Handbook of Case Histories in Failure Analysis
Volume 2

Prepared under the direction of the
ASM International Failure Analysis Committee

Khlefa A. Esaklul, Editor

Co-Editors
  Herman C. Burghard
  George E. Kerns
  William R. Warke

Mary Thomas Haddad, Acquisitions Editor

Grace M. Davidson, Manager, Production Systems
Suzanne E. Frueh, Production Project Manager
Ann-Marie O'Loughlin, Production/Design
Nancy M. Sobie, Production/Design
William W. Scott, Jr., Director, Technical Publications

ASM
INTERNATIONAL
®
**The Materials
Information Society**

Library of Congress Cataloging Card Number: 92-75192
ISBN: 0-87170-495-1

Manager, Production Systems
Grace M. Davidson

Production Project Manager
Suzanne E. Frueh

Production/Design
Ann-Marie O'Loughlin
Nancy M. Sobie

ASM International®
Materials Park, OH 44073-0002

Printed in the United States of America

TABLE OF CONTENTS

TRANSPORTATION COMPONENT FAILURES

Aircraft-Aerospace

Ground Transportation Components

PROCESS EQUIPMENT FAILURES

Heat Exchangers

Pressure Vessels

Pipes and Pipelines

Tanks

Auxilliary Components

ROTATING EQUIPMENT FAILURES

Blades, Disks, and Rotors

ELECTRICAL EQUIPMENT FAILURES

MISCELLANEOUS FAILURES

Biomedical Components

Springs

Tools

Industrial Structures

Nonmetallic Material

High-Temperature

General

PREFACE

Today more than ever, there is a need for a wide base of failure analysis case histories. During economically depressed times, integrity and life extension of damaged or aged plant components are vital to the success of our industries. These analyses require extensive data on field experience, documented failure causes and solutions, and in-service materials properties data base to ensure that a reliable fitness-for-service analysis is attainable.

The Failure Analysis Committee of the Materials Testing and Quality Control Division of ASM International recognized the need for a reliable source of failure analysis case studies, and we thus began the groundwork for this series. Volume 1 of *Handbook of Case Histories in Failure Analysis*, published in 1992, includes over 100 new case histories that emphasize methodology, prevention of failures, and evolving methods for fitness-for-service analysis and integrity evaluation in nearly all branches of industry. In Volume 2, the editorial committee has again attempted to present a diversity of cases that reflect the broad range of interests and the various applications of materials in the engineering profession.

Today we must cross the boundaries of science to gain an in-depth appreciation of the complexity of failures. The failure analyst has to consider all the facts, use proper tools, and exercise judgment before reaching conclusions. The case histories in this series reflect this approach to a large extent. Like any compilation of case histories, this series presents the current knowledge of professionals who believe in sharing their experience and recognize the need to contribute to the success of their profession. It also points out the informational gaps in the field of failure analysis, some of which evolved because of constraints imposed on our profession by the current exploitation of liability litigation. Some companies are not in the position to share their experience because they must protect their business interests.

The editorial committee feels that these two volumes will provide professionals in the field with basic knowledge and with a comprehensive approach for conducting an effective failure analysis. The value of this resource will undoubtedly grow as professionals in the field find the means to use this knowledge. Our committee is considering publishing a third volume that would include more nonmetallic material failures. Another goal is to develop a computer database that would assist engineers in identifying the causes of failures and appropriate measures for prevention, and that could provide the needed data for assessment of component integrity and remaining life. Because each properly performed failure analysis study is unique and a valuable learning tool, we invite potential contributors to submit papers and help expand the literature base. Should Volume 2 be as well received as Volume 1 was, and should the users of these volumes indicate a need for a computer database, the committee will work with ASM International to pursue the development of this source of information.

I wish to acknowledge the contributions of all the authors and reviewers; without their dedicated efforts, this series would not have been possible. Special thanks are due to my co-editors, William R. Warke, Herman C. Burghard, and George E. Kerns, for their efforts in planning, structuring, and reviewing this series. Special thanks are also due to William R. Warke and Gregory S. Gerzen of Amoco Research Center for their assistance with Volume 2 and to the editorial staff of ASM International, particularly Mary Thomas Haddad, for their assistance throughout this project. The support of Amoco Corporation throughout this project is greatly acknowledged. I am grateful to my wife, Vicki, and to my children, Camiella, Ahmed, and Yasmeen, for their patience and for allowing me the time to work on this book.

Khlefa A. Esaklul
Amoco Research Center
Naperville, Illinois
August 1993

REVIEWERS AND CONTRIBUTORS

A.M. Abdel-Latif
Department of National Defence (Canada)

Lynn E. Arnold
Arnold Metallurgical

James H. Arthur, Jr.
Newport News Shipbuilding

Clifford Atkinson
Consultant (U.K.)

W.T. Becker
The University of Tennessee

Valerie J. Berry
Rolls-Royce, Inc.

Edward V. Bravenec
Anderson & Associates/ISI

John E. Brynildson
T. Crane and Associates

S.R. Callaway (retired)
General Motors Electromotive Division

Joseph M. Capus
Consultant

Rafael Castillo
Westinghouse Canada, Inc.

Nicholas E. Cherolis
General Electric Aircraft Engines

J. Ciulik
Radian Corporation

J.A. Clum
State University of New York at Binghamton

Richard L. Colwell
Air Products and Chemicals

E. Philip Dahlberg
Metallurgical Consultants, Inc.

Prabir Deb
Radian Corporation

Peter F. Ellis II
Radian Corporation

Mel J. Esmacher
Betz Laboratories, Inc.

James E. Feather
Exxon Research and Engineering Company

Ross F. Firestone
Ross Firestone Company

Gregory S. Gerzen
Amoco Research Center

Leslie N. Gilbertson
Zimmer, Inc.

George M. Goodrich
Taussig Associates, Inc.

Ronald W. Gruener
Dresser Industrial Valve

Kimberly O. Harding
Failure Analysis Associates

Jerome C. Hill
Coors Brewing Company

Mark Hineman
Taussig Associates, Inc.

Roland Huet
Failure Analysis Associates, Inc.

Joseph C. Jasper
Armco Research, Inc.

Kumar V. Jata
University of Dayton Research Institute

J.M. Johnson
Amoco Corporation

Mitchell P. Kaplan
Willis & Kaplan, Inc.

George Y. Lai
Haynes International, Inc.

Bernard S. Lement
Lement & Associates

Hugo F. Lopez
University of Wisconsin-Milwaukee

Andy Madeyski
Westinghouse Electric Corporation

Michael L. Marx
National Transportation Safety Board

Dennis McGarry
Owens/Corning Fiberglas

Daniel J. Olah
Borg Warner Automotive, Inc.

Michael P. Oliver
University of Dayton Research Institute, USAF

Henry Otto
Failure Analysis Associates

N.S. Palanisamy
Walbar Metals, Inc.

Subhash R. Pati
International Paper Company

Padmanabha S. Pillai
Goodyear Tire and Rubber Company

Robert M. Rose
Massachusetts Institute of Technology

Stuart T. Ross
S.T. Ross & Associates

Ravi Rungta
General Motors Corporation

Larry W. Sarver
Babcock & Wilcox

M. J. Schofield
Cortest Laboratories, Ltd.

David E. Schwab
Allied-Signal, Inc.

Harry Schwartzbart
Consultant

Roch J. Shipley
Engineering Systems, Inc.

A. Kent Shoemaker
Engineering Systems, Inc.

George A. Slenski
U.S. Air Force

Chris Suman
Consultant

G. Mark Tanner
Radian Corporation

Michael A. Urzendowski
DNV Industrial Services

Stuart L. Wilson
Houston Lighting & Power

Daniel A. Wojnowski
Engineering Systems, Inc.

Donald J. Wulpi
Metallurgical Consultant

TRANSPORTATION COMPONENT FAILURES

Aircraft-Aerospace

Ground Transportation Components

Aircraft Accident Caused by Explosive Sabotage

R.V. Krishnan, S. Radhakrishnan, A.C. Raghuram, and V. Ramachandran, National Aeronautical Laboratory, Bangalore, India

Damage to a passenger aircraft that resulted from a midair explosion and subsequent emergency landing was investigated to determine the cause and location of the explosion. Extensive damage had occurred in the front toilet and cockpit areas and to the undercarriage and underside of the aircraft. Fractographic and surface examination of metal fragments (stainless steel and aluminum alloy) from damaged areas indicated that the accident was caused by an explosion in the front toilet. A reconstruction exercise confirmed this conclusion. Damage to the undercarriage and underside resulted from the emergency landing.

Key Words   Explosions   Stainless steels   Aluminum-base alloys
             Aircraft components

Background

Damage to a passenger aircraft that resulted from a midair explosion and subsequent emergency landing was investigated to determine the cause and location of the explosion.

Circumstances leading to failure

The flight of the passenger aircraft was uneventful until 20 min before a scheduled landing, at which time an explosion occurred, causing injuries to the crew and damage to the front toilet area and to instruments in the cockpit. However, the main structure was undamaged. Despite these multiple emergencies, the pilot continued to fly the aircraft and attempted an emergency landing. In process, the aircraft overshot the runway, severing the engines and damaging the underside of the aircraft. Although an in-flight explosion was known to have occurred, it was necessary to provide material evidence to a Court of Inquiry to establish the location and type of explosion.

Visual Examination of General Physical Features

Because the engines and undercarriage were severed, the aircraft was resting on the underside of the fuselage, beyond the end of the runway (Fig. 1). The front toilet and cockpit areas had suffered extensive damage. The paint on the external skin above the front toilet had peeled off, and a hole was found on the skin of the toilet roof, with the metal lip curling outward (Fig. 2a). The panels of the toilet compartment had given way, as did the front bulkhead and the floorboard below the washbasin. Sharp projectiles had penetrated the panels. Fragments were lodged in the toilet panels, in the panel opposite the toilet across the gangway, in the pilot's seat, and in the observer's seat behind the pilot.

Inside the toilet, the stainless steel washbasin and its fittings were severely mangled. The wastepaper receptacle, made of aluminum alloy sheets and kept under the washbasin, had disintegrated into a number of pieces. The toilet bowl was caved in (Fig. 2b), as though deformed by compressive forces directed from the region below the washbasin.

Figure 2(c) illustrates damage to the various instruments in the cockpit. There was a hole in the backrest of the observer's seat, which was situated between the toilet wall and the pilot's seat. The backrest of the pilot's seat also had a hole pierced through it (Fig. 2d).

Fig. 1   Damaged aircraft in which the explosion occurred

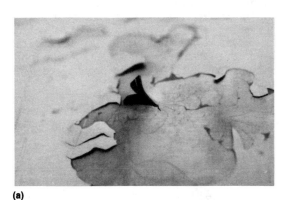

(a)

(b)

Fig. 2   Damage in the front portion of the aircraft. (a) Hole in the external skin on the toilet roof. (b) Caving in of the toilet bowl.

(continued)

(c)

(d)

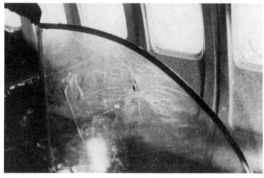

(e)

Fig. 2 Damage in the front portion of the aircraft. (c) Damage to the instrument panel in the cockpit. (d) Hole in the backrest of the pilot's seat, front view. (e) Hole in the plastic panel in the passenger cabin

In the passenger cabin, the plastic panel in front of the left front row of seats had a hole pierced through it (Fig. 2e), with distinct crazing. The cabin crew's folding seat, located in the passage between this plastic panel and the rear wall of the front toilet, had disintegrated, and sharp metal fragments were embedded in the foam of this seat.

After examining the wreckage in the field, a large number of metal fragments were recovered from the front toilet and cockpit areas, as well as from the ground below, for further laboratory examination.

Testing Procedure and Results	

Testing Procedure and Results

Surface examination

Visual. The fragments were curled and twisted. Some of them contained dents and pierced holes. Curling was pronounced in the lips around the holes and along the fracture edges.

Macrofractography/Scanning Electron Fractography. The fragments were examined in a stereomicroscope and in a scanning electron microscope (SEM). The features observed on the fracture edges and on the surfaces of the fragments are shown in Fig. 3. These features were pronounced in the fragments obtained from the disintegrated wastepaper receptacle.

Some of the fragments from the wastepaper receptacle showed a stepwise slant fracture edge, the slope of the slant fracture reversing periodically along the edge (Fig. 3a). Distinct curling was observed on some of the fragments of the wastepaper receptacle, the free fracture end rolling over itself one or more turns (Fig. 3b). A curved surface with a small radius of curvature was another unusual

feature of some of the fragments (Fig. 3c). Dents were observed in some of the pieces of the wastepaper receptacle (Fig. 3d). Some were sharp and others were glancing dents. Figure 3(e) shows an SEM fractograph of metal spall in a fragment. Fracture occurred along the midthickness plane, resulting in flaking. Holes were pieced through some of the fragments. Figure 3(f) shows a typical hole, around which petallike tongues of the metal have curled outward. Sharp spikes were noticed in some pieces along the fracture edge (Fig. 3g). On the surface of the fragments of the wastepaper receptacle, small craters were visible with raised ridges along their rims (Fig. 3h). Many tiny, nondescript, featureless fragments were contained in the debris collected from the damaged areas of the aircraft (Fig. 3i).

Simulation tests

Explosion experiments were carried out in the laboratory on similar aluminum alloy sheet. The fragments thus produced were found to have these characteristics.

Discussion

When the aircraft overshot the runway and hit the approach lamps and the airport fence, the engines, undercarriage, and underside of the aircraft suffered considerable impact damage. However, the damage to the toilet area, cockpit, and the surrounding structures and fittings therein were of an

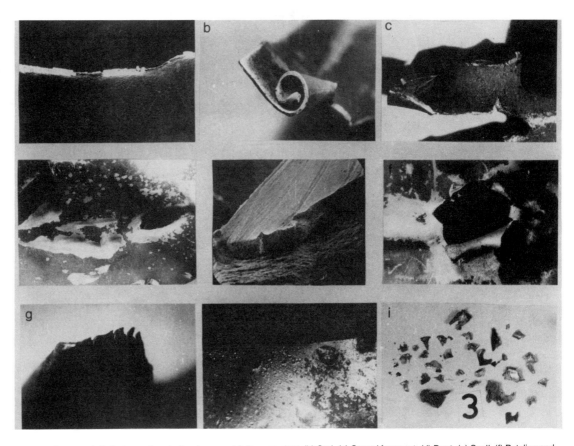

Fig. 3 Characteristic features of explosive damage. (a) Reverse slant. (b) Curl. (c) Curved fragment. (d) Dent. (e) Spall. (f) Petaling and curling around a hole. (g) Spikes. (h) Craters. (i) Nondescript fragments

entirely different nature and could not have been caused by impact on landing.

Features on fragments of sheet metals disintegrated by explosive forces are distinct from those produced by impact alone. These characteristic features, which are pronounced in the primary zone of explosion, survived the subsequent impact forces. Tardif and Sterling (Ref 1) have described in detail the characteristics of explosive damage in aluminum. These have found applications in the detection of explosive sabotage of aircraft (Ref 2-5).

The features described above and illustrated in Fig. 3 are general characteristics of explosive fracture. It has been reported that some features, such as reverse slant in sheet metal, are produced in high-strain-rate deformation and fracture, but craters with raised rims on the surface of sheet metal are produced by gas bubble impingement following a chemical explosion. The craters were visible in the fragments of the wastepaper container.

Spall in sheet metal is a typical shock wave phenomenon. When a compression wave passes through sheet metal, it is reflected from the rear surface of the sheet as a tension wave, and in this process, fracture occurs along the midthickness plane. Some of the larger pieces of the wastepaper container had oriented dents and holes, indicating explosive damage from the inside to the outside. The nature of damage to the structure and fittings in and around the front toilet provided further evidence of this phenomenon. The holes in the roof skin of the toilet (Fig. 2a), in the pilot's backrest (Fig. 2d), and in the plastic panel in the passenger cabin (Fig. 2e) indicated the presence of high-velocity projectiles moving in various outward directions from the toilet. The inward caving of the toilet bowl (Fig. 2b) and the severe deformation of the washbasin pointed to the region under the washbasin as the center of the explosion.

The peeling of the paint from the top skin of the fuselage between the cockpit and the front entrance (Fig. 1) could have been caused only by internal forces from underneath and not during landing. In aircraft accidents caused by explosion, study of the damage to the surrounding structures often leads to determination of the center of explosion by tracing the trajectories of the projectiles in the damaged area (Ref 6,7).

All evidence suggested that the wastepaper receptacle was the center of the explosion. This conclusion is illustrated in the trajectory tracing shown in Fig. 4.

Conclusion and Recommendations

Microscopic examination of the fragments collected from the aircraft clearly established chemical explosion in the front toilet as the cause of the accident. The center of the explosion was found to be the wastepaper receptacle under the washbasin. Further damage to the undercarriage and

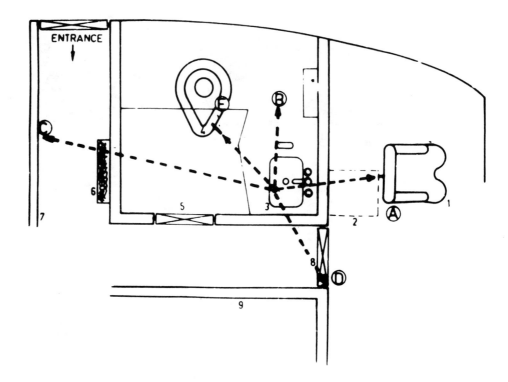

Fig. 4 Trajectories of projectiles from the center of the explosion. 1, pilot's seat; 2, observer's seat; 3, washbasin; 4, toilet bowl; 5, toilet door; 6, cabin crew seat; 7, plastic panel; 8, cockpit door; 9, pantry; A, holes in the backrest of pilot's seat; B, hole in the external skin; C, hole in the plastic panel; D, holes in the door panel; E, dent in the toilet bowl

underside of the aircraft was caused when the aircraft made an emergency landing at high ground velocity and overshot the runway.

References

1. H.P. Tardif and T.S. Sterling, Explosively Produced Fractures and Fragments in Forensic Investigations, *J. Forensic Sci.,* Vol 12 (No. 3), 1967, p 247-272
2. E. Newton, Aircraft Damaged or Destroyed by Deliberate Detonation of Explosives, *Can. Aeronaut. Space J.,* Vol 14, 1968, p 385-392
3. H.P. Tardif and T.S. Sterling, Detection of Explosive Sabotage in Aircraft Crashes, *Can. Aeronaut. Space J.,* Vol 15 (No. 1), 1969, p 19-27
4. R. D. Barer and T.S. Sterling, Investigating an Aircraft Disaster, *Met. Prog.,* Vol 98 (No. 5), 1970, p 84-86
5. A.C. Raghuram, S. Radhakrishnan, R.V. Krishnan, and V. Ramachandran, Death in the Air, *Sci. Age,* Vol 3 (No. 8), 1985, p 32-39
6. V.T. Clancey, Explosive Evidence in an Airplane Accident, *Can. Aeronaut. Space J.,* Vol 14, 1968, p 337-343
7. R.V. Krishnan, S. Radhakrishnan, A.C. Raghuram, and V. Ramachandran, Investigation of an Aircraft Accident by Fractographic Analysis, *Advances in Fracture Research,* Vol 5, S.R. Valluri *et al.,* Ed., Pergamon Press, 1984, p 3677-3684

Cracking in an Aircraft Main Landing Gear Sliding Strut

S.A. Barter and G. Clark, DSTO Aeronautical Research Laboratory, Department of Defence, Melbourne, Australia

Examination of several fighter aircraft main landing gear legs revealed unusual cracking in the hard chromium plating that covered the sliding section of the inner strut. The cracking was associated with cracks in the 35 NCD 16 steel beneath the plating. A detailed investigation revealed that the cracking was caused by the combination of incorrect grinding procedure, the presence of hydrogen, and fatigue. The grinding damage generated tensile stresses in the steel, which caused intergranular cracking during the plating cycle. The intergranular cracks were initiation sites for fatigue crack growth during service. It was recommended that the damaged undercarriage struts be withdrawn from service pending further analysis and development of a repair technique.

Key Words

Military planes	Struts	Hydrogen embrittlement
Nickel-chromium-molybdenum steels	Fatigue failure	Landing gear
Crack propagation		

Alloys

Nickel-chromium-molybdenum steels—35 NCD 16

Background

Examination of several main landing gear legs on a high-performance fighter aircraft revealed unusual bands of cracking in the hard chromium plating that covered the sliding section of the inner strut.

Pertinent specifications

The strut was manufactured from an ultra-high-strength 35 NCD 16 steel (French alloy specification). The strut was shot peened before the sliding section was hard chromium plated.

Performance of other parts in same or similar service

The manufacturer's previous experience indi-cated that this banded cracking could lead to catastrophic failure of the strut well before the calculated safe-life, jeopardizing the safety of the aircraft and crew.

Specimen selection

Information on the cause, type, and growth rate of the cracking was not available from the manufacturer. Therefore, a worst-case example of the cracking was selected from the fleet and sent for laboratory examination.

Visual Examination of General Physical Features

The main undercarriage sliding half-fork assembly (inner strut) consisted of a tube section and an integral lower forged beam into which a press-fitted axle was mounted (Fig. 1). The tube section was designed to slide within a cylinder to provide damping during landings. The lower section of the tube, which normally bears on the outer cylinder seal, was hard chromium plated for a length of approximately 250 mm (10 in.).

Testing Procedure and Results

Nondestructive evaluation

Dye penetrant inspection of the electroplated hard chromium band revealed cracking with a distinct helically banded structure in both the lower (Fig. 2) and upper ends of the plating. The hard chromium plating was removed by chemical dissolution in sodium hydroxide solution, except where strips were required for metallographic examination. The base material was then penetrant tested, and cracking was also found in the steel.

Surface examination

Visual. Generally, the cracking in the steel occurred in a banded pattern similar in appearance and position to the cracking observed in the electroplating. The regular nature and the radial form of this banding strongly indicated grinding damage as the likely cause (Fig. 3). To confirm this, the surface was etched with nital, which resulted in

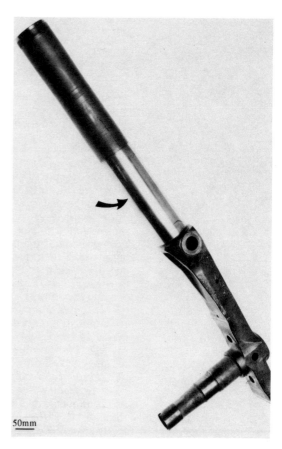

50mm

Fig. 1 Main landing gear sliding strut provided for destructive examination. Arrow indicates region that was hard chromium plated. Wheel axle is at lower end.

Fig. 2 Indications of cracking in the hard chromium plating (arrow) at the axle (lower) end. Note the banded appearance of the cracking.

Fig. 3 Heat-related grinding damage running from the bottom of the chromium plating in evenly spaced bands similar to the bands of cracking. The plating was not removed from several areas, including the shiny area, for metallographic purposes. 2% nital etch

the appearance of helically shaped, dark etching bands.

Close examination of these bands revealed a "mud-flat" cracking pattern (Fig. 4), typified by extensive networks of narrow, randomly oriented cracks. The steel surface below the plating had been shot peened. This peening is usually carried out before plating to eliminate any residual tensile stress in the surface of the part; steels hardened to this strength level are highly susceptible to hydrogen-induced cracking during the plating cycle. Such peening also improves fatigue life by inducing residual compressive stress. The stresses induced by the poor grinding practice apparently were sufficient to cause cracking either before shot peening or despite the beneficial effects of peening.

Macrofractography. An area suspected of containing fatigue cracking was broken open. The exposed surfaces revealed fatigue cracks extending from intergranular crack tips. The maximum fatigue crack depth was approximately 0.13 mm (0.005 in.), which when added to the intergranular cracking from which it grew, gave a total crack depth of approximately 0.25 mm (0.010 in.) into the substrate (Fig. 5). This fatigue cracking extended about 30 mm (1 in.) around the circumference of the tube in a band approximately 3 mm (0.1 in.) wide, centered approximately 3 mm (0.1 in.) from the lower chromium runout. The band was located on the side of the assembly opposite to the

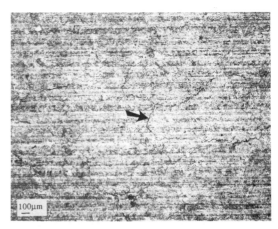

Fig. 4 Closeup view of one of the bands in Fig. 3. Note the network of cracks (arrow), which produces a "mud-flat" cracking appearance.

axle and forward of the centerline; this is an area where high stresses occur during landing and braking of the aircraft.

The combined intergranular/fatigue cracks were more open at the surface than the purely intergranular cracks; this had permitted the ingress of a corrosive medium which led to extensive corrosion of the crack faces (Fig. 6). Thus, although indistinct crack progression marks were visible, determination of the crack growth rate using quantitative fractographic techniques would have been very difficult even if a detailed load history had been available. Examination of the surfaces of cracks in the chromium plating that were associated with cracking in the steel (Fig. 7) revealed evidence of progression marks, indicating that in some cases the chromium cracks grew from the chromium/steel interface.

Metallography

Metallographic sections were taken at different angles through a number of areas where grinding damage had occurred and where the chromium plating had been left intact.

Microstructural Analysis. The general microstructure was that of a low-alloy ultrahigh-strength steel, consisting of a fine-grained tempered martensite with few inclusions. Slight banding (Fig. 8) of the structure was observed, consistent with mild alloy segregation. The plating microstructure was typical of hard chromium plating, with numerous short vertical defects and a columnar grain structure.

Crack Origins/Paths. Sections through the cracking revealed intergranular cracks penetrating approximately 0.13 mm (0.005 in.) into the steel, along the prior-austenite grain boundaries. Most of these cracks were associated with cracking in the chromium plating; some, however, were not. Sections perpendicular to the banding, through the lower region of the chromium-plated band and the plating runout region, revealed straight transgranular cracks growing into the steel from the intergranular crack tips. Straight transgranular cracking of this nature is typical of fatigue in this material.

Chemical analysis/identification

Material. Chemical analysis of the steel indicated that it conformed to the specification for 35 NCD 16.

Coatings or Surface Layers. Chemical analysis of the plating indicated that it was chromium.

Mechanical Properties

Hardness. Testing of several areas of the steel yielded an average value of 51 HRC, which indicates an approximate ultimate tensile strength of 1830 MPa (265 ksi).

Fig. 5 Fracture surface of one of the deeper fatigue cracks. F, fatigue band; I, intergranular cracking; O, overload region. Dotted line indicates the boundary between intergranular cracking and fatigue.

Fig. 6 Closer view of the intergranular and fatigue cracking, showing the marked corrosion damage (pitting) on the surface. O, overload region

Fig. 7 Closer view of the surface of one of the cracks through the chromium plating. This crack was associated with cracking in the steel substrate. Arrows indicate probable progression marks on the chromium crack surface.

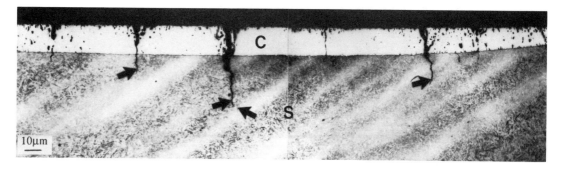

Fig. 8 Section through the cracking adjacent to the chromium plating runout. C, chromium; S, steel. The arrows indicate regions of transgranular cracking produced by fatigue, which grew from the intergranular crack tips. 2% nital etch

Discussion

The presence of dark etching bands on the steel surface beneath the chromium plating clearly indicated that abusive grinding had caused localized overheating of the surface. Such excessive surface heating may transform very thin surface layers of the tempered martensite matrix to austenite (which is then quenched by the surrounding steel to untempered martensite) and/or tempering of the surface layers. Rapid heating and cooling also induce rapid expansion of the surface layers, which in turn can induce plastic flow. When the surface is cooled by the surrounding metal, high residual tensile stresses remain in the surface layers to act on the quenched, untempered, and very brittle martensite or on the overtempered matrix. These stresses (uniform tension in a surface layer) can produce the characteristic "mud-flat" cracking that was found in the strut where extensive networks of narrow cracks were observed. The absence of the white etching layer or patches usually observed when untempered martensite is formed in this manner indicated that the temperature to which the surface layers were raised during the grinding operation was below the transition temperature for this steel, or that a subsequent finish grinding operation removed this white etching layer.

Because several cracks were present in the steel without related cracks in the chromium plating, it is likely that the cracking in the steel preceded the cracking in the chromium. The intergranular cracking was characteristic of hydrogen embrittlement cracking, which probably would have occurred before the strut was baked (de-embrittlement) to remove residual hydrogen.

Although it was not possible to establish the fatigue crack growth rate from observations of the crack surfaces, the very sharp crack tip associated with intergranular cracking would result in a large K_t value for this type of defect, thus allowing the intergranular cracks that were oriented perpendicular to the principal tensile stress direction to initiate fatigue cracks soon after the part was first subjected to significant loads.

Conclusion and Recommendations

Most probable cause

The majority of the cracking in the steel was intergranular and was consistent with the existence of high tensile surface stresses. These stresses resulted from damage produced during grinding of the steel surface prior to the plating cycle. Cracking was probably produced by the absorption of hydrogen during the electroplating process. These intergranular cracks led directly to the growth of fatigue cracks.

Remedial action

It was recommended that undercarriage struts showing this form of unusual cracking in the hard chromium plating be withdrawn from service pending further analysis and development of a repair process.

How failure could have been prevented

Although this part is highly stressed in service, safety can be maintained by using principles based on the results of full-scale fatigue testing and the application of sensible safety factors. The test failure of this main undercarriage leg at a similar location occurred after simulation of 15,000 flights. Applying a safety factor of 3 gives a safe-life of 5000 flights. However, the manufacturer experienced a service failure at 2849 landings, caused by a fatigue crack that initiated from defects similar to those investigated in this study. It is clear that the safe-life of this component is severely compromised by this type of defect, a type which could be avoided by proper grinding procedure and post grinding inspection.

Cracking in an Aircraft Nose Landing Gear Strut

S.A. Barter, N. Athiniotis, and G. Clark, DSTO Aeronautical Research Laboratory, Department of Defence, Melbourne, Australia

A crack was detected in one arm of the right-hand horizontal brace of the nose landing gear shock strut from a large military aircraft. The shock strut was manufactured from a 7049 aluminum alloy forging in the shape of a delta. A laboratory investigation was conducted to determine the cause of failure. It was concluded that the arm failed because of the presence of an initial defect that led to the initiation of fatigue cracking. The fatigue cracking grew in service until the part failed by overload. The initial defect was probably caused during manufacture. Fleet-wide inspection of the struts was recommended.

Key Words	Aluminum-base alloys Landing gear Struts	Fatigue failure Crack propagation Fractography	Scanning electron microscopy Military planes
Alloys	Aluminum-base alloy — 7049		

Background

A crack was detected in one arm of the right-hand horizontal brace of the nose landing gear shock strut from a large military aircraft.

Applications

The shock strut forms a part of the nose landing gear of a large aircraft in service with the Royal Australian Air Force; it carries (and provides damping of) the vertical loads experienced by the wheels during touchdown and ground roll, transmitting those loads to the airframe. It also carries side loads.

Circumstances leading to failure

During a before-flight inspection, a crack was detected in one arm of the right-hand horizontal brace of the nose landing gear shock strut. The crack was in the aft arm of the twin-arm brace approximately 25 mm (1 in.) from the shock strut barrel.

Pertinent specifications

This shock strut was a modified version of the original part, which had a record of stress-corrosion-related problems. The modification involved changing the aluminum alloy from 7079 to 7049.

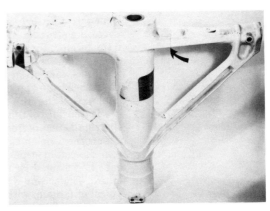

Fig. 1 As-received nose landing gear shock strut. Arrow indicates the crack in one of the support arms. ~.125 ×

Specimen selection

The strut was sent to a laboratory for determination of the cause and nature of the cracking and estimation of the rate at which the defect had grown to failure. This information was required so that inspection intervals could be set to ensure detection of similar cracking in other struts and thus avoid catastrophic failure.

Visual Examination of General Physical Features

The shock strut was manufactured from an aluminum alloy forging in the shape of a delta, with the barrel of the strut running through the center of the delta. The top of the barrel was supported on both sides by braces, one a solid tube and the other consisting of a pair of arms joined by a ligament near the midpoint. The crack severed one of the two arms approximately halfway between the inner edge of the central ligament and the shock strut barrel (Fig. 1). The cross section of the failed arm was approximately trapezoidal, with a thickness of 19 mm (0.75 in.) and a height that varied from an outer face of 24 mm (0.95 in.) to an inner face of 30 mm (1.2 in.), giving a cross-sectional area of approximately 500 mm² (0.8 in.²).

Testing Procedure and Results

Nondestructive evaluation

Liquid Penetrant Inspection. After removal of paint from the area of failure and from the adjacent area of the other arm, these areas were inspected for cracking using a high-resolution dye penetrant. No further cracking was observed.

Surface examination

Visual inspection showed a light gold-colored (chromate-conversion-coated) shot-peened surface after paint removal.

Macrofractography. The cracked region was removed to expose the fracture faces. Failure progressed by a number of modes, as indicated by three distinct areas on the fracture surface (Fig. 2).

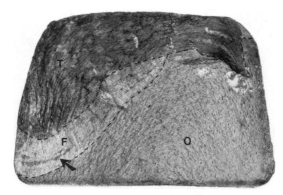

Fig. 2 Fracture surface of the failed support arm. Cracking initiated in the highly textured area (T). Flat area (F) is fatigue cracking with bands of tearing probably caused by heavy landings (arrow). The remainder of the surface was produced by a single overload (O).~2.34×

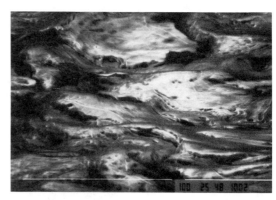

Fig. 3 Surface of the initial cracking. The flattened areas are surrounded by small areas of dimpling. 448 ×

Propagation features in the initial area of cracking revealed a strong association with the metal flow in the forging; these features included high relief, a crack shape that appeared to follow the forging flow lines, and an average plane of propagation that varied from approximately 90° to the longitudinal axis of the arm (initially) to approximately 45°. The surface of this area was dull gray in color, with small areas of fretting damage close to the top outer corner and along the two edges leading away from the corner; this damage suggested that the crack had existed for a considerable time prior to failure. The second area of the fracture face was typical of fatigue, with intermittent bands of stable tearing fracture. Crack progression marks were readily observed, as this area was almost unaffected by fretting or rubbing damage. The final area had features typical of room-temperature overload fracture of high-strength aluminum alloys. The first area of cracking covered approximately 36% of the cross-sectional area, followed by fatigue and intermittent tearing that covered approximately 13%. The remaining area was the overload fracture.

Scanning Electron Microscopy/Fractography. Close examination of the fracture surface using optical and electron-optical microscopy revealed that the initial area of cracking consisted of large areas of platelike structure separated by smaller areas with dimplelike features (Fig. 3). The area near the top corner was covered with fretting product, and some evidence of shallow corrosion was visible. No shear lip was observed along the edges in this region (temperature overload of high-strength aluminum). The fatigue cracking initiated from multiple sites along the face of the initial crack, in a region where the initial crack plane changed direction toward the centerline of the arm (the top of the initial crack and the region on the initial crack surface where the fatigue crack initiated are marked in Fig. 4), and progressed in a plane perpendicular to the longitudinal axis of the arm. Progression marks were clearly visible, and some groupings of striations of similar appearance and spacing were thought to have been caused by separate landings of the aircraft. The fatigue band had not developed along the full extent of the initial crack, but was confined to the area close to the

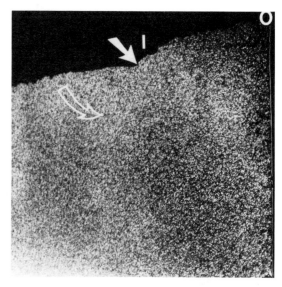

Fig. 4 Macrosection through the failure showing the flow pattern created during the forging operation. O, area at which the initial defect originated; I, the line (visible here as a point) of initiation of the fatigue crack on the inner face of the initial cracking. The approximate point at which the initial cracking ends (open arrow) is shown more clearly in Fig. 8. Acid etched. 4.70×

top corner of the arm, extending along approximately 70% of the length of the initial crack. Numerous broad bands (believed to be single load progressions) and several tear bands were observed on the fatigue surface, perhaps caused by heaver landings. The final overload was typical of room-temperature overload of high-strength aluminum alloys.

Quantitative Fractography. To assist in establishing the time at which the initial cracking occurred, the investigators requested an estimate of the number of landings undergone by the shock strut. The operator reported that this number was unknown, because the strut had been fitted to a number of aircraft, and the exact aircraft tail numbers had not been recorded. However, the number of service flight hours experienced by the component was recorded as 2331.6.

Examination of the records for nine fleet aircraft revealed that the number of flight hours between landings (where the aircraft comes to a full stop) varied from 2.28 to 2.77, with an average of 2.48. The total number of flight hours between touchdowns, which includes touch-and-go landings, ranged between 0.75 and 1.72, with an average of 1.2. The number of loads experienced by the nose landing gear would vary significantly if the nose gear was grounded during a touch-and-go. Assuming that this is not usually the case, the loads from full-stop landings were probably the major cause of crack growth. Based on this information, the number of landings that the strut was expected to have experienced was between 843 and 1022.

Crack progression marks observed on the fatigue area of the fracture and believed to correspond to aircraft landings were counted using an optical microscopy and digital counter. These markings were of a regular nature and consisted of an initial heavy mark (initial touchdown) followed by a series of light, irregularly spaced marks (aircraft ground-roll). These patterns were visible on numerous areas in the fatigue region (Fig. 5).

Measurements were made along a line following the direction of growth of the fatigue crack at approximately the thickest part (4.62 mm, or 0.182 in.) of the fatigue band. After the data were processed, a plot was made (Fig. 6) of the fatigue crack depth (below the point on the initial defect from which the fatigue crack initiated) versus the estimated landing number. Three tear bands observed along this line are evident as steps on the curve (Fig. 6) and are marked 1, 2, and 3, corresponding to tears with widths of 0.14, 0.51, and 0.08 mm (0.0055, 0.020, and 0.0030 in.), respectively. Two large gaps in the data were due to difficulties in observing progression marks in these areas. This procedure gave an estimate of approximately 835 landings prior to failure. This is close to the minimum number of 843 estimated from the operational data.

Metallography

Microstructural Analysis. Metallographic sections taken through the fracture disclosed a number of features. The macrostructure of the arm was typical of a forged high-strength aluminum alloy (Fig. 4). The forging operation had produced a U-shape flow pattern in this area, with the bottom of the U facing the outer face of the arm. The path of the initial cracking was closely related to the structure; the cracking tended to follow the direction of flow of the aluminum, resulting in a crack surface with high relief in some area. Secondary cracking was observed below the surface of the main crack, following the elongated boundaries of the worked grains (Fig. 7). The extent of this secondary cracking was considerably less than that which would be expected if the initial cracking had been caused by stress-corrosion cracking (SCC). The initial cracking extended beyond the point of initiation of the fatigue cracking, continuing to curve toward the longitudinal axis of the arm and clearly following the flow lines (Fig. 8). Several voids were noted along the section of the crack that had not been opened by the failure. The region of fatigue was transgranular, as was the final overload region.

Chemical analysis/identification

Chemical analysis of the shock strut in the area

Fig. 5 Representative sample of fatigue pattern. The heavy marks (arrow) probably represent the landing load, and the finer marks were from ground-roll loads. ~648 ×

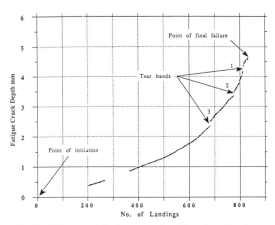

Fig. 6 Fatigue crack depth versus estimated number of landings

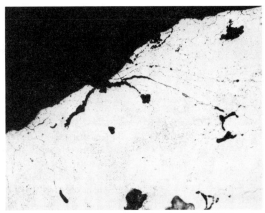

Fig. 7 Secondary cracking below the plane of the initial crack. The cracks are oriented along grain boundaries aligned with the metal flow produced during forging. Etched in Keller's reagent. 320×

of the failure revealed that it was manufactured from an aluminum alloy that conformed to the specification for alloy 7049. Conductivity and hardness measurements were consistent with the T73 temper.

Simulation tests

To determine the true nature of the initial cracking, a number of simulated tests were per-

Fig. 8 Cross section of fatigue (F) and initial cracking (I) surfaces showing the extension of the initial crack (arrow) beyond the point of fatigue initiation. Etched in Tucker's reagent. 7.70×

formed to produce failures in the alloy for comparison purposes.

A section of the alloy was subjected to liquid metal embrittlement by the application of liquid gallium to a mildly stressed C-ring. Failure occurred by intergranular separation, as expected. The crack appearance was considerably different from that of the initial cracking observed in the arm.

A section of the alloy was subjected to conditions considered likely to produce SCC. It was found that the material was highly resistant to SCC in the environment used (surviving 3 weeks of exposure to a standard salt spray without significant cracking); the environment produced only pitting corrosion. Exposure to a more severe environment produced a small amount of cracking that was significantly different from the cracking observed in the initial area of the failure.

Several sections of the alloy were subjected to severe forging at 400 °C (750 °F). The laps and cracks so produced were sectioned and compared with those of the initial cracking in the failed arm. These artificially produced cracks had features that were similar to the initial cracking in the failed component—including tightness, small voids dotted along the main laps, and secondary cracking following flow lines.

Discussion

The evidence indicated that the root cause of the failure was a large crack that was associated with the direction of forging flow rather than oriented perpendicularly to the principal stress direction, as was the case with the fatigue cracking. Other unusual features about this crack were the fact that it extended a considerable distance past the plane on which the fatigue cracking initiated, its gray surface coloration, the strong flow orientation of it and of the secondary cracking, the presence of dimples on its surface, and the "worked" nature of its surface. These features and the similarity to defects produced by forging samples suggested that the initial crack occurred some time before or during the forging process, followed by closing of the crack on subsequent forging and shot peening to produce a very tight defect that was not detected prior to service.

The fatigue band was active for a considerable time and appeared to have progression marks consistent with those expected from landing loads. Counting the marks gave an estimate of the total number of landings that the part experienced. Given the many difficulties in interpretation of this fracture surface, this estimate showed excellent correlation with the number of full-stop landings estimated, indicating that the part contained the initial cracking prior to entering service and that the crack started to grow very early in its service life. Another observation was the large reduction in cross-sectional area prior to failure, indicating that the cracking was prolonged by load transfer to the other arm.

Conclusion and Recommendations

Most probable cause

The arm failed because of an initial defect that led to initiation of fatigue cracking. This fatigue cracking grew in service until the part failed by overload. The initial defect was probably caused during manufacture. Considering the size of the initial defect (36% of the cross-sectional area) and the high probability that it existed from the time of manufacture, the damage tolerance of this brace of the strut was high.

Remedial action

A fleet-wide inspection of the struts was recommended.

How failure could have been prevented

Improved control of manufacturing processes and/or nondestructive inspection might have prevented the failure.

Failed Mixer Pivot Support of An Army Attack Helicopter

Victor K. Champagne, Gary Wechsler, and Marc Pepi, United States Army Research Laboratory, Materials Directorate, Watertown, Massachusetts

A forged, cadmium-plated electroslag remelt (ESR) 4340 steel mixer pivot support of the rotor support assembly located on an Army attack helicopter was found to be broken in two pieces during an inspection. Visual inspection of the failed part revealed significant wear on surfaces that contacted the bushing and areas at the machined radius where the cadmium coating had been damaged, which allowed corrosion pitting to occur. Optical microscopy showed that the crack origin was located at the machined radius within a region that was severely pitted. Electron microscopy revealed that most of the fracture surface failed in an intergranular fashion. Energy dispersive spectroscopy determined that deposits of sand, corrosion and salts were found within the pits. The failure started by hydrogen charging as a result of corrosion, and was aggravated by the stress concentration effects of pitting at the radius and the high notch sensitivity of the material. The failure mechanism was hydrogen-assisted and was most likely a combination of stress-corrosion cracking and corrosion fatigue. Recommendations were to improve the inspection criteria of the component in service and the material used in fabrication.

Key Words		
Nickel-chromium-molybdenum steels, Corrosion	Stress-corrosion cracking	Corrosion fatigue
Stress concentration	Pitting (corrosion)	Intergranular fracture
Helicopters	Notch sensitivity	Hydrogen embrittlement

Alloys   Nickel-chromium-molybdenum steels—4340

Background

A mixer pivot support was found to be broken during a pre-flight inspection.

Applications

The mixer pivot support is a flight safety-critical component and is part of the rotor support as-

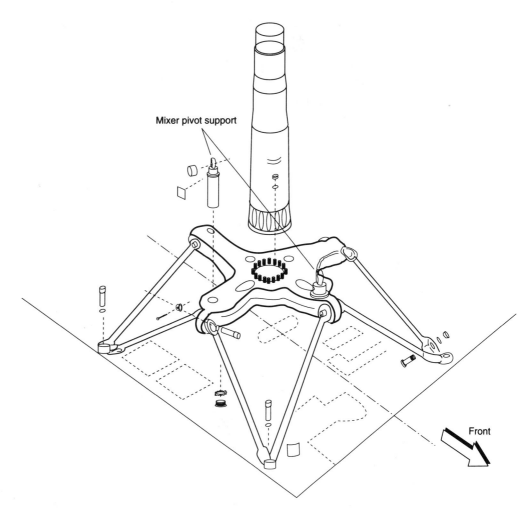

Fig. 1   Schematic of the rotor support assembly. Mixer pivot support is shown with an arrow.

sembly shown in Fig. 1. The mixer pivot support fits through the transmission support. If the part had failed during flight, the helicopter would have crashed.

Circumstances leading to failure

The component was in service approximately 1449 days. The original replacement time for the mixer pivot support was 800 days. However, this period was extended to 1440 days due to a lack of spare parts. A second extension of six months beyond 1440 days was granted for the same reason, as long as no surface corrosion was observed upon visual inspection. According to inspection records, no evidence of corrosion was detected on the part. However, the results of this investigation proved otherwise.

Pertinent specifications

The component was specified to be forged from ESR 4340 steel bar stock. A vacuum-plated cadmium coating was applied according to MIL-C-8837.

Performance of other parts in same or similar service

Mixer pivot supports fabricated from the same manufacturer and from the same heat treat/shipping lots as the broken component had not failed during service.

Specimen selection

The two failed pieces of the mixer pivot support were subject to laboratory failure analysis.

Testing Procedures and Results

Surface examination

Visual. Figure 2 shows the broken component in the as-received condition. The failure occurred at the machined radius. The two broken pieces were labelled Part A (upper half) and Part B (lower half). Part A showed no obvious signs of corrosion or mechanical damage. In contrast, Part B exhibited significant wear on surfaces that contacted the bushing. These regions were characterized by dark stains. The cadmium coating appeared to have been almost entirely worn away during service and severe corrosion pitting had occurred within these areas. The extensive corrosion prevented accurate measurement of the surface finish. However, the radius where the fracture occurred was measured and found to be within requirements (3.3 mm, or 0.130 in.). Figure 3 is an optical fractograph which reveals the crack origin and direction of propagation. The radial lines and chevron patterns indicated that the crack proceeded from the bottom right of the photo-

graph (as designated by the arrow) up along both sides of the central hole. The fracture appears to have originated at the bottom radius.

Scanning Electron Microscopy/Fractography. Figure 4 is a SEM fractograph representative of the morphology found on approximately 90% of the total fracture surface. The mode of fracture was intergranular decohesion, indicative of a brittle fracture. No evidence of fatigue striations were observed, but since these features are difficult to resolve in such high-strength materials, fatigue could not be entirely ruled out as a failure mechanism. Extensive corrosion pitting was found along the edge identified as the crack origin, but the exact point of crack initiation was difficult to resolve due to mechanical damage. Energy dispersive spectroscopy confirmed the by-products of corrosion (Fig. 5). Chlorides and particles consisting primarily of silicon and oxygen (sand) were found within pits. Chlorides have been known to accelerate corrosion of carbon steels.

(a)

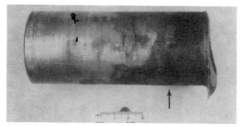

(b)

Fig. 2 Macrographs of the two broken halves of the mixer pivot support as received. Region of significant wear is denoted by arrow.

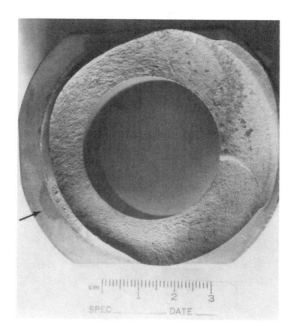

Fig. 3 Optical fractograph of Face A (arrow identifies crack origin). Approximately 0.81×

Metallography

Microstructural Analysis. Samples were prepared to examine the microstructure of the component, as well as determine the extent of pitting. The microstructure of fine martensite was consistent with the heat treatment performed on the component. Unusual material defects or large inclusions were not observed. Cracks were noted originating from the bottom of pits, and extending into the base material parallel to the fracture plane, as shown in Fig. 6. These cracks appeared to extend in an intergranular fashion.

Chemical analysis/alloy identification

Material. Samples from the failed mixer pivot support were chemically analyzed, and the results (Table 1) met the requirements of ESR 4340 steel.

Mechanical properties

Hardness measurements performed across a section of the failed component averaged 55.6 HRC, and conformed to the required hardness of 54 to 57 HRC.

Tensile Properties. Tensile specimens were fabricated from the failed component with the intent of obtaining fracture surfaces which could be compared to the fracture surface under investiga-

tion. The specimens were pulled to failure and the resultant fracture surfaces were examined using a scanning electron microscope. The tensile specimens displayed a typical ductile cup-cone fracture. The dimpled morphology was indicative of a ductile failure.

Fig. 4 SEM showing intergranular mode of failure. Approximately 350×

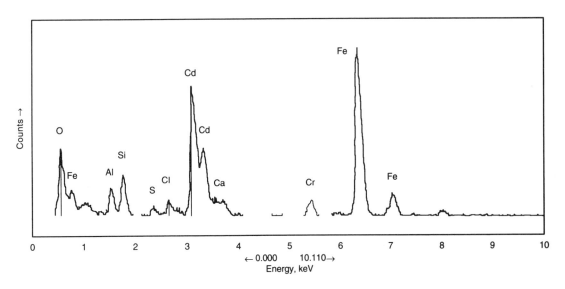

Fig. 5 EDS spectrum from within a pit adjacent to the fracture

Table 1   Chemical analysis

| | Composition, wt% | | | | | | | | | | |
	C	Mn	Si	P	S	Cr	Ni	Mo	Cu	Al	Fe
HMS-6-1121	0.39 to 0.41	0.60 to 0.80	0.20 to 0.35	0.010 max	0.008 max	0.70 to 0.90	1.65 to 2.00	0.20 to 0.30	0.35 max	0.030 max	Remain
Mixer pivot support	0.39	0.69	0.21	0.008	0.005	0.82	1.99	0.28	0.11	0.012	Remain

Discussion

Severe corrosion pitting occurred along the machined radius of the component and served as a crack initiation site. Desert sand taken from Saudia Arabia and analyzed at ARL revealed high concentrations of chlorides which assisted in the corrosion process. Hydrogen diffused into the high strength material (56 HRC) as a result of the corrosion process and migrated into areas of high stress concentration (crack tip). Evidence for this claim is shown by the tensile tests; when a section of material taken from the failed component was pulled to failure, the resulting fracture surface was dimpled, but the failure mode over 90% of the fracture surface under investigation was intergranular. In addition, the final fracture region on the component (shear lip) also displayed a dimpled topography. This evidence indicates that the material can fracture in a ductile fashion. Hydrogen-assisted cracking occurs in an intergranular fashion in this type of material when heat treated to the hardened condition.

Fig. 6 Corrosion pits with cracks extending into the base material parallel to the fracture plane. Approximately 265×

Conclusion and Recommendations

The cracking of the mixer pivot support initiated at the machined radius within a severely pitted region. The fracture did not originate where fretting was most severe. The failure began due to hydrogen charging as a result of corrosion. This condition was aggravated by the stress concentration effects of pitting at the radius and the high notch sensitivity of the material. Failure was hydrogen assisted, and was most likely a combination of stress-corrosion cracking and corrosion fatigue.

It was recommended that the components be removed from service at the first indication of corrosion. Visual inspection with the aid of a magnifying lens could be used to detect corrosion in the field. The component could continue to be utilized when hardened to 54 to 57 HRC as long as there is no visual evidence of corrosion. In this way, the ballistic properties could be maintained. However, a more conservative approach (which would sacrifice some of the ballistic properties of the material) would be to heat treat the component to a softer condition. A softer material would be less notch-sensitive, and the inspection intervals could then be longer since the critical crack size would be increased.

Failure Analysis of a Space Shuttle Solid Rocket Booster Auxiliary Power Unit (APU) Fuel Isolation Valve

J.H. Sanders, IIT Research Institute, Marshall Space Flight Center, Alabama
G.A. Jerman, National Aeronautics and Space Administration, Marshall Space Flight Center, Alabama

A precipitation-hardened stainless steel poppet valve assembly used to shut off the flow of hydrazine fuel to an auxiliary power unit was found to leak. SEM and optical micrographs revealed that the final heat treatment designed for the AM-350 bellows material rendered the AM-355 poppet susceptible to intergranular corrosive attack (IGA) from a decontaminant containing hydroxyacetic acid. This attack provided pathways for which fluid could leak across the sealing surface in the closed condition. It was concluded that the current design is flight worthy if the poppet valve assembly passes a preflight helium pressure test. However, a future design should use the same material for the poppet and bellows so that the final heat treatment will produce an assembly not susceptible to IGA.

Key Words

Valves
Intergranular corrosion,
  Heating effects

Precipitation hardening steels,
  Corrosion
Sensitizing

Stainless steels, Corrosion
Tempering
Booster rockets

Alloys

Precipitation hardening steels—
AM-350

Precipitation hardening steels—AM-355

Background

A poppet valve assembly failed a helium pressure test (Ref 1, 2).

Applications

The poppet valve assembly is part of the fuel isolation valve. This valve isolates hydrazine from an auxiliary power unit that drives the thrust vector control system of the Space Shuttle solid rocket booster (Fig. 1). The poppet valve assembly consists of an AM-350 bellows welded onto an AM-355 poppet that acts as the sealing surface. The valve is normally closed (shut off), so that in an inactive condition no hydrazine fuel is allowed to flow. The hydrazine inlet pressure is normally maintained at about 1.0 MPa (150 psig).

Circumstances leading to failure

Testing of the fuel isolation valve prior to reinstallation led to the use of helium pressure tests to discern the integrity of the poppet valve assembly. The pressure tests were performed at 5.5 MPa (800 psig) and revealed leakage, which was subsequently isolated to the sealing surface of the poppet.

Pertinent specifications

The poppet valve assembly is fabricated from components made out of precipitation-hardened stainless steels AM-355 and AM-350 (see Table 1 for compositions). The bellows is fabricated from AM-350 sheet and welded onto the AM-355 poppet. The AM-355 poppet is machined from bar stock and heat treated as follows:

- Solution heat treat: 1038 °C (1900 °F) for 1 to 3 h, cool to room temperature
- Sub-zero cool: –73 °C (–100 °F) for ≥ 3 h
- Equalized: 774 °C (1425 °F) for ≥ 3 h, cool in air to ≤ 27 °C (80 °F)
- Over-temper: 579 °C (1075 °F) for ≥ 3 h, cool to room temperature

After welding on the AM-350 bellows, the entire assembly is heat treated as follows:

- Solution heat treat: 1038 °C (1900 °F) for 15 min, quick furnace quench

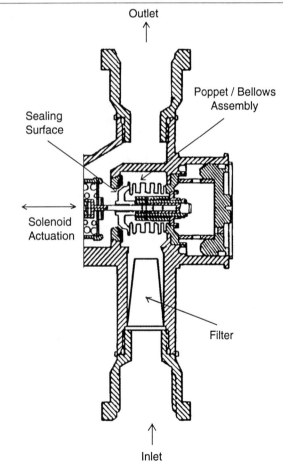

Fig. 1 Schematic of the part of the fuel isolation valve that contains the poppet bellows assembly

- Trigger anneal: 927 to 943 °C (1700 to 1730 °F) for 20 to 35 min., argon quench
- Sub-zero cool: –73 °C (–100 °F) for ≥ 1 h
- Temper: 449 to 460 °C (840 to 860 °F) for 3 h, rapid quench

Table 1   Compositions of AM-355 and AM-350

Alloy	Nominal composition, wt%(a)					
	Cr	Ni	Mo	Mn	C	N
AM-355	15.5	4.5	2.9	0.85	0.12	0.10
AM-350	16.5	4.5	2.9	0.85	0.10	0.10

(a) Balance Fe

This heat treatment was designed to provide the best resistance against stress-corrosion cracking (SCC) in AM-350.

A key aspect of the Space Shuttle is its designed reusability. Therefore, after solid rocket booster recovery, the device is decontaminated with hydroxyacetic acid. Prior to reinstallation for flight, the device is cleaned with Turco® cleaning solution. The assembly is then reinstalled and charged with liquid hydrazine approximately 14 days prior to launch.

Performance of other parts in same or similar service

Other poppet valve assemblies have been helium pressure tested as a consequence of these findings. One other poppet failed the test and was subsequently replaced.

Selection of specimens

Five poppets were analyzed, of which two had failed (#72 and 135) and two had passed (#82 and 129) the pressure test. The fifth poppet was analyzed after the AM-355 heat treatment and before joining of the bellows (#35).

Visual Examination of General Physical Features

The general appearance of all specimens was similar. There were no apparent cracks or visible deformation.

Testing Procedure and Results

Surface examination

Scanning Electron Microscopy. The surface of a failed AM-355 poppet is shown in Fig. 2. The low-magnification micrograph shown in Fig. 2(a) shows that the surface is rough with evidence of intergranular separation and surface cracking. The high-magnification micrograph in Fig. 2(b) shows displacement of entire grains so that some grains appear recessed relative to adjacent grains. Superficial boundary separation is apparent as surface cracks propagate along grain boundaries.

Metallography

Microstructural Analysis. Sections were taken through the vertical rotational axis to produce "pie" slices. Optical photographs of a failed specimen reveal the degraded structure of a poppet sealing surface (Fig. 3a). The degradation extends the length of the sealing surface. At higher magnification (Fig. 3b), the structure shows missing grains on the polished surface. SEM confirms the loss of grains and shows the degraded nature of the grain boundaries (Fig. 4).

Optical photographs and SEM micrographs of the unwelded AM-355 poppet show a uniform appearance with martensitic structure (Fig. 5 and 6). This structure is very different than that of the failed poppet.

Chemical analysis/identification

Material. Electron microprobe analysis using wavelength dispersive spectroscopy was performed to determine if noticeable segregation had occurred in the grains of a failed specimen. Figure 7 shows a representative linescan for carbon, iron and chromium across a grain boundary. From these data, the segregation of carbon and chromium to grain boundaries is clear. Evidence of high concentrations of chromium and carbon were also observed on the surfaces of dislodged grains.

(a)   (b)

Fig. 2 SEM micrographs of an AM-355 poppet valve sealing surface showing extensive surface roughening and surface cracking. (a) 72.8×. (b) 560×

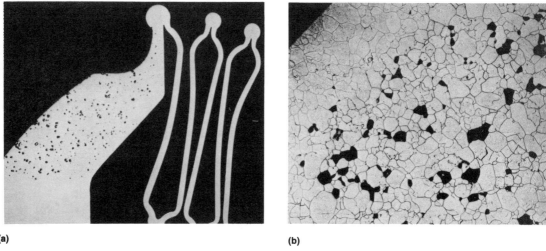

(a)

(b)

Fig. 3 Optical micrographs of a sectioned poppet valve assembly. (a) Porous structure in the AM-355 sealing surface. 10.4×. (b) Porosity determined to be missing grains dislodged during polishing. 65×

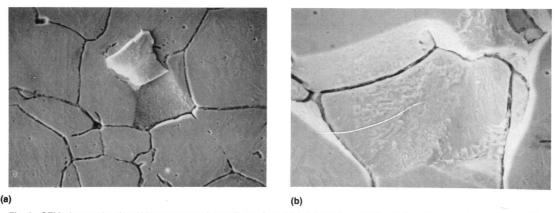

(a)

(b)

Fig. 4 SEM micrographs of an AM-355 poppet valve sealing surface. (a) A dislodged grain. 570×. (b) Sensitized grain boundaries result in grain boundary decohesion. 1140×

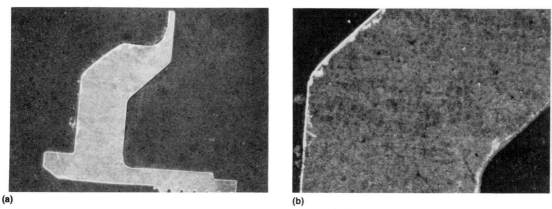

(a)

(b)

Fig. 5 Optical micrographs of a sectioned poppet valve assembly before joining of the bellows showing a uniform appearance and martensitic structure. (a) 5.6×. (b) 21×

These results are indicative of grain boundary sensitization that enhances susceptibility to intergranular attack (IGA).

Mechanical properties

Hardness values for the AM-355 poppets are shown in Table 2. The values for the specimens that were joined to the bellows and given the sec-

ond heat treatment are at least 10 HRC greater than a specimen not subjected to the additional heat treatment.

Simulation tests

From the microstructure, it was apparent that the second heat treatment produced a poppet that experienced IGA. Simulation tests were then nec-

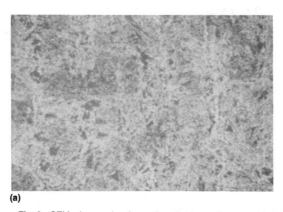

Fig. 6 SEM micrographs of a sectioned poppet valve assembly before joining of the bellows showing the martensitic structure. (a) 285×. (b) 1710×

Table 2 Hardness of investigated poppets

Poppet #	Condition	Hardness, HRC
35	Not welded to bellows	35
72	Failed pressure test	NA
135	Failed pressure test	45 to 47
82	No failure	46
129	No failure	46

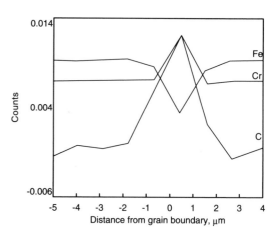

Fig. 7 Carbon, iron and chromium linescans across a grain boundary on a failed AM-355 poppet surface using wavelength dispersive spectrometry. The carbon peak is at the grain boundary. Carbon, 30 counts/division; iron, 6000 counts/division; chromium, 7000 counts/division

essary to isolate the possible corrosive solution(s) so that the appropriate corrective actions could be identified.

Efforts were pursued to determine whether poppet valve assemblies should be considered acceptable for flight if they pass helium pressure tests. Specimens were submerged in hydrazine for a period of 14 days at a temperature of 21 °C (70 °F). These tests revealed that hydrazine has no detrimental effect on the sealing surface under no load (Ref 3). Therefore, if a poppet valve assembly passes the helium pressure test, it will not undergo additional degradation in the charged condition prior to launch.

Exposures to the hydroxyacetic acid decontamination solution and the Turco® cleaning solution were also performed to determine if they were corrosive to the sensitized microstructure of the AM-355 poppet. It was observed that the decontaminant does attack the grain boundaries if left in contact with the poppet for extended periods of time (Ref 1). Therefore, if purging of the decontaminant does not remove all of the solution, degradation of the material may occur prior to reinstallation of the poppet assembly for the next flight. However, helium pressure tests prior to reinstallation will detect excessive degradation and the poppet assembly will be replaced, resulting in an attrition problem which does not affect flight safety.

Discussion   The microstructural investigation revealed that the AM-355 poppet of the poppet valve assembly is in a degraded condition after the bellows joining procedure and subsequent heat treatment. The formation of carbide networks on the grain boundaries creates a chromium depletion zone immediately adjacent to the grain boundary, making it susceptible to IGA. A study by Williamson (Ref 4) revealed that the use of a 538 °C (1000 °F) temper, such as that prescribed in the first heat treatment, produces AM-355 with good resistance to SCC, and a hardness of about 38 to 40 HRC. However, lowering the temper temperature to 454 °C (850 °F), as is the case for the second heat treatment, produced an AM-355 material that is susceptible to SCC with hardness of 46 to 47 HRC.

Hardness values recorded for the poppets corresponded well with the results of the study by Williamson. The degraded microstructure of the poppets given the second heat treatment had hardness values of about 46 HRC and sensitized grain boundaries. Before the second heat treatment was applied to the poppet, the material was in the optimal form for resistance to IGA, having a martensitic structure with hardness of about 35 HRC. The results of this investigation show that although AM-350 and AM-355 are similar in composition, they respond very differently to heat treatment conditions.

The mechanism by which the AM-355 sealing surface fails is a consequence of the sensitized microstructure and decontamination procedure. Residual decontaminant left in contact with the sealing surface for extended periods of time results in IGA that creates a weak grain boundary network. When the extent of attack is sufficient to create a path across the entire sealing surface, fluid may leak through even in the closed position.

Conclusion and Recommendations		
	Most probable cause	*Remedial action*

Most probable cause

The optimal heat treat for the AM-350 bellows transforms the microstructure of the AM-355 poppet to a sensitized structure susceptible to IGA. Residual decontaminant solution attacks the sealing surface in an intergranular fashion, because the sensitized microstructure has a chromium depletion layer immediately adjacent to the grain boundary. This attack provides pathways by which fluid can leak across the sealing surface when the assembly is closed.

Remedial action

The use of current poppet valve assemblies is acceptable if they pass the helium pressure test prior to each launch. Future efforts will focus on using the same material for the bellows and sealing surface. This material should not be susceptible to IGA. One recommendation is the use of AM-350 for the entire poppet valve assembly. This would allow optimum heat treatment of the assembly without sacrificing properties and would require less effort for requalification. Requalification is necessary for design/material changes of flight hardware.

References

1. G. Jerman, "Failure Analysis of Fuel Isolation Valve," Memo EH22-92-222, National Aeronautics and Space Administration, Marshall Space Flight Center, Huntsville, AL, 1993.
2. David Mayo, "Fuel Isolation Valve-Poppet Assembly Failure Investigation," Memo DEM-006-93MP, United Technologies, USBI, Huntsville, AL, 1993.
3. Chris Conway, private communications, United Technologies, USBI, Huntsville, AL, 1992.
4. J.G. Williamson, "Stress Corrosion Studies of AM-355 Stainless Steel," Technical Memorandum TMX-53317, National Aeronautical and Space Administration, August 1965.

Failure of Aircraft Target Towing Cables

G.D.W. Smith, Department of Materials, University of Oxford, Oxford, United Kingdom
K.E. Easterling, *School of Engineering, University of Exeter, Exeter, United Kingdom

Over a period of 2 or 3 years, 40 to 50 premature failures of drawn high-tensile, pearlitic high-carbon (0.8 wt% C) steel wires used as cables for towing targets behind aircraft occurred. Six service failures were examined in detail. Four types of failure characteristics were noted. A close examination of wire that had been flown several times without failure was also made, and dynamic tests were conducted to investigate the fracture characteristics of wire subjected to dynamic loading. It was concluded that dynamic shock loading transmitted by the target during unsteady flight conditions was the major cause of failure. Recommendations emphasized the need for a suitable shock absorber to be fitted at the constant-tensioning device of the winch system.

Key Words

Carbon steels
Military planes

Shock loading
Fracturing

Cables
Military applications

Background

High-tensile, high-carbon steel wire was used to tow targets behind aircraft. Indicated airspeeds of up to 740 km/h (460 mph) and altitudes of up to 10,000 m (32,800 ft) were encountered. The wire (drawn 0.8 wt% C pearlitic steel) was single strand, 1.3 mm (0.05 in.) in diameter, with a breaking strength of 305 to 315 kg (670 to 695 lb) under static loading conditions. Under flight conditions, the target was towed around a figure eight course. Towing cable lengths up to 6 km (3.7 mi) were used. The target was retrieved after use by rewinding the cable onto a winch.

The problem encountered was the premature failure of the cable after a few uses. It was necessary to keep the maximum load on the wire below 200 kg (440 lb) (as measured at the aircraft) and to minimize the towing length to prolong the life of the cable. For a constant load of 200 kg (440 lb), a length effect was observed. The number of failures increased rapidly when the towed length exceeded 3 km (1.85 mi). The static load on the cable was a maximum at the aircraft (200 kg, or 440 lb) and decreased to approximately 50 kg (110 lb) at the target. The major source of static loading was wire drag. The location of the cable failure was generally believed to be within a few meters of the winch, although failures were also observed close to the target and occasionally at other points along the length of the wire.

Over a period of over 2 years involving more than 500 flights, there were 40 to 50 failures. Failure occurred during normal flying conditions, not while the target was being winched in or being let out. The moment of failure frequently occurred after the aircraft emerged from a turn. Under these conditions, the static stress on the cable reached a maximum, and oscillations were observed before failure occurred. However, the moment of failure apparently came shortly after the period of maximum static stress, rather than at the precise instant of maximum load. Under most common conditions, the cable was heavily damped, for transverse oscillation, by the airflow. Large transverse oscillations were observed in the region of the target, however. An investigation was undertaken to (1) explain the reasons for the (apparently premature) failure of the cable and (2) provide suggestions for minimizing the problem.

Visual Examination of General Physical Features

Six service failures were examined in detail metallographically. These showed some variable characteristics, as discussed below.

Failure type 1 showed a tensile fracture of typical cup-and-cone structure. A particular feature of this type of failure was the occurrence of several transverse scratch marks, approximately 0.1 to 0.15 mm (0.004 to 0.006 in.) deep in the vicinity of the fracture. Failure apparently occurred at this defect. Slight necking of the wire was also observed close to the point of fracture. Two specimens exhibited these characteristics, although in one case the fracture appearance was somewhat irregular. Similar secondary markings were observed in each case. The wire appeared flattened or compressed for some distance on one side of the fracture, and a short, deeper mark was observed on the other side.

Failure type 2 showed cup-and-cone fracture similar to specimen 1, with no sign of surface scratching. A different feature was observed on this wire—a pronounced blue color on the wire surface, extending several millimeters from the point of fracture. This is typical of a localized heating effect (e.g., wire being struck by lightning).

Failure type 3 showed a shear-type fracture, oblique to the wire axis. This form of failure can occur when tensile overload conditions are combined with loss of ductility in the wire. This loss of ductility could arise from several causes, such as inclusions in the wire, excessive local work hardening, strain aging, or (possibly) deformation at an excessive strain rate.

Failure type 4 showed a region of severe bending in close proximity to the fracture zone. The fracture itself had apparently occurred by a shearing process, which in this case had taken place almost exactly transverse to the wire axis. This failure is almost certainly due to kinking of the wire. Two examples of this kind of failure were observed.

Metallographic examination was under-

---

*Professor Easterling passed away during the time that this paper was being prepared for publication.

taken of wire that been flown several times without failure. The following points were noted:

- The surface finish of the wire generally remained very good after winding and rewinding.
- Some local necking was observed at regular intervals of a few centimeters, in the region of the wire that was very close to the winch during flight operations.
- The breaking load of the wire increased by 10 to 20 kg (20 to 45 lb) after several cycles of use. This was apparently due to work hardening, possibly occurring during the winding and unwinding process.
- The wire was extremely springy and somewhat lacking in flexibility. This may have contributed to failure—by kinking, for example.

The number of failures analyzed was limited; however, the data suggested strongly that the predominant cause of failure was tensile overload. Occasional accidents such as kinking or local heating were probably not of central importance.

Testing Procedure and Results

Design of laboratory testing program

It was necessary to investigate the possible reasons for the failure of the wire at an (apparent) maximum load of only 200 kg (440 lb) (close to the winch), when the nominal breaking load should be in excess of 300 kg (660 lb). Reasons for breakage at much lower apparent loads (close to the target, or near the midpoint of the wire) also required consideration. Possible reasons that could be investigated in the laboratory included variations in wire quality, fatigue, damage to the wire surface, and localized plastic deformation of the wire. These were investigated systematically.

Unused Wire Quality. Of more than 20 specimens tested, all broke at loads between 305 to 315 kg (670 to 695 lb). There was thus no evidence of faults in the manufacturing process. Failure occurred in a ductile manner (cup-and-cone appearance).

Used Wire Quality. Specimens cut from different sections of used cable were tested. Breaking loads were in the range 315 to 325 kg (670 to 715 lb), i.e., in excess of the original value. No evidence of overall degradation of strength was evident. Some decrease in ductility was observed, however. New wire exhibited 50 to 70% reduction in area at the fracture zone, whereas used wire exhibited only 45 to 50% reduction.

Local Plastic Deformation. Wire was bent through 180° about decreasing radii of curvature, restraightened, and then tensile tested. Reduction in strength was observed only when the bending diameter was less than 45 mm (1.8 in.), a much more severe bend than is likely to be experienced in service (see the discussion below on kinking). When the curve was bent around its own radius, attempts to restraighten it led to cracking and splitting along the axis.

Kinking. Wire was bent into 360° loops of decreasing radii and then tensile tested in loop form. Loops with radii greater than 5 mm (0.2 in.) were removed during the testing process, producing a wire containing a 360° twist. No measurable reduction in breaking load occurred under these conditions. Loops of less than 5 mm (0.2 in.) radius behaved in a drastically different manner. When the load was applied, these small loops tightened until a localized kink was produced in the wire. Failure occurred at the kink at an extremely low load (approximately 50 kg, or 110 lb), and the wire sheared approximately transverse to the axis, leaving a severely plastically deformed and twisted section near the fracture. This correlated well with type 4 service failures.

Scratches and Indentations. Scratches were introduced into the wire surface with the aid of a sharp blade. For scratch depths of up to 0.2 mm (0.008 in.) (in excess of the maximum observed in type 1 service failures), no measurable reduction in breaking load occurred. For scratches in excess of 0.2 mm (0.008 in.) in depth, progressive reductions in strength were observed (for example, 0.3 mm (0.012 in.), 290 kg (640 lb) and 0.39 mm (0.015 in.), 270 kg (595 lb). Fracture occurred by a ductile cup-and-cone mechanism in every case.

Wires were placed crosswise against one another and subjected to compressive loads, such as might be experienced during winding operations. No measurable change in breaking load was detectable after compression loads of up to 315 kg (695 lb) had been applied. Loads in excess of this were unlikely to be obtainable in practice.

Fatigue. To investigate possible effects of cyclic loading on the wire, fatigue experiments were carried out using a combination of a static tensile load of approximately 200 kg (440 lb) and a superimposed pulsed load of 90 kg (200 lb). The following results were obtained:

- New wire under 200 kg (440 lb) static load plus 90 kg (200 lb) pulses: No cracks after 1000 cycles
- New wire containing a 0.08 mm (0.003 in.) deep scratch under 200 kg (440 lb) static load plus 90 kg (200 lb) pulses: No cracks after 1000 cycles
- New wire containing a 0.08 mm (0.003 in.) deep scratch under 224 kg (494 lb) static load plus 90 kg (200 lb) pulses: Breakage after approximately 1000 cycles

Tensile testing of (unbroken) fatigued wire showed that its strength had been increased by cyclic loading. Breaking loads of 325 kg (715 lb) were obtained, compared with 305 kg (670 lb) for the same section of new wire. The ductility of the wire was reduced somewhat by fatigue (45% reduction in area, instead of 50 to 70%). The effect of small surface defects in the wire appeared to be somewhat greater in fatigue than under static loading conditions.

Conclusions from Initial Test Program. The most important conclusion reached from these tests was that, under normal conditions, the wire was extremely tough and resistant to most natural hazards—damage, wear, and plastic deformation—likely to be encountered in service. There-

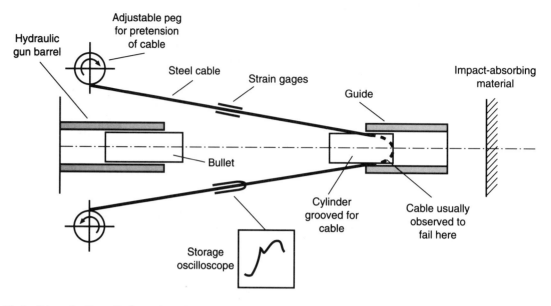

Fig. 1 Schematic of impact testing equipment

fore, apart from the occasional accidental failure, caused by kinking or localized heating, for example, it was not possible to simulate the service failures using conventional laboratory test conditions.

It appeared that an additional factor must be present under flying conditions that did not occur under static test conditions for small lengths of wire. By far the most probable factor was dynamic loading of the wire. This could arise in several ways, for example, because of irregular movements of the plane, target, or cable, or because of oscillations introduced into the cable by the transverse component of the airflow.

Possible effects of stress waves

Two types of waves are possible—transverse and longitudinal. Transverse waves are likely to be heavily damped, particularly when they propagate into the highly stressed region of cable close to the aircraft. Consequently, this type of wave was neglected in the investigation. Longitudinal waves appear to be more important. Assuming for the moment that the origin of the stress wave is at the target, it is immediately evident that stress wave propagation occurs under a set of very unusual conditions, because a static stress gradient exists along the length of the cable. The three main consequences of this are:

- The propagation behavior of the wave will be nonuniform along the length of the cable.
- A wave that starts out as a purely elastic wave may be converted into a plastic wave at some point in the cable where the sum total of static and dynamic load exceeds the elastic limit of the wire.
- A tensile wave reflected at the aircraft will be transmitted back down the wire as another

tensile wave. Therefore, close to the aircraft, where the static load is already at a maximum, the effect of any dynamic tension pulse is doubled. This appeared to be the most likely explanation of the commonest form of cable failure. The speed of propagation of these stress waves would be so high that they would not be detected by the conventional load-measuring device fitted to the aircraft.

Dynamic tests

To investigate the fracture characteristics of wire subjected to dynamic loading, a series of tests was carried out on high-velocity impact loading equipment based at the University of Lulea Applied Mechanics Laboratory (Sweden). For this purpose, a length of the wire was wound 180° around a grooved circular cylinder, with the ends of the wire wound several times around fixing pegs. The pegs could be adjusted to vary the tension in the wire. Shock loading was obtained by firing a "bullet" at the rear of the cylinder, as shown in Fig. 1. The wire was fitted with strain gauges to measure the static and dynamic strains and loads on the wire via a digital oscilloscope.

The results from these tests can be summarized as follows. Even under dynamic loading conditions, a cup-and-cone ductile fracture resulted, with a reduction of area of 50%. Fracture was observed to occur at the shoulder of the grooved cylinder, and the area of fracture on the wire was associated with surface markings and scratches. A study of these markings showed that they were generated by shock-loaded compressive contact between the wire and the surface of the grooved cylinder and that the markings were practically identical with the type 1 service failures.

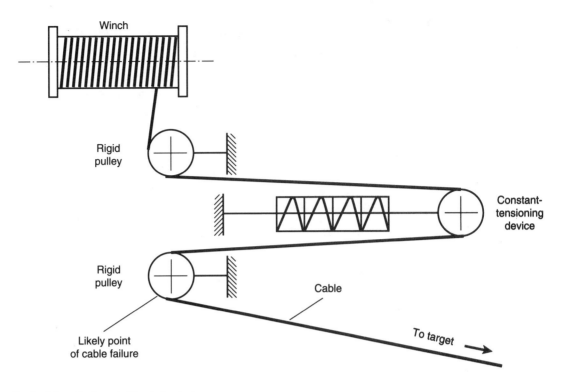

Fig. 2 Schematic of constant tensioning device

Conclusion and Recommendations

Dynamic shock loading was found to be a major cause of failure. The majority of failures appeared to be located at the constant tensioning device on the aircraft winch. This system consisted of a static load absorber and a rigid pulley (see Fig. 2). The shock absorber was unsuitable for damping transient shock loads. Failure occurred at one of the rigid pulley wheels indicated. It should be noted that the size of the pulley used was similar to the grooved cylinder used in the dynamic tests. On this basis, the most important single recommendation was the fitting of a low inertia shock absorber, such as a heavily damped spring, at the constant tensioning device. Additional recommendations that were made included modifications to aircraft flight patterns and winch operating procedures, to minimize the incidence of dynamic loading events.

The origin of the type 2 failures which appeared to be due to local heating of the wire, remained unproven at the end of the study. However, a few years after completion of this study, one of the authors presented the results of this work to the Swedish Air Force at a materials meeting in Linköping. Following the talk, one of the pilots in the audience recounted how he had been one of the trainee pilots using the towing facility about the time of the current investigation. He admitted that to force extra-long weekend leave on the authorities, it became commonplace for the trainees to aim not at the target, but at the towing cable itself. The resulting delay, during which the cable was retrieved from the Lapland forests, was thus used to the pilots' advantage; it was also no mean targeting feat. This piece of supplementary information may well account for these fractures, which the authors had earlier speculated to be the result of lightning strikes. The episode also emphasizes the fact that real-life case studies invariably contain factors that are hard to simulate in a laboratory.

Acknowledgments

The authors are grateful to Karl Erik Wiklund (Svenska Flygtjanst) for many useful discussions and gladly acknowledge the cooperation and assistance of Professor Bengt Lundberg and his group in the dynamic testing work and the help of Thomas Otby and Ove Lindgren in the general experimental work.

Failure of a Transport Aircraft Relay Valve Guide

A.K. Das, Aircraft Design Bureau, Hindustan Aeronautics Ltd., Bangalore, India

A 52000 bearing steel valve guide component operating in the fuel supply system of a transport aircraft broke into two pieces after 26 h of flight. The valve guide fractured through a set of elongated holes that had been electrodischarge machined into the component. Analysis indicated that the part failed by low-cycle fatigue. The fracture was brittle in nature and had originated at a severely eroded zone of craters in a hard, deep white layer that was the result of remelting during electrodischarge machining. It was recommended that the remaining parts be inspected using a stereoscopic microscope and/or a borescope.

Key Words

Bearing steels
Engine components
Electric discharge machining

Aircraft components
Low-cycle fatigue
Airplanes

Control valves
Fatigue failure

Alloys

Bearing steel—52000

Background

A steel relay valve guide component operating in the fuel supply system of a transport aircraft broke into two pieces during flight.

Applications

The valve guide was one of a batch of such components imported for tail plane pitch control. Valve failure introduces a great risk of fire, which may endanger the safety of the aircraft by the escape of fuel to hot zones around the engine bay. The part was reportedly machined to its entire length by conventional machining—except for two sets of elongated holes, which were precision machined by the electrodischarge process (Fig. 1).

Circumstances leading to failure

The valve guide fractured into two pieces after it had served for only 26 h of flight (Fig. 1). The valve was operating at a working load of 20.7 MPa (3000 psi) at the time of failure.

Pertinent specifications

The valve guide was manufactured from SAE 52000, a high-carbon chromium ball bearing steel.

Specimen selection

One of the two broken pieces was selected for fractographic analysis. A section across the origin of failure was metallographically prepared for microstructural studies.

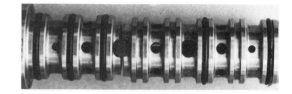

Fig. 1 Rear valve guide, showing transverse fracture through an elongated hole. No plastic deformation was evident around the fracture zone.

Visual Examination of General Physical Features

Visual examination revealed that the valve guide had separated transversely through one of the two sets of elongated holes. The hole bore surface was irregular and discolored.

Testing Procedure and Results

Surface examination

Macrofractography. Examination of the failed elongated hole under a low-power stereoscopic microscope revealed a rough and corroded bore surface, with clear signs of melting and craters (Fig. 2). The fracture had a smooth but rather coarse crystalline surface in a plane at a right angle to the valve axis.

Both elongated holes contained almost identical deep depressions at the bore edge caused by severe burning combined with erosion effects (Fig. 2). The fracture surface exhibited a conspicuous dark-shaded, triangular zone adjoining the sunken hole bore edge. Closer observation revealed several faint arclike markings of rather wide spacing running parallel to the profile of the dark-shaded zone. This indicated that the crack had propagated in stages through several lines of arrest in the crack path, having originated at the crater region in the outer corner of the bore surface. All these features suggested typical brittle cracking under low-cycle fatigue.

Fig. 2 Fractograph of the transverse fracture, showing brittle crack propagation through widely spaced lines of arrest and from the melted crater region of the hole corner

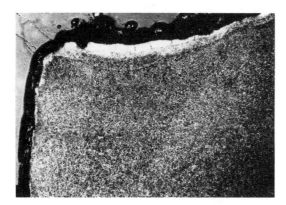

Fig. 3 Low-magnification optical micrograph of a section across the crater zone, showing the presence of a conspicuous hard, white layer that seeped into the cracked zone while in the molten condition during electrodischarge machining. An underlying tempered zone exists between the core and the layer.

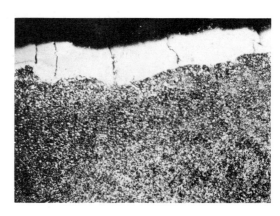

Fig. 4 Higher-magnification view of the zone in Fig. 3, showing parallel transverse cracks in the white layer that occasionally penetrated the tempered martensite matrix

Metallography

Microstructural Analysis. Microexamination of a section across the zone of origin revealed a conspicuous white layer (remelted) in the outer surface of the bore to the maximum depth of 0.035 mm (0.0014 in.). The microstructure also contained a 0.35 mm (0.014 in.) deep overtempered zone just beneath the white layer (Fig. 3). Otherwise, the core structure showed satisfactory tempered martensite with a uniform distribution of numerous tiny globular carbides in the matrix, typical of a high-carbon chromium ball bearing steel.

Microexamination also revealed that the fracture was enveloped with a trace amount of white layer to a considerable depth (Fig. 3). Moreover, the white layer showed several transverse cracks emanating from the outer bore surface and occasionally penetrating the base of the overtempered zone (Fig. 4).

Mechanical properties

Hardness. A microhardness survey using a 500 g load was performed. A maximum hardness value of 1100 HV in the outermost white layer and a minimum value of 550 HV in the overtempered zone were obtained. The core hardness yielded an average value of 800 HV.

Discussion

Fractographic studies established that the failure occurred by low-cycle fatigue, as evidenced by the presence of several lines of arrest marks in the crack propagation path. Close analysis of the fracture path revealed that the brittle crack originated at the severely eroded zone of craters in the outer corner of the bore surface.

The presence of surface defects, such as discoloration, erosion, and craters, indicated a high-temperature melting phenomenon that occurred during electrodischarge machining, forming a very hard, deep white layer. It was evident from the microstructure that, because the white layer had seeped into the fracture surface, crack initiation took place during the high-temperature machining stages.

On the other hand, the development of transverse cracks in the outermost bore surface was a direct consequence of drastic quenching of the hard, brittle white layer. Under this condition, any minor stresses during service would lead to premature brittle failure. However, the depth of the white layer (0.035 mm, or 0.0014 in.) exceeded the limit normally permissible for finished hardenable steel products (0.025 mm, or 0.001 in.). As such, the excess depth was believed to have contributed to the extreme brittleness of the white layer. All these findings suggested that the process parameters adopted for electrodischarge machining could have been faulty, especially with respect to the maintenance of proper current density. It was also possible that short circuiting led to high-temperature melting.

Conclusion and Recommendations

Most probable cause

The brittle cracking of the valve guide was caused by low-cycle fatigue, which was attributed to the presence of a preexisting crack in the outer corner of the bore. This crack resulted from an unduly overheated structure (white layer) formed during machining.

Remedial action

Stereoscopic macroexamination as well as borescopic inspection of the electrodischarge machined bore surfaces should be performed before such components be accepted from the exporter.

Fatigue Fracture of a Helicopter Tail Rotor Blade Due to Field-Induced Corrosion

Richard H. McSwain*, McSwain Engineering, Inc., Pensacola, Florida

A helicopter tail rotor blade spar failed in fatigue, allowing the blade to separate during flight. The 2014-T652 aluminum alloy blade had a hollow spar shank filled with lead wool ballast and a thermoset polymeric seal. A corrosion pit was present at the origin of the fatigue zone and numerous trails of corrosion pits were located on the spar cavity's inner surfaces. The corrosion pitting resulted from the failure of the thermoset seal in the spar shank cavity. The seal failure allowed moisture to enter into the cavity. The moisture then served as an electrolyte for galvanic corrosion between the lead wool ballast and the aluminum spar inner surface. The pitting initiated fatigue cracking which led to the spar failure.

Key Words			
Aluminum base alloys, Mechanical properties	Helicopters	Rotor blades	
Crack initiation	Fatigue failure	Mechanical properties	
	Galvanic corrosion	Pitting (corrosion)	
		Seals	

Alloys   Aluminum base alloys—2014

Background

A helicopter tail rotor blade failed and separated during flight.

Applications

The tail rotor blade spar is the main structural component of the tail rotor blade system and is safety-of-flight critical.

Circumstances leading to failure

During flight, the tail rotor blade failed and the outer section of the blade separated from the aircraft. The crew executed a successful emergency landing. Subsequent inspection revealed that the spar of one tail rotor blade had failed and that the entire section from the spar shank outboard had separated in flight. The departed outboard section was recovered and there was no strike damage evident.

Pertinent specifications

The blade was constructed of 2014 aluminum alloy heat treated to the T-652 condition. The spar shank was hollow and the cavity was filled with lead wool ballast material. The cavity was sealed with a thermoset material.

Visual Examination of General Physical Features

The tail rotor blade spar failure location is shown in Fig. 1. The failure was located approximately 32.38 cm (12.75 in.) from the inboard end of the blade. The fracture was located within the zone of the spar shank where the lead wool ballast filled the spar, as shown in Fig. 2. In addition, the spar cavity was found to contain an unknown fluid. Dimensional analysis revealed no discrepancies.

Fig. 1 Failed rotor blade showing fracture shank of the blade. Approximately 0.31×

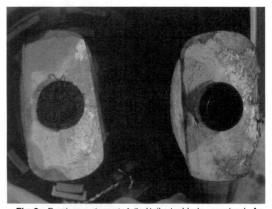

Fig. 2 Fracture surfaces in failed tail rotor blade spar shank. Approximately 0.66×

Testing Procedure and Results

Surface examination

Visual. Visual examination of the spar fracture surface revealed a flat fracture with beach mark features typical of fatigue. The fatigue fracture initiated at the inner surface of the spar cavity and propagated outward until final overload failure occurred.

Macrofractography. Macroscopic examination of the fracture surface revealed a flat area at the origin with a trough-shaped corrosion pit which was 0.43 mm (0.017 in.) long and 0.10 mm (0.004 in.) deep. The crack propagated over half the cross-section of the spar and to the surface of the spar before final overload occurred. Figure 3

---

*Formerly Head, Metallic Materials Engineering, Naval Aviation Depot, Pensacola, Florida

Fig. 3 Scanning electron micrograph of tail rotor blade failure origin. Micrograph shows lead wool ballast in aluminum spar bore cavity and fracture which initiated at bore wall. Approximately 13×

Fig. 4 Scanning electron micrograph of tail rotor blade failure origin. Micrograph shows multiple pits at origin with associated corrosion product. Beach marks are shown emanating from pits, typical of a fatigue failure mode. Approximately 63×

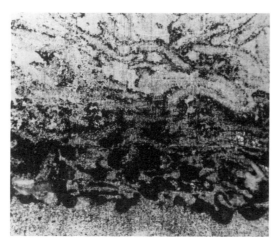

Fig. 5 Light micrograph of tail rotor blade shank bore surface. Patterns remaining on surface were result of electrolytic attack produced by lead wool adjacent to aluminum spar in presence of an electrolyte. Black material was found to be corrosion/reaction product filling spherical pits. Approximately 13×

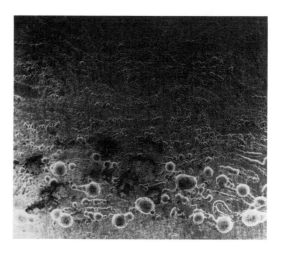

Fig. 6 Scanning electron micrograph of tail rotor blade bore surface after cleaning. Area shown is identical to that shown in Fig. 5. Removal of corrosion/reaction product revealed spherical pits and etched patterns beneath the black material. Approximately 13×

shows the origin area and lead wool fibers in contact with the spar cavity surface. The corrosion pit at the origin is also shown. The origin and origin corrosion pit are shown at higher magnification in Fig. 4. The corrosion product that filled the pits is lighter, due to charging in the scanning electron microscope beam.

Scanning Electron Microscopy Fractography. Fractographic analysis by scanning electron microscopy revealed fatigue striations with varying spacings, caused by the complex load spectrum on the blade during crack propagation. The initial 1.10 mm (0.040 in.) of cracking produced a flat fracture surface typical of high-cycle fatigue. The remainder of the fracture surface was rough with evidence of a higher crack propagation rate.

Corrosion Patterns. Examination of the inner surface of the spar cavity revealed an unusual pattern of corrosion as shown in Fig. 5. The same area with corrosion removed is shown in Fig. 6.

Metallography

Microstructural analysis of the spar revealed no material anomalies that would have contributed to the failure. Sections were taken through the failed spar to determine the extent of the corrosive attack. The sectioning revealed spherical pits at the surface of the cavity wall with isolated intergranular attack.

Chemical analysis/identification

Material. Chemical analysis by X-ray energy spectroscopy revealed a composition consistent with the specified 2014 aluminum alloy.

Mechanical properties

Hardness tests averaged 84 HRB. The typical hardness of 2014-T652 aluminum is 81 to 90 HRB.

The electrical conductivity was also found to be appropriate for the specified material.

Discussion

The fatigue cracking initiated at a corrosion pit on the inner surface of the spar cavity. A fluid substance was found inside the spar cavity and the thermoset plastic seal covering the cavity was broken. The material properties and microstructural features were all consistent with 2014-T652 aluminum. The spar cavity should have been free of moisture or any fluid substance, if the seal had performed as designed.

Conclusion and Recommendations

Most probable cause

The tail rotor blade failed by fatigue which initiated from a trough-shaped corrosion pit located at the inner wall of the spar shank cavity. The pitting was a result of galvanic attack caused by entry of moisture into the spar cavity. The moisture served as an electrolyte for galvanic corrosion between the lead wool ballast and the aluminum spar material. Corrosion pits were formed along the contact points between the lead wool fibers and the spar cavity surface. The moisture entry was due to the failure of a thermoset polymeric seal inside the spar cavity.

How failure could have been prevented

The failure could have been prevented by ensuring the integrity of the thermoset polymeric seal, and by barrier coating the wall of the spar cavity with an insulating material. Alternate methods of encapsulating the lead wool would also have prevented the galvanic corrosion, corrosion pitting, and resulting fatigue failure.

Processing-Induced Fatigue Fracture of a Helicopter Tail Rotor Blade

Richard H. McSwain*, McSwain Engineering, Inc., Pensacola, Florida

A helicopter tail rotor blade spar failed in fatigue, allowing the outer section of the blade to separate in flight. The 7075-T7351 aluminum alloy blade had fiberglass pockets. The blade spar was a hollow "D" shape, and corrosion pits were present on the inner surface of the hollow spar. A single corrosion pit, 0.38 mm (0.015 in.) deep, led to a fatigue failure of the spar. The failure initiated on the pylon side of the blade. Dimensional analysis of the spar near the failure revealed measurements within engineering drawing tolerances. Though corrosion pitting was present, there was an absence of significant amounts of corrosion product and all of the pits were filled with corrosion-preventative primer. This indicated that the pitting occurred during spar manufacture, prior to the application of the primer. The pitting resulted from multiple nickel plating and defective plating removal by acid etching. Post-plating baking operations subsequently reduced the fatigue strength of the spar.

Key Words	Aluminum base alloys, Mechanical properties Pitting (corrosion)	Rotor blades, Mechanical properties Chemical etching	Helicopters Fatigue failure Fatigue strength
Alloys	Aluminum base alloys—7075		

Background

A helicopter tail rotor blade failed and separated during flight.

Applications

The helicopter tail rotor blade provides anti-torque flight control as it reacts against the main rotor system. The tail rotor blade spar is the primary structural component of the tail rotor blade assembly and is safety-of-flight critical.

Circumstances leading to failure.

In flight, the aircrew heard a loud bang followed by severe vibrations. The crew executed a successful emergency landing. Subsequent inspection revealed that the spar of one tail rotor blade had failed and that the entire air foil section had departed the aircraft. There was no strike damage evident. The outboard departed section was never recovered.

Pertinent specifications

The tail rotor blade was constructed of 7075 aluminum alloy heat treated to the T7351 condition. The blade had fiberglass pockets.

Visual Examination of General Physical Features

The residual inboard portion of the failed blade spar is shown in Fig. 1. The fracture was flat and was located approximately 59.7 cm (23.5 in.) from the inboard end of the spar.

Testing Procedure and Results

Surface examination

Visual. The spar fracture surface showed a flat fracture with numerous highly reflective facets present, typical of fatigue, as shown in Fig. 2. Visual examination also revealed that the flat fracture zone extended over approximately 85% of the spar cross-section before final overload. The arrow in Fig. 2 points to a black spot adjacent to the inner surface of the spar where the cracking originated.

*Formerly Head, Metallic Materials Engineering, Naval Aviation Depot, Pensacola, Florida

Fig. 1 Remaining portion of helicopter tail rotor blade after spar failure and outboard section separation. Approximately 0.14×

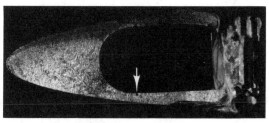

Fig. 2 Tail rotor blade spar failure surface showing flat fracture, reflective facets, and a dark spot at the arrow. Approximately 0.95×

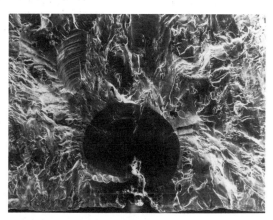

Fig. 3 Fracture origin of tail rotor blade spar. Crack propagated from pit that was 0.38 mm (0.015 in.) deep extending from the inner surface of the spar. Approximately 43.7×

Macrofractography. The black spot was the fracture origin, as shown in Fig. 3. The black spot encircled the bottom of a pit extending from the inner surface of the spar to a depth of 0.38 mm (0.015 in.). The fracture surface exhibited characteristics of fatigue failure for approximately 85% of its cross-section. The remaining portion of the fracture surface, a darker coarser area, exhibited characteristics typical of overload failure.

Scanning Electron Microscopy Fractography. A low-magnification scanning electron micrograph of the fracture origin is shown in Fig. 3. The micrograph shows the corrosion pit located at the fatigue origin. The pit was 0.38 mm (0.015 in.) deep and 0.254 mm (0.010 in.) across at the surface of the spar cavity. No significant amounts of corrosion product were found in the pit. Examination of other spar cavity surfaces revealed numerous smaller corrosion pits. All of the pits were filled with corrosion-preventative primer.

Figure 4 shows a higher-magnification scanning electron micrograph of the spar fracture surface and shows uniformly spaced fatigue striations.

Corrosion Patterns. Though corrosion pitting was present, there was an absence of significant amounts of corrosion product, and all of the pits were filled with corrosion-preventative primer. Moisture entry into the spar cavity that could have caused corrosion damage was not evident.

The lack of corrosion product in the pits, the presence of primer in the pits, and the fact that the spar had never been reworked, all indicated that the pitting occurred during manufacture, prior to application of the primer.

Metallography

Microstructural analysis of the spar revealed numerous large intermetallic alloy particles, some of which were exposed to the surface. Those which were exposed had been chemically attacked, resulting in various sized surface pits.

Chemical analysis/identification

Material and Weld. Chemical analysis of the spar material by X-ray energy spectroscopy re-

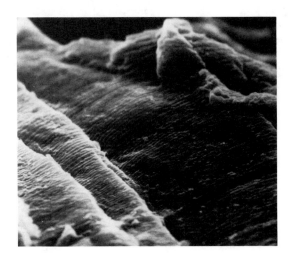

Fig. 4 Scanning electron micrograph of fatigue striations on tail rotor blade spar fracture surface. Approximately 3534×

vealed an alloy composition similar to 7075 aluminum as specified.

Analysis of the fatigue origin pit using X-ray energy spectroscopy showed some residual chlorine and sulfur from the pitting or from later processing.

Mechanical properties

Hardness tests averaged 72 HRB, below the minimum hardness of 77.5 HRB typically specified for 7075-T7351 aluminum.

The conductivity was found to be 41% IACS, on the higher end of the 38.0 to 42.5% IACS range specified for 7075-T7351 aluminum.

Tensile Properties. The ultimate tensile strength was 427 MPa (62 ksi), below the minimum specified of 475 MPa (69 ksi),. The 0.2% offset yield strength was 345 MPa (50 ksi) below the minimum specified yield of 393 MPa (57 ksi). The elongation was 11.3%, satisfactory for the minimum elongation specification of 8%. Hardness, conductivity, and mechanical data all indicated that the spar had been excessively heated after the T7351 condition was established.

Discussion

Further research revealed that the spar had received multiple nickel plating during the manufacturing process. Each plating was removed by acid etching prior to reapplication of the plating. The etching process was the most probable cause for the observed attack of the intermetallic alloy particles and pit formation observed in the microstructural analysis. Records indicated that the failed spar was plated and etched three times before an acceptable plating was achieved. The pits were undetected in the remaining manufacturing

operations and were covered with corrosion-preventative primer. In addition, the spars were given an unauthorized baking operation at 232 °C (450 °F) for 4h following each nickel plating operation. The etch pitting became the site for fatigue crack initiation. The reduction in mechanical properties due to the post-plating baking reduced the fatigue strength of the spar. The fatigue crack propagated through 85% of the cross-section of the spar before final failure occurred.

Conclusions and Recommendations

Most probable cause

The tail rotor blade failed by fatigue which initiated from a 0.38-mm (0.015-in.) deep acid etch pit located at the inner wall of the spar cavity. The pitting was a result of a multiple plating and plating re-

moval etch process during manufacturing. An unauthorized series of baking operations reduced the mechanical properties of the spar. The reduced mechanical properties further contributed to the fatigue process.

How failure could have been prevented

Better processing controls during the etching and plating procedures would have lessened the possibility of defective plating and subsequent stripping. The post-plating baking which reduced the mechanical properties should not have been performed.

Fatigue Fracture of a Transport Aircraft Crankshaft During Flight

A.K. Das, Aircraft Design Bureau, Hindustan Aeronautics Ltd., Bangalore, India

A 4340 steel piston engine crankshaft in a transport aircraft failed catastrophically during flight. The fracture occurred in the pin radius zone. Fractographic studies established the mode of failure as fatigue under a complex combination of bending and torsional stresses. SEM examination revealed that the fracture origin was a subsurface defect—a hard refractory (Al_2O_3) inclusion—in the zone close to the pin radius. Chemical analysis showed the crankshaft material to be of inferior quality. It was recommended that magnetic particle inspection using the dc method be used to check for cracks during periodic maintenance overhauls.

Key Words		
Nickel-chromium-molybdenum steels	Nonmetallic inclusions	Crankshafts
Aerospace engines	Fatigue failure	Airplanes

Alloys   Nickel-chromium-molybdenum steel—4340

Background

A steel piston engine crankshaft in a transport aircraft failed catastrophically during flight.

Applications

The massive, complex-shaped crankshaft (Fig. 1) was machined from a triple alloy steel forging. During flight, the crankshaft experienced a combination of highly complex stresses, including bending and torsion, particularly at the critical pin radius zone. The crankshaft was magnetic particle inspected during periodic major overhauls for the presence of defects or crack indications in the pin radius zone.

Circumstances leading to failure

One of two crankshafts of a twin piston engine catastrophically fractured into two pieces in an area near the pin radius zone (Fig. 2) during flight. The aircraft landed safely with the help of the other engine. The aircraft was maintaining an average flight speed of 2050 rev/min at the time of failure. It had previously undergone three major overhauls involving 3000 h of service.

Pertinent specifications

The crankshaft was manufactured from SAE 4340 steel.

Performance of other parts in same or similar service

Similar crankshafts were reported to have completed their predicted service life satisfactorily.

Specimen selection

The undamaged portion of the crankshaft was selected for detailed fracture analysis under a low-power stereoscopic microscope. A small portion containing the fracture origin zone was sectioned for in-depth analysis by microexamination and scanning electron microscopy (SEM).

Visual Examination of General Physical Features

Visual examination revealed that the failure occurred transversely to the crankshaft axis, with no evidence of plastic deformation around the fracture zone. The fracture features of the broken half on the left side of the crankshaft (Fig. 2) were found to be more or less intact, whereas those of

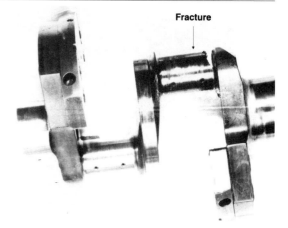

Fig. 1   Both portions of the failed crankshaft, showing fracture through the pin radius zone. 0.02×

Fig. 2   Fractograph of the undamaged fracture portion, showing the distinct eye zone of a fatigue crack. 1.11×

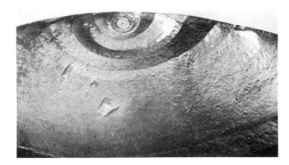

Fig. 3 Closeup view of the eye zone, showing a tiny black subsurface inclusion that acted as the fracture initiation site. 3.92×

Fig. 4 SEM micrograph showing uniformly spaced beach marks around the hard irregular refractory inclusion enclosed in a nonmetallic oxide film

the mating surface were completely obliterated by rubbing that occurred after failure.

Testing Procedure and Results

Surface examination

Macrofractography. Examination of the fracture using a low-power stereoscopic microscope revealed a smooth surface with a distinct "eye zone" formed by several conspicuous beach marks around a tiny subsurface defect that served as the origin (black spot, Fig. 3). The fracture features were indicative of a bending fatigue failure. The smooth zone of the fracture covered slightly more than half the circumference. The remainder exhibited a coarse, irregular surface, typical of overstress failure.

Scanning Electron Microscopy/Fractography. The fracture origin area was examined at higher magnification by SEM to identify the nature of the subsurface defect. The defect was a lone refractory-type inclusion enclosed by a nonmetallic film (Fig. 4). The inclusion appeared to be Al_2O_3, as this type of hard refractory inclusion does not undergo deformation by heavy mechanical working, such as forging. However, actual characterization of the inclusion could not be confirmed owing to a lack of appropriate instrumentation at the time of investigation.

Metallography

Microstructural Analysis. Microexamination of a section across the fracture in the region adjacent to the subsurface defect revealed that the material was very unclean, with numerous globular nonmetallic oxide inclusions (Fig. 5). Otherwise, the microstructure consisted of fine-grained tempered martensite.

Chemical analysis/identification

The chemical composition of the crankshaft material conformed to SAE 4340 specifications (Table 1).

Mechanical properties

Hardness. Microhardness measurements using a 30 kg load produced an average value of 383

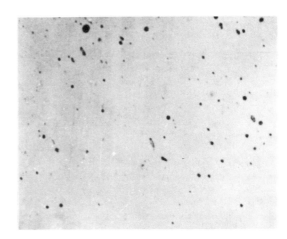

Fig. 5 Micrograph showing nonmetallic globular oxide inclusions

Table 1 Results of chemical analysis

Element	Composition, %	
	Failed crankshaft	SAE 4340 requirements
Carbon	0.41	0.38-0.43
Manganese	0.75	0.65-0.80
Phosphorus	0.015	0.025 (max)
Sulfur	0.025	0.040 (max)
Nickel	1.75	1.65-2.00
Chromium	0.82	0.70-0.90
Molybdenum	0.25	0.20-0.30

HV. The minimum specified hardness requirement was 360 HV.

Discussion

Fractographic studies established the mode of failure as fatigue under a complex combination of bending and torsional stresses. Its origin was traced to a subsurface hard refractory (Al_2O_3) type of inclusion. The subsurface defect acted as a severe notch for the initiation of fatigue cracking under dynamically loaded conditions. The overall cleanliness of the material was marred by a high content of oxide inclusions.

During three successive overhauls, the crankshaft was magnetic particle inspected using the ac method. No crack indication was reported, even though the subsurface crack may have existed in the dormant state. Failure to detect the presence of a defect or crack was attributed to the small size of the inclusion (about 0.01 mm, or 0.0004 in.) and its depth 1 mm (0.04 in.) below the surface. The ac method is insensitive at such depths, and the dc method should be employed instead.

Conclusion and Recommendations

Most probable cause

The catastrophic fracture was attributed to the presence of a subsurface defect in the form of a lone hard refractory inclusion in the critically loaded pin radius zone, which became the initiation site for a hidden fatigue crack. The uncleanliness of the material further reduced its fatigue resistance.

Remedial action

It was recommended that the supplier exercise utmost care to maintain product quality, including the use of the dc method of magnetic particle inspection.

Contact Fatigue Failure of A Bull Gear

G. Mark Tanner, Radian Corporation, Austin, Texas
James R. Harty, The Hartford Steam Boiler Inspection and Insurance Company, Hartford, Connecticut

A bull gear from a coal pulverizer at a utility failed by rolling-contact fatigue as the result of continual overloading of the gear and a nonuniform, case-hardened surface of the gear teeth. The gear consisted of an AISI 4140 Cr-Mo steel gear ring that was shrunk fit and pinned onto a cast iron hub. The wear and pitting pattern in the addendum area of the gear teeth indicated that either the gear or pinion was out of alignment. Beach marks observed on the fractured surface of the gear indicated that fatigue was the cause of the gear failure. Similar gears should be inspected carefully for signs of cracking or mis-alignment. Ultrasonic testing is recommended for detection of subsurface cracks, while magnetic particle testing will detect surface cracking. Visual inspection can be used to determine the teeth contact pattern.

Key Words	Chromium-molybdenum steels, mechanical properties		Bull gears
	Rolling contact	Contact fatigue	Fatigue failure
	Pulverizers	Alignment	
Alloys	Chromium-molybdenum steel—4140		

Background

Applications

The gear is part of a coal pulverizer, built in 1978. The bull gear was driven by a pinion gear at 100 to 200 rpm. The pinion gear was reportedly undamaged. The gear box is sealed and uses extra pressure (EP) lubricant. The bull gear consists of a steel gear ring that is shrunk fit and pinned onto a cast iron hub.

Circumstances leading up to failure

The gear failed in service under normal operation.

No potential problem with the gear was indicated.

Performance of other parts in same or similar service

None of the other coal pulverizers had experienced a failure.

Selection of specimens

Four sections from the bull gear were received for analysis. Three of the sections contained fractures and one section was intact.

Visual Examination of General Physical Features

The four gear sections received for analysis are shown in Fig. 1 through 4. Figures 1 and 2 show the two mating fractured sections. Figure 3 shows a fractured gear tooth and Fig. 4 shows the largest section, which was intact and contained no apparent damage.

(a)

(b)

Fig. 1 One section of the bull gear as received for analysis

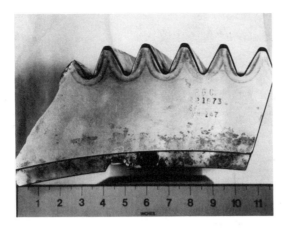

Fig. 2 The mating section of the section in Fig. 1

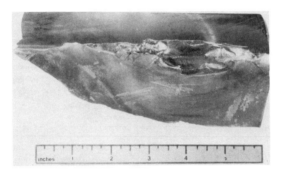

Fig. 3 A gear tooth section as received for analysis

Testing Procedure and Results

Non-destructive evaluation

The largest gear section underwent wet fluorescent magnetic particle testing to check for additional cracks. No additional cracks were observed. The teeth in the gear section were then inspected using ultrasonic testing. Indications were observed in two teeth 51 to 63 mm (2.0 to 2.5 in.) from the edge of the gear. The teeth that contained indications are marked with arrows in Fig. 4.

Surface examination

Beach marks were observed over most of the fracture surfaces on the three fractured gear sections. Figures 5 and 6 show beach marks on two of the fractured sections, as indicated by arrows. Beach marks are small ridges that develop on a fracture surface during cyclic load variations. Beach marks can be used to locate fatigue initiation sites as they are concentric about the origin area. The pattern of beach marks on the fracture surfaces indicated a primary crack initiation site just below the surface of the gear tooth. The initiation site was located approximately 63 mm (2.5 in.) from the edge of the gear. The crack propagated towards the middle of the tooth and then changed direction and propagated radially towards the hub. Several subsurface secondary initiation sites were also observed.

Figure 7 shows a sketch of a gear tooth profile with observed wear and pitting damage. Surface wear, shown in Fig. 8, was observed on the face of the gear teeth in the addendum area and was lo-

Fig. 4 The largest gear section, which was intact, as received for analysis. Arrows point to teeth that contained ultrasonic indications.

cated slightly towards one end of the teeth. The surface wear was on the same side of the gear teeth as the initiation site of the cracks. Figure 9 shows surface pitting observed on the opposite side of the teeth in the addendum area. The pitting started approximately 6.3 mm (0.25 in.) from the edge of the gear and continued for about 25 mm (1.0 in.). A smaller amount of pitting was observed on the gear diagonal to this pitting, as shown in Fig. 7.

Metallography

Three cross sections were cut from the gear, one containing an indication and two without indica-

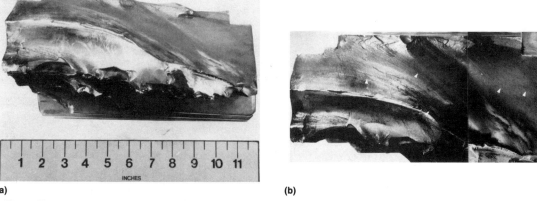

(a)

(b)

Fig. 5 Photographs showing the fracture surface of the gear section shown in Fig. 1. Arrows point to beach marks. (a) 0.65×. (b) 0.30×

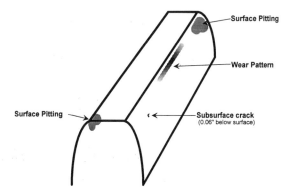

(a)

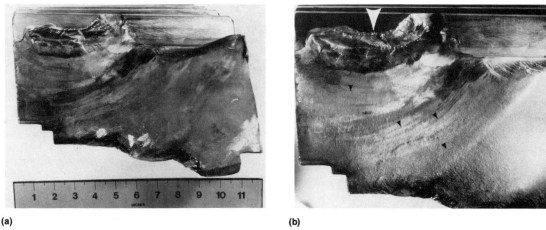

(b)

Fig. 6 The fracture surface of the gear section shown in Fig. 2. Small arrows point to beach marks, while the large arrow points to fatigue crack origin. (a) 0.63×. (b) 0.44×

Fig. 7 Sketch of a gear tooth profile showing observed wear and pitting damage. Approximate location of a subsurface crack is shown.

Fig. 8 Photograph showing surface wear typical of that observed on all of the teeth. 0.22×

tions. The sections without indications were removed from opposite sides of the gear and were ground and macroetched to reveal the hardened case of the gear. Figure 10 shows the macroetched cross section through a tooth, which reveals the location and thickness of the hardened case. The hardened case did not contain any discontinuities; however, the case depth was not uniform on both sides of each tooth. The case depth varied from ap-

proximately 1.8 mm (0.07 in.) on one side of the tooth, to approximately 1.3 mm (0.05 in.) on the other side.

One macroetched section was subsequently prepared using standard metallographic techniques. The microstructure of the gear, shown in Fig. 11, consists of tempered martensite with a few scattered manganese sulfide inclusions, typical of AISI 4140 in the quenched-and-tempered condition. The microstructure in the case-hardened region of the gear tooth, shown in Fig. 12, consists of martensite and dispersed carbides, with a few scattered manganese sulfide inclusions.

The section that contained an indication was ground and polished until a crack was observed. Figure 13 shows a subsurface crack indication on a cross section through a tooth. The crack is located approximately at the same level as the pitch line. Figure 14 shows the crack initiation area. The crack initiated approximately 1.5 mm (0.06 in.) below the surface. The cracking mode is transgranular, characteristic of fatigue. A microstructural feature, termed butterfly wings, was observed in the vicinity of the crack. Butterfly wings are evidence of rolling contact fatigue (Ref 1).

Chemical analysis/identification

The chemical composition of the gear material is given in Table 1, along with the specification for AISI 4140. The gear material meets the chemical requirements of the specification.

Mechanical properties

Microhardness measurements of the base metal and case on all three metallographic sections were taken, using a Vicker's microhardness indenter. The microhardness values were sub-

sequently converted to the more common Rockwell C scale. The hardness ranged from 48 to 60 HRC in

Table 1 Chemical composition of the bull gear (% by weight)

Element	Gear	Chemical requirements per AISI 4140
Carbon	0.40	0.38 to 0.43
Manganese	0.80	0.75 to 1.0
Silicon	0.30	0.15 to 0.30
Phosphorus	0.012	0.35 max
Sulfur	0.017	0.040 max
Nickel	0.08	NS(a)
Chromium	0.82	0.80 to 1.10
Molybdenum	0.15	0.15 to 0.25
Iron	Balance	Balance

(a) NS—Not Specified

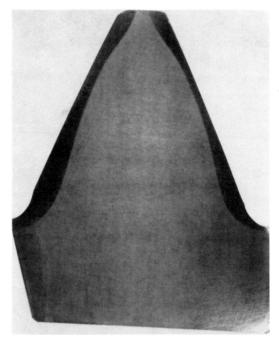

Fig. 10 Macroetched cross section through a gear tooth showing the geometry and thickness of the hardened case. Note the irregular case depth. Nital etchant, 1.86×

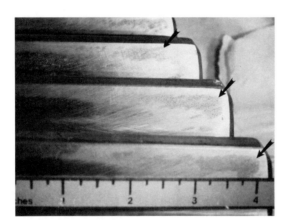

Fig. 9 Surface pitting (arrows) observed on all of the teeth

(a)

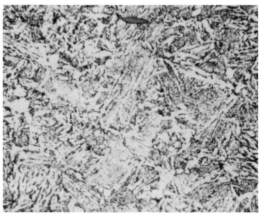

(b)

Fig. 11 Gear core microstructure consisting of tempered martensite with an average hardness of 24 HRC. Nital etchant. (a) 154×. (b) 616×

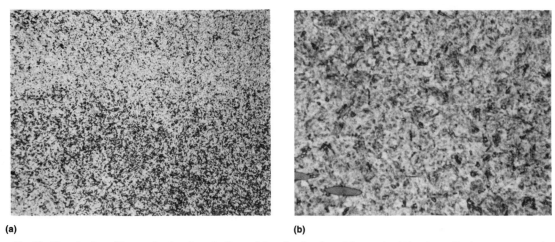

(a)

(b)

Fig. 12 Microstructure of the case-hardened gear teeth, consisting of martensite and dispersed carbides along with a few manganese sulfide inclusions. The average hardness is 55 HRC. Nital etchant. (a) 154×. (b) 616×

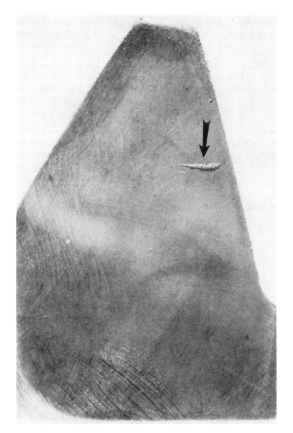

Fig. 13 Cross section through a gear tooth with a crack indication (arrow) revealed by magnetic particle inspection. 2.2×

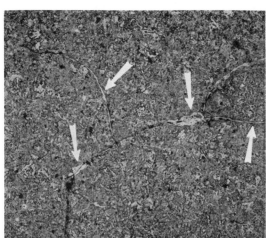

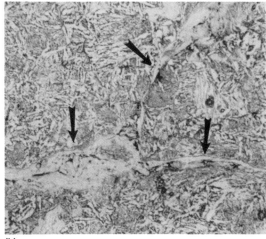

(b)

Fig. 14 Photomicrographs of the crack initiation area. Butterfly wings (arrows) are evidence of rolling-contact fatigue. Nital etchant. (a) 100×. (b) 400×

the case hardened region, and averaged 24 HRC in the core of the teeth without cracks. The hardness for the tooth that contained a crack averaged 55 HRC in the case hardened region and 23 HRC in the core.

Discussion   AISI 4140 steel is a commonly used material of construction for gears. The case and the gear teeth were case-hardened to promote increased wear resistance. The material is appropriate for this service and material. The base metal exhibited a normal microstructure for a quenched-and-tempered low-alloy steel. The core and case hardness values are in the normal range for a gear.

The wear and pitting pattern in the addendum area of the gear teeth indicates that either the gear or the pinion was out of alignment. Misalignment will cause localized overloading of the gear. Since the wear and pitting pattern occurred on all of the teeth, it is unlikely that misalignment occurred as a result of deflection of teeth that were cracked.

The beach marks observed on the fracture surface of the gear indicate that fatigue was the cause of the gear failure. The location of the fatigue initiation site and the microstructural feature of butterfly wings indicate the gear failure started by rolling-contact fatigue.

Rolling-contact fatigue is the result of stresses very near the contact area exceeding the endurance limit of the material. The endurance limit is the minimum stress level required to initiate fatigue. In rolling conditions, the maximum stress is the shear stress located below the surface just ahead of the rolling point of contact. If sliding is occurring in the same direction, the shear stress will increase at the same point. If the shear plane is close to the surface, pitting or spalling will occur (Ref 1). However, if the shear plane is deep due to a continual, heavy-rolling load contact, the crack will propagate inward, as was the case with the bull gear. The crack initiated approximately 1.5 mm (0.06 in.) below the surface in the core; the case depth at this location was 1.3 mm (0.05 in.). Crack propagation was parallel to the surface for a short distance and then turned radially inward.

Conclusions and Recommendations

Most probable cause

The root cause of the rolling-contact fatigue could not be determined from the available information. However, the most likely cause is a continual, localized overloading of the gear. A nonuniform case-hardened surface of the gear teeth may have also contributed to the failure. The wear pattern indicated the gear was overloaded on approximately 5% of the face during service, causing the maximum stress to occur at a greater depth. This depth was located in the core below the maximum case depth on one side of the tooth. A harder material (non-austenitic steel) correlates with a higher tensile strength and a higher endurance limit. The stress level in the gear was above the endurance limit of the core. If minimum case depth on both sides of each tooth had been at least 18 mm (0.07 in.), the maximum stress would have occurred in the case and may have been below the endurance limit. As a result, fatigue failure would not have occurred. Consequently, the nonuniform case-hardened layer on the gear teeth probably contributed to the failure.

Remedial action

Gears in the other coal pulverizers at the utility may also be cracking. All similar gears should be inspected carefully for cracking. Ultrasonic testing is recommended for the detection of subsurface cracks, while magnetic particle testing will detect surface cracking. The gears should be checked for any indication of misalignment, which could cause overloading. Visual inspection of the contact pattern on the teeth is one method which can be used to check for misalignment.

Reference

1. *Metals Handbook*, 9th ed., Vol 11, Failure Analysis and Prevention, American Society for Metals, 1986, p 593

Failure of Four Cast Steel Gear Segments

George M. Goodrich, Taussig Associates, Inc., Skokie, Illinois

Gears in a strip mining dragline failed in service. The material was identified as a low-alloy (NiCrMoV) steel. SEM analysis indicated that the initial fracture and subsequent fractures resulted from impact or a suddenly applied load. Mechanical testing indicated that the gears had low impact strength. Failure was attributed to low toughness caused by the absence of, or improper, heat treatment. Casting defects identified during metallographic examination were determined to be the fracture initiation site, but were considered less significant than the low as-received impact strength of the material. It was recommended that the equipment manufacturer implement an appropriate heat treatment to meet the impact requirements of the application.

Key Words

Nickel-chromium-molybdenum steels
Brittle fracture
Shrinkage

Castings
Cleavage
Mining machinery

Gears
Casting defects

Background

Gears in a strip mining dragline failed in service. The four failed gear segments were submitted for metallurgical analysis.

Applications

The segments reportedly represented one and a half gears from a dragline piece of equipment used for strip mining. Four complete gears were used in the equipment to lift and move the dragline. The total weight lifted by these four gears was 6750 Mg (7450 tons). Each gear was made from two half segments. Each half segment weighed approximately 6800 kg (15,000 lb).

Circumstances leading to failure

In September 1982, the entire bull gear section of the dragline unit was rebuilt. Upon installation of the gears, however, the mounting holes did not line up. This misalignment resulted in the failure of the two rear gears. The front gears were reworked and were certified as acceptable. The rework included opening the counterbore, sleeving previous gear-to-camshaft bolt holes, and redoing the bolt holes. These two reworked gears were then installed at the rear of the dragline unit in April 1983, and new front gears were installed on November of the same year. The rear gears failed the following month.

Pertinent specifications

The gear segments were reportedly manufactured from a cast steel. However, compositional requirements and material specifications were not known.

Visual Examination of General Physical Features

The gear segments were designated as MLR, MRR, MLF, and F. Each was cleaned and visually examined. The MLF segment had 17 gear teeth. These teeth were identified with consecutive numbers beginning at the fracture (Fig. 1). The fracture origin appeared to be coincidental with the gear teeth root. The only other evidence of damage on this gear was at the crown of teeth 2, 3, and 4 (Fig. 2).

The gear segment designated as MRR had 14 teeth. These teeth were numbered consecutively, with the first tooth located at the nonbroken end. The fracture had approximately the same configuration as that observed for segment MLF. These two segments apparently were from the same original gear half. Again, the fracture surfaces suggested that the origin was coincidental with the

Fig. 1 Closeup view of the fracture surface on gear segment MLF

Fig. 2 Damage observed at the crown of teeth 2, 3, and 4 on gear segment MLF

Fig. 3 Closeup view of the fracture at tooth 18 on gear segment F

Fig. 4 Damage observed on tooth 27 of gear segment F at the pitch line

gear teeth root. No other damage was evident.

The MLR gear segment had 28 teeth, which were numbered consecutively beginning from the nonfractured end. The suspected fracture origin was again coincidental with the gear teeth root. No other damage was visible.

Gear segment F had 34 teeth and exhibited no area where complete fracture had occurred across the section. Each gear tooth was numbered consecutively for identification purposes. Two of the teeth (18 and 23) were partially missing. The fracture surfaces that resulted exhibited a faint thumbnail pattern (Fig. 3), indicating that fatigue may have been a factor in this failure. In addition, tooth 24 exhibited visual evidence of fracture at a location that was similar to teeth 18 and 23. The only other visible evidence of damage was observed near the pitch line on teeth 5 and 27 (Fig. 4).

Testing Procedure and Results

Nondestructive evaluation

Liquid Penetrant Inspection. After cleaning, all of the gear segments were subjected to nondestructive testing (NDT) using a dye penetrant. The purpose of this examination was to locate other areas with indications suggesting that additional fractures were present.

Gear segments MLF, MRR, and MLR exhibited the presence of indications in almost every gear tooth root. An example of these indications is shown in Fig. 5. Gear segment F exhibited indications in the roots to a lesser extent. A typical faint indication is shown in Fig. 6.

As a consequence of this investigation, selected teeth were removed. These sections included the fracture surfaces and other areas that contained representative indications.

Magnetic Particle Inspection. The sectioned areas were then subjected to magnetic particle inspection, which produced indication results that were similar to those observed using the dye penetrant technique.

Fig. 5 Indications (dark) of possible cracks on gear segment MLR in the root between teeth 5 and 6

Surface examination

Visual. Each of the sections that were removed from the main gear bodies were examined further using visual techniques and a stereoscopic microscope. The purpose of this examination was to search for evidence of additional damage or other cracks that would be helpful in the failure analysis and to evaluate the fracture surfaces on gear segment F at the suspected fracture origins.

The fracture surface of gear segment MLR (tooth 28) showed the presence of striations that suggested the fracture origin was indeed coincidental with the gear tooth root. The pitch face of this gear tooth exhibited evidence of severe damage (Fig. 7), suggesting that a foreign object had

Fig. 6 Closeup view of a faint indication (center of photograph) found near the root of tooth 8 of gear segment F

Fig. 7 Pitch face and fracture region of tooth 28 of gear segment MLR

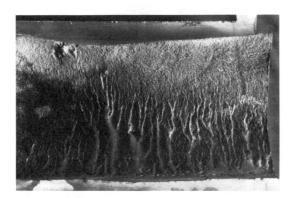

Fig. 8 Thumbnail pattern observed on the fracture surface on tooth 18 of gear segment F

lodged in the gear at this location.

The fracture surfaces on teeth 18 and 23 of gear segment F exhibited similar thumbnail patterns that appeared to emanate from a single location on each fracture surface. Figures 8 and 9 show the thumbnail pattern and the suspected fracture origin, respectively, for tooth 18. Stereomicroscopic examination of both teeth revealed that the suspected fracture origins exhibited evidence of porosity.

Scanning Electron Microscopy/Fractography. The fracture surfaces of tooth 28 of gear segment MLR (MLR-28) and of teeth 18 and 23 of segment F (F-18 and F-23) were each examined further using scanning electron microscopy (SEM) to search for evidence of defects or discontinuities that could have caused or contributed to the failure. Determination of the fracture mode was also a goal.

Examination of tooth MLR-28 at selected locations along the gear root, suspected as being the fracture origin, revealed that the fracture was transgranular and exhibited characteristics commonly associated with cleavage (Fig. 10 and 11), which indicated that the segment fractured as the result of a suddenly applied load. Cleavage also can be an indication of low toughness. No evidence of defects or discontinuities was found along the fracture surface. Also, no evidence of beach marks was observed, which indicated that the fracture was not caused by fatigue. Dimples or microvoid

Fig. 9 Suspected fracture origin on tooth 18 of gear segment F

coalescence was not observed, which ruled out ductile overload. Finally, because the fracture occurred transgranularly, rather than intergranularly, the material could not have been in an embrittled condition at the time of failure.

Examination of tooth F-18 revealed that the suspected fracture origin exhibited evidence of a dendritic pattern—an indication of shrinkage porosity (Fig. 12 and 13). Further examination of the fracture surface in the immediate vicinity of the dendrite and porosity revealed characteristics

Fig. 10 SEM micrograph showing the fracture surface of tooth MLR-28. A location near the center is shown at higher magnification in Fig. 11. ~20×

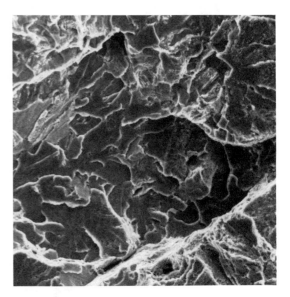

Fig. 11 SEM micrograph showing the cleavage detected on the fracture surface of tooth MLR-28. ~500×

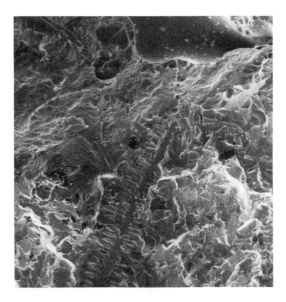

Fig. 12 SEM micrograph showing an apparent dendrite near the suspected fracture origin of tooth F-18. ~150×

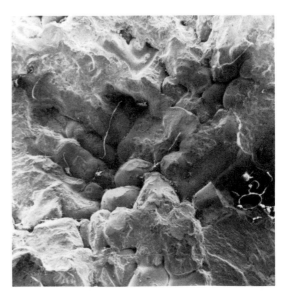

Fig. 13 SEM micrograph showing shrinkage porosity detected at the suspected fracture origin on tooth F-18. ~15×

typically associated with cleavage.

Evidence of cleavage was also present in the immediate vicinity of the fracture origin on tooth F-23. Neither of the teeth from gear segment F exhibited beach marks, microvoid coalescence, or intergranular fracture.

Although visual examination of segment F revealed the presence of a possible thumbnail pattern around the suspected fracture origins, the presence of this pattern at a microscopic level was not confirmed by SEM. This analysis did show that the fracture surfaces had cleavage rather than beach marks. It is also important to note that the suspected fracture origins on both teeth from segment F were associated with shrinkage porosity.

Metallography

Several gear teeth were selected for metallographic analysis: MLR-28, which had an apparent fracture origin at the root; MRR-4, F-8, and F-9, which had NDT indications in the root; F-5, with visual pitch line damage; and F-18 and F-23, which had fracture origins at shrinkage cavities. Cross sections were prepared in accordance with standard metallographic techniques. They were examined unetched and etched at magnifications up to 1000×.

Gear tooth MLR-28 was examined at the suspected fracture origin and along the pitch face. In the unetched condition, the fracture surface appeared relatively flat immediately adjacent to the suspected fracture origin, which was indicative of the smeared condition observed by SEM. This con-

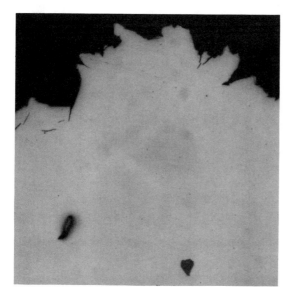

Fig. 14 Cross section of the MLR-28 fracture surface, showing the irregular nature of the surface and secondary cracking. Unetched. 200×

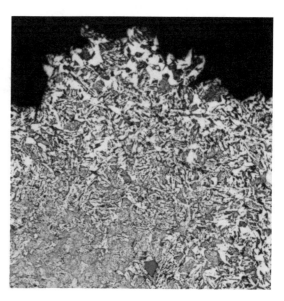

Fig. 15 Same region shown in Fig. 14, but in the etched condition. A pearlite and ferrite structure was revealed. Nital etch. 200×

Fig. 16 Deformed region along the pitch face of tooth MLR-28. Nital etch. 50×

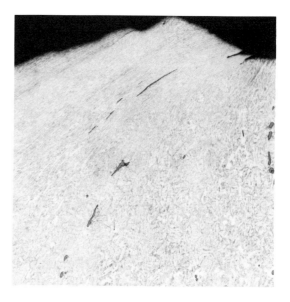

Fig. 17 Cross section through the damaged pitch line area of tooth F-5. The damage was the result of cold working, suggesting that a foreign object had become temporarily wedged at this location. Unetched. 50×

dition indicated that the surface at this location may have sustained additional damage after fracturing occurred. Farther along the fracture surface, the fracture had an irregular contour and exhibited secondary cracks (Fig. 14)—indicative of a cleavage-type fracture. In the etched condition, the tooth exhibited a mixture of ferrite and pearlite, the appearance of which suggested that the material may have been normalized and tempered (Fig. 15).

The damage that was observed along the pitch face of tooth MLR-28 exhibited strong evidence of cold working (Fig. 16). The presence of cold working is indicative of damage caused by the wedging of a segment against the pitch face.

Tooth 4 from gear segment MRR was metallographically examined to investigate the NDT indications of a possible crack in the root. No discontinuities were evident, suggesting that nondestructive testing had indicated machine tool marks, not cracks.

Gear tooth F-5 was chosen to further investigate the damaged condition observed at the pitch line. Evidence of surface cold working was found (Fig. 17), suggesting that a small piece of material may have become temporarily wedged at this location and caused the damage.

A sample from the suspected fracture origin on tooth F-18 revealed characteristics typically associated with shrinkage (Fig. 18). Foreign constitu-

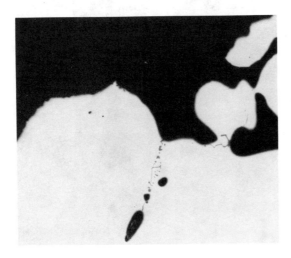

Fig. 18 Shrinkage porosity observed at the suspected fracture initiation site on tooth F-18. Unetched. 50×

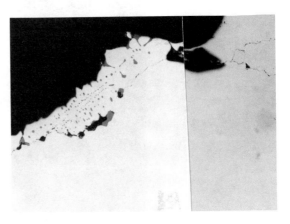

Fig. 19 Closeup view of slag particles detected along the fracture surface near the fracture origin (upper right) shown in Fig. 18. Unetched. 122×

Table 1 Results of chemical analysis

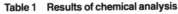

| | Composition, % | | |
Element	Gear MLR	Gear MRR	Gear F
Carbon	0.31	0.26	0.31
Manganese	1.00	1.27	1.01
Phosphorus	0.019	0.021	0.020
Sulfur	0.028	0.028	0.027
Silicon	0.28	0.36	0.53
Nickel	0.77	0.81	0.77
Chromium	0.63	0.70	0.81
Molybdenum	0.21	0.22	0.43
Copper	0.11	0.12	0.07
Lead	<0.02	<0.02	<0.02
Aluminum	0.02	0.04	0.02
Tin	<0.02	<0.02	<0.02
Vanadium	0.02	0.02	0.01
Niobium	<0.05	<0.05	<0.05
Boron	0.0021	0.0021	0.0028
Titanium	<0.03	<0.03	<0.03
Hydrogen	0.0002	0.040	0.001
Oxygen	0.002	0.005	0.010
Nitrogen	0.008	0.008	0.007

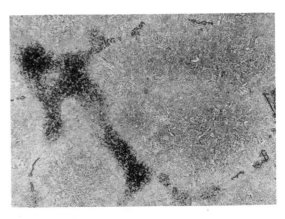

Fig. 20 Detailed view of microsegregation observed on tooth F-23. Nital etch. 14.8×

ents, probably slag inclusions, were present in the metal (Fig. 19). Farther along the fracture surface were irregularities and secondary cracking, again typical of cleavage-type fracture. The etched structure at this location reflected the presence of ferrite and pearlite.

Shrinkage and slag segregation were also found in the immediate vicinity of the suspected fracture origin on tooth F-23. In the etched condition, the structure at this location showed evidence that segregation occurred during solidification (Fig. 20).

The presence of shrinkage, slag segregation, and microsegregation at the suspected fracture origins on teeth F-18 and F-23 indicated that a shrinkage condition or an area representative of the last region to solidify in the vicinity served as fracture origins on both broken teeth. This analysis did not identify additional cracks on any other

gear teeth, except the one observed visually on tooth 24. Although NDT evidence suggested the possible presence of other cracks, the use of metallographic techniques was unable to confirm that these indications actually existed.

Chemical analysis/identification

The results of chemical analysis of gear segments MLR, MRR, and F are presented in Table 1. This analysis showed that the three segments had basically comparable compositions, although the F gear did have slightly higher silicon and molybdenum contents than the other two gears. Based on these results, the composition was determined to be similar to grades 4, 6, and 7 in ASTM A487 (however, only grade 7 contains boron). Basically, the carbon contents of the gear segments were higher than specified in those three grades. None of the gears contained elements at levels considered detrimental to performance.

Mechanical properties

Tensile Properties. Tensile bars were machined from teeth MLR-28, MRR-3, MRR-4, and F-

Table 2 Results of tensile testing

Gear segment	Tensile strength		0.2% offset yield strength		Elongation in 50 mm (2 in.), %	Reduction of area, %
	MPa	ksi	MPa	ksi		
MLR	738	107	607	88	6.0	11.9
MRR	880	127.8	695.7	100.9	7.5	15.2
F	769	111.5	640.5	92.9	2.0	3.1

Table 3 Results of impact testing in the as-received condition

Gear segment	Test temperature		Impact energy (average)	
	°C	°F	J	ft · lbf
MLR	–20	0	11.1	8.2
	0	32	18.0	13.3
		RT	24.1	17.8
MRR	–20	0	12.2	9.0
	0	32	12.6	9.3
		RT	14.5	10.7
F	–20	0	5.0	3.7
	0	32	7.1	5.2
		RT	7.9	5.8

Table 4 Results of impact testing after quenching and tempering

Gear segment	Test temperature		Impact energy (average)	
	°C	°F	J	ft · lbf
MLR	–20	0	30.8	22.7
	0	32	50.6	37.3
		RT	61.0	45.0
MRR	–20	0	68.5	50.5
	0	32	73.2	54.0
		RT	76.6	56.5
F	–20	0	54.0	39.8
	0	32	54.2	40.0
		RT	63.0	46.5

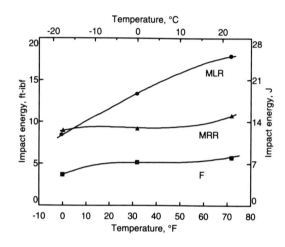

Fig. 21 Results of Charpy impact testing on as-received gear segments. See also Tables 2 and 3.

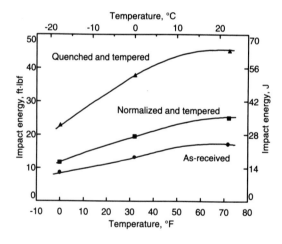

Fig. 22 Impact energy of heat-treated gear segment MLR in as-received gear segment. See also Tables 3 to 5.

8. The results of tensile testing are shown in Table 2. The three gear segments had approximately comparable tensile strengths and yield strengths, with the strength of segment F intermediate between the two other segments. The F segment had the lowest elongation and reduction of area values.

Impact Toughness. Charpy impact bars were machined from the three gear segments and tested at –20 °C (0 °F), 0 °C (32 °F), and room temperature—representative of possible operating conditions for the strip mining dragline. Figure 21 and Table 3 show the data obtained by tests of the three gears in the as-received condition.

The low impact energy measured for gear segment F, combined with its low elongation and reduction of area values, suggested that under any impact situation where all three gear segments were used, segment F would fracture first. Because initial impact testing indicated that segment F, as well as segment MRR, had relatively low impact strengths compared with expectations for material to be used under the conditions described, additional impact tests were performed. This time, however, samples of the three segments were heat treated by quenching and tempering

and by normalizing and tempering. An austenitizing temperature of 845 °C (1550 °F) was used in all cases.

Gears that were quenched and tempered generally exhibited impact strengths greater than 40 J (30 ft · lbf) and, in most instances, greater than 55 J (40 ft · lbf) (Table 4). A normalizing and tempering heat treatment produced impact strengths slightly higher than the as-received condition (Table 5). Figures 22 to 24 summarize the information that was developed on impact energy for the MLR, MRR, and F gear segments, respectively.

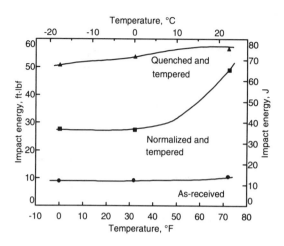

Fig. 23 Impact energy of heat-treated gear segment MRR relative to as-received gear segment. See also Tables 3 to 5.

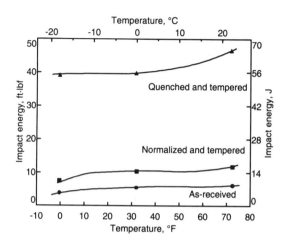

Fig. 24 Impact energy of heat-treated gear segment F relative to as-received gear segments. See also Tables 3 to 5.

Discussion

Based on the results of this analysis, gear segment F was the initial segment to fracture in service. The teeth that became dislodged during this fracturing incident were carried over by the drive pinions or timing pinion and ultimately caused the other two gear segments to fracture. The initial fracture and subsequent fractures were the consequence of impact or a suddenly applied load in a material with low impact strength.

SEM examination of the fracture surfaces, which revealed the presence of cleavage, formed the basis of this conclusion. The fact that segment F had lower impact strength in the as-received condition further substantiated that the F gear broke first.

Although visual evidence suggested that fatigue may have been a factor in this failure, closer examination with the SEM and metallographic analysis produced results that indicated otherwise. In segment F, imperfections were found at the suspected fracture origins. These imperfections appeared as shrinkage resulting from the casting operation and were to be expected for a casting of this size. These discontinuities served as the fracture initiation site, but did not actually cause the fracture to occur.

Sites such as shrinkage porosity serve to initiate fracturing under many conditions. When im-

Table 5 Results of impact testing after normalizing and tempering

Gear segment	Test temperature °C	°F	Impact energy (average) J	ft · lbf
MLR	−20	0	15.6	11.5
	0	32	26.2	19.3
		RT	34.6	25.2
MRR	−20	0	37.3	27.5
	0	32	37.7	27.8
		RT	66.7	49.2
F	−20	0	10.6	7.8
	0	32	14.0	10.3
		RT	15.6	11.5

pact is a factor, the mode of failure becomes more important than the fracture initiation site. Had the fracture been caused by fatigue, more significance would have been attached to the presence of shrinkage.

Mechanical property testing of three gear segments revealed relatively low impact strengths compared with expectations for this application. The ability of the material to withstand impact in this application should be a primary requisite. The selected material exhibited high impact strengths in the quenched and tempered condition.

Conclusion and Recommendations

Most probable cause

Because the impact strength of the steel decreased with temperature, this failure probably occurred on the first day that ambient temperature was low enough to cause fracture when the load was applied. The temperatures on the day of failure were approximately 1 to 5 °C (35 to 40 °F). Based on test results, the impact energy required to cause failure was approximately 8 J (6 ft · lbf). The combined ef-

fect of low impact strength, low temperatures, and high impact loads caused the failure.

Remedial action

The user should work with the gear supplier to determine the necessary heat treatments to achieve maximum impact resistance. Considering the large size of these gears, induction heat treatment may be advantageous.

Failure of Trailer Kingpins Caused by Overheating During Forging

T.M. Maccagno, J.J. Jonas, and S. Yue, McGill University, Department of Metallurgical Engineering, Montreal, Quebec, Canada
J.G. Thompson, Atlas Specialty Steels, Welland, Ontario, Canada

Two forged AISI 4140 steel trailer kingpins fractured after 4 to 6 months of service. Fractographic and metallographic examination revealed that cracks were present in the spool-flange shoulder region of the defective kingpins prior to installation on the trailers. The cracks grew and coalesced during service. Consideration of the manufacturing process suggested that the cracks were the result of overheating of the kingpin blanks prior to forging, which was exacerbated during forging by deformation heating in the highly-strained region. This view was supported by results of two types of tensile tests conducted near the incipient melting temperature at the grain boundaries. All kingpins made by the supplier of the fractured ones were ultrasonically inspected and six more anticipated to fail were found. It was recommended that the heating of forging blanks be more carefully controlled, especially with respect to the accuracy of the optical pyrometer temperature readout. Also, procedures must be developed such that forging blanks that trigger the over-temperature alarm are reliably and permanently removed from the production line.

Key Words

Chromium-molybdenum steels, mechanical properties		Kingpins
Forging defects	Overheating	Crack propagation
Fracturing		

Alloys

Chromium-molybdenum steels—4140

Background

Applications

The trailer kingpin, shown schematically in Fig. 1, fits into the fifth wheel of a tractor to provide the link in a tractor-trailer unit.

Circumstances leading to failure

The kingpins of two trailers fractured after the trailers had been in service for approximately 4 to 6 months. The first failure occurred during hook-up of the tractor to the trailer, and the second occurred while the tractor-trailer unit was maneuvering in a yard. These two kingpins were traced to a single supplier. All the kingpins made by this supplier were ultrasonically inspected; both those in service (*i.e.* on trailers in use), and those held in inventory. Six more kingpins considered likely to fail were found: two in service, and four in inventory.

Pertinent specifications

This supplier produced the kingpins by forging AISI 4140 steel bar stock which had been cut into blanks. The blanks were heated in an induction furnace for about 5 minutes, by which time the temperature was normally about 1177 to 1232 °C (2150 to 2250 °F). At this point, the temperature is monitored by an optical pyrometer near the end of the heating line, and if the temperature exceeds 1260 °C (2300 °F), a warning light and buzzer are activated and the blank is taken out of the production line and scrapped. A warning is also given if the temperature of the blank is below about 1149 °C (2100 °F). Such blanks are also set aside, but can be re-inserted into the heating line after cooling.

This forging was made in two steps. After the blank left the induction coil, it dropped through a chute and an operator placed it in a hammer forge roughing die, where it was given a single large deformation at a high strain rate. The partially-formed kingpin was then moved to the finishing die and given a smaller deformation at a slower strain rate. The final piece was trimmed of flash and dropped in a steel bin to cool. It was then re-austenitized in a controlled atmosphere, quenched, and tempered to 380 to 420 HB.

Performance of other parts in same or similar service

Of the approximately 15,000 kingpins that were supplied by the forging shop to the trailer manufacturer, eight were found to be seriously defective. Four of these were in service, and four were in inventory.

Selection of specimens

Fractographic and metallographic evaluation was carried out on all eight defective kingpins. Mechanical testing was also carried out on specimens taken from these kingpins, and compared with specimens taken from other heats of 4140 steel that did not give problems.

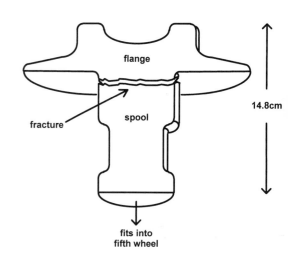

Fig. 1 Schematic of a trailer kingpin, showing where fracture occurred.

Visual Examination of General Physical Features

The two kingpins that failed in service had fractured completely across the shoulder between the flange and spool regions (Fig. 1). The failure surfaces exhibited an oxidized and soiled dark region that covered about 40% of the failure surface (Fig. 2). The oxidation indicated that this region had been exposed to air at high temperature.

Adjacent to the dark surface was a clean uncorroded light region, where final fracture occurred. The shear lip around the light region edge of the failure surface provided support for this view. Gross plastic deformation was not evident anywhere on the kingpins.

Testing Procedure and Results

Fracture surface examination

Because the dark regions of the failure surfaces were judged to be where the failure originated, these regions were the focus of the examination.

Macro (optical) Fractography. Under a low-power microscope, the dark region of the failure surface consisted of flattened step-like features that appeared to be worn, but with areas dispersed throughout that appeared smooth and unworn. The smooth unworn areas were especially evident on the failure surfaces of the second failed kingpin.

No obvious failure origin could be identified on any of the surfaces.

Scanning Electron Microscope (SEM) Fractography. In an SEM, the smooth unworn areas seen in the optical microscope exhibited a faceted appearance, characteristic of intergranular cracking (Fig. 3). The size of the individual facets indicated a large grain size (0.5 mm or 0.02 in. or more), and suggested that the cracking was the result of high temperature at some point during the processing. Moreover, the highly equiaxed appearance of these facets suggested that the grains formed *during or after* the high deformation roughing step.

SEM examination at higher magnification revealed dendritic MnS precipitates on many of the intergranular facets (Fig. 4). The facets themselves appeared to be smooth and rounded at the edges. These features suggested that very high temperatures had been reached and that grain boundary liquation (melting) had occurred.*

Sectioning along the lengths of the kingpins revealed many cracks away from the actual failure surfaces. SEM fractography of cracks that were subsequently opened up in the lab clearly showed intergranular facets with features similar to those seen on the failure surfaces.

Metallography

Microstructural Examination. Sectioning and polishing along the length of the kingpins revealed many cracks away from the actual failure surfaces, and etching additionally revealed the flow lines due to forging (Fig. 5). Cracking and failure took place in the region of the most intense deformation.

The cracks open to the outer surface appeared to be somewhat oxidized, and the outer surface itself also showed evidence of decarburization. Cracks not connected to the surface, however, showed little evidence of oxidation or decarburiza-

Fig. 2 The lower part of a failed kingpin.

Fig. 3 SEM micrograph of the fracture surface in the dark region of a failed kingpin. Note the large facets characteristic of intergranular cracking at high temperature.

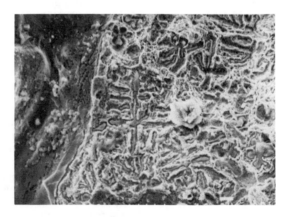

Fig. 4 Higher magnification SEM micrograph of the same fracture surface, showing dendritic MnS precipitates on the facets. This is an indication of incipient melting at the grain boundaries.

* For a more detailed discussion of grain boundary melting, see L.E. Samuels, *Optical Microscopy of Carbon Steels*, American Society for Metals, Metals Park, Ohio, 1980, p 257-291.

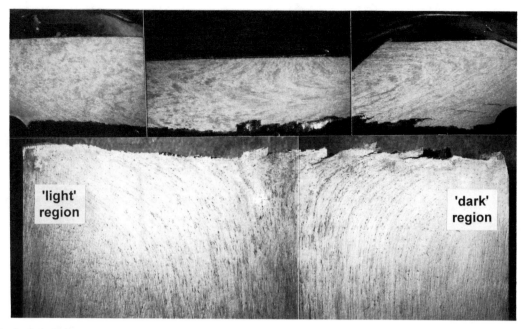

Fig. 5 Optical micrograph of axial section through the shoulder region of a failed kingpin. This polished and etched view reveals the flow pattern due to forging, and shows cracking in the dark region in addition to the main fracture.

tion. This last point indicated that the internal cracks did not originate from folding of the metal surface (say, during hot rolling of the starting blanks) but were developed at a late stage in the processing.

It was also found that the morphology of MnS precipitates differed considerably according to location. In the spool region, far away from the failed surfaces, the MnS appeared as elongated stringers (Fig. 6a) often seen in wrought steel. However, at locations closer and closer to the failure surfaces, the MnS appeared more globular, and more associated with voids identified as solidification shrinkage cavities (Fig. 6b). This morphology is characteristic of material that has undergone local melting and re-solidification. These voids do not appear to be highly deformed, also indicating that the melting occurred at a late stage during processing.

Crack Origins/Paths. The failure surface profiles corresponding to the dark and light regions of the failure surfaces are marked in Fig. 5. Cracking is associated with the deformed flow lines in the dark region.

In many instances, the shrinkage cavities and globular MnS inclusions were associated with a large prior grain structure (Fig. 7), which corresponds with the large intergranular facets observed in the section on SEM fractography. Note also that the globular MnS inclusions in Figs. 6b and 7 are probably sections through dendritic precipitates similar to those seen on grain facets in Fig. 4.

Chemical Analysis of the Bulk Material. The chemical compositions for the four heats from which the defective kingpins were made were compared with the specification for AISI 4140 steel, and the chemical compositions for other heats which did not give problems. All heats, including the ones which resulted in defective kingpins, were within

(a)

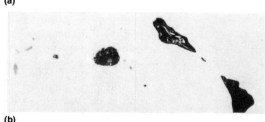

(b)

Fig. 6 Optical micrographs of the spool region of a failed kingpin. (a) Away from the site of the failure. This view shows MnS stringers typical of those seen in wrought steel products. Not etched. (b) Close to the site of the failure. This view shows the globular MnS inclusions associated with solidification shrinkage cavities. Not etched

the specification for 4140 steel, and contained low levels of residual elements. Moreover, no obvious relationship exists between the levels of residuals and the defective kingpins.

Mechanical properties

Tensile Properties. Room-temperature tensile testing was carried out on specimens cut from inventory kingpins from the same heat as the two service failures. The specimens were machined such that the gage lengths were oriented along the axis of the kingpin, in the spool-flange shoulder region. The yield stress, ultimate tensile stress, and elongation at room temperature were all found to be well within the normal ranges.

High-temperature tensile testing was also performed (on specimens cut from an inventory kingpin from the same heat as the first failed kingpin), using a technique whereby the specimen was slowly pulled while the temperature was simultaneously increased.* The results of this test (Fig. 8) established that the temperature where the ductility dropped to zero, due to incipient melting at the grain boundaries, was about 1408 °C (2566 °F). The increasing-temperature test also revealed a sharp strength drop in flow stress at about 1320 °C (2408 °F) due to dynamic recrystallization.

Impact Toughness. Charpy V-notch impact tests were performed at room temperature on specimens cut from an inventory kingpin from the same heat as the first failed kingpin, and from an inventory kingpin from a heat that did not give problems. The failure heat exhibited impact energy values virtually identical to those exhibited by the non-failure heat, and all the results were within the normal range.

Simulation Tests

Tensile Simulation Tests. Efforts were made to produce, in the laboratory, fracture surfaces that exhibited features similar to those exhibited by the failed kingpins. This was achieved by pulling the specimens at a high strain rate (about 500%/s), at a temperature a few °C below the incipient melting point. This procedure led to fracture surfaces (Fig. 9) that exhibited large intergranular facets remarkably similar to those exhibited by the actual failed kingpins (compare Fig. 9 with Fig. 3).

Introduction of Artificial Defects Prior to Forging. In order to determine whether it was possible that defects present in the blanks *before* the forging operation (*i.e.* developed during hot rolling of the bar stock) could be responsible for the cracking observed in the defective kingpins, some starting blanks containing saw cuts were forged at the forging shop. These artificial rolling defects did result in 'cracks' in the final forged pins, but the locations of these cracks depended sensitively on the location of the saw cut in the starting blanks. The artificial cracks were only occasionally found in the critical spool-flange shoulder region, in contrast with the actual defective pins, where cracking was observed *only* in the shoulder region.

Moreover, the surface morphology of the artificially-introduced cracks was quite different from the cracks observed in the defective kingpins. The artificial cracks exhibited smooth, non-granular

Fig. 7 Optical micrograph of the spool region of the failed kingpin, close to the site of the failure. This view shows the globular MnS precipitates along large prior grain boundaries.

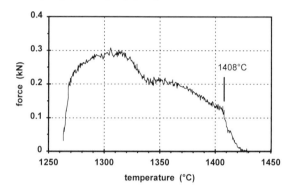

Fig. 8 Force vs. temperature for a specimen machined from a kingpin from the same heat as the first failed kingpin, and pulled in an increasing temperature test. The specimen was pulled at 2 mm/min. (0.08 in./min.) while the temperature was increased at 1 °C/s (1.8 °F/s).

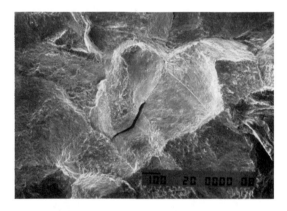

Fig. 9 SEM micrograph of the fracture surface of a specimen tested about 10 °C (18 °F) below the incipient melting temperature, at a strain rate of about 500%/s. This view shows large intergranular facets similar to those seen in Fig. 3.

surfaces with evidence of significant oxidation and decarburization, whereas the service cracks were intergranular, and the fully internal ones showed no evidence of oxidation or decarburization.

---

* See F. Hassani, T.M. Maccagno, J.J. Jonas, and S. Yue, "Behaviour of steels near the incipient melting temperature," *Metall. Trans. A.* in press, 1993.

Discussion

The fractography and metallography clearly indicated that cracks were present in the spool-flange shoulder region of the defective kingpins prior to their installation on the trailers. The shoulder is a highly stressed region of the kingpin, and these cracks grew and coalesced during the 4 to 6 months of service. At some point, the load bearing capability of the kingpin was insufficient and a final catastrophic burst of crack propagation occurred, as shown by the clean light regions of the failure surfaces.

The cracks were found in the region which undergoes the most deformation during forging. Moreover, the cracks exhibited rounded large intergranular facets associated with localized melting at the grain boundaries. Throughout the shoulder regions there were solidification shrinkage cavities and MnS inclusions. The latter showed evidence of having been melted and re-solidified (sometimes along grain boundaries) at some point after hot rolling of the forging blanks. The evidence indicated that the cracks arose as a result of overheating of the blanks before forging, exacerbated by the deformation heating developed in the highly strained shoulder region during forging. Because the processing was conducted in air, the cracks open to the surface of the kingpin oxidized, causing the 'dark' regions seen on the failure surfaces.

The likely sequence of events was reconstructed as follows. If a blank is initially overheated to 1360 °C (2480 °F) or more, the additional deformation heating produced during forging in the shoulder region can raise the temperature by some 20 to 30 °C (36 to 54 °F). Further straining becomes concentrated in this region, resulting in even more deformation heating, all the way up to the incipient grain boundary melting temperature. Upon cooling, shrinkage cavities occur, and the previously elongated MnS particles become globular and precipitate out at the boundaries of the large grains. More importantly, though, the re-solidified grain boundaries have little strength, and if a tensile stress is acting across the grain boundaries at this point, the boundaries open up to form cracks.

This explanation for the cracking is supported by the results of the tensile simulation testing.

Conclusions and Recommendations

Most probable cause

The kingpins probably failed as a result of the propagation of cracks that had formed during the forging operation. These cracks developed in kingpin blanks overheated prior to forging, a situation which was exacerbated during forging by deformation heating in the highly strained regions.

How failure could have been prevented

Optical pyrometers of the type used to measure temperature in heating lines can be notoriously imprecise; the temperature readout is significantly affected by oxide scale on the surface of blanks, calibration errors, and smoke or steam. Moreover, examination of the production line raised questions about the effectiveness of the over-temperature alarm, and of the procedures applied to the blanks that trigger the alarm.

The heating of forging blanks needed to be more carefully controlled, especially with respect to the accuracy of the temperature readout given by the pyrometer in the heating line. Similarly, procedures needed to be developed so that forging blanks that trigger the over-temperature alarm were reliably and permanently removed from the production line.

Acknowledgements

The authors wish to thank Mr. J.R. Kattus of AMC-Vulcan, Inc., Birmingham, Alabama, for permission to reproduce Fig. 5. Thanks are also due to Prof. T. Gladman of the University of Leeds, U.K., for permission to reproduce Fig. 6 and 7.

Failure of a Sprocket Drive Wheel in a Tracked All-Terrain Vehicle

W.A. Pollard, Metals Technology Laboratories, CANMET, Ottawa, Canada

A sand-cast LM6M aluminum alloy sprocket drive wheel in an all-terrain vehicle failed. Extensive cracking had occurred around each of the six bolt holes in the wheel. Evidence of considerable deformation in this area was also noted. Examination indicated that the part failed because of gross overload. Use of an alloy with a much higher yield strength and improvment in design were recommended.

Key Words	Aluminum-base alloys	Sprockets	Motor vehicles
	Sand castings	Cracking (fracturing)	Wheels
	Chain drives		
Alloys	Aluminum-base alloy—LM6M		

Background

A sprocket drive wheel in a tracked all-terrain vehicle (ATV) failed.

Pertinent specifications

The part was sand cast in alloy LM6M of British Standard 1490 (alloy A413.0).

Visual Examination of General Physical Features

The wheel, showing failure locations, is illustrated in Fig. 1. It was about 380 mm (15 in.) in diameter, and the section thickness was about 13 mm (½ in.) throughout. There was extensive cracking around each of the six bolt holes and although not shown clearly in Fig. 1, evidence of considerable deformation in this region of the wheel.

Testing Procedure and Results

Metallography

Microstructural Analysis. Sections were taken from the cracked region and prepared for metallographic examination. The structures were typical of modified aluminum-silicon alloys. Some variations in the fineness of the silicon eutectic were apparent (see Fig. 2 and 3), presumably as a result of various chilling rates.

The structure also contained a relatively high concentration of a thin, platelike constituent (appearing as needles in the cross sections shown in Fig. 2 and 3). Attempts were made to establish its composition, but met with limited success. The usual platelike constituent found in these alloys is known as β(Al-Fe-Si), and the amount present is

Fig. 1 General view of sprocket drive wheel, showing extensive cracking around each bolt hole. ~0.21 ×

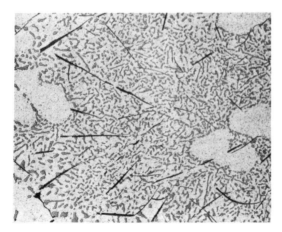

Fig. 2 Section showing comparatively coarse eutectic structure and (dark) platelike constituent. Etched in 20% H_2SO_4 at 70 °C (160 °F). 152.5×

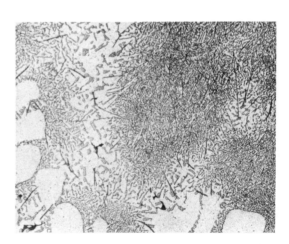

Fig. 3 Section showing fine eutectic and platelet structure. Etched in 20% H_2SO_4 at 70 °C (160 °F). 152.5×

proportional to the iron content. In this case, the etching reaction of the needles did not correspond to that of β(Al-Fe-Si), and it is presumed that other minor impurities, such as copper and manganese, were also present. In fact, many of the plates had a duplex structure, as shown at higher magnification in Fig. 4, which is from a section through a crack.

The casting was reasonably sound in the area of the flange. The moderate porosity present would not have seriously affected the strength of the part.

Crack Origins/Paths. Figure 4 also shows the strong influence that the plates had on the path of the fracture. Most of the fracture surfaces shown are either through a plate or at the interface between a plate and the aluminum matrix. Also, several incipient cracks are associated with plates.

Chemical analysis/identification

Drillings were taken from the wheel and submitted for chemical analysis. The results are shown in Table 1.

The composition of the wheel was within specification limits. The iron content, although below the specified maximum, was higher than usual for this class of material and, as will be detailed later, may have had a deleterious effect on the ductility of the alloy.

Mechanical properties

Tensile Properties. Four tensile test pieces, 75 mm (3 in.) long and 6.4 mm (¼ in.) in diameter,

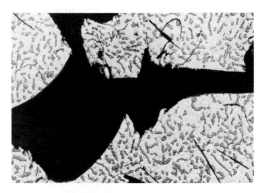

Fig. 4 Section through fracture showing influence of platelike constituent. Etched in 20% H_2SO_4 at 70 °C (160 °F). 235×

were taken from the flange between the bolt holes, as indicated in Fig. 1. The results of tensile tests are shown in Table 2, together with the minimum specified properties for separately cast test bars of alloy LM6M.

There is no provision in BS 1490 for testing bars from castings but, in North America, it is common for the minimum specifications in the casting to be 75% of the ultimate strength and 25% of the elongation of the corresponding separately cast test bars. Using that criterion, the properties of the sprocket wheel would be considered satisfactory in alloy LM6M.

Discussion

Tensile testing showed that the casting was as strong as could be expected for this alloy. Somewhat better ductility might have resulted if the iron content had been lower and, consequently, less of the embrittling, platelike constituent had been present. However, the extensive cracking and evidence of substantial deformation in the region around each of the six bolt holes indicated that failure of the part resulted from gross overload. More extensive deformation before fracture would have been of little or no value.

Conclusion and Recommendations

Most probable cause

Examination of the failed sprocket drive wheel showed that gross overload resulted in considerable deformation and ultimate failure.

Table 1 Results of chemical analysis

	Composition, %	
Element	Sample from sprocket wheel	Aluminum alloy LM6M (A413.0)
Silicon	10.99	11.0-13.0
Copper	0.25	0.6 (max)
Magnesium	0.10	0.1 (max)
Iron	0.52	1.3 (max)
Manganese	0.28	0.35 (max)
Nickel	0.04	0.5 (max)

Table 2 Results of tensile testing

| Specimen | Ultimate tensile strength | | Yield strength | | | | Elongation (4D), % |
| | | | 0.2% | | 0.1% | | |
	MPa	ksi	MPa	ksi	MPa	ksi	
A	154.4	22.4	85.5	12.4	84.1	12.2	1.0
B	172.4	25.0	93.8	13.6	80.7	11.7	2.0
C	156.5	22.7	91.7	13.3	82.0	11.9	2.0
D	155.8	22.6	86.9	12.6	75.8	11.0	2.0
Average	160.0	23.2	88.3	12.8	80.7	11.7	2.0
BS 1490:1963 (LM6M) minimum specification	162.0	23.5	...		53.8	7.8(a)	5.0

(a) Minimum not specified; for information only

How failure could have been prevented

Possible avenues for improvement include use of a premium-strength aluminum alloy, such as C-355 heat treated to the T6 condition. Strength and ductility in the critical bolt circle area would be optimized by the use of chills in the mold to maximize the freezing rate and to minimize the size and adverse morphology of silicon plates. Such techniques should develop a yield strength of 140 MPa (20 ksi) or higher, a tensile strength of 210 MPa (30 ksi), and true ductility of 1.5 to 2.0%. Plaster-mold castings of this alloy weighing 14 kg (30 lb) have consistently met specification minimum values of 303 MPa (44 ksi) tensile strength, 228 MPa (33 ksi) yield strength, and 1.5% elongation at a 95% statistical confidence level.

If space permits, increasing the thickness of the wheel web in the bolt circle area with a generous fillet blending into the web could lower the effective stress level. Stress around the bolt holes would be reduced roughly proportional to the increase in cross section. The weight penalty for such a change would be minimal (an addition of 2.8 g/cm^3, or 0.1 $lb/in.^3$).

Introducing residual compressive stress at the bore of each bolt hole might be very effective. The technique used is essentially "autofrettage." An oversize plug is forced through the hole, stretching the surrounding metal until it yields plastically. Once the plug has passed through, nonyielded material closes in elastically to create a residual compressive stress in the metal near the bore. The residual compression offsets tension service stress and thereby tends to increase resistance to fatigue failure.

Acknowledgment   Sincere thanks to Samuel R. Callaway for his comments and constructive suggestions.

Failure Analysis of an Exhaust Diffuser Assembly

K. Sampath*, Concurrent Technologies Corporation, Johnstown, Pennsylvania
M.D. Chaudhari, Columbus Metallurgical Services, Inc., Columbus, Ohio

An exhaust diffuser assembly failed prematurely in service. The failure occurred near the intake end of the assembly and involved fracture in the diffuser cone (Corten), diffuser intake flange (type 310 stainless steel), diffuser exit flange (type 405 stainless steel), expansion bellows (Inconel 600), and bellows intake flange (Corten). Individual segments of the failed subassemblies were examined using various methods. The analysis indicated that the weld joint in the diffuser intake flange (type 310 stainless steel to Corten steel) contained fusion-zone solidification cracks. The joints had been produced using the mechanized gas-metal arc welding process. Cracking was attributed to improper control of welding parameters, and failure was attributed to weld defects.

Key Words			
Austenitic stainless steels	Welded joints	Low-cycle fatigue
Corrosion-resistant steels	Air intakes	Transition joints
Exhaust systems	Diffusers	Fatigue failure
Weld defects		

Alloys			
Austenitic stainless steel—310	Martensitic stainless steel—405	Corrosion-resistant steel—Corten
Nickel-base superalloy—Inconel 600		

Background

An exhaust diffuser assembly that handled air at about 450 °C (850 °F) failed prematurely in service after approximately 5000 h of operation. The failure occurred near the intake end of the assembly and involved fracture in the diffuser cone (Corten), diffuser intake flange (type 310 stainless steel), diffuser exit flange (type 405 stainless steel), expansion bellows (Inconel 600), and bellows intake flange (Corten).

Pertinent specifications

The exhaust diffuser assembly consisted of two subassemblies: a diffuser cone assembly and an overlapping bellows assembly (Fig. 1). The diffuser cone assembly consisted of a diffuser cone with circumferential intake and exit flanges. The diffuser cone was produced by longitudinal welding of a formed steel sheet (Corten). The two circumferential joints in the diffuser cone assembly—one between the Corten and the intake flange (type 310 stainless steel) and the other between Corten and the exit flange (type 405 stainless steel)—were produced by mechanized gas-metal arc welding (GMAW) using type ER309 filler metal (Ref 1) and 98Ar-2O₂ shielding gas. The bellows assembly consisted of an expansion bellows (Inconel 600) bolted to circumferential intake and exit flanges (Corten). The field failure occurred near the intake end of the exhaust diffuser assembly and involved

material fracture in the diffuser cone, diffuser intake and exit flanges, expansion bellows, and bellows intake flange.

Fig. 1 Exhaust diffuser assembly. Arrow indicates direction of air flow.

Visual Examination of General Physical Features

Visual examination of the failed components indicated that the fracture path near the intake end of the diffuser cone was either random or irregular (Fig. 2). The random orientation of the fracture path suggested that failure was not associated with any specific metallurgical mechanism or phenomenon. Although the zigzag fracture path

---
*Formerly at Columbus Metallurgical Services, Inc.

Fig. 2 Diffuser cone with intake side up. Note the zigzag fracture path along intake end (large arrow) and separation (small arrows) along exit side.

in the diffuser cone appeared to coincide with the longitudinal weld in the diffuser cone, the fracture in this region appeared to be secondary. Furthermore, selected areas on the diffuser cone showed blisters and rust spots on the aluminum-painted surface (Fig. 3). However, the external surface did not exhibit significant evidence of any kind of macroplastic deformation. The external surface of the diffuser assembly also showed evidence of abrasion (rubbing) along the exit flange of type 405 stainless steel and an appreciable reduction in thickness across the weld seam. These abrasion marks were observed only at a select location rather than along the circumference of the exit diffuser flange, which suggested that abrasion had occurred at a later stage, following the complete fracture of the diffuser support assembly and perhaps immediately prior to shutdown. Alternatively, these abrasions might have occurred during transportation subsequent to failure.

Visual examination of the fracture paths in the diffuser intake and exit flanges showed a clear association with the weld seam. The fracture path in the diffuser intake end exhibited cracking principally along the weld interface with the type 310 stainless steel diffuser intake flange and occasionally along the center of the weld and the Corten diffuser cone. The fracture path in the diffuser exit end exhibited cracking principally along the weld seam with the type 405 stainless steel diffuser exit flange and occasionally in the diffuser cone.

The fracture in the expansion bellows was localized near the intake end (Fig. 4), whereas the exit end did not show any evidence of fracture. The fracture on the intake end appeared to be associ-

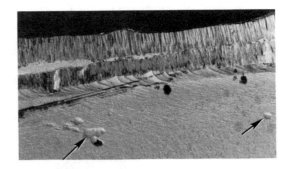

Fig. 3 Fracture edge of diffuser cone near exit side, showing abrasions across weld and blisters (arrows) in diffuser cone

Fig. 4 Partial circumferential fracture in expansion bellows. Arrows indicate partial tearing along intake end.

ated with a specific location. Ductile tearing at this location indicated possible failure by mechanical overload.

Individual segments of the failed subassemblies were examined using liquid penetrant testing to enable detailed characterization of the failure and to develop an overall approach for failure analysis. Multiple specimens were then obtained from representative transverse sections of the failed subassemblies, resin mounted, and prepared for macroscopic and microscopic examination and stereomicroscopy and light microscopy. The metallographic specimens were etched using equal volumes of HCl, HNO_3, and acetic acids. Representative fracture sections were also examined using scanning electron microscopy (SEM) to determine specific failure modes and to develop suitable correlations with the observed microstructures. Prior to SEM examination, the fracture surfaces were repetitively cleaned with acetone and methanol, followed by ultrasonic cleaning in a bath containing 5% phosphoric acid solution.

The individual chemistries of the diffuser cone, diffuser intake flange, diffuser exit flange, expansion bellows, and bellows intake and exit flanges were also determined by inductively coupled plasma spectrometry to identify possible deviation in chemistry from material specifications.

Nondestructive evaluation

Liquid penetrant inspection showed that the diffuser intake flange exhibited a continuous fracture front, traveling along the type 310 stainless steel flange adjacent to the weld seam and extending a substantial length in a zigzag (irregular) manner into the diffuser cone. Several secondary cracks were also revealed.

The diffuser exit flange also exhibited a continuous fracture adjacent to the weld seam along the type 405 stainless steel flange, which occasionally propagated into the diffuser cone. The shift or transition in the propagation of the crack appeared to coincide with the weld stops and restarts. Again, several secondary cracks were also revealed.

Although liquid penetrant inspection did not enable identification of fracture initiation sites in the intake and exit diffuser flanges, it provided additional support to the findings of the visual examination. These findings indicated that the fracture initiated principally in the intake flange and probably caused cracking/separation along the exit diffuser flange at later stages of fracture propagation. Consequently, further detailed metallographic and fractographic examination was performed on the diffuser intake and exit flanges, whereas limited analysis was performed on the failed regions of the diffuser cone and the expansion bellows.

Metallographic and fractographic examination

Diffuser Intake Flange. Stereoexamination of transverse sections of the diffuser intake subassembly revealed a single-pass weld made from the external side of the diffuser assembly. Examination of multiple sections showed several microcracks near the fusion boundary with type 310 stainless steel and incomplete weld penetration at selected locations. Interestingly, locations exhibiting complete weld penetration showed the fracture occurring either through type 310 stainless steel (Fig. 5a) or the Corten diffuser cone (Fig. 5c), whereas locations exhibiting incomplete penetration showed the fracture occurring through the weld center (Fig. 5b).

Subsequent microscopic examination of the transverse section of the diffuser intake flange revealed two weld beads, one made from the external side and the other made from the internal side. The internal weld bead was not readily evident in stereomicroscopy because of a postweld machining operation. Despite a two-pass welding procedure, both the external and internal weld beads exhibited a dendritic microstructure.

Optical microscopic examination of representative failed specimens of the diffuser intake subassembly indicated cracks near the weld interface with the type 310 stainless steel flange (Fig. 6). Furthermore, the weld metal exhibited a primary austenite mode of solidification, especially at regions close to the weld interface ("unmixed zone") with the type 310 stainless steel flange. This indicated the possibility of fusion-zone solidification cracks at these locations. Examination of the weld microstructure at higher magnification confirmed the presence of microcracks in the weld fusion boundary. The weld fusion-zone microstructure also exhibited tiny fragments of Corten steel from the diffuser cone, indicating possible inadequate joint preparation (i.e., deburring prior to welding) and insufficient melting during welding.

SEM examination of the failed regions adjacent to the fusion zone of the type 310 stainless steel diffuser intake flange showed minimal evidence of macroscopic deformation, indicating that fracture in these regions occurred unaided by deformation or mechanical loading—possibly by the propagation of the fusion-boundary cracks. To evaluate the nature of the fusion-boundary cracks, a transverse section containing the crack was slit open using a precision diamond saw.

Subsequent SEM examination of the opened region revealed an intergranular type of separation along columnar dendrite boundaries (Fig. 7), which is typical of fusion-zone solidification cracks. The occurrence of fusion-zone solidification cracks in the weld between the diffuser cone and the type 310 stainless steel intake diffuser flange suggested that local variations in welding conditions probably produced sluggish fluid flow and promoted the occurrence of a weld metal chemistry (with a Cr_{eq}/Ni_{eq} ratio of less than 1.35) susceptible to solidification cracking, especially in the "unmixed zone" of type 310 stainless steel (Ref 2). The occurrence of solidification cracks in the unmixed zone of type 310 stainless steel was attributed to improper control of welding parameters—specifically, weld travel rate. Lower weld travel rates along with the weld outer edge probably resulted in sluggish fluid flow and higher base metal dilution and promoted weld solidification to primary

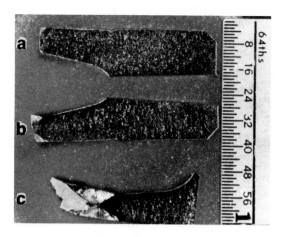

Fig. 5 Representative transverse sections of diffuser intake subassembly. (a) Fracture through fusion boundary with type 310 stainless steel. (b) Fracture through welds. (c) Fracture through Corten

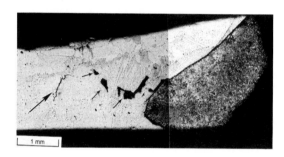

Fig. 6 Transverse section of weld between intake flange (left) and diffuser cone (right). Large arrow indicates cracking near fusion boundary with type 310 stainless steel. Small arrows indicate fragments of Corten steel in fusion zone.

Fig. 7 Fracture surface in type 310 stainless steel flange, showing separation along columnar dendrite boundaries typical of fusion-zone solidification cracks

austenite near the fusion boundary with type 310 stainless steel.

SEM examination of the failed regions along the Corten showed failure by mechanical loading and did not reveal any unusual features. SEM examination of the failed regions through the weld showed features intermediate between those in the fusion boundary with type 310 stainless steel and those along the Corten.

Diffuser Exit Flange. Stereoexamination of the failed sections of the diffuser exit assembly showed two weld passes between the diffuser cone and the exit flange. Analysis of the weld bead shapes indicated that one of the welds was made from the internal side and the other was made from the external side (Fig. 8). Examination of multiple sections indicated complete penetration (tie-in) between the external and internal weld beads. Optical microscopic examination of the two weld beads revealed different etching responses: the internal bead showed a transformed microstructure, and the external bead showed an as-solidified microstructure. Microstructural analysis of the transformed weld metal microstructure (internal bead) showed appreciable grain-boundary martensite. The two-pass weld also showed a coarse-grained heat-affected zone (HAZ), with the type 405 stainless steel suggesting a probable drop in ductility in this region.

Standardless quantitative SEM/EDS (energy-dispersive spectroscopy) analysis was performed to determine the chemistries of the internal and external weld beads. The chemical analysis (Table 1) indicated no appreciable variation in chemistries of the two beads, further suggesting that the microstructural transformation observed in the internal bead was caused by the combined effect of low nickel content (resulting from appreciable base metal dilution) and the thermal cycling associated with the external bead.

Optical microscopic examination of the fracture edge of the type 405 stainless steel flange showed significant shear-type plastic deformation (Fig. 9), indicating mechanical overload followed by shearing as the principal mode of failure. SEM examination of the failure in the diffuser cone exit flange showed evidence of ductile failure (Fig. 10), which confirmed that these regions failed by deformation caused by mechanical overloading.

Diffuser Cone. SEM examination of the failure in the diffuser cone near the intake end showed minimal evidence of macroscopic deformation, which confirmed that these regions failed unaided by either deformation or mechanical loading. SEM examination at a higher magnification revealed

laminations and microcracks in the Corten (Fig. 11), thereby suggesting that the observed zigzag

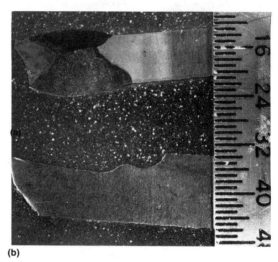

(b)

Fig. 8 Representative transverse sections of diffuser exit subassembly. (a) Fracture in Corten. (b) Fracture in type 405 stainless steel adjacent to the weld

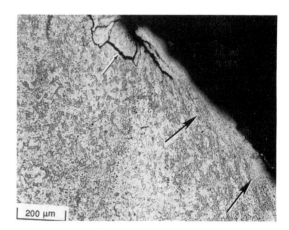

Fig. 9 Fracture edge of exit diffuser flange of type 405 stainless steel, showing cracking (small arrow) and shear-type plastic deformation of grains (large arrows)

Table 1 Chemistries of two-pass weld between Corten and type 405 stainless steel, wt%

Component	Cr	Mn	Fe	Ni
Internal bead	15.35	0.41	80.27	3.97
External bead	15.39	0.37	80.25	3.98

Table 2 Chemical analysis of individual components, wt%

Sample ID	C	Mn	P	S	Si	Cu	Ni	Cr	Mo	Al	V	Nb	Co
Diffuser													
Intake flange	0.04	1.63	0.027	0.001	0.52	0.14	18.9	25.9	0.11	0.009	0.06	0.013	0.18
Exit flange	0.17	0.34	0.020	0.002	0.43	0.06	0.13	11.9	0.04	0.002	0.06	0.006	0.02
Cone	0.089	0.35	0.09	0.009	0.32	0.33	0.15	0.88	0.019	0.005	0.001	0.000	0.000
Expansion bellows													
Intake flange	0.13	1.3	0.006	0.027	0.32	0.40	0.12	0.63	0.019	0.000	0.057	0.000	0.009
Exit flange	0.14	1.0	0.008	0.024	0.35	0.31	0.033	0.54	0.010	0.000	0.030	0.000	0.000
Bellows	0.04	0.18	0.010	0.002	0.25	0.08	Remainder	15.41 (Fe)	8.0	...	...	...	...

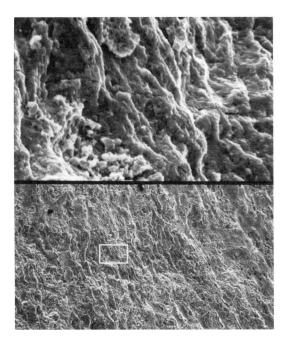

Fig. 10 Fracture surface of diffuser cone near exit end, showing ductile dimples in type 405 stainless steel

Fig. 11 Fracture surface of diffuser cone near intake end, showing laminations and microcracks in Corten steel

fracture was associated with "weak" interfaces in the Corten.

Expansion Bellows. SEM examination of the failure regions of the expansion bellows intake end revealed fatigue striations (Fig. 12). The wide spacing of these striations indicated that the failure propagated by low-cycle thermal and/or mechanical fatigue prior to complete breakdown.

Chemical analysis/identification

Table 2 shows the complete chemical analysis of the individual components. Detailed analyses did not reveal any significant deviation from material specifications, except for a marginally higher carbon content in the case of the diffuser exit flange versus the corresponding type 405 stainless steel specification. Interestingly, the intake diffuser flange showed a high phosphorus content (0.027%). A combined sulfur and phosphorus content higher than 0.02% in austenitic weld deposits is known to promote weld solidification cracking (Ref 3).

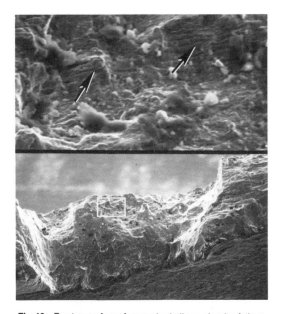

Fig. 12 Fracture surface of expansion bellows, showing fatigue striations (arrows)

Conclusion   The failure analysis procedure clearly indicated that the weld joint in the diffuser intake flange (type 310 stainless steel to Corten steel) contained fusion-zone solidification cracks. These cracks were promoted by a high phosphorus content in the type 310 stainless steel flange and a primary austenite mode of solidification in the unmixed zone of type 310 stainless steel. The occurrence of solidification cracks in the unmixed zone of type 310 stainless steel was attributed to improper control of welding parameters (specifically, weld travel rate), which in turn affected base metal dilution and promoted weld solidification to primary austenite near the fusion boundary with type 310 stainless steel.

Furthermore, at selected locations along the weld circumference in the diffuser intake flange, the two-pass welds did not tie in, and led to the occurrence of discontinuities, which may have served as additional sites for fracture initiation/propagation in the diffuser intake assembly. The failure analysis procedure showed that the failure of the exhaust diffuser assembly primarily occurred from preexisting solidification-related cracks near the weld seam with the diffuser intake flange, whereas the eventual in-service failure of other components occurred by shear or tear caused by mechanical overload and/or fatigue.

Acknowledgement   Adapted from an article that appeared in Welding Journal, Nov 1991, p 45-50. Used with permission of the American Welding Society.

References   1. "Specification for Corrosion Resisting Chromium and Chromium-Nickel Bare and Composite Metal Cored and Stranded Welding Electrodes and Rods," AWS A5.9-81, American Welding Society

2. S. Kou, *Welding Metallurgy*, John Wiley & Sons, 1987, p 179-185
3. V.P. Kujanpaa, N.J. Suutala, T.K. Takalo, and T.J. Moisio, *Met. Construct.*, Vol 12 (No. 6), 1980, p 282-285

Fatigue Failure of Steering Arms Due to Forging Defects

Roy G. Baggerly, Kenworth Truck Company, Kirkland, Washington

Several heavy truck Cr-Mo steel steering arms in service less than three years fractured during station-ary or low-speed turning maneuvers that required power-assisted steering. Metallographic examination of the cracked AISI 4135 arms, heat treated to a hardness of 285 to 341 HB, revealed that fatigue crack initiation occurred from the tip of oxide scale inclusions forged into the U-shaped arm at the inside radius. Corrective action involved redesigning the steering arm to increase the minimum forging radius and reduce the stress level at the inner-bend radius, and reducing the level of power assistance to the wheels to encourage the driver to put the vehicle in motion prior to turning.

Key Words Chromium-molybdenum steels, mechanical properties Steering arms
Fatigue failure   Crack initiation   Forging defects
Stress concentration

Alloys Chromium-molybdenum
steel—4135

Background

Several steering arms in service less than three years fractured during a dry park turn or slow speed turn when a high load was applied from hydraulically assisted power steering.

Applications

Maneuvering of heavy trucks during parking or backing into loading/unloading docks is greatly facilitated by power-assisted steering. The load applied to the steering system is maximum at this time due to the scrubbing friction of the tires across a surface, often dry pavement. The geometry of the steering system is such that the steering arm usually has a bend, the inside radius of which can vary depending on the particular front axle arrangement. Since this component is forged, potential forging discontinuities can be introduced when the forged radius of the bend is too small. A schematic of the steering arm is shown in Fig. 1.

Circumstances leading to failure

A steering arm failed when a fully loaded truck was maneuvering in a parking lot in preparation for unloading. The vehicle was equipped with a hydraulic power-assisted steering gear, and maneuvering required full turns of the tires from the left

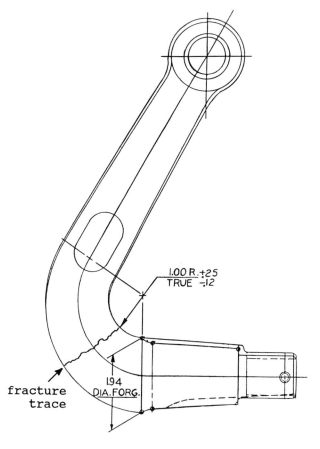

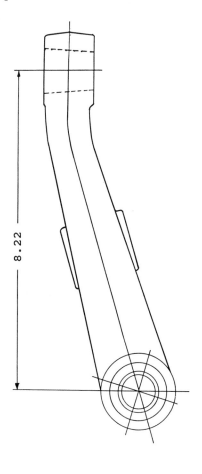

Fig. 1 Schematic of steering arm

to the right. Subsequently several other steering arms failed under similar circumstances, *i.e.* stationary or low-speed turning maneuvers that required power-assisted steering.

Pertinent specifications

The manufacturer's material specification for the steering arm required an AISI 4135 forged steel part, heat treated to a Brinell hardness of 285 to 341 HB.

Performance of other parts in same or similar service

The same steering arm is also used in manual steering gear applications. The manual resistance to turning the steer axle wheels while the truck is stationary is too great, and the vehicle is generally placed in motion in order to turn. No failures of steering arms occurred when used with manual steering gears, although small cracks have been observed at very high vehicle mileages in manual gear steering arms. These cracks initiated at oxide scale forged into the inner bend of the steering arm. The number of fatigue cycles at these high applied loads from manual steering gears, however, was not sufficient to cause failure during the lifetime of the vehicle.

Selection of specimens

The result of these early failures prompted a federal recall campaign and many returned steering arms were metallographically sectioned and examined microscopically to determine the statistical distribution of the forged-in oxide scale and the propensity to initiate fatigue cracks. Several steering arms that were unused were also sampled and examined for forged-in oxide scale.

Visual Examination of General Physical Features	Complete fracture of the steering arm could occur along a plane that passed through the inside radius. A macrophotograph of a characteristic steering arm fracture is shown in Fig. 2 and the failure location is also shown in Fig. 1.

Fig. 2 Fractured steering arm

Testing Procedure and Results	

Nondestructive evaluation

Magnetic Particle. The location on the steering arm where fatigue cracks initiated occurred over a very restricted area of the inside radius, and most of the steering arms that were returned in the recall did not require detailed inspection prior to sectioning through the suspect area. Several unused steering arms were subjected to magnetic particle inspection, and Fig. 3 shows a typical example of the magnetic indications associated with forged-in oxide scale.

Surface examination

Visual. The fracture surfaces of the failed steering arms revealed characteristic beach marks indicating a fatigue mechanism with either single or multiple origins emanating from the inside radius of the steering arm (Fig. 4). Occasionally fatigue initiation would occur from both the outside and inside of the bend, and the fatigue crack would propagate from both sides of the steering arm simultaneously in a reverse bending mode.

Metallography

Microstructural Analysis. The microstructure of the AISI 4135 forged steering arm exhibited a tempered martensitic structure that was indicative of a quench and temper heat treatment. This was consistent with the hardness specification and hardness measurements and showed that the steering arms were in compliance with the heat treatment specification.

Crack Origins/Paths. The metallographic examination of several cracked steering arms revealed that fatigue initiation occurred from the tip of the oxide scale inclusions forged into the arm. Typical oxide scale inclusions that were present in every steering arm investigated are shown in

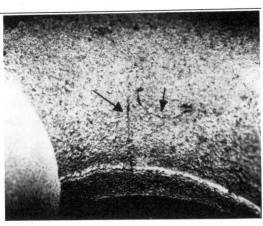

Fig. 3 Magnetic particle indications of oxide scale forged into the steering arm

Fig. 4 Fracture surface of steering arm showing characteristic beach marks

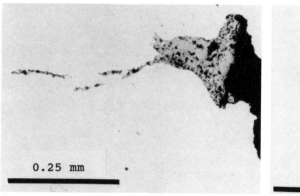
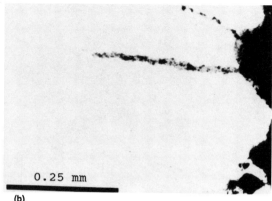

(a)   (b)

Fig. 5   Oxide scale inclusions forged into the inner radius of the steering arm

Fig. 5.

Chemical analysis/identification

The chemical composition of the failed steering arms was determined using optical emission spectroscopy for most of the alloy elements and a Leco carbon/sulfur analyzer for carbon and sulfur content. The latter analysis required the use of drill flakes for ignition and the subsequent analysis of the combustion gases. Table 1 lists the chemical analysis for a characteristic failed steering arm. The composition of the material complies with the standard composition for AISI 4135 alloy steel.

Mechanical properties

Hardness was the only mechanical property determined. The hardness of the failed steering arms were consistently within the required hardness range of 285 to 341 HB.

Table 1   Chemical analysis results

| | Composition, wt% | |
Element	Steering arm	AISI 4135 steel
Carbon	0.38	0.32 - 0.38
Silicon	0.240	0.15 - 0.30
Manganese	0.75	0.60 - 1.00
Chromium	0.94	0.75 - 1.00
Molybdenum	0.15	0.15 - 0.25
Nickel	0.15	...
Phosphorus	0.008	0.035 max
Sulfur	0.023	0.040 max
Vanadium	0.004	...
Copper	0.17	...
Aluminum	0.018	...

Discussion

Forging slugs were blanks are prepared from annealed 4135 alloy bar material by shearing to the appropriate length. The process of forging the steering arm required repeated hits of the forging billet using a series of three separate die cavities. The arm was forged into a U-shape in the second die cavity, during which there was a distinct possibility of entrapping surface oxide scale in the region of the inner bend radius of the part. A forging die is normally designed to expel the surface oxide scale with the flash at the parting line. The inner bend radius was too sharp, however, and some of the oxide scale was entrapped within the forging instead of being expelled with the flash. Many steering arms, both new and unused, as well as those returned from the field, were metallographically sectioned and the depth of the oxide scale measured. The data was subsequently analyzed using Weibull statistics. Figure 6 shows the best fit of the data corresponding to a Weibull slope of 2.11 and a characteristic oxide depth of 0.6 mm (0.02 in.) (63.2% of the population). The expected oxide flaw distribution depth is shown in Fig. 7, which indicates a finite probability of oxide scale extending to depths greater than 1 mm (0.04 in.).

The failed steering arms would be expected to be from the population that had the greater depth of oxide scale. The maximum stress on the steering arm occurs during a dry park, right hand turn which results in scrubbing the tires over dry pavement. A stress as high as 903.2 MPa (131 ksi) was measured during this maneuver. The maximum stress intensity present at the tip of the oxide flaw, assuming that it acts as a sharp crack, can be calculated using standard fracture mechanics methodology. The relationship between stress intensity

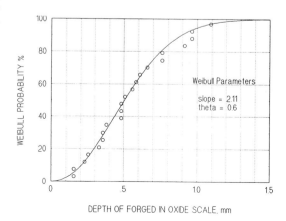

Fig. 6   Weibull analysis of the depth of forged-in oxide scale

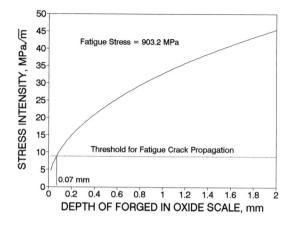

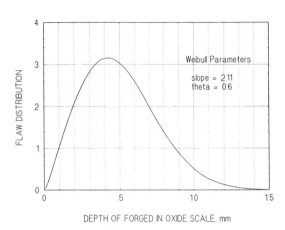

Fig. 7 Statistical frequency distribution of the depth of forged-in oxide scale

and depth of oxide scale is shown in Fig. 8. An approximate threshold for fatigue crack propagation of 8.8 MPa√m (8.0 ksi√in.) is also shown (Ref 1). A flaw larger than 0.07 mm (0.003 in.) will propagate by fatigue crack growth under repeated cyclic stressing at a stress amplitude of 903.2 MPa (131 ksi).

The number of dry park maneuvers that can be accommodated by these steering arms before they fail by fatigue is therefore of interest. The failures occurred with steering arms that were less than three years old. The manufacturer indicated that if the arm was subjected to an average of 12 cycles per day for three years, approximately 13,000 cycles would accumulate. At this level of fatigue cycles and stress, a steering arm with a flaw approximately 0.9 mm (0.035 in.) or larger would be expected to fail. Figure 9 shows the relationship between the integrated Paris equation, *da/dn*, us-

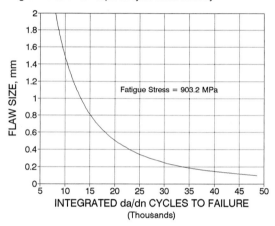

Fig. 8 Effect of flaw depth on cyclic stress intensity

Fig. 9 Effect of flaw size on integrated fatigue life

ing fatigue crack growth data for AISI 4140 steel of similar strength (Ref 1) and a final crack size of 15.2 mm (0.6 in.) to define failure. If the steering arms had not been removed from service, arms with smaller flaws would have continued to fail with time, since the average lifetime of heavy trucks is greater than 15 years.

Conclusion and Recommendations	***Most probable cause***

Most probable cause

The most probable cause of failure of the steering arms was the presence of forged-in oxide scale at a location along the inner radius of the arm subjected to a high stress level. Tire scrubbing movement, facilitated with hydraulic power assist when the vehicle is stationary (*i.e.* the dry park condition) imparts a high stress level to the steering arm, resulting in fatigue crack propagation from pre-existing flaws.

Remedial action

The steering arm was redesigned to reduce the stress level at the inner bend radius. The level of hydraulic power assistance was also reduced to en-

courage the driver to place the vehicle in motion prior to turning the steering axle wheels.

How failure could have been prevented

An appropriate design analysis to lower the overall operating stress on the steering arm should have been performed. Insuring an adequate inspection procedure prior to release of the design for manufacture could also have prevented failures. The use of metallographic techniques would have shown the presence of oxide scale in the critical section, and an effective NDT inspection technique such as magnetic particle inspection would have prevented the deployment of flawed steering arms.

Acknowledgements   The author gratefully acknowledges the Kenworth Truck Company for encouraging publication of this failure analysis investigation.

Reference   1. Paul Thielen and Morris Fine, "Fatigue Crack Propagation in 4140 Steel," *Metall. Trans.*, Vol 6, Nov. 1975, p. 2133.

Fracture of a Train Wheel Due to Thermally Induced Fatigue and Residual Stress

Carmine D'Antonio, Department of Metallurgy and Materials Science, Polytechnic University, Brooklyn, New York

An ASTM A504 carbon steel railway car wheel that was used on a train in a metropolitan railway system failed during service, causing derailment. The wheel was completely fractured from rim to hub. Macrofractography of the fracture surface showed road grime, indicating that the crack had existed for a considerable time prior to derailment and initiated in the flange. Failure propagated from the flange across the rim and down the plate to the bore of the hub. Two zones that exhibited definite signs of heating were observed. The fracture initiation site was typical of fatigue fracture. No defects were found that could have contributed to failure. The wheel conformed to the chemical, microstructural, and hardness requirements for class A wheels. Failure was attributed to repeated severe heating and cooling of the rim and flange due to brake locking or misapplication of the hand brake. It was recommended that the brake system on the car be examined and replaced if necessary.

Key Words

Carbon steels
Railroad wheels
Brakes

Thermal fatigue
Railroads

Brake shoes
Fatigue failure

Background

A railway car wheel failed during service, causing derailment. The wheel was completely fractured from rim to hub.

Applications

The wheel was part of a train in a metropolitan railway system and thus was subjected to repeated stops.

Pertinent specifications

The wheel was specified to be a wrought, class A (ASTM A504) type. Class A wheels are required to be quenched and tempered to 255 to 321 HB. Wheels of this type are intended for high-speed service with severe braking loads.

Specimen selection

A cut was made parallel to the fracture to allow removal of a section containing the fracture surface. Two additional samples were cut from the rim and used for metallographic examination, scanning electron microscope (SEM) examination, and hardness testing. Chemical analysis was conducted using drillings machined from the rim.

Visual Examination of General Physical Features

Initial examination of the wheel revealed a crack that traversed radially from flange to hub. The crack was 4.8 mm (3/16 in.) wide at the tread. The opening decreased linearly to only 1 mm (0.05 in.) at the hub bore. This degree of taper indicated that high tangential residual stresses were present in the rim at the time of failure.

Testing Procedure and Results

Surface examination

Macrofractography. Figure 1 is a photograph of the fracture surface as it appeared when it was cut from the wheel. The dark areas are road grime, indicating that the crack had existed for a considerable time prior to derailment. The surface was cleaned with benzene and carbon tetrachloride and examined a second time. Figure 2 shows the fracture after cleaning. It is evident that fracture initiated in the flange at a circular area approximately 19 mm (3/4 in.) in diameter. Failure propagated from this area across the rim and down the plate to the bore of the hub. The texture of the fracture surface indicated that final failure took place in three or four episodes of tensile overload.

Although not evident in Fig. 2, two zones exhibited definite signs of heating. The entire hub fracture surface was tinted blue, indicating it reached temperatures of approximately 315 °C (600 °F). This heating occurred after the wheel fractured

Fig. 1 Fracture surface removed from the train wheel by making a parallel radial cut. Dark areas are road grime, indicating that the crack had existed prior to derailment.

Fig. 2 Fracture surface shown in Fig. 1 after cleaning. Fracture initiated in the flange and propagated across the rim, plate, and hub.

Table 1 Results of chemical analysis

Element	Composition, %	
	Wheel material	Specified (ASTM A504)
Carbon	0.54	0.57 (max)
Manganese	0.61	0.60-0.85
Phosphorus	0.009	0.05 (max)
Sulfur	0.032	0.05 (max)
Silicon	0.18	0.15 (min)

Fig. 3 Closeup view of the fracture initiation site. Fracture surface in the flange is typical of fatigue.

and was caused by the axle rotating relative to the wheel. The fracture surface at the rim was also tinted. Its color suggested lower temperatures caused by brake friction. The web fracture surface was clean.

Figure 3 shows a closeup view of the fracture initiation site. This zone was typical of fatigue fracture; it was brittle in nature and exhibited classical beach marks. SEM examination of a specimen subsequently cut from this zone verified that it was a fatigue zone.

Metallography

Microstructural Analysis. Metallographic examination of a sample cut from the wheel rim indicated that the structure was tempered martensite with a few small patches of ferrite. The wheel thus met the heat treatment requirement for class A wheels.

Chemical analysis/identification

Table 1 presents the results of wet chemical analysis conducted on material removed from the rim, along with the specified composition requirements. The wheel material met specification.

Mechanical properties

The hardness of the wheel rim was 265 to 268 HB, which met the requirements for class A wheels and was consistent with the observed chemistry and microstructure.

Discussion

Cyclic tensile stresses induced in the wheel flange were high enough to cause fatigue initiation and progression. On subsequent cycles of high tensile stress, the fracture extended over large areas of the rim, culminating in a final event that compromised the remainder of the rim, web, and hub. The stresses in the rim were amplified by the existing crack.

Repeated tangential tensile stress typically stems from intermittent heating and cooling of the wheel rim and flange due to brake locking or misapplication of the hand brake. Either of these events can cause severe frictional heating of the rim and flange zones, resulting in hoop compressive stresses high enough to exceed the yield strength of the wheel material at the elevated temperature. It is significant that the rim and flange are normally in residual compression because of their manufacture; the compressive stresses caused by thermal expansion are added to these stresses. Upon release of the brake or stopping of the train, the rim and flange cool and contract. This contraction is opposed by the web, which is not heated in the process, thus causing the rim and flange to be placed in residual tension.

Conclusion and Recommendations

Most probable cause

No defects were found that could have contributed to failure. The wheel conformed to the chemical, microstructural, and hardness requirements for class A wheels.

Failure of the wheel was caused by repeated severe heating and cooling of the rim and flange, which initiated fatigue fracture in the flange. The heating and cooling also reverted the normal residual compressive hoop stresses to residual tension stresses, culminating in complete radial fracture of the wheel and subsequent derailment.

Remedial action

The brakes were the sole source of rim and flange heating. It was recommended that the brake system on the car be examined and replaced if necessary.

In-Service Failure of SAE Grade 8.1 Wheel Studs

Roch J. Shipley, Peter C. Bouldin, and Edward W. Holmes, Engineering Systems, Inc., Aurora, Illinois

Failure of carbon-manganese steel wheel studs caused by improper tightening of the inner wheel nuts resulted in separation of a dual wheel assembly on a heavy truck. The beachmark pattern observed on the fracture surfaces of the studs evidenced fatigue cracks emanating from multiple origins around the circumference. There was no indication that any microstructural characteristics of the material contributed to the failure. Inclusions that were present were small and relatively few in number. Failure to check the torque of the inner wheel nuts as per the manufacturer's recommendation caused the inner wheel nuts to loosen during break-in and lose the required clamping force. The development and promotion of educational programs on proper wheel tightening procedures was recommended.

Key Words			
	Carbon manganese steels, Mechanical properties	Fasteners, Mechanical properties Fatigue failure	Automotive wheels Bolted joints, Mechanical properties

Alloys   Carbon manganese steels—1541

Background

Application

This case study concerns the failure of specialty alloy fasteners used in the mounting of dual wheel assemblies in medium and heavy trucks.

Circumstances leading up to failure

The dual rear wheel assembly on the left side of a single rear axle heavy truck completely separated from the hub. The wheel assembly was mounted to the hub by 10 left hand thread wheel studs and twenty specialty nuts designed to match the wheels. The truck was relatively new at the time of this incident, having travelled approximately 20,000 total miles.

Pertinent specifications

The wheel studs were specified to be Grade 8.1 under SAE Standard J429, "Mechanical and Material Requirements for Externally Threaded Fasteners." Medium-carbon alloy steels in general and alloy 1541 in particular meet the requirements of this standard.

The torque requirement for both the inner and outer nuts was 746 N · m (550 ft · lbf).

Performance of other parts in same or similar service

Currently, over 1.5 million medium/heavy trucks log more than 90 billion miles annually in the United States. In a 1992 report, the National Transportation Safety Board estimated the incidence of wheel separations at 750 to 1,050 per year (Ref 1). The portion of these separations due to the fracture of wheel studs is unknown. Failure of bearings and failure of the wheel itself can also result in wheel separation.

Visual Examination and General Physical Features

Due to the design of the assembly, none of the pieces of the fractured studs were lost. The head pieces of all ten studs were retained in the hub and the remainder of the stud and the nut assemblies were retained within the dual wheel assembly. After the left wheel separated, the studs on the right wheel were inspected and one fractured stud was discovered.

Figure 1 shows a side view of one of the fractured studs and specialty nuts. The nut shown in the figure was used to clamp the inner wheel of the dual assembly. The second nut, used to clamp the outer wheel, was not involved in the failure and is not shown.

Figure 2 shows the fracture surface of one of the studs, both the head side and the portion still in the nut. The fracture appearance of this stud was typical of those studs which exhibited nearly all fatigue fracture. Some studs appeared to have some areas of overload fracture, but it was difficult to classify all areas of all fractures due to damage of the fracture surface. While there was some corrosion and damage in some areas, the beachmark pattern is clear evidence of fatigue cracks emanating from multiple origins around the circumference. The wheel studs are a classic example of fatigue in rotational bending at high nominal stress with mild stress concentration.

Although the inner wheel associated with the fractured studs was not available for inspection, it was reported some of the mounting holes appeared to be elongated.

Fig. 1   Side view of fractured wheel stud and nut which fastened inner wheel

(a)

(b)

Fig. 2 Fracture surfaces of wheel stud shown in Fig. 1. The shank end of the stud was not removed from the nut. Both surfaces were slightly damaged. (a) Head end. (b) Shank end still in nut

Testing Procedure and Results

Metallography

Microstructural Analysis. Figure 3 shows the microstructure of the wheel studs. The microstructure is consistent with the requirements of Grade 8.1, SAE Standard J429, for an elevated-temperature drawn steel. The microstructure is ferrite and very fine pearlite and/or bainite.

There was no indication that any microstructural characteristics contributed to the failure. The microstructure was uniform from the edge to the center of the cross section. No decarburization, laps, or other types of surface discontinuities were observed. The inclusions that were present were small and relatively few in number.

Chemical analysis/identification

Material. A sample of one of the wheel studs was analyzed to check the steel chemical composition. The results are given in Table 1. This product analysis is consistent with AISI 1541, one of the alloy possibilities mentioned for grade 8.1 wheel studs in SAE Standard J429.

Mechanical properties

Hardness. An average hardness of 36.5 HRC was measured on both transverse and longitudinal cross sections. This value is consistent with the requirements of SAE J429 for Grade 8.1 studs.

Stress analysis

Low-cycle Fatigue Analysis. As previously mentioned, the fracture characteristics are consistent with a relatively high stress fatigue process. The fatigue cracks in the stud typically initiated at the roots of the first few threads engaged with the nut. These threads experienced the highest stress and the thread root created a stress concentration.

As seen in Fig. 2, virtually the entire fracture surface of this particular stud exhibits fatigue characteristics. Although the stress was relatively

Fig. 3 Microstructure of wheel stud. 2% nital etch. 500×

Table 1 Chemical analysis of Grade 8.1 wheel stud

			Element, wt%					
C	Si	Mn	Ni	Cr	Mo	B	P	S
0.36	0.23	1.66	0.03	0.08	0.01	<0.0005	0.013	0.014

high at the time of crack initiation, the stress decreased as the cracks grew. This is an example of load shedding; because there are redundant load paths, the other studs in the assembly would have taken a greater portion of the load. Studs which exhibited a greater amount of overload fracture would have been the last to fail.

Discussion

Synthesis of evidence

For bolted connections which experience cyclic loading, control of clamping force is critical to avoid fatigue failure of the fasteners. If the clamping force is insufficient, the fasteners will experience the full separating force applied to the joint. If the clamping force is excessive, the fasteners are

stressed above their yield strength and the joint is loosened.

Clamping force is typically controlled in the trucking industry by means of torque specifications. Therefore, calibration and proper application of torque wrenches is critical.

The stretching of the fastener associated with

the clamping force is typically a few thousandths of an inch or less. Thus, even a slight change in the dimensions of the clamped surfaces can result in relaxation of the fasteners. This can occur with new parts, where mating surfaces may not match perfectly at assembly and may "wear-in" during initial operation. Vibration can also cause the joint to loosen under certain conditions.

For the above reasons, manufacturers often recommend that torques of critical fasteners be checked periodically, especially when equipment is relatively new or after fasteners have been removed and reinstalled. In the case study of this ar-

ticle, the manufacturer specified that the torque of the wheel nuts should be checked after a short break-in period, typically 85 to 170 km (50 to 100 miles). In order to check the torque of the inner nuts, the outer nuts should have been removed. Either the torque check was not done at all, or else only the torques of the outer nuts were checked.

At least some of the fasteners were able to relax, leading to initiation of fatigue cracks. Once a few of the fasteners began to crack, the remaining fasteners would also fail, whether they had relaxed or not, due to increased loads resulting from load shedding.

Conclusion and Recommendations	***Most probable cause*** The most probable cause of the fracture of the wheel studs was failure to check the torque of the inner wheel nuts as per the manufacturer's recommendation. The inner wheel nuts loosened during break-in and the required clamping force was lost. ***Remedial action*** As a result of failures of this type, the National	Transportation Safety Board (NTSB) has recommended the development and promotion of educational programs on proper wheel tightening procedures for the trucking industry. ***How failure could have been prevented*** Failure could have been prevented by maintenance procedures in accord with the manufacturer's recommendations for wheel studs.

Acknowledgements

Mr. Robert Franzese assisted with the preparation of the photographs. The permission of the truck manufacturer to publish this report is also gratefully acknowledged.

Reference

1. National Transportation Safety Board, Special Investigation Report, "Medium/Heavy Truck Wheel Separations," Report No. PB92-917004, NTSB/SIR-92/04, Sept 1992.

Mechanical Failure of a Repair Welded Ferritic Malleable Cast Iron Spring Hanger

Alan A. Johnson and Joseph A. von Fraunhofer, University of Louisville, Louisville, Kentucky

The right front spring hanger on a dual rear axle of the tractor of a tractor-trailer combination failed, causing the vehicle to roll-over. The hanger was made from malleable cast iron that had been heat treated to produce a decarburized surface layer and a pearlitic transition layer. It had been repair welded after breaking into two pieces longitudinally in a prior incident, using cast iron as weld metal. The repair weld bead on both surfaces missed the fracture over 15 to 20% of their lengths. The incomplete repair weld and brittleness of the weld metal and heat-affected zones led to the failure.

Key Words

Mechanical properties,
 Malleable cast iron
 Heat affected zone
 Weld metal

Cracking (fracturing)
Brittleness

Repair welding
Cast irons

Background

The right front spring hanger on the dual rear axle of a tractor-trailer combination failed, causing the vehicle to roll over. The hanger had fractured longitudinally into two pieces in a prior incident and had been repair welded. Failure occurred at this repair weld.

Applications

Spring hangers on each side of the vehicle secure springs to the frame. In service they are subjected mostly to compressive stresses, except when braking.

Circumstances leading to failure

A tractor-trailer combination loaded with wood chips was approaching an intersection on a rural highway. When the driver applied the brakes he heard a snap and found that he had lost control of the vehicle. The vehicle veered to the left and then rolled over. The driver suffered a broken neck.

Specimen selection

Three samples were cut from the fracture surface and one from a location remote from the fracture. These were mounted to provide sections normal to the front and back surfaces of the hanger.

Visual Examination of General Physical Features

An exemplar tractor is shown in Fig. 1, and the right front rear axle spring hanger is indicated with an arrow. Figure 2 shows the relationship between the hanger and the stack of leaf springs on the right side of the vehicle. One half of the broken spring hanger, together with the corresponding hanger from the left side of the vehicle, is shown in Fig. 3. The two show the same asymmetry because, in fact, both are hangers intended for use on the left side of the vehicle. This is evidenced by the fact that the broken hanger removed from the right side of the vehicle had the letter "L" cast into it. The photograph in the figure was taken at a point in the investigation when two metallographic specimens had been cut from the broken hanger and one had been cut from the unbroken hanger

Fig. 1 An exemplar truck showing the location of the spring hanger that broke (arrow)

Fig. 2 The relationship between the right front spring hanger and the stack of leaf springs on the right side of the vehicle

from the other side of the vehicle.

Examination of the fracture revealed that it had occurred at a repair weld. Weld beads were present from front to back on both the inside and outside surfaces. On both surfaces the weld beads had missed the fracture along 15 to 20% of their lengths. Thus, the repair was incomplete. Consistent with this observation, it was found that the corrosion products on the part of the fracture not repaired were dark red-brown, while those on the freshly created fracture were bright orange. The owner of the tractor had just purchased it in a used condition, and the repair may have been made just before it was purchased.

Metallography

One of the three metallographic specimens from the fracture is shown after polishing and etching in nital in Fig. 4. The weld bead on the outside surface of the hanger, A, had the effect of mending the fracture. That on the inside surface, B, missed the fracture surface, C, altogether. Thus at this location only a partial repair of the fracture was accomplished. It is estimated that both weld beads missed the fracture over 15 to 20% of their lengths. The hanger's microstructure remote from the surface is shown in Fig. 5. It shows that the hanger was made from malleable cast iron.

The hanger was found to have a decarburized surface layer approximately 0.15 mm (0.006 in.) in depth and, below this, a transition zone about 0.40 mm (0.016 in.) in depth. Examination of this transition zone at a higher magnification revealed the presence of pearlite. In two of the three metallographic specimens, a layer of martensite was found at the weld metal/base metal interface. An example of such an interface region is shown in Fig. 6. The martensite is shown at a higher magnification in Fig. 7. The microstructure of the weld metal is shown in Fig. 8.

EDAX measurements

The three metallographic specimens were subjected to electron dispersive analysis by x-ray (EDAX) in a scanning electron microscope. The results, which are semiquantitative, are shown in Table 1. The instrumentation used was not capable of detecting carbon. The EDAX data, in conjunction with the metallographic examination, indicate that both the weld metal and the hanger are cast iron.

Microhardness measurements

Microhardness measurements were carried out on the specimen shown in Fig. 4 using a Vickers microhardness tester with a 200 gm (7 oz) load.

Fig. 3 The broken right front spring hanger and the unbroken left front spring hanger

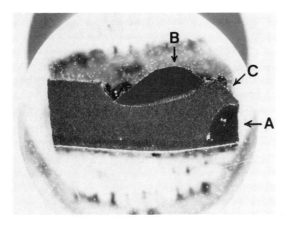

Fig. 4 A metallographic section through the fracture surface. 2.2×

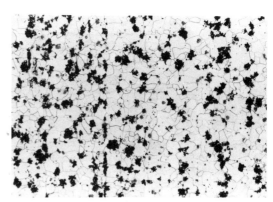

Fig. 5 The microstructure of the broken spring hanger. 62×

Table 1 EDAX results obtained on the broken spring hanger and its weld metal

Specimen No.	Location	Composition, %		
		Fe	Mn	Si
1	Weld metal	97.7	0.5	1.8
1	Weld metal	96.8	0.7	2.5
1	Weld metal	89.7	0.7	9.6
1	Base metal	91.7	0.6	7.69
2	Weld metal	97.9	0.6	1.5
2	Weld metal	98.7	0.6	0.7
3	Weld metal	97.5	0.7	1.8
3	Weld metal	95.8	0.6	1.1
3	Base metal	95.9	0.4	3.7

The results are shown in Table 2. Standard deviations are not given because of hardness gradients in the weld beads.

Table 2 Microhardness measurements made on the metallographic section shown in Fig. 4

Location	Hardness, VPN
Base metal	153, 185, 141, 146, 156, 129
	155, 143, 130, 140, 131
Weld bead on inner surface	430, 426, 401
Weld bead on outer surface	366, 385, 397

Discussion

The spring hanger was made from ferritic malleable cast iron. This is, of course, produced by the heat treatment of a white iron casting in a neutral atmosphere (Ref 1). Decarburization is limited and produces a shallow decarburized surface layer and a thin associated pearlitic transition zone. The rest of the casting has an ordinary malleable cast iron microstructure. Sometimes the term "blackheart" malleable cast iron is used to describe this kind of material.

The weld metal exhibits a microstructure characteristic of rapid solidification. The white phase is cementite, and its presence is reflected in its high microhardness (Table 2). A somewhat similar microstructure can be found in the ASM *Metals Handbook* (Ref 2). The martensite found at the weld metal/base metal interface was formed as the heat-affected zone was austenitized and then rapidly cooled. Both the cementite and the martensite are, of course, brittle.

Conclusion

This failure was caused by a combination of two factors. First, the repair weld was incomplete. This subjected the region that had been partially repaired to complex stresses, even though the overall stress on the spring hanger when in service was mainly compressive. Secondly, the weld was brittle because it contained cementite and martensite. Apparently the weld was not preheated to prevent these from forming.

The safety of the driver of the tractor-trailer combination depended on the integrity of this spring hanger. Once broken it should have been discarded and replaced by a new one. It is not possible to achieve ductility in a malleable cast iron repair weld using a cast iron welding rod even with preheating. In the present case, the problems associated with such a repair were compounded by the carelessness of the welder.

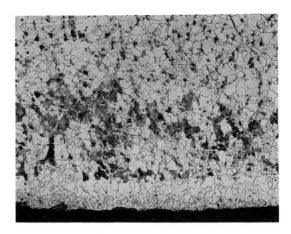

Fig. 6 A region of weld metal/base metal interface that exhibits martensite. 38×

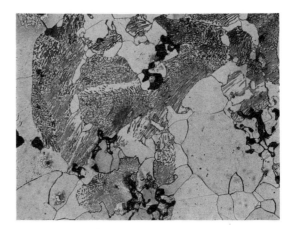

Fig. 7 The interface martensite that can be seen in Fig. 8, at a higher magnification. 304×

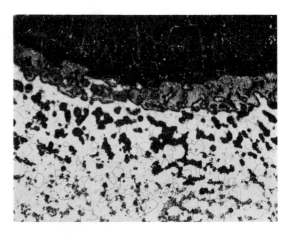

Fig. 8 The microstructure of the weld metal. 304×

References

1. H.T. Angus, *Cast Iron: Physical and Engineering Properties*, Butterworths, 2nd ed., 1976
2. *Metallography and Microstructures, Metals Handbook*, 9th ed., Vol 9, American Society for Metals, 1985, p 247

Quench Cracking in a Turntable Rail

Tina L. Panontin, Test Engineering and Analysis Branch, NASA Ames Research Center, Moffett Field, California

Persistent cracking in a forged 1080 steel turntable rail in a wind tunnel test section was investigated. All cracks were oriented transverse to the axis of the rail, and some had propagated through the flange into the web. Through-flange cracks had been repair welded. A section of the flange containing one through-flange crack was examined using various methods. Results indicated that the cracks had initiated from intergranular quench cracks caused by the use of water as the quenching medium. Brittle propagation of the cracks was promoted by high residual stresses acting in conjunction with applied loads. Repair welding was discontinued to prevent the introduction of additional residual stresses. Finite-element analysis was used to show that the rail could tolerate existing cracks. Periodic inspection to monitor the degree of cracking was recommended.

Key Words	Carbon steels	Stress cracking	Forgings
	Repair welding	Flanges	Wind tunnels
Alloys	High-carbon steel—1080		

Background

A turntable rail in a wind tunnel test section had been experiencing cracking since its installation in the 1970s. The rail was forged from SAE 1080 steel.

Applications

Wind tunnel testing often requires model orientation adjustment. This is accomplished by rotating the carriages that support the test section floor, the model support, and the model around a turntable rail (see Fig. 1).

Stresses in the rail result from the weight of the carriage, test section floor, and other elements, as well as from loads transferred from the model during testing. These stresses are typically small. However, large, residual, tangential stresses are believed to exist in this particular rail from fit-up distortion, flame-hardening treatment, and weld repairs.

Circumstances leading to failure

The first crack in the rail appeared immediately following installation and heat treatment. The rail flange separated with a loud report; no load had yet been applied. This and several subsequent cracks were repaired by welding.

A recent scheduled inspection of the turntable rail revealed two new, through-flange cracks and a number of small, surface-connected cracks spaced randomly along the bottom rail corners (Fig. 2). An investigation was requested to discern the cause of the repeated cracking and to evaluate the weld repair procedure.

Pertinent specifications

The rail was forged from high-carbon SAE 1080 steel (0.75 to 0.88% C, 0.6 to 0.9% Mn, 0.040% [max] P, 0.040% [max] S, balance iron) according to rail specification 90 lb ARA-A. The top and bottom of the rail flange (Fig. 2) were flame hardened to improve wear resistance. The specification for hardening required a minimum hardness of 52 HRC to a depth of

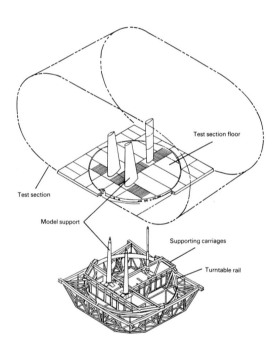

Fig. 1 Wind tunnel model orientation adjustment system, showing model support, test section floor, supporting carriages, and turntable rail

6.4 mm (0.25 in.) on the top surface and to whatever depth could be achieved on the bottom surface. Heat treatment of the rail was performed after installation. Oxyacetylene torches and a water-quench system mounted on a carriage were used to harden the contact surfaces. The torches were passed over the rail a second time to temper the hardened layer.

Specimen selection

A section of rail flange containing one through-flange crack was cut from the rail. The saw cut was then prepared as a V-groove joint and a full-penetration weld repair of the rail was performed.

Visual Examination of General Physical Features

The crack had split the rail flange such that the crack plane was perpendicular to the tangential direction of the turntable rail. There was no visible evidence of plastic deformation associated with the crack. The contact surfaces of the rail appeared worn and contained many shallow wear craters. Slight corrosion of the contact surfaces was evident.

Testing Procedure and Results

Surface examination

Macrofractography. The section of the rail flange was pried apart after cooling to liquid nitrogen temperature to reduce distortion of the fracture surface. The mating fracture surfaces are shown in Fig. 3. Two distinct regions (not including the laboratory-induced fracture) were evident on the surface. Regions Ia and Ib appeared dull and smooth, whereas region II had the shiny, crystalline appearance of cleavage fracture. No features indicating fatigue crack growth, such as beach marks, were evident in region I. River patterns in region II indicated that the direction of crack propagation was from the edge of the hardened case across the interior of the flange, as shown in the bottom photograph of Fig. 3.

Scanning Electron Microscopy/Fractography. Microscopic examination of the fracture surface in region I showed that the crack had propagated intergranularly. The "rock candy" appearance of region Ia near the top surface of the rail is shown in Fig. 4. High magnification of the region (Fig. 5) revealed large, exposed grains nearly devoid of features. Corrosion products were evident on the fracture surface at the top surface of the rail (Fig. 6).

In region II, the fracture path was transgranular and, as expected, the entire surface was composed of cleavage facets (Fig. 7).

Metallography

Microstructural Analysis. The microstructure of the steel was examined after polishing and etching with a 2% nital solution. Boundaries of the hardened case were clearly delineated by the etchant and coincided with the boundaries of regions Ia and Ib. The case depths were approximately 6.4 mm (0.25 in.) on the top surface and averaged 3.3 mm (0.13 in.) on the bottom surface.

The observed microstructures—tempered martensite in the hardened case (Fig. 8) and pearlite/ferrite in the core (Fig. 9)—were consistent with the specified material requirements. No material anomalies, such as segregation at the grain boundaries, were observed.

Mechanical properties

Hardness. A microhardness traverse of the section showed that the hardness changed from 22 HRC in the flange core to 52 HRC in the case. These values correlated well with the corresponding microstructures.

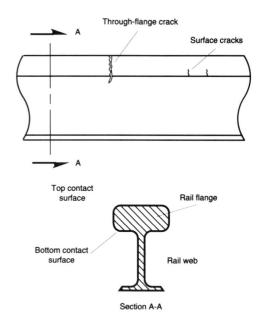

Fig. 2 Turntable rail, showing type and location of cracking

(a)

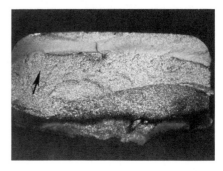

(b)

Fig. 3 Mating fracture surfaces of rail flange. Regions Ia and Ib are dull and relatively smooth. Region II has the shiny, crystalline appearance of cleavage fracture. River patterns in region II indicate that brittle cracking propagated from the edges of the hardened case across the interior of the flange, as shown.

Fig. 4 SEM micrograph of region Ia. "Rock candy" appearance indicates intergranular crack propagation. 120.6×

Fig. 5 SEM micrograph of region Ia. Exposed grains possess few distinguishing features. 670×

Fig. 6 SEM micrograph of region Ia near top of rail flange. Note presence of corrosion products. 308.2×

Fig. 7 SEM micrograph of region II. Cleavage facets evident throughout indicate a transgranular fracture path. 670×

Discussion

Transgranular, brittle crack propagation, such as that seen in region II (Fig. 7), is often associated with an existing defect or crack. Defects and cracks produce high hydrostatic stresses, which inhibit plastic deformation and promote failure by cleavage (separation of atomic planes). Therefore, it is believed that region II was the final fracture area and that cracks existed in the rail (regions Ia and Ib) prior to complete fracture of the flange.

The type of intergranular cracking seen in regions Ia and Ib (Fig. 4) can be produced either subcritically, as with fatigue cracking, stress-corrosion cracking, or hydrogen-induced cracking, or in an unstable manner, as with grain-boundary embrittlement or quench cracking. In the case of the rail flange, the evidence suggests that the intergranular cracks were created during or immediately following quenching when the rail was heat treated. The observations that the regions of intergranular fracture coincide with the flame-hardened regions, and that these regions are discolored

Fig. 8 Micrograph of section of hardened case from rail flange. Microstructure is tempered martensite, as expected from specified heat treatment. Nital etch. Note that no anomalies, such as grain boundary segregation, are extant. 268×

as if they had been exposed to high temperatures, support this conclusion.

No evidence was found to support other causes of intergranular cracking. Fatigue and environmentally assisted cracking were ruled out based on the operating conditions of the rail and the fractographic analysis of the crack—the cyclic loads on the rail were negligible, the environment benign, and the corrosion products seen on the fracture surface were judged by their appearance and location to have developed some time after crack formation. Examination of the steel microstructure provided no indication of weakened grain boundaries, again indicating that fatigue and grain boundary embrittlement were not factors in the cracking.

Quench cracks in steels result from residual stresses created by dimensional changes during cooling, large temperature gradients between the surface and center, and the volume change accompanying the austenite-to-martensite transformation. Ductile steels will plastically flow to relieve these stresses. However, heat-treated high-carbon steels, such as the rail steel, deform very little. Therefore, residual stresses can build to sufficiently high levels to form cracks. Because the surface of a quenched part is the first to transform, it experiences tensile residual stresses. Hence, quench cracks typically start from an external surface at stress concentrations, such as sharp corners or deep tool marks.

The factors controlling the formation of quench cracks include the existence of a stress raiser, steel composition and grain size, quenching medium, and

Fig. 9 Micrograph of section of core from rail flange. Microstructure is predominantly pearlite with small amounts of grain-boundary ferrite, as expected from near-eutectic composition. Nital etch. 268×

the time interval between quench and temper. For the turntable rail, the use of water as a quenching medium was believed to be the primary cause of cracking. Water provides a very rapid cooling rate, which in turn produces high residual stresses. For medium- to high-carbon steels, it is generally recommended that oil quenching or air hardening be used to minimize the potential for cracking.

Conclusion and Recommendations

Most probable cause

Quench cracks initiated during the flame-hardening treatment of both the top and bottom surfaces of the rail flange and propagated through the hardened case. Because of the increased toughness of the core, the cracks became stable at that size and depth. Applied stresses and additional residual stresses from welding, acting in conjunction with existing residual stresses, caused the crack to propagate across the remaining flange area. The direction of crack propagation was from the tip of the quench crack at the bottom surface across the interior of the rail flange. The through-flange crack then propagated down to the rail web until the crack-driving stresses (residual stresses in the tangential direction) were relieved.

Remedial action

Because the cracks initiated from a heat treatment given to the entire rail, other quench cracks probably existed. Nondestructive inspection of the rail did indeed reveal numerous small surface cracks. To minimize the growth of these cracks, it was recommended that welding repair be curtailed; welding could exacerbate the cracking problem by introducing additional residual stresses into the rail. Existing through-flange cracks were analyzed by a finite-element method to show that the cracks did not compromise the integrity of the rail. However, because additional cracks could be propagated through the flange during further operation, it was recommended that periodic inspections be continued and that, if additional through-flange cracks were found, the rail be replaced.

How failure could have been avoided

Cracking probably could have been avoided by using oil or air, rather than water, as a quenching medium for the rail. Alternatively, a lower-carbon steel, such as 1050 or 1060, or an alloy steel, such as 4130, could have been substituted for the 1080 steel to minimize cracking from water quenching and still have met the hardness requirements of the turntable rail.

Acknowledgment

The assistance of Dan Dittman in metallographic preparations and examinations is gratefully acknowledged.

Three-Wheel Motorcycle Frame Failures and Redesign

Thomas A. Knott, Willis and Kaplan, Inc., Buffalo Grove, Illinois

Bending fatigue caused crack propagation and catastrophic failures at several locations near the welds on the low-carbon steel tubular cargo box frame of police three-wheel motorcycles. ANSYS finite element analysis revealed that bending stresses in some of the frame members were aggravated by poor detail design between vertical and horizontal tubes. Stresses observed in the ANSYS analysis were not sufficient to cause the onset of fatigue. However, when compounded by stress concentration factors and in-service dynamic loading, the frame could have been regularly subjected to stresses over the fatigue limit of the material. A strain gage static loading test verified FEM results, and finite element techniques were applied in the design of reinforcing members to renovate the frames. Material properties were determined and welding procedures specified for the reinforcing members. Inspection intervals were devised to avoid future problems.

Key Words	Low-carbon steels, mechanical properties	Tubing
	Weldments	Motorcycles	Fatigue failure
	Bending fatigue	Crack propagation	Stress concentration
	Finite element method		
Alloys	Low-carbon steel—1008		

Background

A significant percentage of the approximately 300 1982-model-year three-wheel motorcycles used by a police department exhibited cracking in several locations on the cargo box frame members. These cracks, which initiate adjacent to or at welds on the frame, propagate and result in total separation of the frame members. The failed tube members not only render the motorcycle inoperable, but also fail in a sudden and catastrophic manner, which may cause the operator to lose control of the vehicle. A different but equally undesirable failure occurs at the frame attachment between the original motorcycle and the two-wheeled cargo box. A third failure location on this vehicle is between the axle and rear bumper. This failure is a result of impact loading on the bumper.

Visual Examination of General Physical Features

A stock 1982 500-cc motorcycle was modified with a two-wheeled cargo box attached in place of the single rear wheel. The cargo box is a fiberglass unit on a tubular steel frame, as shown in Fig. 1. For the conversion to a three-wheeled vehicle, the swing arm on the original motorcycle was cut and the cargo box frame was spliced to form a longer swing arm. The swing arm is connected to the main motorcycle frame through spring/shock units, and forms the rear suspension.

Upper and lower tubes were formed in a "horseshoe" configuration. The apex of each horseshoe was welded to the top and bottom surface of the swing arm stub. The left and right tubes of the upper horseshoe are parallel to the ground and extend rearward to the bumper. The tubes of the lower horseshoe are slanted towards the ground. The lower tubes curve 90°, intersect with the upper tubes, and turn inwards to connect across and form a crossmember behind the rear axle that supports the fiberglass cargo box. A similar cargo box support crossmember is forward of the axle, but originates only from the upper tubes. A third crossmember (called the shock crossmember) is located in front of the other two, and connects the cargo box frame to the main motorcycle frame through three spring/shock units. At the location where the shock crossmember joins the main frame tubing, short vertical members connect the upper and lower tubes on each side.

The cargo box frame is constructed of 25.4-mm (1.0-in.) steel tubing with a wall thickness of 3.2 mm (0.125 in.). The bumper is made of rectangular tubing, 25.4 mm (1.0 in.) wide by 50.8 mm (2.0 in.) high with a 3.2 mm (0.125 in.) wall. The swing arm

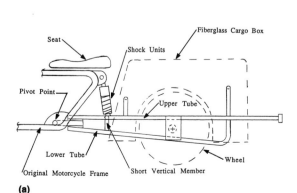

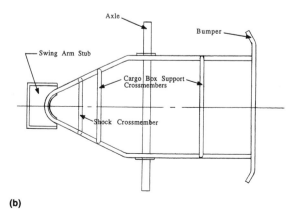

Fig. 1 Frame nomenclature

stub is constructed of both tubes and plates approximately 3.2 mm (0.125 in.) thick, while the plates that locate the rear axle are 4.8 mm (0.188 in.) thick.

Cracks in the cargo box frame occurred in several locations as shown in Fig. 2. At one location, marked "A," the upper and lower tubes are welded to the swing arm plates. A second failure location on the upper and lower tubes, marked "B," is behind the short vertical member that joins them together. From field reports on failures, this location had the highest frequency of cracking. Towards the rear of the frame, the lower tube fails behind the

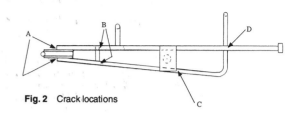

Fig. 2 Crack locations

axle plate ("C"), and the upper tube cracks behind the rear cargo box crossmember ("D").

<table><tr><td>**Testing Procedure and Results**</td><td></td></tr></table>

Surface examination

Examination of the fractured surfaces was hindered by corrosion. Several examples of partially cracked tubes were found, and uncorroded areas of these tubes exhibited beach marks characteristic of fatigue.

The fractures on the tubes followed a vertical path, diverting only at locations induced by the welds (beads of filler, etc.). The fractures generally appeared to start at undercuts in the welds or at the top fillet. No shear lips characteristic of torsional failure were noted.

Chemical analysis/identification

A chemical analysis was conducted on a sample of frame tubing to identify the type of steel used in manufacturing the frame. The composition (wt%) 0.01 Si, 0.40 Mn, 0.08 C, 0.006 P, 0.028 S, was found. This composition corresponds with SAE/AISI 1008, a low-carbon steel.

Mechanical properties

The hardness of the base tubing material was found to be 31 HRB, or an approximate tensile strength of 235 MPa (34 ksi). The hardness in the weld area was found to peak at 74 HRB, with a corresponding tensile strength of 450 MPa (65 ksi).

Stress analysis

Analytical. A significant portion of this study was the finite element analysis of the cargo box

frame structure using the ANSYS computer program. The locations analyzed are shown in Fig. 3.

The finite element analysis indicated stresses in the frame as high as 7.38 MPa (1070 psi) from a 44.5-N (100-lb.) load (location 9). The highest

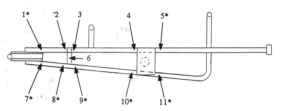

(a)

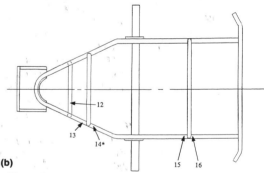

(b)

Fig. 3 Selected analysis locations

Table 1 Local stresses resulting from an applied 44.5-N (100-lb.) load

| | Experimental(b) | | Finite element analysis | | | |
| | | | Baseline | | After reinforcement | |
Location(a)	kPa	psi	kPa	psi	kPa	psi
1	414	60	304.7	44.2	455.7	66.1
2	...	...	3208.1	465.3	1234.8	179.1
3	...	...	6346.6	920.5	104.8	15.2
4	...	...	4199.6	609.1	609.5	88.4
5	620	90	768.8	111.5	166.8	24.2
6	...	...	590.2	85.6	348.2	50.5
7	1241	180	1115.6	161.8	1043.2	151.3
8	4757	690	5448.9	790.3	2142.2	310.7
9	8894	1290	7378.1	1070.1	900.5	130.6
10	4137	600	4700.2	681.7	131.7	19.1
11	2482	360	1470.0	213.2	111.7	16.2
12	...	...	5067.0	734.9	4769.1	691.7
13	...	...	3776.9	547.8	199.3	28.9
14	1655	240	2352.5	341.2	20.7	3.0
15	...	...	226.8	32.9	104.1	15.1
16	...	...	226.8	32.9	172.4	25.0

(a) See Fig. 3 for locations. (b) From strain gage reading. (c) Using ANSYS

stresses were found on the upper and lower tubes behind the short vertical members (locations 9 and 3). These locations exhibited the highest frequency of cracking. Other locations with relatively high stresses (in the range of 3.45 to 6.20 MPa, or 500 to 900 psi) were found in front of the short vertical member on the lower tube (location 8), in front of the forward cargo box support crossmember on the upper tube (location 13), on both the upper and lower in front of the axle mounting plates (locations 4 and 10), and in the shock crossmember (location 12). The stresses in the tubes between intersections were moderate (less than 2.07 MPa, or 300 psi). The lowest stresses (less than 0.69 MPa, or 100 psi) were found in frame members behind the rear axle (locations 15 and 16). Figure 3 and Table 1 show the analysis locations and selected ANSYS stress data.

The stresses observed in the ANSYS analysis are not sufficient to cause the onset of fatigue, but it should be noted that those stresses are due to a 44.5-N (100-lb.) load. The effective load of the motorcycle and rider is close to 445 N (1000 lb.), so in actual service the highest stresses would be over 68.9 MPa (10 ksi). When compounded by the stress concentration factors at the tube joints and the dynamic loading seen in service, the frame could be regularly subjected to stresses over the fatigue limit of the material.

Besides being used as an analytical tool, the ANSYS program was used as a design aid to develop the frame reinforcements. Various configurations of reinforcements were added to the baseline model and run with ANSYS. The prospective reinforcements consisted of plates that originated from the swing arm stub and continued back to three different locations, and a flat plate that was added to the top of the structure. The selected redesign configuration consisted of a 4.8-mm (0.188-in.) plate on the side of the frame from the swing arm stub to the axle. This configuration, shown in Fig. 4, effectively reduced stresses in the upper and lower tubes. The highest stress was 2.14 MPa (310 psi), and was located on the lower tube in front of the short vertical member (location 12). Table 1 also contains stress data due to a 44.5-N (100-lb.) load for a reinforced frame.

Experimental. A strain-gaged static loading test was performed to verify the ANSYS results. Several strain gages were mounted on an actual cargo box frame in selected locations, including areas where cracks have occurred, as 1.6-mm (0.062-in.) strain gages were used to measure the strain along the axis of the tube.

To simulate the loading pattern of a frame in service, the test frame was supported at the front pivot and on both sides of the axle. The applied load was suspended from the shock crossmembers to simulate the downward load of the spring/shock units. Strain gage readings were taken at various applied loads. The load was incremented to confirm that the strains indicated by the instrumentation were proportional to the load applied, and that no geometric non-linearities (gaps between members) were present.

Stresses obtained experimentally and the stresses theoretically determined from the finite element (ANSYS) model at the same locations showed acceptable correlation. In addition, a stress analysis using handbook analysis techniques was performed on the shock crossmember (location 12). The result of 5.24 MPa (760 psi) was in close agreement with the result of 5.07 MPa (735 psi) indicated by the ANSYS analysis.

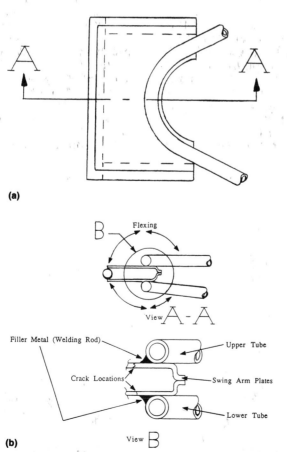

(a)

(b)

Fig. 4 Design of reinforced motorcycle frame

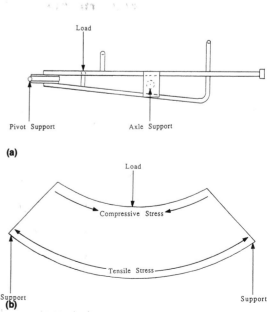

(a)

(b)

Fig. 5 Failure loading, swing arm plates

Discussion

The material in this application, AISI 1008, was probably selected for its ease of manufacture in both cutting and welding. In this case, the fatigue limit of this material appears to have been exceeded due to both material selection and component design.

Conclusions and Recommendations

Most probable cause

Swing Arm Plates ("A"). The failure at the swing arm plates is directly attributable to poor engineering design. Using the cargo box and frame as a swing arm results in high stresses in the swing arm plates, owing to the short length of the connection with the upper and lower tubes. This area becomes a "hinge" when loaded, as shown in Fig. 4, and fails from bending fatigue due to the flexing encountered as the motorcycle experiences road bumps. The failures in this area are consistent with plate flexing since the weld metal remains intact with the tubes after separation.

A second reason for failure at the swing arm plates is that the design does not follow basic vehicle dynamics precepts. One goal of vehicle dynamics is to minimize the amount of unsprung weight, i.e., the weight not carried by the springs. A vehicle with a lower unsprung mass can have a lighter frame, since reduced forces are transmitted through the springs. A significant portion of the total vehicle weight of this motorcycle is unsprung, since the cargo box and its frame is connected below the spring/shock units. This modified design allows unnecessary and excessive loading at the swing arm plates.

Behind Short Vertical Members ("B"). The cracking in the upper and lower tubes adjacent to the short vertical members occurs because of the load transfer patterns from the shock crossmember to the pivot, on the swing arm in front and the axle at the rear. The frame can be represented as a simple beam, as shown in Fig. 6, with both ends supported and a vertical load applied to the center. Classic beam theory indicates that the upper surface would have a compressive stress while the lower surface would have a tensile stress. Though the "beam" on the cargo box frame consists of two tubes separated by a finite distance, the resulting behavior is similar to that of a composite beam: the upper tube compresses and the lower tube elongates from the loading, resulting in a bending moment resisted by the short vertical member. The 90° angle formed by the intersection of the short vertical members with the upper and lower tubes results in high stresses behind the members, and provides a location for cracking.

Fatigue was the apparent failure mechanism of the upper and lower tubes. The maximum stress occurs in the frame from the loadings as the motorcycle is deflected by a bump, and conversely the minimum stress occurs when the motorcycle descends from the bump.

Behind Rear Axle ("C" and "D"). The frame cracking that appears behind the axle is due to impact loads on the bumper. Two long tubes, relatively far apart, support the bumper, with the structure appearing from above as shown in Fig. 7. The dimensions of the frame members themselves are disproportionate. The tubes are 25.4 mm (1.0 in.) in diameter and 610 mm (24 in.) apart, resulting in a structure that lacks the rigidity to resist side loads. A side load or partial side load causes the bumper tubes to bend about the union with the rear cargo box support crossmember. Those locations fail as a result of the rigidity of the joint. The lower tube behind the rear axle plate also exhibited cracking from a similar overload mechanism.

Frame Member Intersection Methods. The joining methods for the frame tubes compound both the fatigue and impact loading problems. Where tubes intersected, one was cut straight and placed on top of the other. The resulting gaps were filled with welding rod, a practice that should be avoided. The proper method of joining the tubes would have been to "fishmouth" the end of one tube, as shown in Fig. 8a, so that it conforms to the round surface of the other tube. This practice also minimizes the amount of welding filler required to complete the weld.

The sharp 90° angles found at many of the tube

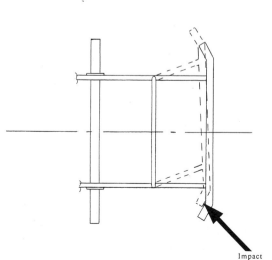

Fig. 6 Failure loading, behind short vertical members

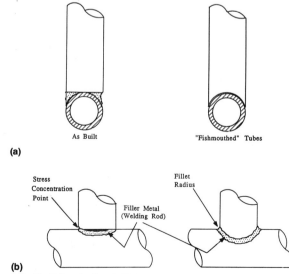

Fig. 7 Failure loading, behind the rear axle

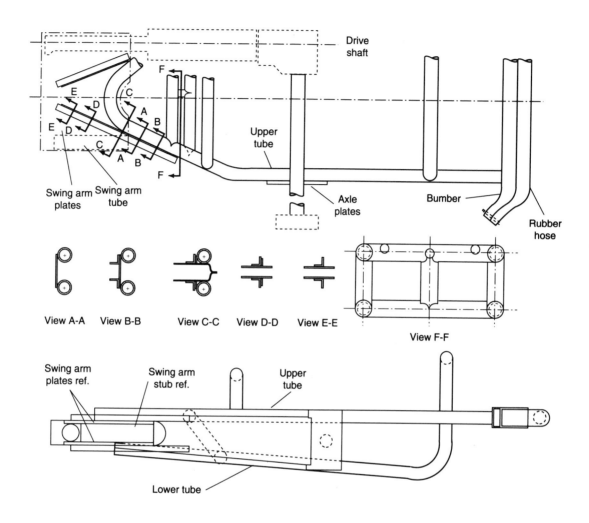

Fig. 8 Methods of tube joining. (a) Fishmouthing. (b) Reduction of stress concentration using fillets

joint locations intensified the stresses in the tubes. On some of the welds, the gap between the intersecting tubes was concave, producing stress concentration points. These stress concentration points could have been avoided by smooth transitions (fillets) between intersecting members, as shown in Fig. 8b.

Remedial action

Welded reinforcements were designed and implemented, as shown in Fig. 4. The repair design was complicated because no two frames were built alike. Statistics were collected, and the repair reinforcements were designed to fit the most frames with a minimum of fitting.

Maintenance operations incorporated an inspection to check for any problems in the future. Based on the stress levels in the redesigned structure and an estimated fatigue spectrum, incorporating a crack examination into a currently scheduled yearly inspection offers an adequate factor of safety to find cracks before reaching critical proportions.

How failure could have been prevented

The design of vehicle structures should follow engineering principles and proven manufacturing procedures to ensure safety. Dynamic effects (inertial loads) of vehicle components increase the loadings beyond the peak static load.

The method of joining the tubes certainly accelerated the process of crack growth. Had the tubes been properly manufactured (fishmouthed and filleted), the same patterns of failure would have occurred, but later in the life of the vehicle. The magnitude of bending stresses was more of a causal factor in these failures than manufacturing details.

The designer should design for infinite life, or specify inspection intervals and procedures if there is a possibility of fatigue loading which may lead to catastrophic cracking.

PROCESS EQUIPMENT FAILURES

Heat Exchangers

Pressure Vessels

Pipes and Pipelines

Failure of Admiralty Brass Condenser Tubes

H.S. Khatak, J.B. Gnanamoorthy, and P. Rodriguez, Metallurgy Programme, Indira Gandhi Centre for Atomic Research, Kalpakkam, India

Leaks developed in 22 admiralty brass condenser tubes. The tubes were part of a condenser that was being used to condense steam from a nuclear power plant and had been in operation for less than 2 years. Analysis identified three types of failure modes: stress-corrosion cracking, corrosion under deposit (pitting and crevice), and dezincification. Fractures were transgranular and typical of stress-corrosion cracking. The primary cause of the corrosion deposit was low-flow conditions in those parts of the condenser where failure occurred. Maintenance of proper flow conditions was recommended.

Key Words

Admiralty metal, corrosion
Condenser tubes, corrosion
Transgranular fracture

Pitting (corrosion)
Crevice corrosion

Nuclear reactor components
Stress-corrosion cracking

Background

Leaks developed in 22 admiralty brass condenser tubes.

Applications

The tubes were parts of a condenser that was being used to condense steam from a nuclear power plant and had been in operation for less than 2 years. Cooling was done with river water.

Specimen selection

One failed tube was submitted for failure analysis. The specimens were cut to suit the stage of the scanning electron microscope (SEM). Care was taken so that the attacked regions were not altered. Figure 1 shows a schematic of the specimen locations in the failed condenser tube. Samples outside the failed regions were used for chemical analysis of the tube material.

Visual Examination of General Physical Features

A black deposit with white patches and a number of pits covered about 75% of the inside circumference of the 150 mm (6 in.) long tube specimen. The top side of the inner surface was clean. Two circumferential through-cracks (locations 6 and 7, Fig. 1) were visible in the deposit region. The outside surface (steam side) looked fairly bright and

uncorroded.

Figure 1 shows the wall thickness measured at different locations along the circumference of the fractured surface with respect to the location of the cracks. The lowest thickness measured was 0.76 mm (0.03 in.). Region 1 in Fig. 1 did not show any loss of thickness.

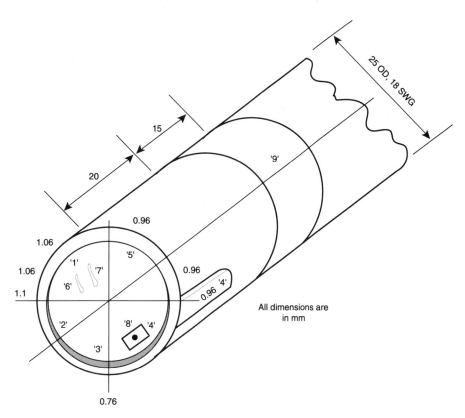

Fig. 1 Schematic of relative locations of samples taken for examination.

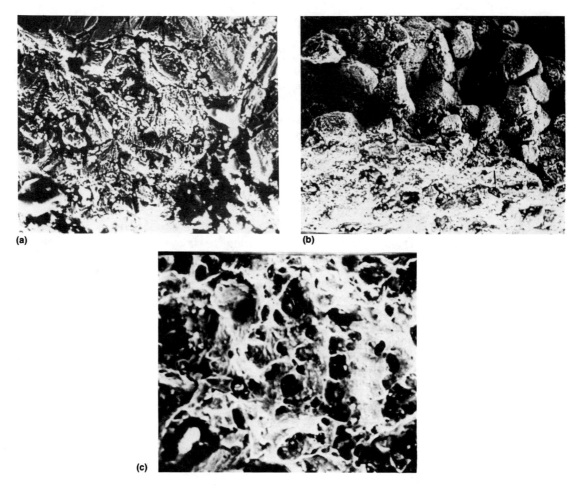

Fig. 2 SEM fractographs of the major fracture surface. (a)Typical transgranular SCC fracture. (b)Intergranular attack followed by transgranular cracking. (c)Typical ductile fracture

Testing Procedure and Results

Surface examination
Scanning Electron Microscopy/Fractography. For SEM examination of the major fracture surface, the entire circumference was sectioned into five parts (Fig. 1). Section 1 indicated a typical transgranular stress-corrosion cracking (SCC) surface (Fig. 2a). Intergranular attack followed by transgranular fracture was seen at one of the locations (Fig. 2b). All other sections, including the thinnest, exhibited ductile failure. This fracture was caused by tensile pulling during removal of the tube. A typical ductile fracture is shown in Fig. 2(c).

A small through-crack (location 6, Fig. 1) was opened under tension and the fracture surface examined by SEM. The fracture morphology of this surface indicated transgranular SCC and was similar to that observed at section 1. It was also observed that the length of the crack inside the tube exceeded that on the outer surface, indicating that the crack had initiated from the inside. Furthermore, when the sample was tilted and examined from the inside, intergranular attack was seen throughout the length of the crack. At one of the locations, the intergranular attack changed to intergranular cracking.

Corrosion Patterns. Examination of the crevice at location 9 in Fig. 1 (Fig. 3a) showed a porous mass inside the crevice and intergranular at-

tack at the bottom (Fig. 4).

Metallography
Crack Origins/Paths. Based on the corrosion morphology and the dimensions of the crack, it was concluded that the attack/crack originated on the inside surface.

A sample from location 7 in Fig. 1, which contained a crack, was polished and etched in a solution containing 30 g $FeCl_3$, 90 mL HCl, and 360 mL H_2O, then observed in an optical microscope. The crack was transgranular. Two smaller transgranular cracks were also seen near the larger one.

A transverse section of location 4 (Fig. 1) was also examined. It showed the same type of transgranular through-thickness crack.

Chemical analysis/identification
Metal chips taken from an area some distance from the failed region were analyzed chemically, and the composition was found to be 69.5% Cu, 29.1% Zn, 1.33% Sn, and 0.046% As.

The corrosion deposits on the inner surface of the tube and the fracture surface were analyzed using the energy-dispersive X-ray (EDX) technique. Copper, zinc, arsenic, tin, silver, sulfur, chlorine, calcium, iron, sodium, and potassium were detected in the corrosion product. EDX analysis of

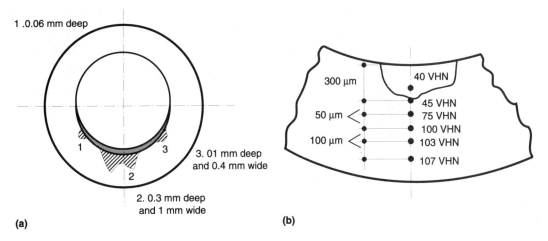

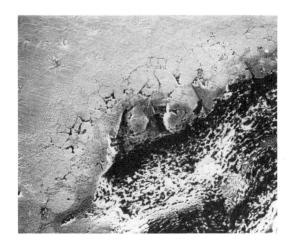

Fig. 3 Schematic of sample taken from location 9 in Fig. 1. (a) Crevice under the deposit. (b) Enlargement of region 2 in (a), showing hardness measurements along the crevice

the intergranularly attacked portion showed that the copper-zinc ratio had increased. It indicated the selective loss of zinc. Analysis of the porous mass in the crevice indicated a complete loss of zinc. Dezincification was observed up to 50 μm in the tube material from the bottom of the crevice.

Mechanical properties

Hardness. Microhardness was measured on the cross section of location 9 (Fig. 1), which was some distance from the failed region, and on a sample from location 4 (Fig. 1), which was closer to the failed region. The hardness values ranged from 95 to 107 HV. A value of 75 HV is typical for annealed admiralty brass tubing.

Microhardness was measured on thickness along the creviced region (Fig. 3a and b). The tube material indicated lower hardness values up to about 100 μm from the bottom of the crevice.

Fig. 4 Micrograph showing crevice attack, porous mass, and intergranular attack at the bottom of the crevice

Discussion

Chemical analysis and the microstructure indicated that the material conformed to specifications. Three types of failure modes were identified: SCC, corrosion under the deposit (pitting and crevice), and dezincification.

Stress-Corrosion Cracking. Examination of the fracture surface of the major crack and a smaller crack showed typical transgranular SCC failure, with evidence of crack branching and no sign of plastic deformation. Stress-corrosion cracking requires the presence of stresses and a suitable environment. The hardness obtained (95 to 107 HV, compared with 75 HV for an annealed tube) indicated the presence of cold working, which resulted in residual stresses. Thermal stresses due to a localized increase in temperature resulting from the low-conducting corrosion deposit must have contributed to the total stresses. In addition, during the installation of the tubes in the tube sheet by rolling, excessive expansion may have resulted in high stresses, not just at the end of the tubes but also through their entire length. Such

stresses are caused by the increased distortion of the tubes.

An environment conducive to SCC was provided by the concentration of salts in the corrosion deposit. In addition to ammonia, a number of environments can cause cracking, such as oxygen and carbon dioxide with ammonia, hydrogen sulfide, air contaminated with sulfur and nitrogen oxides, nitrates, phosphates, alkalis, chlorides, hydrogen fluoride, and nitric acid (Ref 1, 2). The EDX analysis detected the presence of sulfur, silicon, chlorine, calcium, sodium, potassium, and iron in addition to the components of the base material.

The profile of crack (location 6, Fig. 1), the greater quantity of cracks on the inside surface, and the absence of corrosion on the outside surface indicated that the crack initiated from the inside. The intergranular attack leading to intergranular cracking, along with the intergranular attack at the bottom of the crevice on the inside surface, confirmed the above conclusion.

Corrosion Under the Deposit. At one of the

locations, a minimum wall thickness of 0.76 mm (0.03 in.) was noted, compared with 0.9 to 1.1 mm (0.035 to 0.043 in.) at the other areas. This was caused by a differential aeration cell. Crevice-type attack (Fig. 2 and 5) was caused by the same effect.

Dezincification. The EDX analysis at the intergranularly attacked surface (Fig. 3b) showed evidence of dezincification. In the crevice (Fig. 3 and 4), complete loss of zinc was observed. Dezincification was also observed up to about 50 µm be-yond the bottom of the crevice. Hardness measurements on the cross section (Fig. 3b) indicated low values up to about 100 µm. Whatever the mode of attack, it was caused by the corrosion deposit.

The main cause of the corrosion deposit appeared to be the low-flow conditions in those portions of the condenser where failure occurred. Low-flow conditions resulted in the corrosion deposit and in an increase in temperature, which in turn further aggravated the problem.

Conclusion and Recommendations	**Most probable cause**	**How failure could have been prevented**
	Low-flow conditions in some portion of the condenser resulted in corrosion deposits and concentration of the corrosive environment. This led to SCC in the presence of residual and thermal stresses.	Proper operating conditions and equal flow in all tubes could have prevented the concentration of the corrosive environment and thus the failure.

References

1. *The Corrosion of Copper, Tin, and Their Alloys,* John Wiley & Sons, 1971, p 148

2. *8th Int. Cong. Metallic Corrosion,* Vol 1, 1981, p 479

Failure of Nickel-Base Superalloy Heat-Exchanger Tubes in a Black Liquor Heater

J. Robert Kattus, AMC–Vulcan, Inc., Birmingham, Alabama

Several nickel-base superalloy (UNS N06600) welded heat-exchanger tubes used in processing black liquor in a kraft paper mill failed prematurely. Leaking occurred through the tube walls at levels near the bottom tube sheet. The tubes had been installed as replacements for type 304 stainless steel tubes. Visual and stereoscopic examination revealed three types of corrosion on the inside surfaces of the tubes: uniform attack, deeper localized corrosive attack, and accelerated uniform attack. Metallographic analysis indicated that pronounced dissimilar-metal corrosion had occurred in the base metal immediately adjacent to the weld seam. The corrosion was attributed to exposure to nitric acid cleaning solution and was accelerated by galvanic differences between the tubes and a stainless steel tube sheet and between the base metal of the tubes and their dendritic weld seams. A change to type 304 stainless steel tubing made without dendritic weld seams was recommended.

Key Words

Nickel-base alloys, corrosion
Heat-exchanger tubes, corrosion
Nitric acid, environment

Superalloys, corrosion
Galvanic corrosion
Papermaking

Welded joints, corrosion
Heat-affected zone

Alloys

Nickel-base superalloy—N06600

Background

Several nickel-base superalloy (UNS N06600) heat-exchanger tubes used in processing black liquor in a kraft paper mill failed prematurely.

Applications

In the heat-exchanger saturated steam is used to heat kraft black liquor as it circulates out of and

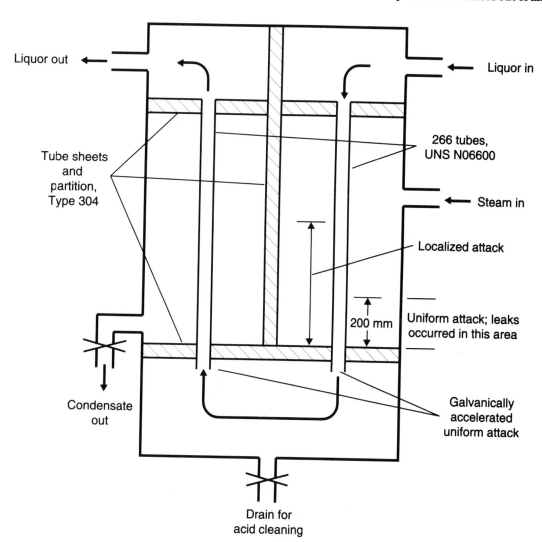

Fig. 1 Schematic of heat-exchanger system, showing the areas of various types of corrosive attack on the inside surfaces of the tubes

back into a digester (Fig. 1). The maximum steam temperature is about 190 °C (375 °F).

Circumstances leading to failure

Premature leaking occurred through the walls of several tubes at levels near the bottom tube sheet about 1 year after new UNS N06600 superalloy tubes were installed. The heat exchanger operates continuously except when the black liquor side is cleaned biweekly by circulating 6% nitric acid at 60 °C (140 °F) and a pH of 1.5 to 2 for several hours. The pH levels of the black liquor and steam condensate are maintained at about 13 and 7.8, respectively—degrees of alkalinity that are relatively benign with respect to corrosive attack on nickel-base superalloys and type 304 stainless steel. Additionally, the boiler water is treated with amines, which are corrosion inhibitors. Nitric acid is much more corrosively aggressive than alkaline solutions, particularly with respect to the UNS N06600 superalloy (Ref 1-3).

Pertinent specifications

The original specifications for the heat exchanger called for the use of type 304 stainless steel for both the tube sheets and tubes, in accordance with ASME SA240 and SA249, respectively. The tubes that failed were made of UNS N06600 nickel-base superalloy and had been installed as replacements for type 304 stainless steel tubes. A total of 266 tubes, 31.75 mm (1.250 in.) in diameter with a 1.52 mm (0.060 in.) wall, were used in the heat exchanger.

Performance of other parts in same or similar service

The type 304 stainless steel tubes had performed satisfactorily for 10 years; the superalloy replacements failed after only 1 year of service.

Specimen selection

After disassembly of the heat exchanger, nine of the tubes representing various degrees of corrosive deterioration, were selected for examination and testing.

Visual Examination of General Physical Features	The outside surfaces of the tubes generally were intact, except for holes in some tubes, which had allowed leakage to occur near the bottom tube sheet. The wall thicknesses were normal in the upper portions of the tubes, but decreased because of corrosive attack on the inside surfaces in the lower portions.

Testing Procedure and Results

Surface examination

Corrosion Patterns. Visual and stereomicroscopic examination revealed three types of corrosion on the inside surfaces of the tubes, as shown in Fig. 1. First, uniform corrosive attack, concentrated in the bottom 200 mm (8 in.) of the tubes, resulted in wall thickness reductions ranging from more than 50% in the most corroded samples to less than 10% in the least corroded specimens. Second, deeper localized corrosive attack occurred along the edges of the longitudinal weld seam (Fig. 2), causing leaking in several of the tubes. The localized attack extended higher up the tubes than the uniform attack, but not all the way to the top. Third, accelerated uniform attack occurred at the bottom of the tubes adjacent to their galvanic couplings with the tube sheet. Corrosion penetrated the full wall thickness in this area; however, leakage was prevented by the surrounding tube sheet.

The relatively undamaged condition toward the top of the tubes, in contrast to the severe corrosion toward the bottom, is believed to be the result of differences in exposure to the dilute nitric acid cleaning solution. For example, rings of various colors on the insides of the tubes indicated that the nitric acid was allowed to stand for periods of time at various intermediate levels and was therefore in contact with the lower levels longer than with the top levels.

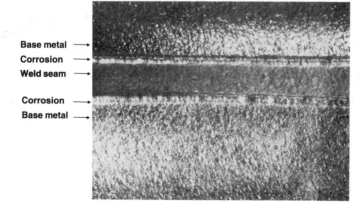

Fig. 2 Typical localized corrosion on inside surface of a tube along the edges of the weld seam. 3.04×

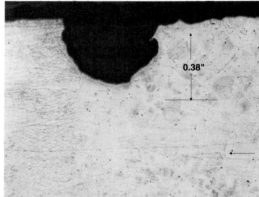

Fig. 3 Typical section through the inside surface of a tube in an area above the level of uniform corrosion, showing localized dissimilar-metal corrosion to a depth of 0.38 mm (0.015 in.) in the base metal at an interface with the weld seam. Note the dendritic microstructure of the weld. Electroetched in oxalic acid. 49.4×

Metallography

Microstructural Analysis. Transverse tube sections through the weld seams and base metals in areas of severe and negligible uniform corrosion were examined metallographically. The typical microstructure (Fig. 3) of a section through an interface between a weld seam and the base metal, at intermediate levels above the area of uniform corrosion, shows that the weld was dendritic, indicating either improper control of the flash-welding process or the use of filler metal. Pronounced localized dissimilar-metal corrosion occurred to a depth of about 0.38 mm (0.015 in.) in the base metal at the interface on the inside surface. This corrosion corresponded to the corrosion that appeared along both edges of the weld seam shown in Fig. 2. In sections near the bottom of the tubes where substantial uniform corrosion occurred (Fig. 4), the greatest depths of corrosive attack and ultimate failure (leaking) occurred in the base metal immediately adjacent to its interfaces with the weld seam. This was caused by a combination of both general and localized corrosion.

Chemical analysis/identification

Chemical analyses on two samples representing the most and least corroded of the tubes provided almost identical results: 0.03% C, 0.33% Mn, <0.005% S, 0.21% Si, 74.80% Ni, 15.75% Cr, <0.01% Cu, and 8.42% Fe. This composition complies with the chemical requirements of ASTM B516 (UNS N06600), which prohibits the use of

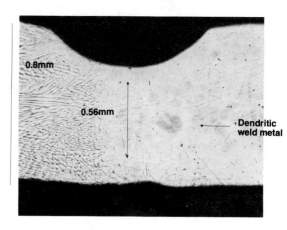

Fig. 4 Tube section near the bottom tube sheet, showing an area at the weld/base metal interface where a combination of both uniform and localized corrosion occurred. In this particular section, the wall thickness has been reduced to 0.8 mm (0.032 in.) by the uniform attack and to 0.56 mm (0.022 in.) by the combined attack; in the tubes that leaked, of course, the combined attack reduced the wall thickness to the point of breakthrough at the weld/base metal interfaces. Electroetched in oxalic acid. 49.4×

filler metal in the weld seam.

Mechanical properties

Hardness. Tests on several tubes showed hardnesses of 67 to 71 HRB in the base metals and 75 to 82 HRB in the weld seams.

Discussion

Three types of corrosion occurred on the inside surfaces of the tubes: localized attack along the interfaces between the weld seam and the base metal, uniform attack, and accelerated uniform attack. The localized attack, which occurred at levels periodically exposed to dilute nitric acid cleaning solutions, was caused by galvanic differences between the base metal and the dendritic weld metal in the presence of the nitric acid (Ref 4). Uniform attack, which was confined primarily to the bottom 200 mm (8 in.) of the tubes, was the result of the natural aggressiveness of dilute nitric acid toward the UNS N06600 tube alloy (Ref 2). Accelerated nitric acid attack occurred at the bottom ends of the tubes, immediately adjacent to the bottom of the tube sheet, because of the anodic galvanic properties of the tubes with respect to the tube sheet (Ref 3, 5). However, this particular attack on the inside

surfaces of the tubes was not a significant factor in the leakage through the tube walls, because the areas where the accelerated attack occurred were enclosed on the outside diameter by the bottom tube sheet.

Marked differences in the severity of the corrosive attack among different tubes, despite similar chemical compositions and microstructures, are probably the result of variations in thickness of the natural oxide surface skins that impart passivity—that is, resistance to galvanic corrosion—to alloys of this type. It is likely that the more corroded tubes had relatively thin or nonuniform skins that broke down quickly during the acid cleaning operations, whereas the skins were thicker and longer lasting on the less corroded tubes.

Conclusion and Recommendations

Most probable cause

The leaking of the tubes, which occurred near the bottom tube sheet, was caused by the combined effects of localized and uniform corrosion induced on the inside surfaces by dilute nitric acid cleaning solutions. Galvanic differences between the base metal and the dendritic weld seam were responsi-

ble for the localized attack. The uniform attack was caused by the limited resistance of the UNS N06600 tube alloy to nitric acid. Accelerated uniform galvanic attack at the bottom ends of the tubes in the areas surrounded by the tube sheet, although relatively severe, was not a factor in the leakage.

Remedial Action

The principal recommendation to correct leakage was to replace the UNS N06600 tubes with type 304 stainless steel tubes made without dendritic weld seams in accordance with ASTM A249.

Matching of the tube alloy with the tube sheet eliminates one source of dissimilar-metal galvanic corrosion, and use of a nondendritic weld seam eliminates the other source of potential dissimilar-metal corrosion. In addition, type 304 stainless steel is reported to have a greater inherent resistance than the UNS N06600 superalloy to nitric acid (Ref 1-3). An additional recommendation was to control cleaning cycles to prevent prolonged exposure to nitric acid.

References

1. "Corrosion Resistance of the Austenitic Chromium-Nickel Stainless Steels in Chemical Environments," Bulletin 2M 8-75 (3M 1-73) 3846(8), International Nickel Co., 1975
2. "Resistance to Corrosion," Bulletin 3M 8-88 S-37, Inco Alloys International, 1988, p 29
3. *Metals Handbook*, 9th ed., Vol 13, ASM International, 1987, p 235, 645
4. "Joining," Bulletin 10M 12-78 T-2, Huntington Alloys, 1978
5. M.G. Fontana and N.D. Greene, *Corrosion Engineering*, McGraw-Hill, 1967, p 31-33

Failure of a Copper Condenser Dashpot

Rakesh Kaul, N.G. Muralidharan, N. Raghu, K.V. Kasiviswanathan, and Baldev Raj, Indira Gandhi Centre for Atomic Research, Kalpakkam, India

A copper condenser dashpot in a refrigeration plant failed prematurely. The dashpot was a long tubular component with a cup brazed at each end. Stereomicroscopic examination of the fracture surface at low magnification revealed a typical ductile mode of failure. The failure was attributed to insufficient component thickness, which made the dashpot unable to withstand internal operating pressure, and to extensive annealing in the heat-affected zones of the brazed joints. It was recommended that the condenser dashpot design take into account the annealing effects of brazing. Hydrostatic testing at a pressure 1.25 times greater than the maximum operating pressure prior to placing the component in service was also suggested.

Key Words

Copper
Condensers (liquefiers)

Ductile fracture
Refrigerating machinery

Condenser tubes

Background

A copper condenser dashpot in a refrigeration plant failed prematurely.

Applications

The refrigeration plant is equipped with two refrigeration units. Each unit operates for 24 h, followed by a 24 h shutdown, during which time the standby unit is put into operation.

The condenser unit of the refrigeration plant was composed of a 350 mm (14 in.) long, 62 mm (2.4 in.) diam dashpot with a wall thickness of 1.2 mm (0.05 in.). A copper cup was brazed at each end. The dashpot was surrounded by a helical copper tube, and the entire assembly was housed in a galvanized steel enclosure filled with polyurethane foam (PUF) insulation. Figure 1 shows a diagram of the condenser unit. The condenser dashpot carried R-503 refrigerant chilled by a cascade-type refrigeration unit (using R-502 refrigerant) to a temperature of –80 °C (–110 °F). The component was operated at a nominal pressure of 1.6 to 1.8 MPa (230 to 260 psi). During shutdown, temperature rises caused evaporation of refrigerant, leading to an increase in the internal pressure. R-503 refrigerant has an equilibrium pressure of 4.2 MPa (609 psi), so the possibility of internal pressure reaching this value during a prolonged shutdown period cannot be ruled out (Ref 1). The condenser unit is not equipped with a pressure measurement instrument.

Circumstances leading to failure

After about 6 months of operation, during which time each condenser unit had undergone about 200 cycles of operation, one unit blew off, ripping open the surrounding PUF insulation and galvanized steel enclosure. The accident took place when the concerned unit was shut down.

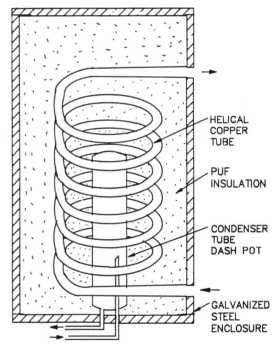

Fig. 1 Diagram of the condenser unit

HELICAL COPPER TUBE

PUF INSULATION

CONDENSER TUBE DASH POT

GALVANIZED STEEL ENCLOSURE

Performance of other parts in same or similar service

No failure of this kind has thus far been encountered in the other condenser unit. The condenser dashpot is housed in a galvanized steel enclosure, so it was not possible to determine whether the other component has undergone plastic deformation.

Visual Examination of General Physical Features

Figure 2 shows the failed condenser dashpot. The failure resulted in the fragmentation of the component into three parts: an upper cup portion, an intermediate tubular portion, and a lower tubular portion. The upper cup portion and the intermediate tubular portion had been thrown out of the enclosure, while the lower tubular portion remained intact with the assembly. The intermediate tubular portion had been completely flattened during the explosion. Hammer marks on the outer surface of this part were introduced by plant personnel who tried to reshape it. It was clear that the failure of the component was preceded by extensive plastic deformation.

Testing Procedure and Results

Nondestructive evaluation

Dimensional Measurement. Approximate dimensional profiles of the component before and after failure are presented in Fig. 3. The two end cups and central tubular region of the dashpot did not show any plastic deformation, whereas the tubular regions adjacent to the two brazed ends exhibited extensive swelling.

Dimensional measurements made on the flattened region showed that this portion had bulged to a maximum diameter of about 86 mm (3.4 in.) (Fig. 3). Diameter and thickness measurements were also carried out on the lower tubular portion (see Fig. 2). Ultrasonic thickness measurements (accuracy of ±1 μm) were made at various locations on the condenser tube component. Diameter and thickness variations of the lower tubular portion of the failed dashpot with respect to the distance from the lower brazed end are presented in Fig. 4.

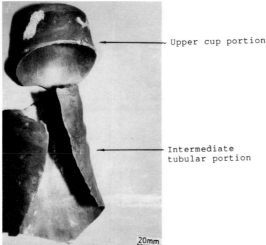

Upper cup portion

Intermediate tubular portion

20mm

(a)

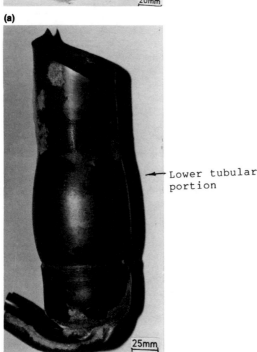

Lower tubular portion

25mm

(b)

Fig. 2 Failed dashpot in the as-received condition

Surface examination

Visual. All of the fracture edges exhibited extensive wall thinning.

Macrofractography. Stereomicroscopic examination of the fracture surface at low magnification showed a typical ductile mode of failure.

Scanning Electron Microscopy/Fractography. Examination of the upper and lower circumferential fracture surfaces by scanning electron microscopy (SEM) revealed the presence of elongated dimples (Fig. 5a), whereas the longitudinal fracture surface exhibited equiaxed dimples (Fig. 5b).

Metallography

Microstructural Analysis. Optical metallographic examination of the longitudinal section of an undeformed region of the lower tubular portion revealed an elongated grain structure with a high degree of cold work (Fig. 6). *In situ* metallographic examination of the bulged region of the lower tubular portion (see Fig. 2) using a portable microscope revealed that the region had undergone annealing. A region near the brazed joint exhibited very large equiaxed grains; grain size diminished with distance from the brazed joint. Figure 7(a) shows the change in grain size with distance from the lower brazed end. Similar grain size variation was noticed in the heat-affected zone (HAZ) of the upper brazed joint (Fig. 7b). Extensive stretching of material near the upper circumferential fracture surface is also evident in Fig. 7(b). Metallographic examination of the inner surface of the intermediate tubular portion, which was opened as a result of the failure, revealed evidence of impingement attack (Fig. 8) (Ref 2).

Chemical analysis/identification

The chemical composition of the condenser dashpot material conformed to commercial-grade copper, with 0.04 wt% Sn, 0.02 wt% Al, and traces of other elements such as iron, zinc, and nickel. No chemical specification details of this material were known by plant personnel.

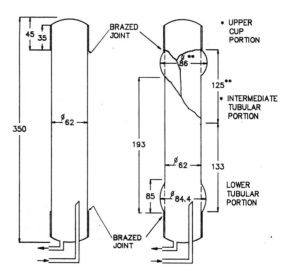

Fig. 3 Dimensional profiles of the dashpot before (left) and after failure. Dimensions given in millimeters. *Portions blown off during the accident. **Measurements of the flattened tubular portion

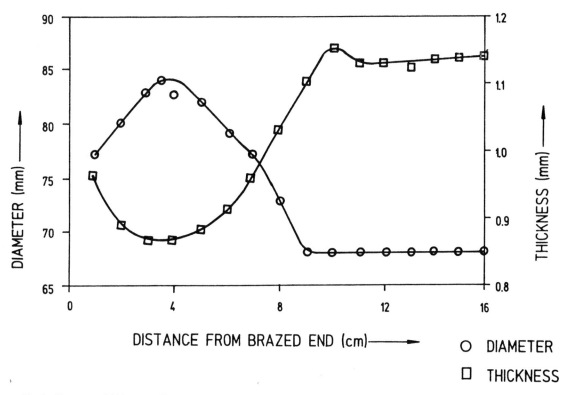

Fig. 4 Diameter and thickness profiles on the lower tubular portion of the failed dashpot (see Fig. 2). Original dimensions: diameter, 62 mm (2.4 in.); thickness, 1.2 mm (0.05 in.)

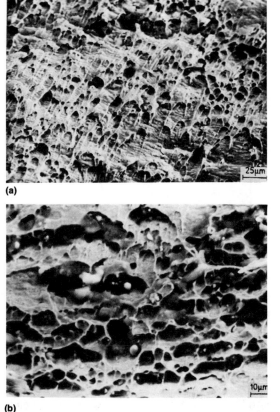

(a)

(b)

Fig. 5 SEM micrographs of the fracture surfaces. (a) Upper and lower circumferential surface exhibited elongated dimples. (b) Longitudinal surface exhibited equiaxed dimples.

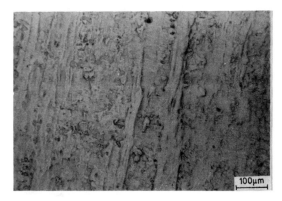

Fig. 6 Original microstructure of the tubular component

Mechanical properties

Hardness. Due to the uneven surface of the failed component, hardness values obtained by *in situ* measurements using a portable ultrasonic hardness tester on the lower tubular portion did not exhibit repeatability. Moreover, these values were markedly different from the microhardness values obtained on mounted and polished specimens. In order to arrive at the correct hardness values, cut portions of the failed component were separately mounted and polished. Microhardness values obtained on these specimens are presented in Fig. 9.

Tensile Properties. To study the effect of brazing on the mechanical properties of copper tube in the HAZ, miniaturized tensile test speci-

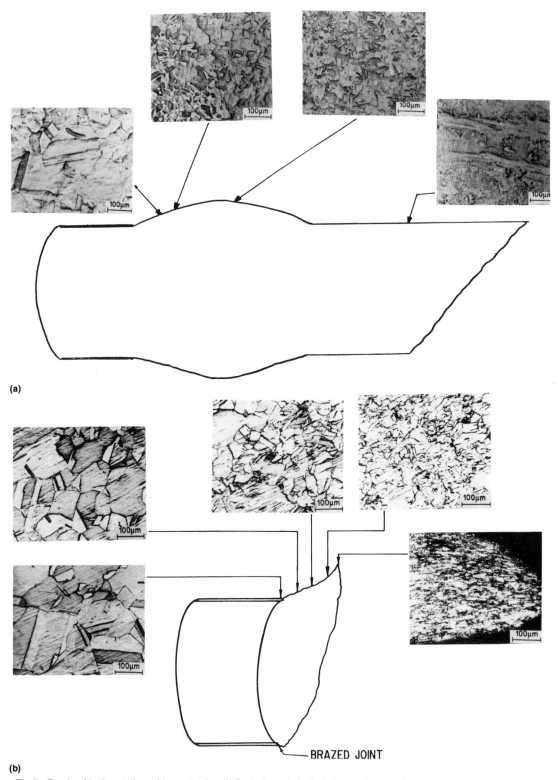

Fig. 7 Results of *in situ* metallographic examination. (a) Grain size variation in the lower tubular portion. (b) Grain size variation in the HAZ of the upper brazed joint

mens were taken from the annealed region (near the upper cup portion) as well as from an undeformed region of the failed component. The results of these tensile tests, along with the stress-strain behavior of an annealed copper specimen (with an average grain size of about 0.08 mm), are presented in Fig. 10. Note that the deformation be-

havior exhibited by specimens taken from the HAZ of the failed component included the effects of work hardening induced by plastic deformation that the component had undergone prior to failure. This can be understood by comparing the deformation behavior of a tensile test specimen from the HAZ with that of an annealed copper specimen.

Stress analysis

Analytical. The internal pressure of refrigerant in the condenser tube component during operation is about 1.6 to 1.8 MPa (230 to 260 psi). During the shutdown period, the pressure increases due to evaporation of refrigerant. During a period of prolonged shutdown, there is a possibility of internal pressure reaching 4.14 MPa (600 (psi), the equilibrium pressure of R-503 refrigerant.

The initial diameter and wall thickness of the condenser tube component were 62 and 1.2 mm (2.4 and 0.05 in.), respectively, as observed from the undeformed central tubular region. The lower tubular portion of the failed component was found to have bulged to a diameter of 84.4 mm (3.3 in.), accompanied by a reduction in wall thickness to 0.865 mm (0.034 in.) (Fig. 4). The region near the upper brazed end had bulged to the maximum diameter of 86 mm (3.4 in.), with accompanying wall thinning to 0.6 mm (0.02 in.). Values of hoop stresses developed on the outer surface of the condenser tube component of initial dimensions as well as on the outer surface of the bulged condenser tube component (considering a circular cross section of maximum diameter), as a result of internal pressurization to various levels, are presented in Fig. 11.

As seen in Fig. 7(a) and 10, two HAZs of the brazed joints underwent extensive annealing, while the central tubular region (unaffected by the brazing process) retained its original strength. The junctions of the HAZs and the central tubular region can be considered as dissimilar joints of two metals having different stress-strain characteristics. As the hoop stress on the outer surface of the dashpot exceeded the yield strength of an-

nealed copper, the HAZs of the brazed joints would have undergone plastic deformation in the form of swelling, whereas the central tubular region would have remained undeformed because of its high yield strength. The two end cups (reinforced by double wall thickness) also would not have undergone any plastic deformation because of the low hoop stress generated on their outer surfaces. Differential deformation of the two annealed HAZs of the tubular component with respect to the relatively stronger central tubular region and the two end cups would have generated local discontinuity stresses (shear stress and bending moments, as shown in Fig. 12) at the junctions of the deformed and undeformed regions to preserve the continuity of the tube wall. In fact, these stresses are relatively local in extent and self-limiting in nature, because once the differential deflection is satisfied by plastic flow of material, a more favorable stress distribution results (Ref 3).

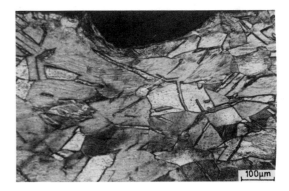

Fig. 8 Impingement pitting on the inner surface of the tubular component

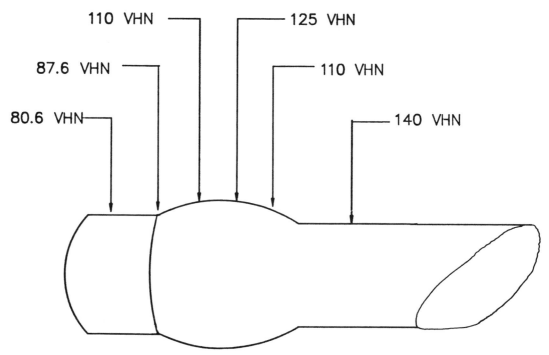

110 VHN

125 VHN

87.6 VHN

110 VHN

80.6 VHN

140 VHN

Fig. 9 Typical microhardness variations in the lower tubular portion

Discussion   Metallographic observations and hardness measurements carried out on the failed condenser tube component revealed that brazing at the two ends of the tubular component had resulted in the annealing of material in the HAZs of the brazed joints while the central tubular region remained unaffected (i.e., in the cold-worked condition). The annealing effect is associated with a substantial increase in the ductility of the material in the HAZ. Due to the high thermal conductivity of copper, the HAZ was large (about 85 mm, or 3.3 in., long at the lower brazed joint). Because of the lack of manufacturing details, it is not known whether the original brazing was repaired during manufacture.

Heat input due to repair would further influence the microstructure and mechanical properties of the HAZ.

Considering the initial dimensions of the tubular component (D = 62 mm, or 2.4 in.; t = 1.2 mm, or

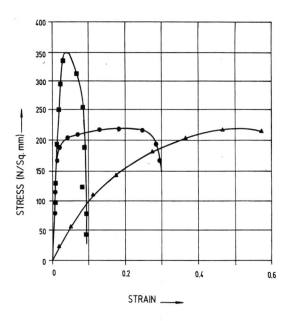

ANNEALED COPPER WITH AVERAGE GRAIN SIZE OF 80 μm
UNDEFORMED REGION OF LOWER TUBULAR PORTION
NEAR UPPER BRAZED END

Fig. 10   Deformation behavior of tensile test specimens taken from the HAZ of the upper brazed joint as well as from the undeformed region of the lower tubular portion. The deformation behavior of an annealed copper specimen with an average grain size of 0.08 mm is included for comparison.

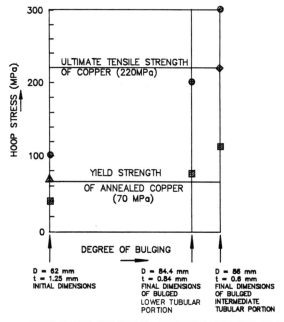

HOOP STRESS FOR NORMAL OPERATING PRESSURE OF 1.6 MPa
HOOP STRESS FOR EQUILIBIRIUM PRESSURE OF 4.14 MPa
HOOP STRESS FOR AN INTERMIDIATE PRESSURE OF 2.8 MPa
HOOP STRESS FOR AN INTERMIDIATE PRESSURE LEVEL OF 3 M

Fig. 11   Hoop stresses generated on the outer surface of the tubular component

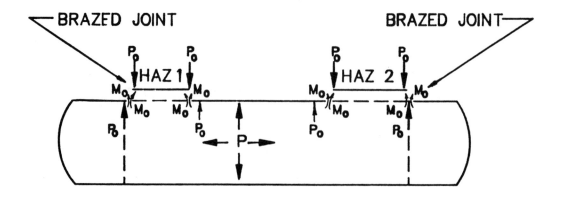

P = INTERNAL PRESSURE   P₀ = SHEAR STRESS   M₀ = BENDING STRESS

Fig. 12   Discontinuity stresses developed in the dashpot as a result of the differential deformation of the HAZs of two brazed joints

0.05 in.), at a normal operating pressure of 1.6 to 1.8 MPa (230 to 260 psi), hoop stresses generated at the outer surface of the tubular component would lie in the range of 40 to 45 MPa (5.8 to 6.5 ksi), which is much below the yield strength of annealed copper (70 MPa, or 10 ksi). During the shutdown period, internal pressure rises due to the evaporation of refrigerant, and hoop stresses developed on the outer surface of the tube reach the yield strength of annealed copper at an intermediate pressure level of about 2.8 MPa (405 psi) (Fig. 11).

The hoop stresses may reach the yield strength of annealed copper even at lower pressure levels at some localized zones where wall thickness would have been less than the nominal wall thickness (e.g., at the locations where impingement-type pitting had formed as a result of the flow of refrigerant or at any other defect on the inner side of the condenser tube wall). This would result in the differential deformation of the two annealed HAZs of the condenser tube component with respect to the relatively stronger central tubular region and the two end cup portions, which were reinforced by increased wall thickness. Plastic deformation of the HAZs of the two brazed joints in the form of swelling (increase in diameter) and wall thinning elevates hoop stress (as well as longitudinal stress), causing further swelling and wall thinning. This increase in hoop stress as a result of plastic deformation is accompanied by work hardening of material, raising the stress required to cause further deformation.

These two competing processes continue in tandem, and the response of the material is decided by the dominant process. During normal operation, internal pressure drops down with an accompanying drop in hoop (and longitudinal) stress, and the reduced hoop stress may not be sufficient to cause further plastic deformation. Thus, incremental deformation would have taken place during each cycle (where high pressurization up to 4.14 MPa, or 600 psi, may occur) and culminated in final longitudinal failure when the wall thickness was not sufficient to sustain the stresses.

It should be noted that the maximum swelling that the component had undergone in upper HAZ before failure was on the order of 40%, which is very close to the percentage elongation that an annealed tensile test specimen can undergo before failure. The implication is that longitudinal rupture following swelling was the primary mode of failure. It can also be seen that both HAZs on the dashpot had undergone an almost equal amount of swelling; longitudinal failure took place in the upper HAZ, where hoop stress approached ultimate tensile strength earlier. Because most of the fracture surface was damaged and was not well protected after the failure, scanning of the fracture surface even at high magnification did not provide any clue regarding the initiation site of failure.

The longitudinal crack would have propagated unhindered along the length of the upper HAZ. At the junctions of the annealed HAZ and the undeformed central tubular region/upper cup portion, however, crack growth in the longitudinal direction would have been arrested because of the low hoop stress (small diameter and thicker wall) available to drive forward the crack, and the crack would have tended to follow the easiest path for propagation. Here the reaction force of the longitudinal burst would make the crack bend (rather than make it propagate in the longitudinal direction) and propagate in two circumferential directions along the interface. A failure of an almost identical nature is described in Ref 4.

Due to the differential deformation of the two HAZs of the tubular component with respect to the undeformed central tubular region and the two end cups, local discontinuity stresses (shear stress and bending moments) would have developed at the junctions of the deformed and undeformed regions to preserve the continuity of the tube wall. Any plastic deformation of material due to these localized discontinuity stresses in the form of wall thinning would have facilitated bending of the longitudinal crack in the circumferential direction at the junctions of the annealed HAZ and the undeformed central tubular region/upper cup portion.

According to the ASME Code for design of pressure vessels, the thickness of shells under internal pressure should not be less than that computed by the following formula (Ref 5):

$$t = \frac{PR}{SE - 0.6P}$$

where t is the minimum required thickness of the shell (inches), P is the design pressure (psi), R is the inside radius of the shell course under consideration (inches), S is the maximum allowable stress value (psi), and E is joint efficiency (taken as 1). Considering a design pressure of 4.14 MPa (600 psi) (the most severe condition of pressure expected in normal operation) and a maximum allowable stress of 46 MPa (6700 psi) (for annealed copper), the minimum thickness of the tubular component would have to be about 2.93 mm (0.12 in.). The actual thickness of the component is only 1.2 mm (0.05 in.). Clearly, there is a design deficiency.

Conclusion and Recommendations

Most probable cause

The failure of the condenser dashpot can be attributed to insufficient component thickness, which made it unable to withstand internal operating pressure, and to extensive annealing in the HAZs of the two brazed joints. In fact, it is quite clear that the effects of brazing had not been taken into account during design of the condenser tube component.

Remedial action

The dashpot of the condenser unit must be designed in accordance with the ASME Code (Ref 5), while taking into account the annealing effect of brazing. The brazing process must be standardized to minimize the degree of annealing.

The component should be subjected to hydrostatic pressure testing at a pressure 1.25 times greater than the design pressure before being placed in service. Hydrostatic testing would have helped to detect the design deficiency at the manufacturing stage. The same components in other operating plants should be examined to avoid similar failures.

References

1. *ASHRAE Handbook of Fundamentals*, American Society of Heating, Refrigerating and Air-Conditioning Engineers, 1972, p 613
2. R.D. Barer and B.F. Peters, *Why Metals Fail*, Gordon and Breach, 1974, p 111-112
3. J.F. Harvey, *Theory and Design of Modern Pressure Vessels*, Van Nostrand Reinhold, 1974, p 134, 151
4. *Metals Handbook*, 8th ed., Vol 10, *Failure Analysis and Prevention*, American Society for Metals, 1975, p 528-529
5. "ASME Boiler and Pressure Vessel Code," Section VIII, Division 1, American Society of Mechanical Engineers, 1989, p 20-24

Stress-Corrosion Cracking of a Brass Tube in a Generator Air Cooler Unit

G. Mark Tanner, Mechanical and Materials Engineering Department, Radian Corporation, Austin, Texas

An arsenical admiralty brass (UNS C44300) finned tube in a generator air cooler unit at a hydroelectric power station failed. The unit had been in operation for approximately 49,000 h. Stereomicroscopic examination revealed two small transverse cracks that were within a few millimeters of the tube end, with one being a through-wall crack. Metallographic examination of sections containing the cracks showed branching secondary cracks and a transgranular cracking mode. The cracks appeared to initiate in pits. EDS analysis of a friable deposit found on the inside diameter of the tube and XRD analysis of crystalline compounds in the deposit indicated the possible presence of ammonia. Failure was attributed to stress-corrosion cracking resulting from ammonia in the cooling water. It was recommended that an alternate tube material, such as a 70Cu-30Ni alloy or a titanium alloy, be used.

Key Words

Admiralty metal, corrosion
Pitting (corrosion)
Cooling systems

Heat exchanger tubes, corrosion
Ammonia, environment

Stress-corrosion cracking
Hydroelectric power generation

Alloys

Brasses—C44300

Background

A brass finned tube in a generator air cooler unit failed.

Applications

The generator air cooler tube was one of eight cooler units at a hydroelectric power station. The unit had operated for approximately 49,000 h. The cooling medium for the tubes is water from a river. Air flows over the finned exterior of the tubes, while water circulates through the tubes.

Circumstances leading to failure

The cooler unit had been experiencing tube leaks with increasing frequency for 3 years. It was extremely difficult to determine the location of the leaks without destroying the cooler. Two previous leaks had been determined to have been caused by poor-quality rolling between the tube and tube sheet. Plant personnel suspected the possibility of fatigue cracks caused by flow-induced vibration.

Specimen selection

The cooler unit was sent back to the original equipment manufacturer (OEM) in France to determine the cause of the leaks. The tube was removed from the cooler by the OEM and returned to the plant for an independent assessment of the failure mode. The tube reportedly contained two leaks, one at a distance of approximately 4 mm (0.16 in.) from the marked end of the tube and the other approximately 50 mm (2 in.) from the marked end.

Visual Examination of General Physical Features

The as-received tube (Fig. 1) was visually examined. The tube was approximately 1.5 m (5 ft) in length, with rectangular external fins to improve heat transfer. No cracks were observed in the tube, but examination of the external surface was limited because of the fins. The internal surface contained a friable deposit.

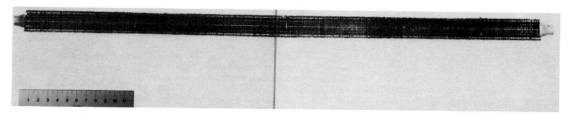

Fig. 1 As-received cooler tube

Testing Procedure and Results

Nondestructive evaluation

The internal surface of the tube was examined using a borescope. No cracks were visible. Subsequently, the cooler tube was filled with water and the ends plugged. The tube was then pressurized to approximately 410 kPa (60 psi) to help locate any leaks. None was observed.

Surface examination

The end of the tube that reportedly contained the leaks was removed from the remainder of the tube, cleaned of internal deposits, and leak tested; no leaks were found. Subsequently, the small tube section was cut longitudinally along the minor axis of the tube. The internal surface was examined using a stereomicroscope at magnifications of 7 to 45×. Several small cracks were observed. One small longitudinal crack, propagating from the inside diameter to the outside diameter, was observed at the very end of the tube. Two small transverse cracks were observed within a few millimeters of the end, with one being a through-

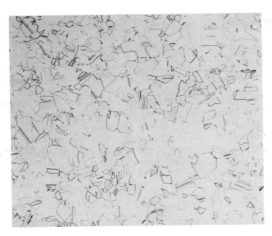 

Fig. 2 Representative photomicrographs of the microstructure of the cooler tube. The structure consists of equiaxed alpha grains with annealing twins. Etched in potassium dichromate. (a) 91×. (b) 364×

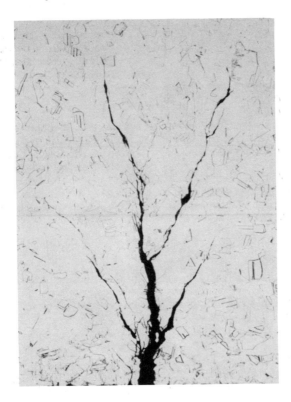

Fig. 3 Photomicrographic montage of the longitudinal crack. The inside diameter of the tube is at the bottom. Etched in potassium dichromate. 100×

Table 1 Results of chemical analysis

| Element | Composition, wt% | |
	Cooler tube	Chemical requirements for UNS C44300 brass(a)
Copper	70.86	70.0-73.0
Tin	0.95	0.9-1.2
Arsenic	0.03	0.02-0.10
Phosphorus	<0.01	NR
Antimony	<0.05	NR
Iron	<0.05	0.06 max
Lead	<0.05	0.07 max
Zinc	bal	bal

(a) NR, no requirement

The typical microstructure of the cooler tube is shown in Fig. 2; it consisted of equiaxed alpha grains with annealing twins. This microstructure is normal for an admiralty brass in the annealed condition.

The transverse through-wall crack contained branching secondary cracks. The mode of cracking was transgranular. Branched transgranular crack paths are characteristic of stress-corrosion cracking (SCC). The cracks appeared to initiate in pits. Figure 3 shows a photomicrographic montage of the longitudinal crack. This crack, like the other, was transgranular. Figure 4 shows a transgranular crack that was observed in the remote ring section. The crack had propagated through approximately 95% of the wall thickness. This crack also appeared to initiate from a pit. Very small cracks were observed emanating from two pits next to this crack. Figure 5 shows a typical pit containing a very small crack initiating at its base.

Chemical analysis/identification

An additional segment was removed from the tube and chemically analyzed. The composition of the tube is presented in Table 1, along with the chemical requirements for UNS C44300, an arsenical admiralty brass. The cooler tube met the chemical requirements of the specification.

A sample of the internal tube deposits that had been collected prior to hydrotesting was analyzed using energy-dispersive X-ray spectroscopy (EDS). This analytical technique is used in conjunction

wall crack. This through-wall crack had been covered by the plug during leak testing.

Metallography

The tube was sectioned into the following pieces: a longitudinal section containing the through-wall transverse crack, a transverse section containing the longitudinal crack, and a transverse ring section remote from the end exhibiting the cracks. All three sections were ground, polished, etched, and then examined using a metallurgical microscope to evaluate the microstructure, crack morphology, and inside and outside diameter surface conditions.

with a scanning electron microscope and can detect elemental components with an atomic number of 9 (fluorine) or greater. Therefore, it cannot identify light elements such as carbon, hydrogen, nitrogen, and oxygen. Results of such an analysis are semiquantitative and indicate the relative amounts of elemental constituents pres-ent. Crystalline compounds in the deposit were identified using powder X-ray diffraction (XRD) techniques.

The EDS spectrum and tabulated XRD results for the internal deposits removed from the cooler tube are shown in Fig. 6. The internal deposits consisted of silicon oxide (SiO_2), potassium aluminum silicate ($KAlSiO_4$), calcium carbonate ($CaCO_3$), and ammonium copper sulfite hydrate $[(NH_4)_7Cu(SO_3)_4 \cdot 5H_2O]$. In addition to the elements contained in these compounds, EDS detected minor to trace amounts of titanium, manganese, iron, and zinc. Most of these deposits are common minerals found in water. Ammonium copper sulfite hydrate indicated the possible presence of ammonia. Therefore, approximately 0.5 g of internal deposit were analyzed for ammonia by the distillation/nesslerization method. The analysis

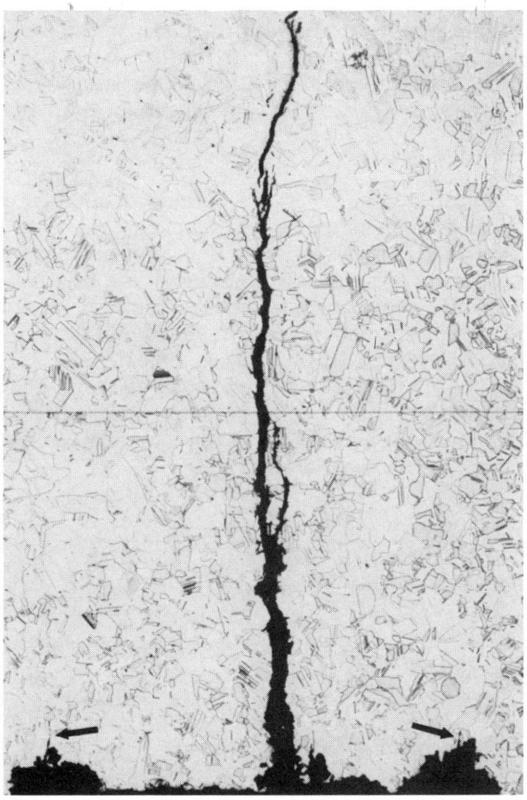

Fig. 5 Photomicrograph of a typical pit containing a very small crack initiating at its base (arrow). The inside diameter of the tube is at the bottom. Etched in potassium dichromate. 152×

Fig. 4 Photomicrographic montage showing a transgranular crack in the remote ring section. The crack appears to initiate in a pit. Very small cracks can be seen emanating from two pits next to the crack (arrows). The inside diameter of the tube is at the bottom. Etched in potassium dichromate. 91×

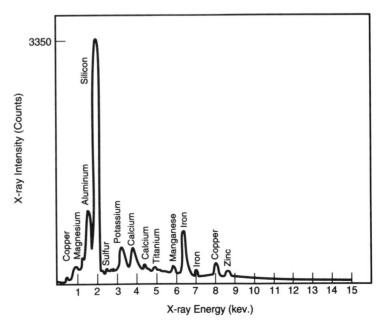

Fig. 6 Summary of the results of EDS elemental analysis and XRD compound identification of the internal deposits removed from the cooler tube

confirmed the presence of ammonia in a concentration of 368 µg/g.

| **Discussion** | The branched transgranular cracking in the brass cooler tube was a result of SCC. Stress-corrosion cracking requires sustained tensile stresses, a corrosive agent, and a susceptible material. The stresses in the tube were probably a combination of hoop stresses from the internal pressure, residual stresses from manufacturing, and possibly | flow-induced stresses. The corrosive agent was ammonia, which was found in the internal deposits in the form of ammonium copper sulfite hydrate, a corrosion product of ammonia and copper in the brass. The technical literature indicates that copper alloys such as brass are easily stress cracked in an ammonia environment (Ref 1, 2). |

Conclusion and Recommendations

Most probable cause

The cause of the tube leaks was ammonia-induced SCC. Because the cracks initiated on the inside surfaces of the tubes and because the river water is not treated before it enters the coolers, the ammonia was likely present in the river water and probably concentrated under the internal deposits.

Remedial action

Two possible options are available to alleviate the cracking: either the ammonia can be eliminated or an alternate material can be used that is resistant to ammonia corrosion as well as to chlorides and sulfur species. Unless the ammonia is biologically produced, the first option is not practical; the ammonia could originate from any company upstream of the plant, locating the source would be very difficult, and continuous monitoring of the river water would be required. The second option is more viable. Alternate tube materials include a 70Cu-30Ni alloy or a more expensive titanium alloy.

References

1. *Metals Handbook*, 9th ed., Vol 11, *Failure Analysis and Prevention*, American Society for Metals, 1986, p 220

2. H.L. Logan, *The Stress Corrosion of Metals*, John Wiley & Sons, 1966, p 157

Thermal Fatigue Failure in a Vaporizer

W.T. Becker, University of Tennessee, Knoxville, Tennessee

A gas-fired, ASTM A-106 Grade B carbon steel vaporizer failed on three different occasions during attempts to bring the vaporizer on line. Dye penetrant examination indicated the presence of multiple packets of ductile cracks on the inside of the coil radius at the bottom of the horizontal axis coils. Visual examination of the inside of the tubing indicated the presence of a carbonaceous deposit resulting from decomposition of the heat-exchanging fluid. Subsequent metallographic examination and microhardness testing indicated that the steel was heated to a temperature above the allowable operating temperature for the fluid. The probable cause for failure is thermal fatigue due to the localized overheating. Flow conditions inside the tubing should be reexamined to ensure suitable conditions for annular fluid flow.

Key Words		
Heat exchangers,	Vaporizers,	Carbon steels,
Mechanical properties	Mechanical properties	Mechanical properties
Thermal fatigue	Overheating	Cracking (fracturing)
Transgranular fracture		

Alloys   Carbon steels—A106 Grade B

Background

A vaporizer installed in a foodstuffs industry processing plant failed three times in attempts to bring the system on line.

A schematic flow chart illustrates the location of the vaporizer in the system (Fig. 1). The vaporizer was thermally cycled depending upon demand of the system, typically cycling 2 to 4 times per hour during normal operation. After the second and third attempts to bring the system on line, studies were done to determine cause for failure. The most complete analysis was done after the third failure, which is discussed here.

Applications

The vaporizer is a common component in a continuous flow heat exchanger system. It was designed as a horizontal axis helical coil system enclosed in a sealed tank which was internally gas fired. Fluid entered the vaporizer through 32 coils of nominal 6-inch diameter schedule 80 steel pipe (6.25 in. OD, 0.432 in. wall) wound around the horizontal axis of the tank. A ring gas burner was installed at the center line of tank axis at the entry end of the tank. The uniformity of the flame distribution inside the tank is unknown. Entry and exit of the heat exchange medium was along the tank axis.

Entry and exit connections to the tank were relatively inflexible due to their geometry. As fabricated, individual tubing sections were cold formed to the required radius and butt welded together. All but the final three coils towards the exit were randomly stitch welded together. These stitch welds were not all visible when examined after failure of the system. (For purposes of discussion, coils are numbered from the exit towards the entry to the tank.)

The fluid used in the system as a heat exchanging medium had a normal working range of 60-750 °F. Use of the fluid above 800 °F was not recommended, and it was subsequently determined that the fluid decomposed rapidly at temperatures above 900 °F, leaving a carbon rich sooty deposit on the tube wall.

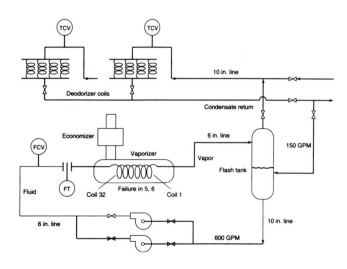

Fig. 1 Schematic flow chart for system showing location of vaporizer

Fig. 2 Photograph showing cracking revealed by the dye penetrant on the tube OD at the coil ID in coil 5. Numbers on the photograph refer to the coil number and the distance from one end of the removed section. Macrophotograph. Tube OD is 6 11/16 in.

Fig. 3 Photograph showing cracking revealed by the dye penetrant on the tube OD at the coil ID in coil 6

(a)   (b)

Fig. 4 Macrophotograph showing cracking at two locations in coil 6. Note branched and jagged nature of cracks.

The design operating temperatures were 650° ± 10 °F for the fluid and 720 °F for the steel tubing.

Circumstances leading to failure

Two shut downs occurred prior to the investigation described here. The first failure was a rupture of the tubing wall in the stitch welded area near the exit of the tank. This crack was ground out and repair welded, and attempts were made to again bring the system on line. A second failure then occurred due to rupture of the tubing at coil 5. Approximately two coils of tubing were cut out of the system, and new tubing installed. The third failure, later identified as a through-wall rupture in coil 6, occurred as attempts were again made to bring the system on line.

During operation, burners were ignited on demand, causing the system to thermally cycle approximately two to four times per hour from a low thermal demand condition (nominal 2 million BTU/hr) to a high thermal demand condition (6 million BTU/hr). The final failure occurred after a total of 6200 hours of operation over approximately nine months, corresponding to a maximum of 12000-24000 cycles. The system was expected to have a life of at least ten years; clearly it was never successfully brought on line.

Pertinent specifications

The vaporizer coils were fabricated from Grade B ASTM A 106 tubing ("Seamless Carbon Steel Pipe for High-Temperature Service"). The system was presumably designed according to ASME boiler and pressure vessel code Section 1, Power Boilers.

Performance of other parts in same or similar service

The system was a new design. Vaporizers of this type are designed with both a vertical and horizontal coil axis. Larger tubing designs (e.g., 6-inch diameter) have been successfully operated with a vertical coil axis, while smaller (e.g., 4-inch diameter) have been successfully used with a horizontal coil axis.

Selection of specimens

Field NDE dye penetrant examination revealed extensive cracking in coils 5 and 6, and no cracking in coils 1-4 (see Fig. 1). Coils 5 and 6 were removed from the vaporizer for additional testing, including a more complete NDE examination. It is unknown if there was any cracking in coils 7 and higher (i.e., closer to the burner).

Testing Procedure and Results

Examination of the coils indicated the appearance of only one through-wall crack. There was an obvious dark sooty deposit on the inside of the coil. This deposit could be partially rubbed off, but a layer was tightly bonded to the metal surface.

Surface examination

Scanning Electron Microscopy Fractography. No scanning electron microscopy was done on the fracture surface, although it had been hoped that it would be possible to use this additional tool in analysis of the failure. As is visible in the optical micrographs, there is an extensive amount of oxide on the fracture surface. Attempts to remove this oxide resulted in the loss of detail on the fracture surface and the inability to draw any conclusions from a study of the fracture surface.

Metallography

Coils were sectioned both perpendicular and parallel to the coil axis (Fig. 5). Metallographic specimens were removed from these larger sections in regions corresponding to the inside and

outside diameter of the coil. The inside coil diameter specimens were chosen so as to contain cracks identified from the dye penetrant examination. These same specimens were subsequently utilized for microhardness testing. Orientations A and D correspond to longitudinal specimens on the inside (A) and outside (C) of the coil diameter. Orientations B and D are sections transverse to the tube axis, with B on the ID of the coil and D on the OD of the coil (see Fig. 5).

Metallographic examination revealed several features.

1. General features of the microstructure at the OD and ID of the tube at the coil OD and ID are shown in Figs. 6, 7, 8, and 9. Cracking on the surface is shown in Figs. 9 and 10.
2. Visual and NDE examination revealed only one through-wall crack. No cracks were found metallographically which initiated on the ID of the tube wall at the coil ID. No cracks were found on either the ID or OD of the tube at the coil OD.

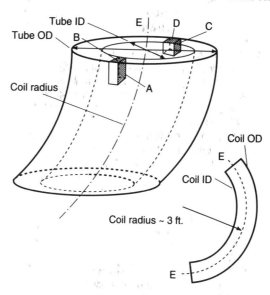

Fig. 5 Schematic diagram showing location of metallographic sections. Sections taken parallel and perpendicular to the tube axis at the inside (ID) and outside (OD) coil diameter. (TID, tube inside diameter; TOD, tube outside diameter; CID, coil inside diameter; COD, coil outside diameter)

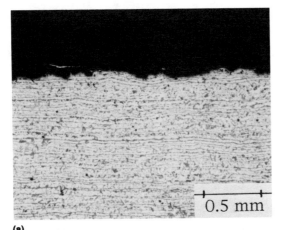

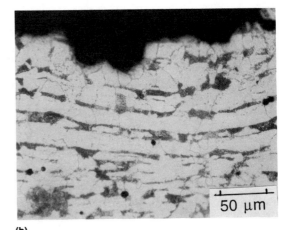

(a)   **(b)**

Fig. 6 Section A in Fig. 5. Microstructure of a crack-free region on the tube OD at the coil ID. Tube axis is horizontal in the figure. Microstructure consists of pearlite and ferrite. Compare coarse microstructure at surface to Fig. 9. Surface roughening where oxide not present. 2% Nital etch

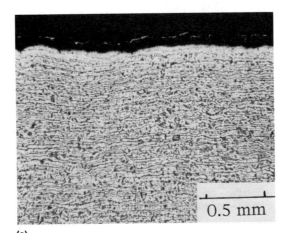

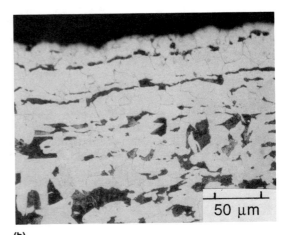

(a)   **(b)**

Fig. 7 Section D in Fig. 5. Microstructure at the tube ID at the coil OD. Tube circumference is horizontal in the figure. Small amount of decarburization visible. Nital etch

3. Cracks were transgranular, often containing a deposit which was contiguous with a discontinuous deposit on the ID coil surface of the tube. This deposit is assumed to be iron oxide (Fig. 10).
4. Surface roughness along the OD at the coil ID varied from relatively rough (Fig. 6) to relatively smooth (Fig. 9). Smooth surfaces were associated with the presence of what was presumed to be iron oxide. Interior tube wall surfaces were smooth.
5. There was considerable variation in the microstructure at the tube OD at the coil ID, which when examined at high magnification was a varying extent of spheroidization of the pearlite (Figs. 11a-d). No spheroidization was observed at the tube ID at the coil ID, or at any location on the tube at the coil OD. No cracks were found in regions where there was not some tendency of the pearlite to be spheroidized.
6. In some regions at both the tube ID and OD surface, ferrite grains appeared to be more darkly outlined suggesting the possibility of the presence of an intergranular film (Fig.

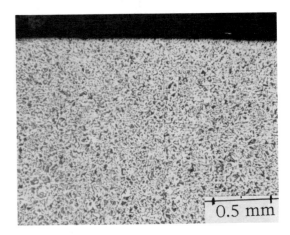

Fig. 8 Section D. Microstructure at the tube OD at the coil OD. 2% Nital etch

14). However, when these same regions were examined unetched, the grain boundary outlining disappeared. This indicates that no intergranular deposit was present on the tube ID due to decomposition of the fluid and

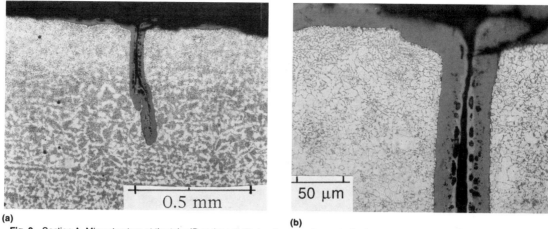

Fig. 9 Section A. Microstructure at the tube ID at the coil ID showing a crack penetrating from the surface. Some oxide is visible on the surface which continues inside the crack. Compare refined microstructure at surface to Fig. 6. 2% Nital etch

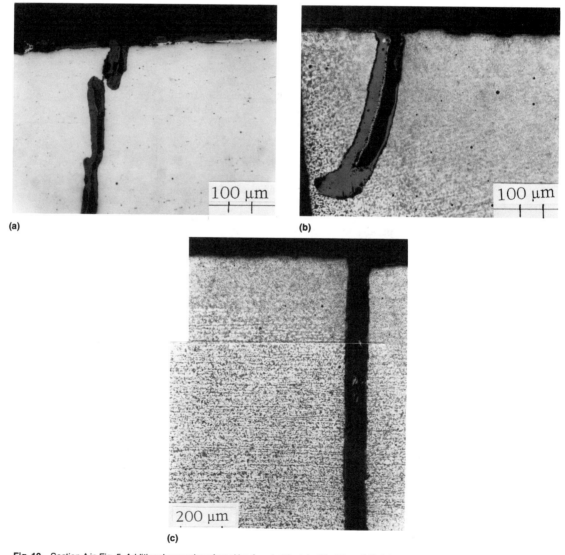

Fig. 10 Section A in Fig. 5. Additional examples of cracking found at the tube ID at the coil ID. (a) Unetched. (b and c) 2% Nital etch. Note the alterted surface microstructure as in Figs. 9(a) and (b).

subsequent diffusion of decomposition products into the ID wall. Such a deposit had been identified after the second failure.

7. Note the reduction in grain size in Fig. 9 compared to Fig. 6. This would indicate that the material in Fig. 9 was heated to a tem-

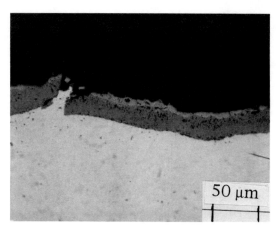

Fig. 11 Section A in Fig. 5. Note the radial bulging of the material and the extrusion of the metal between the fractured oxide. Unetched

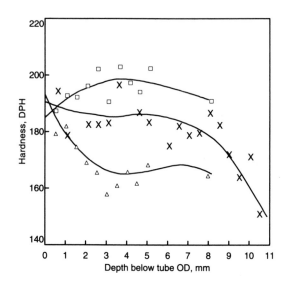

Fig. 12 Section A in Fig. 5. Microstructures showing partial spheroidization of pearlite at various depths below the tube OD at the coil ID in a region containing cracks. Compare structure at surface to Fig. 6. Depth below the surface increases from (a) to (d). 10(d) located approximately 75% through the wall. Nital etch

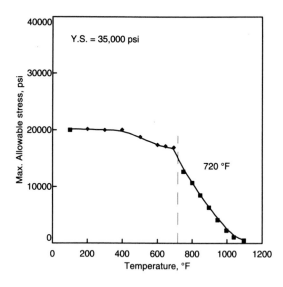

Fig. 13 Section A. Microstructure at a region showing outlining of ferrite grains. 2% Nital etch

perature greater than A1. It was not possible to determine with certainty whether the temperature also exceeded the A3 temperature due to the variation in grain size observed.

8. When Figs. 10a-c are examined closely, it is apparent that there is a radial displacement of material on each side of the partial thickness crack. Figure 11, also on surface A, does not show any cracking but does show two other interesting features. There is a radial bulging of material. The oxide has been ruptured and the base material appears to be extruded between the two sides of the oxide fracture surface. Bulging of the base material will occur due to a comprehensive stress parallel to the oxide surface. The offset of the cracks in Figs. 10a-c can also occur due to circumferential compressive stress. In all three figures, there is a positive evidence of local plastic deformation.

Mechanical properties

Hardness. Microhardness surveys (1 Kg load, 136 diamond) were taken on the metallographic specimens (Fig. 14). Traverses were taken on a crack-containing specimen (i.e., at the coil ID in both a crack containing region and in a crack-free region). For comparison purposes, a traverse was also made across the tube wall at the coil OD (surface D in Fig. 5). The hardness in the crack-containing region at the coil ID is lower than that in the coil ID crack-free region. The hardness variation across the wall in the crack-containing region was also lower than the hardness across the wall at the coil OD. The drop in hardness from OD to ID at the coil OD is consistent with the decarburization observed at the tube ID in Fig. 7, which may be associated with original fabrication of the tube.

The average hardness across the wall of the section containing cracks was 169 ± 7 DPH, while that of the crack-free material was 195 ± 5 DPH, while that of the crack-free material was 195 ± 5 DPH. The average hardness across the wall on the coil OD, neglecting the small decarburized region, was 184 DPH ± 6.

Dye Penetrant Examination. A detailed dye penetrant examination was done in coils 5 and 6 under laboratory conditions (Figs. 2, 3, 4). This examination revealed several features:

1. Significantly more cracks were observed in addition to the one through-thickness crack visible to the naked eye.
2. Cracking was concentrated at a location corresponding to the bottom of the coil (at 6 o'clock looking down the axis of the coil), decreasing towards the 3 and 9 o'clock positions. No cracking was identified from 9 o'clock to 3 o'clock position.*
3. All cracking initiated on the inside of the coil diameter at the tube OD. No cracks were

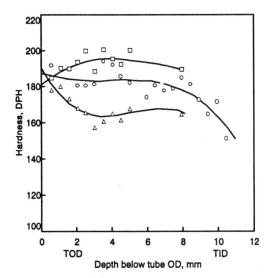

Fig. 14 Microhardness (1 Kg load, 136 diamond) traverses through the material at different locations. ± At coil ID in crack free region. ± At coil OD. ± At coil ID in crack containing region

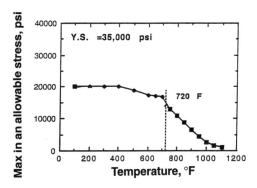

Fig. 15 Maximum allowable stress versus service temperature for ASTM A-106 Grade B material. Adapted from Ref 2. Design operating temperature indicated

found on the OD of the coil.

4. Cracks typically occurred in packets of 2-5 or more cracks, randomly spaced along the tube near the 6 o'clock position.

5. Crack appearance was typically jagged and branched, although there were a few isolated single cracks which were straight with no branches.

Discussion

The vaporizer was designed to operate under conditions for which a sufficient volume of fluid circulates at a velocity to create annular flow conditions in the coil. That is, the interior wall surface of the tubing is to be coated with a continuous liquid layer surrounding vaporized fluid. If annular flow conditions are not established, fluid is not available to absorb heat, causing a temperature rise in the material itself. If sufficient fluid is present to maintain the annular flow, pooling of liquid in the bottom of the coils near the exit of the tank is expected to occur. This pooling results in one and possibly two conditions in these coils which leads to the development of thermal stresses:

1. Pooling of liquid will cause a dry wall condition at the bottom of the coil and the tube OD and coil ID. As noted above, however, no liquid exists at the tube ID to absorb the heat from the gas flame. Heat dissipation from the hot coil ID occurs by conduction through the metal but is less by convection through the vapor than through the liquid. The presence of any sooty deposit on the inside of the tube is expected to further hinder heat transfer.

The net result is a temperature rise in the

tube wall to a level that can exceed design limits.

A plot of maximum allowable stress versus service temperature (Fig. 15) suggests that the yield strength of the tubing decreases rapidly above 700-750 °F. The presence of the black sooty deposit on the inside of the coil as well as the presence of spheroidized pearlite on the tube ID indicates that this material was exposed to temperatures above 750 °F, and as evidenced by the sooty deposit, to temperatures above 900 °F. There is indication of still higher temperatures being obtained as revealed by the reduction in ferrite grain size in Fig. 9 compared to Fig. 6. At such excessive service temperatures, the yield strength of the material at the coil ID is low, permitting plastic deformation to occur. Material at the coil ID is then plastically deformed while material adjacent to the liquid pool is elastically deformed, which sets up thermally induced stress gradients when the total system cycles. This is one of the criteria used to associate the failure with thermal fatigue or creep-thermal fatigue.

2. A second more localized condition may develop if boiling of the liquid pool occurs such that droplets of fluid spatter onto the dry wall. There, the droplets are vaporized and the resulting heat transfer quenches the heated metal at localized regions, again setting up thermally generated biaxial tensile stress gradients, but on a more localized scale (see Ref 2). This is a second source of cyclic plastic strain.

---

* For purposes of discussion, orientation of the failure with respect to the coil geometry is that 12 o'clock corresponds to the top of the tank and 6 o'clock to the bottom of the tank. With reference to Fig. 5, cracking and metallographic specimens are also oriented with respect to the inside (CID) and outside of the tube coil diameter (COD), and the inside (TID) and outside of the pipe diameter (TOD).

Some concern was directed towards the influence of the stitch welding on the failure mode. The geometry of the tubing at the exit from the tank is relatively stiff, and the stitched tubing section is also relatively stiff. Any gross movement of the tubing near the exit would then be expected to occur in the three floating coils. If movement became excessive, cracks would be generated at either the 90° bend or near the first stitch welded section of the tubing. No cracks were observed at either location. Thus, although there was considerable misfit at the exit from the tank during fabrication, any stresses created by this misfit would probably be relaxed during firing of the system.

A second issue considered was whether any residual stresses created from the cold forming operation of the tube coils could have been wholly or partially responsible for the failure. While there may have been a residual tensile stress parallel to the coil axis on the inside diameter of the coil after cold forming, this stress would have been relaxed during initial firing of the system. Additionally, ASTM A-106 requires a stress relief anneal after a cold forming operation.

Conclusions and Recommendations

There is no direct fractographic evidence of fatigue. However, it is believed that the thermal stress gradients produced by slug operation and/or the self-quenching due to liquid droplets splattering make it highly likely that failure was caused by thermal fatigue or creep-thermal fatigue. These conditions developed due to inadequate flow conditions in the tubing, which in turn caused the development of thermal stresses. One might ask what direct information could have been obtained to prove the presence of thermal fatigue. Such information probably cannot come directly from SEM fractographic analysis since: 1) The presence of fatigue striations are poorly developed in the steels, and typically only appear for long-life conditions (i.e. no or small plastic strains); 2) At larger plastic strains the fracture might well be by microvoid coalescence for fatigue loading and at higher stresses, perhaps even intergranular fracture if wedge cracking associated with creep failures is involved. (No wedge cracks were found metallographically.) 3) Oxide scale on the surface prevents the direct observation of the fracture surface. Attempts to remove this scale were unsuccessful without destroying necessary fracture surface information if it had not already been destroyed by the oxidation process.

However, the nominal macroscopic appearance of the cracks on the tubes does support the assumption of thermal fatigue (see Ref 3, for example.). There is also strong evidence that local differential plastic strains occurred in the material (Figs. 10, 11). Thus, with information concerning the thermal cycling of the system in conjunction with the macroscopic crack appearance and evidence of plastic strain, it seems reasonable to conclude that the likely failure mechanism is thermal fatigue. A one-time plastic strain overload event as the cause for failure is ruled out because the system was thermally cycled some 1200 to 2400 times before cracking penetrated the wall and caused a sharp temperature rise due to combustion of the heat exchanger fluid.

How failure could have been prevented

Flow conditions inside the tube were inadequate to prevent a dry wall condition. Flow conditions inside the tubing should be re-examined to ensure suitable conditions for annular flow.

References

1. Perry and Green, *Perry's Chemical Engineers Handbook*, 6th ed., McGraw-Hill, 1984.
2. Franco, "Failures of Heat Exchangers," *ASM Metals Handbook*, 9th Ed., Vol 11, Failure Analysis and Prevention, ASM International, 1980, p 633.
3. Wright, *et al.*, "Root Cause of Circumferential Cracking in the Water Walls of Super Critical Boilers: A Report on Work in Progress," EPRI International Conference on Boiler Tube Failures in Fossil Plants, Nov 1991, San Diego, CA.

Failure of AM350 Stainless Steel Bellows

T. Jayakumar, D.K. Bhattacharya, and Baldev Raj, Indira Gandhi Centre for Atomic Research, Kalpakkam, India

AM350 stainless steel bellows used in the control rod drive mechanism of a fast breeder reactor failed after 1000 h of service in sodium at 550 °C (1020 °F). Helium leak testing indicated that leaks had occurred at various regions of the welded joints between the convolutes in the bellows. The weld failure was attributed to poor quality assurance during fabrication, which resulted in cracklike openings at the fusion zone. The openings extended during tensile loading. Use of proper welding procedures and quality control measures were recommended to prevent future failures.

Key Words

Austenitic stainless steels
Nuclear reactor components

Welded joints
Fast reactors

Welding parameters
Breeder reactors

Alloys

Austenitic stainless steels—AM350

Background

AM350 stainless steel bellows used in the control rod drive mechanism of a fast breeder reactor failed after 1000 h of service in sodium at 550 °C (1020 °F).

Applications

The bellows act as a barrier to sodium, preventing it from contacting the moving parts of the control rod drive mechanism. Figure 1 shows the dimensional details of the bellows. The details of the position of the weld joint between the convolutes are also shown.

Circumstances leading to failure

There were a large number of tension and compression cycles during the operating life of the bellows. Failure was detected in the form of sodium leakage from the outside to the inside of the bellows.

Performance of other parts in same or similar service

It is generally reported that the lives of similar bellows operating under similar conditions are unpredictable. The experience of the authors has also shown prediction of the lives of these types of bellows to be unreliable.

Visual Examination of General Physical Features

No specific features could be identified from visual examination of the bellows.

Testing Procedure and Results

Nondestructive evaluation

Helium leak testing of the bellows indicated leaks of greater than 10^{-7} cm^3/s (6×10^{-8} in.^{3}/s) at different locations in the weld joints between the convolutes (detail Y, Fig. 1).

Metallography

Crack Origins/Paths. Optical metallography of the autogenous weld profiles indicated nonuniform fusion. The extent of fusion was found to vary from 150 to 275 µm (6 to 11 mils) (Fig. 2). There was considerable variation in the shape of the fused region as well. A sharp cracklike opening at the center of the fusion line was observed in 4 of the 12 welds examined (Fig. 3). In one such case, the crack extended throughout the fusion zone and

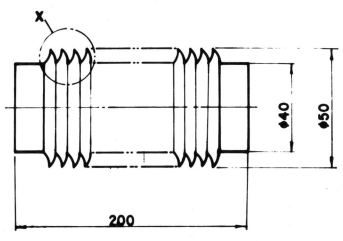

Fig. 1 Dimensional details of the AM350 bellows. Dimensions are given in millimeters. **(continued)**

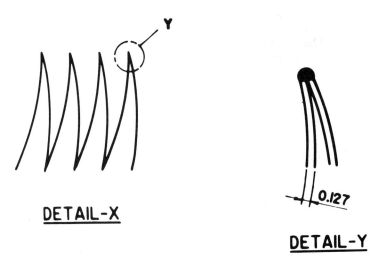

DETAIL-X

DETAIL-Y

0.127

Fig. 1 (continued) Dimensional details of the AM350 bellows. Dimensions are given in millimeters.

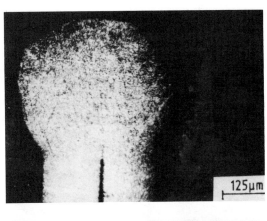

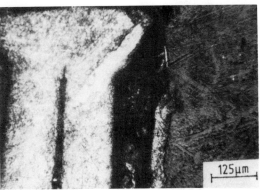

(b)

Fig. 2 Optical micrographs showing weld profiles and extent of fusion

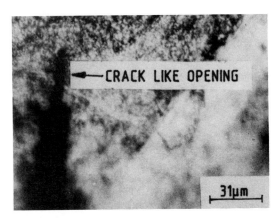

CRACK LIKE OPENING

31μm

Fig. 3 Optical micrograph showing a sharp cracklike opening at the fusion line

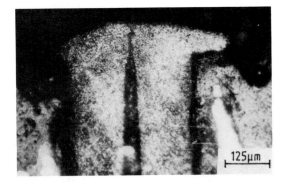

125μm

Fig. 4 Optical micrograph showing a cracklike opening that extends throughout the fusion zone. It is on the verge of separation.

was on the verge of separation (Fig. 4). Extension of the cracklike opening was observed primarily on a portion of the bellows that had been under a tensile load (indicated by increased pitch).

Discussion When the bellows operated under tensile loading conditions, the cracklike opening at the center of the fusion line (Fig. 3) opened up further (Fig. 4). A succession of such openings during the life of the bellows ultimately led to failure. The life of a bellows depends on the combined effect of the tensile load, the extent of the fusion zone, and the presence of cracklike openings. The somewhat unpredictable nature of the combination of these parameters explains the unpredictable life of such bellows, as reported by others (Ref 1) using similar bellows in similar applications.

The weld failure was attributed to poor quality assurance during fabrication of the bellows. The unpredictable life of bellows under similar operating conditions also suggests that the failure was related to variations in fabrication quality rather than to operating conditions.

Conclusion and Recommendations	

Most probable cause

Poor quality assurance during fabrication led to inadequate fusion zone size and cracklike openings at the fusion lines of the bellows. In the presence of a tensile load, the cracklike openings grew, leading to failure.

Remedial action

It is recommended that destructive examination of the welded convolute joints be performed on a sample basis as part of the fabrication procedure qualification. However, regardless of quality control measures during fabrication or random destructive testing after fabrication, failure of the bellows in service appears to be inevitable with this type of welded construction.

Reference

1. P. Allegre, R. Jacquelin, and J.L. Carbonnier, *Bellows for Sodium Systems*, IAEA Specialists Meeting, Tokyo, IWGFR/32, 1979, p 9

Failure of Three Cyclone Separators

George M. Goodrich, Taussig Associates, Inc., Skokie, Illinois

Three 1006 carbon steel steam / water separators failed in a boiler installation after several years of service. Annual inspection had revealed no evidence of deterioration until the last inspection, when they were removed from service. Metallurgical investigation determined that the separators had deteriorated because of erosion corrosion. Further analysis of the boiler operation revealed that operational changes made in the last year of service caused an increase in velocity of the water / steam mixture. It was recommended that the operating parameters for the boiler be reevaluated and prior levels of operation be reinstituted.

Key Words	Carbon steels, corrosion Boilers	Separators, corrosion	Erosion corrosion
Alloys	Low-carbon steel—1006		

Background

Three low-carbon steel steam/water separators failed in a boiler installation after several years of service. Annual inspection had revealed no evidence of deterioration until the last inspection, when they were removed from service because of significant sidewall metal loss.

Applications

The units separated steam and water in a multiple-fuel-fired boiler operating at 227,000 kg/h (500,000 lb/h) of steam. Steam conditions were reported to be 6.2 MPa (900 psig) and 480 °C (900 °F). Information regarding the water treatment indicated that the boiler water chemistry included a chelant-polymer program with softened makeup water. Analysis of the boiler operation revealed that operational changes made in the last year of service caused an increase in the velocity of the steam/water mixture.

Testing Procedure and Results

Surface examination

Visual. All of the separators were examined visually and using a stereoscopic microscope to identify the locations exhibiting metal loss and to evaluate overall appearance. The first separator exhibited significant metal loss from the top portion, as shown in Fig. 1. The metal loss had completely consumed a portion of the wall. Figure 2 shows separator 2 and illustrates the configuration without significant metal loss.

All three separators exhibited a dark, tenacious scale on the outside and inside diameter surfaces. All three had a rough-textured surface on the inside diameter, and separator 2 also had a rough outside surface. The amount of rough texturing was least severe on separator 3. None of the rough texturing exhibited a distinguishable orientation.

Scanning Electron Microscopy/Fractography. Crescent-shaped, dished-out metal loss features observed metallographically (see next section) on the inside surfaces of the three separators were examined using a scanning electron microscope (SEM). Examination revealed that the metal loss features had a smooth-surfaced, horseshoe-shaped appearance (Fig. 3 to 5). The orientation of the horseshoe pattern was most obvious on the inside surface of separator 2 (Fig. 4). This pattern of metal loss is characteristic of erosion.

The inside surfaces were studied further using energy-dispersive spectroscopy (EDS) to identify the elements associated with the oxidized inner surfaces that may have caused corrosion. For all three separators, the following elements were detected in decreasing order of content: iron, aluminum, oxygen, carbon, silicon, sodium, and sulfur. Trace amounts of magnesium, chlorine, potassium, calcium, chromium, and manganese were also identified. Figures 6 to 8 show the EDS spectra obtained. No evidence of a corrosion accelerant was observed.

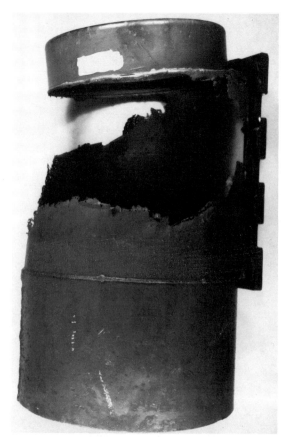

Fig. 1 Overall view of separator 1

Fig. 2 Overall view of separator 2

Fig. 3 SEM micrograph showing horseshoe pattern of metal loss on the inside surface of separator 1. 27.9×

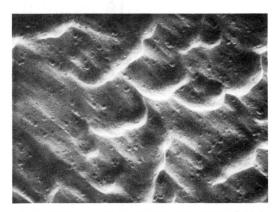

Fig. 4 SEM micrograph showing horseshoe pattern of metal loss on the inside surface of separator 2. 27.9×

Table 1 Results of chemical analysis

| | Composition, % | | |
Element	Separator 1	Separator 2	Separator 3
Carbon	0.05	0.05	0.04
Manganese	0.28	0.31	0.27
Phosphorus	0.013	0.012	0.008
Sulfur	0.014	0.015	0.014
Silicon	<0.05	<0.05	<0.05
Nickel	<0.05	<0.05	<0.05
Chromium	<0.05	<0.05	<0.05
Molybdenum	<0.03	<0.03	<0.03
Copper	<0.03	<0.03	<0.03
Lead	<0.02	<0.02	<0.02
Aluminum	<0.01	0.04	<0.01
Tin	...	<0.02	<0.02
Vanadium	<0.01	<0.01	<0.01
Niobium	...	<0.01	<0.01
Boron	...	<0.0005	<0.0005
Titanium	<0.03	<0.03	<0.03

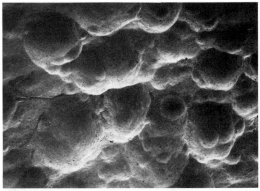

Fig. 5 SEM micrograph showing horseshoe pattern of metal loss on the inside surface of separator 3. 27.9×

Metallography

Longitudinal cross sections were removed from the sidewalls of the three separators. In separator 1, samples were removed immediately adjacent to the opening in the sidewall and 180° from the opening. In separators 2 and 3, samples were removed 180° from the opening only. The four samples were metallographically prepared and examined in the unetched and etched conditions.

Each separator displayed evidence of metal loss in shallow, dished-out regions on the inside diameter surface. Sample 2 also exhibited evidence of minor metal loss on the outside surface. Figure 9 illustrates the features associated with the metal loss. All of the separators exhibited an oxide scale on the outside surface but none on the inside. This oxide scale was most prevalent on samples 2 and 3.

The etched structures revealed a uniform, equiaxed ferrite grain structure interspersed with carbides, indicative of cold-rolled and annealed AISI 1006 steel. The metal loss did not appear to follow metallographic features. Figure 10 illustrates the etched structure for separator 1.

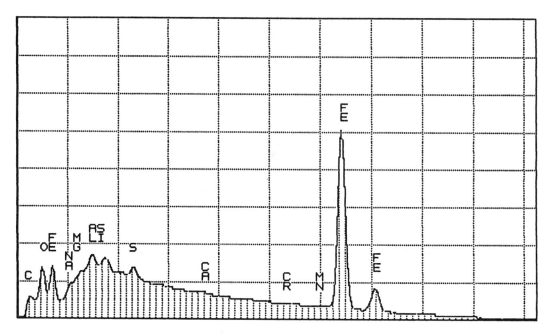

Fig. 6 EDS spectrum of elements representing the inside surface of separator 1

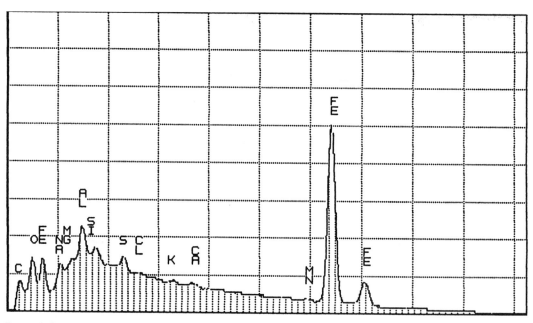

Fig. 7 EDS spectrum of elements representing the inside surface of separator 2

Chemical analysis/identification

Sections from the sidewalls of the three separators were subjected to quantitative chemical analysis. The results are shown in Table 1. Based on test results, the material was identified as AISI 1006 steel. The specified material was not known.

Discussion

Erosion is the consequence of turbulent flow eroding the protective oxide scale and base metal during operation. In boiler systems, an oxide scale is established to protect the components from further metal loss due to oxidation. This scale was visible on the outside surfaces of separators 2 and 3. The scale was not present on any of the inside surfaces. If this protective scale is removed, the oxidation process will reestablish the scale. When turbulent flow erodes the scale, a horseshoe pattern is established on the affected surface. This pattern has characteristics that indicate the direction that the steam and water mixture flow and generally points in the direction of flow. In this instance, the turbulent flow of the water/steam mixture eroded the sidewall from the inside diameter and caused the severe deterioration observed in separator 1.

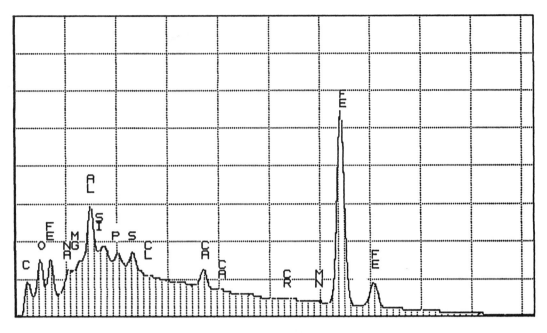

Fig. 8 EDS spectrum of elements representing the inside surface of separator 3

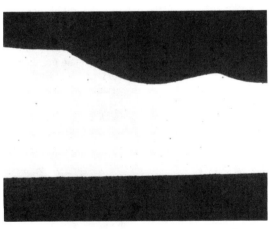

Fig. 9 Longitudinal cross section through separator 1 adjacent to the opening. A crescent-shaped pattern of metal loss is evident on the inside diameter (top). Unetched. 62×

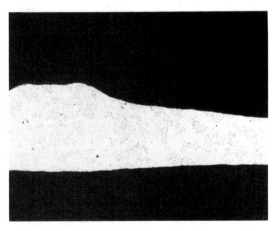

Fig. 10 Longitudinal cross section through separator 1 180° from the area shown in Fig. 9. Note the equiaxed ferrite grain structure. Nital etch. 62×

Conclusion and Recommendations

Most probable cause

The three separators sustained metal loss on the inside diameter surface as a consequence of erosion. This mechanism was most obvious on separator 1. SEM examination of all three separators revealed evidence of a smooth horseshoe pattern on the inside surfaces, indicative of erosion. This erosion was the result of changes in boiler operations that increased the velocity of the water/steam mixture.

Remedial action

It was recommended that boiler operating parameters be reevaluated and that prior water/steam velocity be reestablished.

Intergranular Stress-Corrosion Cracking Failure in AISI Type 316 Stainless Steel Dished Ends Near Weld Joints

D.K. Bhattacharya, T. Jayakumar, and Baldev Raj, Indira Gandhi Centre for Atomic Research, Kalpakkam, India

Two AISI type 316 stainless steel dished ends failed through the formation of intergranular stress-corrosion cracks (IGSCC) within a few months of service. The dished ends failed in the straight portions near the circumferential welds that joined the ends to the cylindrical portions of the vessel. Both dished ends were manufactured from the same batch and were supplied by the same manufacturer. One of the dished ends had been exposed to sodium at 550 °C (1020 °F) for 500 h before failure due to sodium leakage was detected. The other dished end was used to fabricate a second vessel that was kept in storage for 1 year. Clear evidence of sensitization was found in areas where IGSCC occurred. Sensitization was extensive in the dished end that had been exposed to sodium at high temperature, and it occurred in a narrow band similar to that typical of weld decay in the dished end that had been kept in storage. Solution annealing was recommended to relieve residual stress, thereby reducing the probability of failure. It was also recommended that the carbon content of the steel be lowered, i.e., that a 316L grade be used.

Key Words	Austenitic stainless steels, corrosion	Stress-corrosion cracking	Dished ends, corrosion
	Sensitization	Intergranular corrosion	Weld decay
Alloys	Austenitic stainless steels—316, 316L		

Background

Two AISI type 316 stainless steel dished ends that were supplied by the same manufacturer from a single batch failed within a few months of service. Both failures occurred by the formation of intergranular stress-corrosion cracking (IGSCC) in the straight portions of the dished ends near the circumferential welds joining the dished ends to the cylindrical portions of the vessels.

Applications

The dished ends are used in the fabrication of cylindrical vessels. They typically are either cold spun or cold pressed. In the present case, they were manufactured by cold spinning. Figure 1 shows a schematic of a dished end. The end of the straight portion of the dished end is joined to the cylindrical shell by butt welding.

Circumstances leading to failure

One of the dished ends (specimen 1) was exposed to sodium at 550 °C (1020 °F) for 500 h. The temperature was increased to this level in stages. Failure was detected through sodium leaks in the vessel in the straight portion of the dished end. This vessel, which had only one dished end, was positioned vertically with the dished end at the bottom. Subsequent to the detection of this leak, another vessel (not yet put in service) was examined. It was fabricated using two dished ends from the same batch as the failed dished end. Cracks were detected by liquid penetrant testing on the outside surfaces of the straight portions of both

dished ends of this vessel.

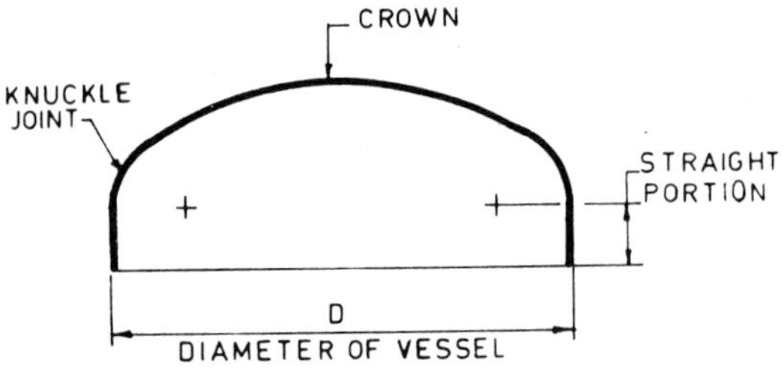

Pertinent specifications

The dished ends were fabricated from 6.0 mm (0.24 in.) thick AISI type 316 stainless steel plate by cold spinning. Stress relieving of the dished ends was not specified. The dished ends were joined to the cylindrical shell portion by tungsten inert gas welding for the root pass and manual metal arc welding for the subsequent passes. Control of interpass temperature and a dye penetrant check after each pass were specified. The vessel had successfully undergone soap bubble testing, ammonia leak testing, and pressure testing.

Performance of other parts in same or similar service

There is an element of unpredictability in the performance of dished ends, even if they are from the same supplier. However, dished ends from the same batch typically appear to behave similarly. In the present case, the failed dished ends were from the same batch. Dished ends from other batches have performed without problems.

Specimen selection

Specimen 1 was removed by cutting along the circumferential weld joint (retaining some portion of the weldment with the dished end) after the sodium was drained off and the internal hardware was removed. Similarly, one of the two dished ends from the other vessel (specimen 2) that had been in storage was removed. Portions of the straight areas of the dished ends containing the cracks were cut out for optical microscopy. Rectangular strips containing part of the circumferential weldment and the straight portion of the dished end were removed, with the longer axes of the strips parallel to the weld to assess the degree of sensitization per ASTM Standard A 262, Practice E. To examine the crack surfaces by scanning electron microscopy (SEM), several small, square-shaped pieces, which had cracks positioned in the middle of the specimen, were cut out. The areas flanking the cracks were

Fig. 1 Schematic of the cross section of a dished end

then torn open by placing the material in a tension testing machine and pulling the specimens apart.

Visual Examination of General Physical Features

No abnormal features were observed visually that had bearing on the failures.

Testing Procedure and Results

Nondestructive evaluation

Liquid penetrant testing was carried out to determine the extent of the cracking. Testing was also helpful in evaluating repair of the cracked areas of dished end 1, which was in service. It was ultimately decided that the dished end could not be repaired. The repair procedure involved grinding the areas containing cracks from the outside, leaving about 1 mm (0.04 in.) of thickness, and depositing weld metal on the areas that had been ground. After each pass during the repair, liquid penetrant testing was carried out. No cracks on the outside surface were detected during this test. However, cracks did develop on the inside surface due to repair welding (see the section on ultrasonic testing below).

In the dished end in storage (dished end 2), the greater number of indications on the inside surface than on the outside and the increased coarseness of these indications pointed to initiation of the cracks from the inside.

Ultrasonic Examination. The objective of ultrasonic examination was to determine whether the cracks originated from the inside or outside of the straight portion of the dished ends. This was particularly important in the case of the dished end in service, because the inside surface was not accessible during the initial stages of the investigation when the decision to abandon repair of the dished end had not yet been reached.

Repair welding was found to be futile when ultrasonic testing showed initiation of new cracks from the inside in areas near the weld deposits. As indicated earlier in the section on liquid penetrant testing, repair of dished end 1 was attempted by grinding the cracked areas and depositing weld metal to fill these areas.

Thickness measurements obtained by ultrasonic testing indicated that the thickness in the straight portion was greater (6.8 mm, or 0.28 in.) compared to the nominal value of 6.0 mm (0.24 in.) for the starting material, indicating significant material flow in this region with consequent cold work and retention of residual stress.

Surface examination

Scanning Electron Microscopy/Fractography. Figure 2 is a typical SEM fractograph of an intergranular stress-corrosion crack from dished end 2 (in storage). Fractographs of crack faces from dished end 1 (in service) showed similar IGSCC, but the surfaces were smeared with sodium oxide.

Metallography

Microstructural Analysis. Figures 3 and 4 illustrate the inside surface crack morphology for dished ends 1 and 2, respectively. The cracks are

Fig. 2 Scanning electron micrograph of an intergranular stress-corrosion crack

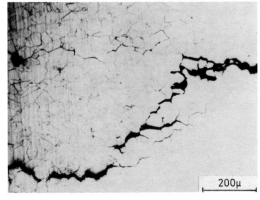

Fig. 3 Optical micrograph of IGSCC in dished end 1 (in service)

Fig. 4 Optical micrograph of IGSCC in dished end 2 (in storage)

typical examples of IGSCC. The microstructure, however, does not show typical features that are normally associated with sensitization, i.e., dark, wide grain boundaries in the etched condition that indicate precipitation of chromium carbides and decreased passivation in these regions. Sensitization typically occurs in steels such as types 304 and 316, particularly when exposed at temperatures from 500 to 750 °C (930 to 1380 °F). However, because the crack features greatly resembled IGSCC, confirmation of the extent of sensitization and the associated tendency for IGSCC was obtained by using ASTM Standard A262-70, Practice E.

The test consists of exposing rectangular strips of the material to boiling and acidified (added sulfuric acid) copper sulfate solution mixed with copper turnings for 24 h. After exposure, the strips are bent into a U-shape configuration. The nature and extent of cracking that develops reflects the tendency for IGSCC. Figure 5 shows numerous cracks at the top of such a U-shape strip from dished end 1 (in service at 550 °C (1020 °F) for 500 h), and Fig. 6 is of dished end 2 (in storage). In Fig. 6, the narrow band of orange peel structure in dished end 2 indicates a typical case of weld decay corresponding to the narrow zone near the weld/parent metal boundary at which the temperature range for sensitization prevails for a sufficiently long time. In this dished end, cracks were observed in the orange peel region.

Crack Origins/Paths. Cracks originated at

Fig. 5 Numerous cracks at the tip of a U-shape strip from dished end 1. Testing was conducted per ASTM Standard A262, Practice E, for assessment of the extent of sensitization.

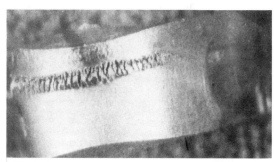

Fig. 6 Orange peel structure on the tip of a U-shape strip taken from dished end 2 per ASTM Standard A262, Practice E

several locations in the inside surface of both the dished ends. This was confirmed by liquid penetrant and ultrasonic testing. Some of the cracks propagated through the wall thickness, reaching the outside surface.

Discussion

All of the evidence indicates that both of the dished ends had been sensitized in the area where the intergranular cracks were observed. The cracks originated at the inside surfaces. Significant cold working in the straight portion of the dished ends, as evidenced by an increase in wall thickness, implies the retention of residual stress. Cold work levels on the order of 10% were estimated for both dished ends. Although no rigorous analytical and experimental stress analyses were carried out, consideration of the nature of expected material flow at the inside and outside surface of the dished ends during cold spinning and subsequent to the completion of the cold spinning process indicate that the nature of the stress in the inside surface should be tensile.

The occurrence of IGSCC in dished end 2 is easily understandable. The work site where these SCC failures occurred is a coastal area with average monthly temperatures in the range of 25 to 32 °C (77 to 90 °F). The relative humidity ranges from 70 to 80% and the chloride contents in the atmosphere, evaluated by the wet candle method, ranged from 8 to 45 mg/m^2 per day of sodium chloride. Because the vessel was not stored in an air-

conditioned atmosphere, the sensitized structure, residual stress, and chloride content in the environment effectively combined to produce the observed IGSCC.

For dished end 1, in which IGSCC was observed during service originating from the inside surface that was in contact with pure sodium, interpretation of these findings is less straightforward. Pure sodium is not known to lead to IGSCC. If cracks had already initiated in the inside surface at the time the vessel was put into service, the cracks would have grown during service due to thermal operational stresses. It should be noted that the vessel operated below the maximum temperature of 550 °C (1020 °F). The occurrence of thermal stresses was very possible during each temperature increase that was undertaken to reach the maximum temperature. Ultrasonic inspection clearly showed that even the thermal stresses generated during repair welding were sufficient to initiate new cracks on the inside surface. Therefore, even if there were no cracks in the vessel at the beginning of service, the cracks could have been generated during service.

Conclusion and Recommendations

Most probable cause

There were two failures—one in a dished end in service in a vessel containing sodium at 550 °C (1020 °F) and the other in two dished ends in a vessel that was kept in storage for 1 year. The dished ends were not stress relieved after fabrication by cold spinning. In the straight end portion of one of the dished ends kept in storage, sensitization occurred in a narrow

band a slight distance from the weld joint that joined the dished end to the cylindrical portion of the vessel. In the other dished end, sensitization was extensive, most probably due to exposure at 550 °C (1020 °F) for more than 500 h.

For the dished end in storage, the most probable cause of failure was the combined effect of a sensitized microstructure, residual stresses, and a humid environment having a high chloride content. For the dished end in service, the extensive sensitization and thermal stresses during operation combined to lead to failure.

How failure could have been prevented

Failure could have been avoided by the use of low-carbon austenitic stainless steel, such as type 316L or 304L, and by relieving the residual stresses in the dished ends by solution annealing.

Liquid Metal Embrittlement of Bronze Rupture Discs

William R. Watkins, Jr., Air Products and Chemicals, Inc., Allentown, Pennsylvania

Failure of three C22000 commercial bronze rupture discs was caused by mercury embrittlement. The discs were part of flammable gas cylinder safety devices designed to fail in a ductile mode when cylinders experience higher than design pressures. The subject discs failed prematurely below design pressure in a brittle manner. Fractographic examination using SEM indicated that failure occurred intergranularly from the cylinder side. EDS analysis indicated the presence of mercury on the fracture surface and mercury was also detected using scanning auger microprobe (SAM) analysis. The mercury was accidentally introduced into the cylinders during a gas-blending operation through a contaminated blending manifold. Replacement of the contaminated manifold was recommended along with discontinued use of mercury manometers, the original source of mercury contamination.

Key Words			
Bronzes	Rupturing		Intergranular fracture
Embrittlement	Mercury		Brittle fracture
Safety equipment			

Alloys Bronzes—C22000

Background

Applications

Commercial bronze rupture discs (C22000) are used in a safety device integral with valves used on flammable gas containing cylinders. C22000 composition limits are 89.0 to 91.0 wt% Cu, 0.05 wt% max Pb, 0.05 wt% max Fe, and remainder Zn. Burst pressure is normally 26 MPa (3775 psi). Disc dimensions for this pressure are nominally 1.44 cm (0.565 in.) in diameter and 0.165 mm (0.0065 in.) thick. Overpressurization typically causes ductile blowout of the disc with subsequent gas release to the surrounding environment.

Circumstances leading to failure

Several cylinders had been filled with a hydrogen/ethylene gas mixture through a gas-blending manifold. Two days later, after normal working hours, one cylinder released its contents into a closed area, precipitating an explosion. Also, two cylinders that had been loaded onto a truck released their contents into the atmosphere with no further consequences. Investigation of the cylinder valves indicated that their rupture discs had failed.

Selection of samples

The three rupture discs from the cylinders that had leaked were examined along with unfailed discs from other cylinders filled at the same time. Also, a rupture disc that had failed to overpressurization was examined.

Testing Procedure and Results

Surface examination

Visual. Surface appearance of all the discs exposed to the hydrogen/ethylene mixture was similar. The atmosphere side was unaffected; however, the cylinder side had a dark gray appearance.

Macrofractography. Discs that ruptured prematurely (Fig. 1) had an atypical appearance compared to a disc that had ruptured due to overpressurization (Fig. 2). Normal burst failures initiate at the center of the disc, causing a serrated fracture profile. Fracture in the prematurely failed discs initiated off-center at a point corresponding to the inside edge of a shear ring (similar in shape to a washer) that supports the rupture disc. Fracture progressed circumferentially in both directions from the initiation point. On one disc (the one that caused the explosion) the center portion was missing; on the other two discs the center remained attached.

Scanning Electron Microscopy Fractography. Using scanning electron microscopy, surface deposits were detected on only the cylinder side of both failed and unfailed discs that had been exposed to the hydrogen/ethylene mixture (Fig. 3 and 4).

Fracture surface morphology of the three failed discs were, for the most part, intergranular (Fig. 5 and 6). A ductile fracture morphology was found in areas that were either the last area of a disc to fail or were proximate to the atmosphere side of the disc (Fig. 7). Secondary intergranular cracking was found on the cylinder side of all three discs (Fig. 8 and 9); however, no secondary cracking was found on the atmosphere side. For comparison purposes, a ductile fracture mode produced by overpressurization is shown in Fig. 10.

Metallography

Crack Origins/Morphology. Sections near the fracture edge were examined on the three failed discs. Fracture was intergranular from the cylinder-side surface to approximately midwall of the disc thickness, and then changed to ductile fracture (Fig. 11). Secondary intergranular cracks that had not propagated through the thickness were found adjacent to the primary fractures on the cylinder side of the discs (Fig. 11 and 12).

Chemical analysis/identification

EDS analysis in the SEM was used to chemically characterize surface deposits on the discs that had been exposed to the hydrogen/ethylene gas and the fracture surfaces of the failed discs. Only Cu, Zn, Fe and S were initially detected (Fig. 13). After very extensive examination, trace quantities of Hg were detected. An isolated indication of a low-energy Hg peak in the form of a broadening of a sulfur peak was found on the cylinder side of

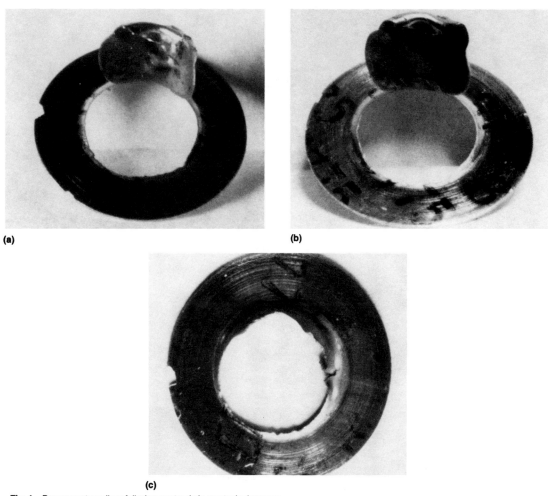

Fig. 1 Bronze rupture discs failed prematurely in an atypical manner

Fig. 2 Typical appearance of a bronze disc ruptured due to overpressurization

Fig. 3 SEM micrograph showing surface deposits on the cylinder side of a disc exposed to hydrogen/ethylene blend. 271×

an exposed, but unfailed, disc (Fig. 14). Subsequently, a similar indication along with higher-energy Hg indications barely above background energy was found on or near the fracture surfaces of two of the three failed discs (Fig. 15).

X-ray photoelectron spectroscopy (XPS) and scanning auger microprobe (SAM) analysis were used to corroborate the EDS detection of Hg or to detect nitrogen, since ammonia stress corrosion cracking was initially suspected as a failure

mechanism. XPS was performed on the unfailed disc on which trace Hg was detected by EDS. Nitrogen, as nitrate, and iron, as iron oxide, were the only elements found that were extraneous to the findings on a disc that had not been exposed to the gas blend.

Since the minimum XPS analysis area of 4 mm^2 (0.016 in.^{2}) was inappropriate for analyzing the narrow fracture surface of the failed discs, SAM was used. After ion sputtering to remove

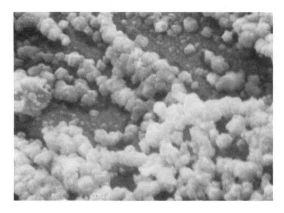

Fig. 4   Same as Fig. 3 at higher magnification. 3087×

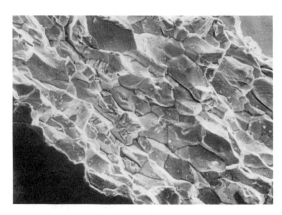

Fig. 5   SEM micrograph showing intergranular fracture morphology of a prematurely failed disc. 630×

Fig. 6   SEM micrograph showing an area similar to Fig. 5 at higher magnification. 1953×

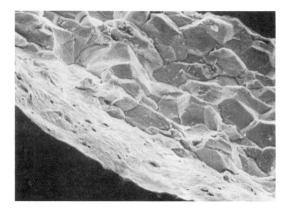

Fig. 7   SEM micrograph showing transition from intergranular to ductile fracture morphology. The atmosphere side of the disc is the lower left of the micrograph. 756×

Fig. 8 SEM micrograph showing secondary intergranular cracking on the cylinder side of the failed disc. Arrow indicates primary fracture surface. 239×

Fig. 9   Same as Fig. 8 at higher magnification. 945×

carbon, which was initially detected along with oxygen, trace quantities of mercury were found on one fracture surface.

Discussion   Since initial analyses on the disc samples revealed no indications of mercury, stress-corrosion cracking by ammonia, nitrogen compounds or sulfur compounds was at first thought to have caused premature failure of the discs. This conclusion was supported by the intergranular nature of the fractures and the fact that the failures occurred two days after filling of the cylinders. Mercury embrit-

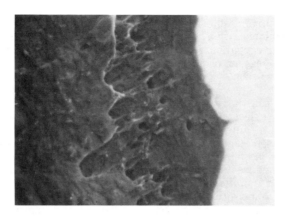

Fig. 10 SEM micrograph showing a ductile fracture morphology typical of overpressurization. 756×

Fig. 11 Photomicrograph of a cross-section of a failed disc. Intergranular fracture was evident on the cylinder side of the disc (lower right) changing to ductile fracture on the atmosphere side (upper right). 315×

Fig. 12 Photomicrograph showing a high magnification view of secondary intergranular crack at the bottom of Fig. 11. Crack is on the cylinder side of the failed disc. 630×

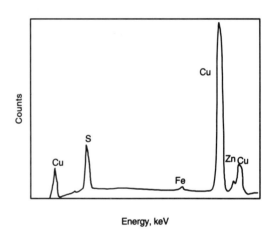

Fig. 13 EDS spectrum obtained from deposits on the cylinder side of a disc exposed to the hydrogen/ethylene blend.

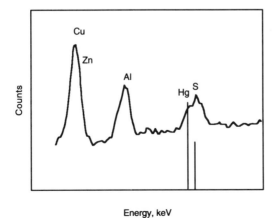

Fig. 14 EDS spectrum (low-energy band) obtained from deposits on the cylinder side of an exposed disc. An isolated indication of Hg was detected as evidenced by the broadening of a sulfur peak.

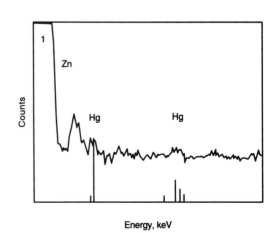

Fig. 15 EDS spectrum (high-energy band) representative of areas on or near fracture surfaces of two failed discs. Hg indications are barely above background.

tlement failures typically occur almost instantaneously once mercury amalgamates with a host metal under stress. Further investigation revealed, however, that no evidence of oxygen or moisture, both necessary for stress corrosion, was found in the cylinders.

SEM examination indicating that fracture had initiated on the cylinder side at least narrowed the search for a contaminant to the internal environment of the cylinder or to some agent that may have contaminated the disc prior to installation. More extensive EDS analysis was subsequently performed. Copper, zinc and iron indications are the result of the base alloy composition. Some of the iron detected, especially on the cylinder side, may have been the result of iron oxide from the cylinder inner diameter. Sulfur was thought to be a contaminant that originated in the gas-blending manifold that had previously been used for hydrogen sulfide.

Ironically, the energy level at which the primary sulfur peak occurs on an EDS scan is almost coincident with a low-energy mercury peak. This made detection of mercury very difficult. Only after extensive and repeated EDS examination was sufficient evidence to confirm mercury contamination was collected. The high vacuum environment in the SEM combined with electron bombardment during examination may vaporize most of the mercury originally present on fracture surfaces.

XPS and SAM analyses provide evidence that corroborated EDS results indicating the presence of mercury. Final corroboration occurred when the failure analysis results prompted the disassembly of the gas-blending manifold at the site. A large amount of mercury had accumulated in the manifold. Apparently, mercury manometers had previously been used in the system and had contaminated the manifold.

Conclusion and Recommendation	The rupture discs failed prematurely due to mercury embrittlement. The hydrogen/ethylene gas blend had been contaminated with mercury from the gas-blending manifold. It was recommended that the gas-blending manifold be replaced and the use of mercury manometers be discontinued.

Transgranular Stress-Corrosion Cracking Failures in AISI 304L Stainless Steel Dished Ends During Storage

D.K. Bhattacharya, J.B. Ghanamoorthy, and Baldev Raj, Metallurgy and Materials Programme, Indira Gandhi Centre for Atomic Research, India

Several type 304L stainless steel dished ends used in the fabrication of cylindrical vessels developed extensive cracking during storage. All of the dished ends had been procured from a single manufacturer and belonged to the same batch. When examined visually, several rust marks were observed, indicating contamination by rusted carbon steel particles. Liquid penetrant testing was used to determine the extent of the cracks, and in situ metallographic analysis was performed over the cracked region. The morphology of the cracks was indicative of transgranular stress-corrosion cracking (TGSCC). Conditions promoting the occurrence of the TGSCC included significant tensile stresses on the inside of the dished ends, the presence of surface contamination by iron due to poor handling practice using carbon steel implements, and storage in a coastal environment with an average temperature of 25 to 32 °C (77 to 90 °F), an average humidity ranging from 70 to 80%, and an atmospheric NaCl content ranging from 8 to 45 mg/m^2/day. Recommendations preventing further occurrence of the situation were strict avoidance of the use of carbon steel handling implements, strict avoidance of cleaning practices that cause long-term exposure to chlorine-containing cleaning fluid, and solution annealing of the dished ends at 1050 °C (1920 °F) for 1 h followed by water quenching to relieve residual stresses.

Key Words			
Austenitic stainless steels, corrosion	Stress-corrosion cracking	Dished ends, corrosion	
Contaminants	Transgranular corrosion	Marine environments	

Alloys   Austenitic stainless steel—304L

Background

Several type 304L stainless steel dished ends developed extensive cracking during storage. All of the dished ends had been procured from a single manufacturer and belonged to the same batch.

Applications

The dished ends are used to fabricate cylindrical vessels. The straight portions of the dished ends are welded to cylindrical shells.

Circumstances leading to failure

Customary visual inspection of the dished ends before welding revealed extensive zigzag cracks in the inside surfaces of many of them. Liquid penetrant testing revealed the cracks clearly. There were no cracks on the outside surfaces.

Handling procedures used by the manufacturer of the dished ends were not known. The fabricator of the cylindrical vessels took no special care in handling the ends.

Pertinent specifications

The material used for the dished ends was to conform to AISI 304L stainless steel. They were to be manufactured by cold spinning; no stress relieving was specified.

Performance of other parts in same or similar service

There is always an element of uncertainty in the occurrence of this type of transgranular stress-corrosion cracking (TGSCC), because an effective combination of (residual) stress, microstructures, and environment must be present. However, the widespread occurrence of TGSCC in dished ends from the same batch was striking.

Specimen selection

Specimens were selected for chemical analysis to determine whether the material used to make the dished ends conformed to AISI 304L. The specimens were obtained by drilling at regions away from those where cracks were observed.

Microscopic examination of the surface region was carried out *in situ*. A specimen was extracted from the crown area of one dished end by interconnecting drills in order to observe the crack morphology across the wall thickness.

Visual Examination of General Physical Features

Several rust marks were visible, indicating contamination by rusted carbon steel particles.

Testing Procedure and Results

Nondestructive evaluation

Liquid penetrant testing was used to determine the extent of the cracks. This technique was also used to determine whether any new cracks were generated when the dished ends that had not yet exhibited cracks were stress relieved by solution annealing.

Metallography

Metallographic inspection of the cracked regions was carried out *in situ* by using a fixture to position a portable microscope vertically over the cracked region (Fig. 1). Surface preparation was performed using standard commercial equipment. Figure 2 shows the morphology of a portion of a crack, and Fig. 3 shows a mosaic of another crack.

Fig. 1 Setup for *in situ* metallographic inspection of a dished end

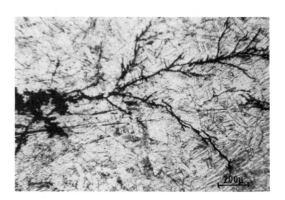

Fig. 2 Transgranular stress-corrosion crack on the inside surface of a dished end

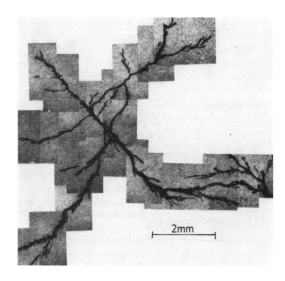

Fig. 3 Mosaic of a crack on the inside surface of a dished end

The crack morphology across the wall thickness is illustrated in Fig. 4. Figures 2 to 4 indicate that the cracks were classical transgranular stress-corrosion cracks.

Crack Origins/Paths. Figure 4 shows that the cracking only partially penetrated the wall. This behavior was observed in every dished end. Cracking appeared to prefer to spread over the inside surfaces and subsurface regions.

Stress analysis

Experimental. The presence and qualitative nature of the residual stress in the dished ends were determined by making two saw cuts, as shown in Fig. 5. There was considerable movement of portion B toward the axis of rotation of the dished end, indicating that the nature of the residual stress at the inside surface of the straight-edge portion of the dished end was tensile. It was expected that the stresses in the inside surfaces of the crown areas would also be tensile.

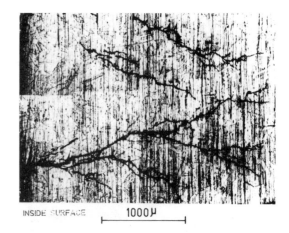

Fig. 4 TGSCC morphology across the wall thickness of a dished end

Discussion	*Synthesis of evidence*

The TGSCC cracks were heavily branched with macro- and microbranches, and the main cracks were wide, indicating that the operating stress-intensity factor was high. Based on this evidence and on Fig. 5, it is highly probable that the residual stresses at the inside surfaces of the crown areas of the dished end were tensile in nature.

Rusting of the carbon steel contaminants (Fig. 6) was observed in all the dished ends that developed cracks. The contamination could have occurred at the manufacturer of the dished ends or at the fabricator of the cylindrical vessels.

The common causative factors for SCC of the stainless steel components were determined to be (1) rusting of the iron contaminants, (2) the presence of chloride in the coastal environment, and (3) stresses arising from fabrication or fit-up. The work site where these SCC failures occurred is a coastal area with average monthly temperatures in the range of 25 to 32 °C (77 to 90 °F). Relative humidity ranges from 70 to 80%, and the NaCl content in the atmosphere, evaluated by the wet candle method, ranges from 8 to 45 mg/m^2/day. In such a hot, humid, coastal atmosphere, long exposures of stainless steel components containing iron surface contaminants could result in the rusting of iron debris. Such rust particles absorb moisture and chlorides from the humid environment, finally leading to SCC of the components.

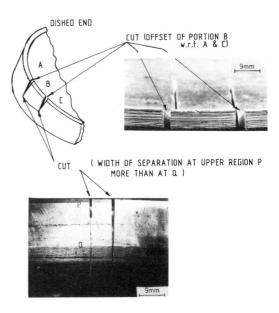

Fig. 5 Reshuffling of residual stresses after hacksaw cutting at a location in the straight portion of a dished end

Conclusion and Recommendations

Most probable cause

Conjoint action of residual stress and chloride ions in the environment contributed to the cracking. The rusting of the carbon steel contaminants enhanced the effect of the chloride species absorbed at the rust debris.

Remedial action

Extreme care was needed to avoid pickup of iron contaminants by the stainless steel components during fabrication and subsequent handling. Pickling and passivation of the stainless steel components prior to final dispatch or storage were also required. Otherwise, unexpected SCC failures of stainless steel components in the unsensitized condition may be encountered even in ambient-temperature atmospheres.

How failure could have been prevented

Failure of the dished ends could have been prevented by avoiding carbon steel contamination, by stress relieving through solution annealing, and by keeping the components dry (by limiting their exposure to humid atmosphere and to water).

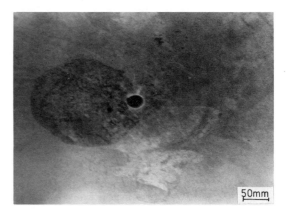

Fig. 6 Rust marks at sites of carbon steel contamination on the inside surface of a dished end

Acidic Pitting in Waterwall Tubing

Wendy L. Weiss, Radian Corporation, Austin, Texas
John C. Jones, Hartford Steam Boiler and Insurance Company, Atlanta, Georgia

Severe pitting was found on the internal surfaces of SA-210 Grade C waterwall tubing of a coal-fired boiler at a cogeneration facility. Metallographic examination showed the pits to be elliptical, having an undercut morphology with supersurface extensions; a type of pitting characteristic of acidic attack. Energy-dispersive X-ray spectroscopy revealed the presence of chlorine in the pit deposits, indicating that the pitting was promoted by underdeposit chloride attack. The presence of copper in deposits on the internal surface of the tubing may have acted as a secondary factor. Acidic conditions may have formed during a low-pH excursion that reportedly occurred several years prior. To prevent future failures, severely damaged tubing must be replaced. Internal deposit buildup must be removed by chemical cleaning to prevent further pitting. Water quality needs continued monitoring and maintenance to ensure that another low-pH excursion does not occur.

Key Words	Carbon steels, Corrosion Chlorides, Environment	Boiler tubes, Corrosion	Pitting (corrosion)
Alloys	Carbon steels—SA-210 Grade C		

Background

Severe pitting was found on the internal surface of waterwall tubes at a southeastern cogeneration facility. The areas most affected were waterwall tubes extending from the lower waterwall headers (at the gate level) approximately 1.2 m (4 ft) above the burner grates. The target wall and both side walls were affected. These sections experienced higher heat intensity due to proximity of fuel combustion products. A window sample was received for analysis (Fig. 1).

Applications

The tube sample was from a Foster Wheeler coal-fired boiler with a stoker grate burner system that has been in service since 1986. The boiler has a steam capacity of 71400 kg/h (157000 lb/h) and is designed to operate at a pressure of 10.3 MPa (1500 psi) with a final steam temperature of 513 °C (955 °F).

A coordinated pH-phosphate system is used for water treatment. Well water is the source for the make-up water. Reportedly, all water treatment control limits, including the demineralizer, are rigorously maintained. However, a demineralizer excursion occurred on one occasion approximately six years prior.

Circumstances leading to the failure

The boiler had reportedly suffered from a low pH excursion (the result of a water quality demineralizer excursion) several years prior to observation of the pitting. The problem occurred approxi-

Fig. 1 The window section received for analysis

mately six years ago and has not reoccurred.

Pertinent specifications

The tube sample came from a 7.62-cm (3-in.) outer diameter SA-210 Grace C tube section, with a minimum wall thickness of 5.2 mm (0.203 in.). The tubing is reportedly original.

Performance of other parts in similar service

Another boiler at the same location also experienced internal pitting, but it was less severe.

Testing Procedure and Results

Surface examination

Visual. The internal surface of the sample was covered with large, deep pits (Fig. 2). The maximum pit depth observed was 1.5 mm (0.06 in.). A significant amount of dark internal deposit was also present in some locations on the window sample.

Metallography

A cross section from the tube was prepared for metallographic examination. The typical morphology of the pitting is shown in Fig. 3. The pits are el-

liptical and have an undercut morphology with subsurface extensions. This type of pitting is characteristic of acidic attack. Some pits were deposit-filled and appeared to be active. The most severe pits were more than one-third through the thickness of the tube wall.

The deposits on the internal surface contained a large amount of copper (Fig. 4). The presence of copper indicates corrosion of pre-boiler water circuit tubing and is of concern because of its possible stimulation of corrosion under certain conditions.

Fig. 2 The internal surface of the window section showing severe pitting

Table 1 Composition of the tube metal and requirements for ASTM SA-210 Grade C steel (wt%)

Element	Tube metal	ASTM SA-210 Grade C
C	0.24	0.35 max
Mn	0.63	0.29 to 1.06
P	0.005	0.048 max
S	0.011	0.058 max
Si	<0.05	0.10 min
Ni	<0.05	ns
Cr	<0.05	ns
Mo	<0.05	ns

ns, not specified

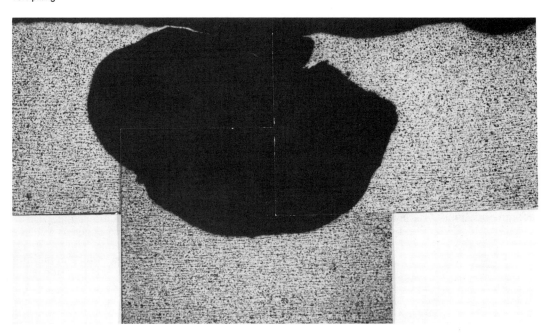

Fig. 3 Cross-section of a typical pit observed on the window section. Notice the undercut morphology indicative of acidic attack. Nital etch, 32×

Fig. 4 Deposit found on the internal surface of the sample. The bright speckles seen through the deposit are copper. Nital etch, 61×

Chemical analysis/identification

Material. A portion of the sample was analyzed for chemical composition (Table 1). The tube material did not meet the requirements for SA-210 Grade C steel due to a low silicon content. The low silicon content was not considered to have contributed to the failure.

Corrosion or Wear Deposits. The transverse section removed from the window sample was dry ground to expose pit deposits and was analyzed using energy dispersive X-ray spectroscopy (EDS) to determine the elements present. The pit deposits were found to contain major amounts of chlorine and potassium, a moderate amount of iron, and minor to trace amounts of sulfur, silicon, sodium, and aluminum (Fig. 5). The presence of chlorine (chlorides) in the pit deposits indicates that the pitting was promoted by underdeposit chloride attack.

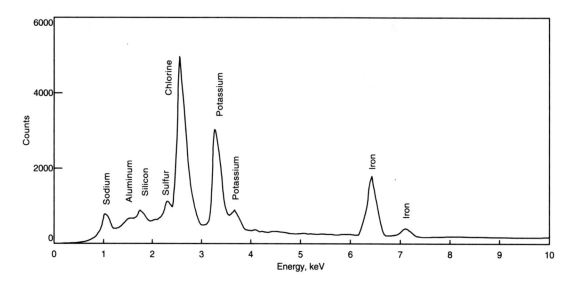

Fig. 5 EDS results from a pit deposit

Discussion	The pitting found on the internal surface of this tube sample results from underdeposit acidic attack, promoted by the presence of chlorine. Certain sulfur compounds can add to pit acidity. Acidic conditions can occur underneath built-up internal deposits and attack the tube metal. The copper in the internal deposits may have acted as a secondary factor by accelerating the pitting corrosion (by setting up localized galvanic cells).

Conclusion and Recommendations

Most probable cause

The acidic conditions may have formed during a low pH excursion that reportedly occurred several years ago.

Remedial action

Severely damaged tubing must be replaced. The internal deposit buildup must be removed by means such as a chemical clean to prevent further pitting. The water quality must be continually monitored and maintained to ensure that another low pH excursion does not occur.

Alkaline-Type Boiler Tube Failures Induced by Phosphate Water Treatment

Moavinul Islam, CC Technologies Inc., Columbus, Ohio

Tube failures occurred in quick succession in two boiler units from a bank of six boilers in a refinery. The failures were confined to the SAE 192 carbon steel horizontal support tubes of the superheater pack. In both cases, the failure was by perforation adjacent to the welded fin on the crown of the top tubes and located in an area near the upward bend of the tube. The inside of all the tubes were covered with a loosely adherent, black, alkaline, powdery deposit comprised mainly of magnetite. The corroded areas, however, had relatively less deposit. The morphology of the corrosion damage was typical of alkaline corrosion and confirmed that the boiler tubes failed as a result of steam blanketing that concentrated phosphate salts. The severe alkaline conditions developed most probably because of the decomposition of trisodium phosphate, which was used as a water treatment chemical for the boiler feed water.

Key Words	Carbon steels, Corrosion pH	Boiler tubes, Corrosion Water chemistry	Corrosion Alkaline corrosion
Alloys	Carbon steels—SAE 192		

Background

Unscheduled shutdowns occurred in boilers No. 1 and No. 3 from a bank of six boiler units in a refinery, due to failure of the horizontal support tubes (Type SAE 192 carbon steel) of the superheater pack. Figure 1 shows a detailed sketch of the support tubes and the location of the failures. The support tubes are arranged in three rows with horizontal and vertical sections. Each row has two tubes fixed one on top of another with welded support plates. The bottom of the superheater pack is welded with fins to the top tubes of the horizontal support.

In Boiler No. 1, failure occurred on the top tubes of the first and the third row of the horizontal support; in Boiler No. 3, the top tube of the first row failed. In both cases, the failure was by perforation adjacent to the welded fin on the crown (or roof) of the top tubes, located in an area near the upward bend of the tube.

Applications

The boilers have a maximum continuous rating of 163 Mg/h (180 ton/h) with an average operating load of 109 Mg/h (120 ton/h) and an average operating pressure of 6.2 MPa (900 psig). The internal temperature of the main bank tubes is 275 °C (527 °F) while the superheater steam temperature is 450 °C (842 °F).

Demineralized make-up water/condensate return is deaerated and then injected with hydrazine and amine before entering the boiler feed pumps. Trisodium phosphate (Na_3PO_4) is injected into the boiler drum. Continuous and intermittent blowdown facilities are installed in the boiler drum.

Specimen selection

The tube sections selected for failure analysis were: Boiler No. 1, Support No. 1 (failed); Boiler No. 3, Support No. 1 (failed); and Boiler No. 3, Support No. 3 (not failed but with corrosion from inside). For purposes of identification, these tubes are referred to here as B1S1, B3S1, and B3S3, respectively.

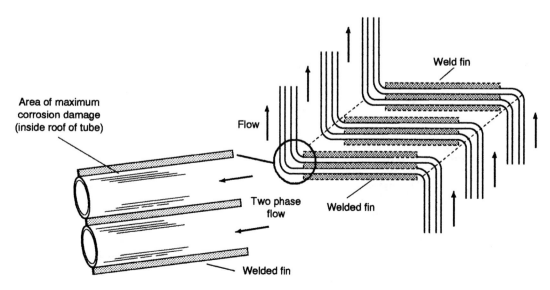

Fig. 1 Schematic of superheater support tubes showing location of maximum corrosion damage

Visual Examination

The average diameter of the tubes was 56.7 mm (2.23 in.), and the average wall thickness was 5.9 mm (0.23 in.). No outside bulging of the tubes was apparent. The outer part of the tubes had a thin adherent oxide, dark brown to black in color.

All three sections were cut longitudinally for closer examination. Samples B1S1 and B3S1 had the maximum corrosion damage which was confined to the crown (inside top or roof) of the tube in the form of gouging and a deep trench following the line of the welded fin. The bottom half of the tubes did not show any corrosion damage. Figure 2 shows the type of attack observed on both tubes. The corrosion trench had actually gone through

the thickness of the tube and into the welded fin. This type of severe attack was confined to an area about 30 cm (11.8 in.) near the upward bend of the horizontal tube with less attack in the adjacent area on the right (Fig. 3). The horizontal tubes are approximately 2.5 m (8.2 ft) long. No attack was observed at the end where these tubes bend downwards (see Fig. 1). Sample B3S3 showed less severe attack, shown in Fig. 4.

The inside of all the tubes were found to be covered with a very loosely adherent, black, powdery deposit. However, the corroded areas had relatively less amounts of deposit.

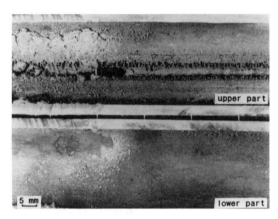

Fig. 2 Deep gouging (trenching) type of attack in the inside upper part of Tubes B1S1 and B3S1

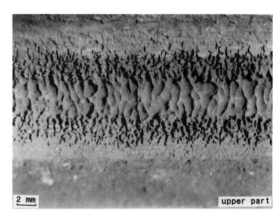

Fig. 3 Less severe corrosion damage in an adjacent part to the right of the area shown in Fig. 2

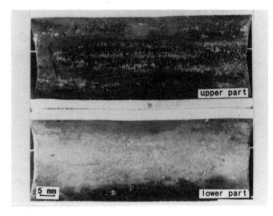

Fig. 4 Gouging type of attack in Tube B3S3 on the inside top part

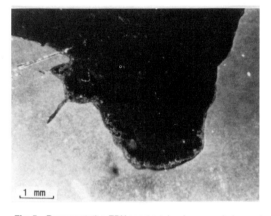

Fig. 5 Representative EDX spectra taken in a corroded area at the boiler tube

Testing Procedure and Results

Surface examination

Scanning electron microscopy was used to study the morphology of the corroded surface, while EDX analysis was carried out on the corroded areas to identify various elements in the corrosion products. Figure 5 shows a representative EDX spectrograph. The main peaks were due to Fe but minor peaks due to Cu, Ca and P were also seen.

Metallography

Cross-sectional and longitudinal sections were cut from Samples B1S1, B3S1, and B3S3 near the

corroded areas. After preparation and etching with nital, they were examined under the optical microscope. Figure 6 shows a low-magnification cross-sectional view of Sample B3S1 showing the corrosion trench which had gone through the tube material and into the welded fin. The same type of attack was observed in sample B1S1. Figure 7 shows a longitudinal section of Sample B3S1 from a less severely attacked area. The gouging attack is apparent.

Except for some lack of fusion observed between the weld bead and the parent metal (Fig. 6),

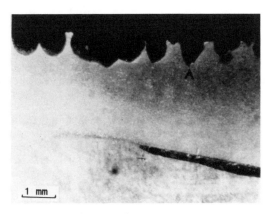

Fig. 6 Deep trench-like attack, which had gone through the tube material and into the welded fin, representative of type of attack in Tubes B1S1 and B3S1

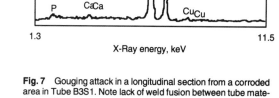

Fig. 7 Gouging attack in a longitudinal section from a corroded area in Tube B3S1. Note lack of weld fusion between tube material and fin.

no microstructural abnormalities were observed in the area of corrosion damage or in other sections of the tube.

Chemical analysis/identification

Corrosion Deposits. The black powdery deposit was scraped from the tubes and analyzed with the help of X-ray diffraction (XRD) which in-dicated that the deposit was mainly composed of Fe_3O_4 (magnetite).

Some of the scraped products (about 1 g, or 0.04 oz.) was leached with about 25 ml of hot double dis-tilled water. The pH of this water extract was then measured. An increase in the pH value from 6.5 (pH of the distilled water) to 8.4 was observed, in-dicating that the deposit was alkaline.

Discussion

The literature on boiler corrosion failures was reviewed to rationalize the findings of the present investigation and to suggest a plausible explana-tion for the observed severe corrosion damage of the boiler tubes.

As mentioned earlier, the most severely cor-roded part of the tubes had a trench-like appear-ance on the roof (or crown) of the tube which also coincided with the bottom of the welded fin. Since the tubes are horizontal and there is two-phase flow, the corrosion had occurred in the steam phase. An identical example of boiler tube corro-sion with photographic documentation could not be found in the literature. However, it is men-tioned that in horizontal or inclined tubing, insuf-ficient mass velocity can allow the flow of steam bubbles along the top of the tube (steam blanket-ing) and cause a strip along the top of the tube to slowly thin due to caustic attack, ultimately lead-ing to perforation (Ref 1).

Under proper operating conditions, boiler tubes develop a two-layer magnetite (Fe_3O_4) layer (Ref 2) according to the reaction:

$$3Fe + 4H_2O) \leftrightarrow Fe_3O_4 + 4H_2 \qquad (Eq\ 1)$$

Ferrous hydroxide, $Fe(OH)_2$, is believed to be the in-termediate in this process, converting to magnetite above 100 °C (212 °F) according to the Schikorr reac-tion (Ref 3):

$$3Fe(OH)_2 \leftrightarrow Fe_3O_4 + H_2 + 2H_2O \qquad (Eq\ 2)$$

The inner layer is relatively thick, compact, ad-herent and continuous while the outer layer is thinner, porous and powdery in texture. Counter currents of ions set up between the metal surface and the aqueous environment are responsible for the growth of the oxide layers (Ref 2). Reactions at the metal/oxide interface result in the growth of the compact Fe_3O_4. The outer porous oxide is formed when Fe^{+2} ions migrating outward react with OH^- ions at the oxide/solution interface to give $Fe(OH)_2$, which then converts to Fe_3O_4 as indi-cated above. The different conditions under which the two magnetite layers are formed result in their different morphologies. Porous magnetite is formed under alkaline conditions. Strong alkaline conditions can lead to dissolution of the magnetite and enhanced corrosion of the iron.

The porous and powdery magnetite deposit in-dicates that alkaline conditions must have existed in the boiler tubes. This observation is also sup-ported by the fact that the water extract of the magnetite deposit was found to be slightly alka-line.

The morphology of the corrosion damage in the present case is apparent from Fig. 2 to 6. The fea-tures may be considered typical of alkaline corro-sion or chelant attack. However, since no chelates (e.g., EDTA) were used in these particular boilers, chelant attack can be discounted. Alkaline attack most likely occurred due to steam blanketing in an area of high heat flux, in this particular case, the welded fin on the top tubes. The bottom tubes do not have a similar area of high heat flux because of the geometry and location as well as the presence of a liquid phase, hence the absence of corrosion at-tack on the bottom tubes. A similar case of severe corrosion has been cited for studded tubes of baker's ovens (Ref 4) where the attack was ob-served to occur at the base of the welded studs. The cause was attributed to an abnormally high heat flux through the studs leading to local boiling and concentration of boiler salts, in particular NaOH. Alkaline attack then occurred.

The source of the NaOH causing alkaline attack is the trisodium phosphate (Na_3PO_4), used as one of the boiler water chemicals. It has been reported (Ref 1, 5) that when a solution of trisodium phosphate is concentrated at elevated temperature, crystals of disodium phosphate (Na_2HPO_4) precipitate out and the supernatant liquid is rich in NaOH.

$$Na_3PO_4 + H_2O \leftrightarrow Na_2HPO_4 + NaOH \qquad (Eq\ 3)$$

To eliminate the formation of free caustic, a mole ratio of Na to PO_4 of 2.6 is recommended. This is accomplished by adding a mixture of Na_3PO_4 and Na_2HPO_4. A ratio of 65% Na_3PO_4 to 35% Na_2HPO_4 corresponds to a mole ratio of 2.6 (Ref 5).

The mole ratio of Na to PO_4 in Na_3PO_4 is 3. Judging from the occurrence of alkaline corrosion in the present case, boiler operating conditions have led to the decomposition of trisodium phosphate to free caustic according to the equation above. Based on the existing literature, a coordinated phosphate or a congruent method of water treatment (Ref 1, 5) under the present circumstances would have possibly avoided the problem. It may be recalled that the refinery used only trisodium phosphate in its boiler water.

Although the pH of the bulk water may be in the safe range (i.e. between 9.5 to 10.5), the occurrence of hot spots and steam blanketing leads to the concentration of salts, so that highly alkaline conditions can be established for boilers that use Na/PO_4 ratios of 2.6 or less.

The occurrence of hot spots and areas of high heat flux in boiler tubes can lead to the formation of thermogalvanic cells where hotter areas act as the anode and suffer accelerated corrosion. Such a situation has been reported by some authors in the corrosion of boilers. Thermogalvanic effects would certainly enhance the alkaline corrosion observed in the present case.

Small amounts of copper were detected by EDX on the corroded surface (Fig. 5). The presence of copper and copper oxides are known to create anodic sites on iron (Ref 6, 8). In the present case, the presence of copper possibly contributed to the overall corrosion process. Although copper was detected by the EDX technique, XRD analysis of the deposits failed to pick up any copper compounds, possibly because the amount present was below the detection limit.

Conclusion

The configuration of the superheater support tubes with its welded support plates, gave rise to hot spots and areas of high heat flux, resulting in steam blanketing. The morphology of the corrosion damage confirmed that the boiler tubes failed as a result of steam blanketing that concentrated the phosphate salts resulting in alkaline attack.

The powdery non-adherent magnetite deposit on the tube internals as well as its alkaline characteristics indicate that too severe an alkaline condition must have existed in the tubes. Severe alkaline conditions developed most probably due to the decomposition of trisodium phosphate to free NaOH because of steam blanketing in areas of high heat flux.

References

1. *Principles of Industrial Water Treatment*, Chapters 11 to 14, Drew Chemical Corporation, Boonton, NJ, 1981.
2. E.C. Potter and G.M.W. Mann, "The Fast Linear Growth of Magnetite on Mild Steel in High Temperature Aqueous Conditions," *Corrosion*, Vol 21, 1965, p 57-67.
3. U.R. Evans, *The Corrosion and Oxidation of Metals*, Chapter XII, Secondary Supplementary Volume, Edward Arnold, London, 1976.
4. F.R. Hutchings, "On-load corrosion in Tubes of High Pressure Boilers," in *Failure Analysis: The British Engine Technical Reports*, American Society for Metals, Metals Park, OH, 1981, p 28-48.
5. C.G. Bozeka and F.J. Pocock, "Waterside Corrosion Control in Industrial Boilers," *CORROSION/85*, Paper No. 249, NACE Corrosion Conference, Boston, MA, 1985.
6. D.R. Holmes and G.M.W. Mann, "A Critical Survey of Possible Factors Contributing to Internal Boiler Corrosion," *Corrosion*, Vol 21, 1965, p 39-48.
7. R.H. Bailey, M.J. Esmacher and T.L. Warner, "Aqueous Corrosion in Boilers: Analysis and Mitigation," *Power Engineering*, Vol 34, 1985, p 33-40.
8. J.P. Engle and C.E. Fox, "The Role of Copper and Physico-chemical Factors in Steam Generator Tube Failures," SE Electric Exchange, Beloxi, MS, April 1958.

An Unusual Case of Hydrogen-Induced Failure in a Refinery Boiler Tube

Moavinul Islam, CC Technologies Inc., Columbus, Ohio

A failed SAE-192 carbon steel tube from a 6.2-MPa (900-psig), 200-Mg/h (180-ton/h) capacity refinery boiler was analyzed to determine its failure mode. Optical and SEM examination results were combined with knowledge of the boiler operating conditions to conclude that the failure was hydrogen-induced. The hydrogen was probably generated by the steam-iron reaction. The source of steam on the flue gas side could be traced to a cracked fillet weld in the boiler. The failure mode was unusual in that the attack was found to originate from the flue gas side of the tube rather than the steam side.

Key Words	Boiler tubes, Corrosion	Carbon steels, Corrosion	Hydrogen embrittlement
	Weld cracking	Refineries	
Alloys	Carbon steels—SAE-192		

Background

A refinery boiler was shut down due to ruptured tubes and consequent flooding of the furnace chamber. Inspection of the inside of the boiler revealed that there were three failed tubes:

- Tube 1, a ruptured vertical superheater support tube positioned approximately in the middle of the furnace
- Tube 2, a vertical main bank tube with a small split, located near the side wall
- Tube 3, a horizontal superheater support tube with a cracked fillet weld

Some of the tubes adjacent to the cracked fillet weld had suffered erosion damage caused by high-pressure steam escaping from the cracked weld and impinging on the adjacent tubes. Judging from the extent of the erosion damage, the cracked weld had probably existed for some time before the shutdown of the boiler.

Figure 1 shows the locations in the boiler of the three failed tubes. Table 1 gives the operating conditions of the boiler. The failure analysis carried out on tube 2, the failed vertical main bank tube, will be described here.

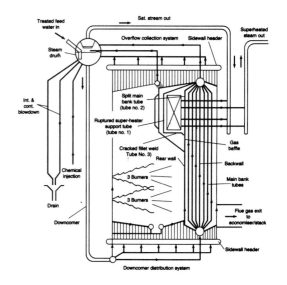

Fig. 1 Schematic diagram of boiler showing the locations of the three failed tubes

Table 1 Boiler operating conditions

Boiler rating (maximum continuous)	200 Mg/h (180 ton/h)
Average operating load	130 Mg/h (120 ton/h)
Average operating pressure	6.2 MPa (900 psig)
Internal temperature of main bank tubes	275 °C (527 °F)
Tube material	SAE-192
Superheater steam temperature	450 °C (842 °F)

Visual Examination of General Physical Features

A section from tube 2, approximately 800 mm (31.5 in.) long, was cut out and subjected to careful visual and macroscopic examination. The tube had a split about 50 mm (1.97 in.) long and 2 mm (0.08 in.) at the widest part of the crack as shown in Fig. 2. The tube had an outer diameter of 57.5 mm (2.26 in.) and a wall thickness of 4.5 mm (0.18 in.).

There were no deposits inside the tube and the inner surface had a tenacious black film approximately 150 μm (0.006 in.) thick. The outer surface had a dark brown-black oxide film approximately 90 μm (0.0035 in.) thick, that was not as adherent as the inner layer. One or two secondary cracks were visible on the outer surface adjacent to the main crack. The inner surface showed no secondary cracks when the tube was cut open and inspected.

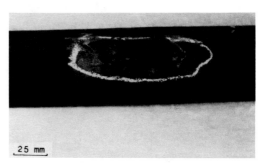

25 mm

Fig. 2 Tube 2 showing the 50-mm (1.97-in.) split

Testing Procedures and Results

Surface examination

Scanning Electron Microscopy Fractography. Several specimens from the fracture area were examined on the scanning electron microscope (SEM). Figure 3 is a low-magnification SEM fractograph which shows the whole fracture surface and the outer surface of the tube (top of the picture) while Fig. 4 shows a close-up of the area marked "a" in Fig. 3. As can be seen in Fig. 4, the fracture mode is predominantly characterized by the familiar river patterns associated with cleavage and quasi-cleavage type failures. The river lines converge from the outer surface of the tube towards the inner surface, indicating that crack propagation was from the outer surface to the inner surface. Some intergranular features are also apparent in Fig. 4 in addition to the river patterns.

Metallography

Metallographic samples were prepared from the fracture area and the end of the tube section, about 300 mm (11.8 in.) away from the fracture. The sample from the fracture area (Fig. 5) showed a high incidence of intergranular cracks, most of which were on the outside of the tube. Grain boundary fissuring was also evident, but there was hardly any crack branching. The microstructure showed only ferrite and was decarburized with no pearlite phase. The sample from the end of the tube section showed a spheroidized microstructure almost devoid of pearlite and consisting of ferrite with grain-boundary cementite, Fe_3C (Fig. 6).

Chemical analysis/identification

Material and Surface Products. Chemical analysis of the tube material showed that it corresponded to ASME grade SAE-192 carbon steel as per specification. X-ray diffraction analysis of the surface layers from both the inside and outside of the tube indicated that the major constituent was magnetite (Fe_3O_4) with some hematite (Fe_2O_3). The internal condition of the tube indicated that the water treatment was functioning properly.

Mechanical properties

Microhardness measurements near the fracture area gave an average value of 92HV, while the end of the tube section averaged 98HV. The recommended value for SAE-192 carbon steel is 137HV.

Fig. 3 Low-magnification SEM fractograph of fracture area of tube 2. Area marked "a" is shown magnified in Fig. 4.

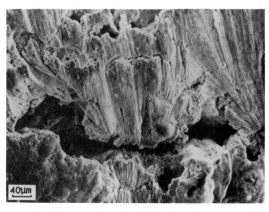

Fig. 4 Higher-magnification SEM fractograph of area marked "a" in Fig. 3. Fracture surface shows intergranular facets and quasi-cleavage river patterns. Note convergence of river lines from outer side of tube to inner side.

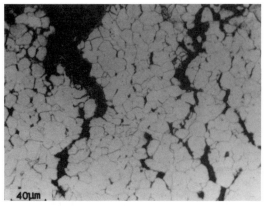

Fig. 5 Optical micrograph of section of tube 2 from fracture area showing intergranular cracks and fissuring. Note the decarburized microstructure.

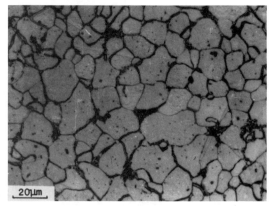

Fig. 6 Optical micrograph of section of tube 2, 300 mm (11.8 in.) away from fracture area showing ferrite and grain boundary cementite.

Discussion

The normal microstructure of SAE-192 carbon steel contains ferrite and pearlite phases. However, the microstructure of tube 2 in the fracture area (as well as at some distance from the rupture), consisted of ferrite with grain boundary cementite or spheroidal carbides (Fe_3C) and no pearlite phases. Occurrence of such a microstructure indicates that the alloy was overheated to temperatures in the range of 538 to 677 °C (1000 to 1250 °F) (Ref 1), which results in the transformation of pearlite to cementite. The formation of spheroidal carbides by the decomposition of pearlite is a time- and temperature-dependent reaction. Transformation at 677 °C (1250 °F) is about 100 times faster than at 538 °C (1000 °F) (Ref 1). However, in the present case, it is difficult to assess the duration of exposure and exact temperature to which the tube was subjected. The microstructure indicates that prior to the failure the tube had been exposed to local temperatures in the range 538 to 677 °C (1000 to 1250 °F).

The mode of failure of tube 2 as revealed by metallographic examination was found to be primarily intergranular with hardly any branching. Grain-boundary fissuring was also evident (Fig. 5). Almost all the cracks were on the outside of the tube. The nature of cracking and the fact that there was hardly any corrosion damage on the inside of the tube suggested that the failure was probably due to hydrogen attack rather than stress corrosion cracking. SEM examination of the fracture surface confirmed that the failure was most like due to hydrogen attack. The source of hydrogen was probably from the steam-iron reaction.

$$3Fe + 4H_2O \leftrightarrow Fe_3O_4 + 4H_2 \qquad (Eq\ 1)$$

The source of steam was from the cracked fillet weld on tube 3. As mentioned before, the fracture surface was characterized by river patterns encountered in cleavage and quasi-cleavage failures (Ref 2), which are often associated with hydrogen induced cracking (Ref 3,4). The direction of propagation of cracking in such failures is indicated by the convergence of the river line (cleavage steps) in the direction of local crack propagation (Ref 2), which in this case was from the outside of the tube to the inner side.

Optical and SEM examinations showed that the failure mode in tube 2 was by a mixture of cleavage and intergranular fracture. Both these fractures are characterized by low energy mechanisms, and simultaneous operation of both can oc-

cur when the resolved stresses for grain boundary separation and for transgranular cleavage are approximately equal. Also, if the preferred grain-boundary fracture path is not continuous and if the cleavage stress is relatively low, the regions that do not fracture intergranularly can fracture by cleavage (Ref 5).

The sequence of events that resulted in the failure of the tube is suggested to be as follows:

- Production of hydrogen by the steam-iron reaction according to Eq 1
- Diffusion of the hydrogen into the metal through the magnetite layer
- Reaction between the diffused hydrogen and the grain-boundary cementite (Fe_3C) present in the alloy, to generate methane (CH_4) according to the reaction below (Ref 6)

$$Fe_3C + 2H_2 \leftrightarrow 3Fe + CH_4 \qquad (Eq\ 2)$$

The formation of CH_4 would result in grain-boundary fissuring and cracking. The above steps involving production of hydrogen, its diffusion and subsequent reaction with Fe_3C, are all temperature-dependent reactions, but to differing extents. In the case of the normally protective magnetite, it is considered that the rate of production of hydrogen is never sufficient to result in significant embrittlement, whatever the temperature, and the rate tends to fall off with time (Ref 6).

When magnetite of the non-protective variety develops, as in the present case, the production of hydrogen is greatly increased and enhances the rate at which embrittlement proceeds. Assuming non-protective magnetite is produced, the chief factor in the embrittlement process is the effect of temperature on the steam-iron reaction. As a consequence, high service temperatures would raise the incidence of embrittlement.

The type of hydrogen attack encountered here, with the cracking initiating from the flue gas side, is rather unusual. Hydrogen attack is generally attributed to the action of high-temperature acidic solutions on the water side of the tube (Ref 7).

Conclusions

The cracked fillet weld was probably the first to fail. This caused loss of circulation and consequently, overheating in tubes 1 and 2, but to different extents. The steam released from the cracked weld interacted externally with tube 2, owing to its conducive microstructure, to produce hydrogen damage and finally rupture. Tube 2 was probably the second to fail. The overheating in tube 1 was more severe and localized, which led to sudden rupture. Tube 1 was the last to fail, resulting in the shut down of the boiler

References

1. G.W. Mann, "History and Causes of On-load Waterside Corrosion in Power Boilers," *Combustion*, Vol 50, No. 2, 1978, p 29-37.
2. *Metals Handbook, Vol 10: Failure Analysis and Prevention*, 8th ed., American Society for Metals, Metals Park, OH, 1975, p 93.
3. Ibid., p 236.
4. I. Ingel and H. Klingele, in *An Atlas of Metal Damage*, Wolfe Science Books, 1981, p 127.
5. *Metals Handbook, Vol 9: Fractography and Atlas of Fractographs*, 8th ed., American Society for Metals, Metals Park, OH, 1975, p 72.
6. F.R. Hutchings, in *Failure Analysis: The British Engine Reports*, F.R. Hutchings and P.M. Unterweiser, Ed., American Society for Metals, Metals Park, OH, 1981, p 46.
7. V.K. Gouda, M.M. Nassrallah, S.M. Sayed and N.H. Gerges, "Failure of Boiler Tubes in Power Plants," *Br. Corros. J.*, Vol 16, 1981, p 25-31.

Caustic Gouging and Caustic-Induced Stress-Corrosion Cracking of Superheater Tube U-Bends

Allen W. Hearn, CF Industries, Inc., Plant City, Florida
Harold Rawlinson, The Hartford Steam Boiler Inspection and Insurance Company, Maurepas, Louisiana
G. Mark Tanner, Radian Corporation, Austin, Texas

Original carbon steel and subsequent replacement austenitic stainless steel superheater tube U-bend failures occurred in a waste heat boiler. The carbon steel tubes had experienced metal wastage in the form of caustic corrosion/gouging, while the stainless steel tubes failed by caustic-induced stress-corrosion cracking. Sodium was detected by EDS in the internal deposits and the base of a gouge in a carbon steel tube and in the internal deposits of the stainless steel tube. The sodium probably formed sodium hydroxide with carryover moisture and caused the gouging, which was further aggravated by the presence of silicon and sulfur (silicates and sulfates). It was recommended that the tubes be replaced with Inconel 600 or 601, as a practical option until the carryover problem could be solved.

Key Words	Carbon steels, Corrosion Environmental effects Sodium hydroxide, Environment	Austenitic stainless steels, Corrosion Alkalies, Environment Boiler tubes, Corrosion	Stress-corrosion cracking, Environmental effects
Alloys	Carbon steels—SA106 Grade B	Austenitic stainless steels—316	Inconel 600, 601

Background

Applications

The superheater tube U-bends were part of a waste heat boiler system that produced steam at 2.07 MPa (300 psi) and 274 °C (525 °F). The steam was used in various processes in a phosphate fertilizer chemical plant.

Circumstances leading up to failure

The new superheater had been installed during a recent outage. In less than six weeks, several superheater tubes started leaking. All of the leaks occurred in the first bend after steam entered the inlet header.

Pertinent specifications

The original superheater tube U-bends were specified to be SA-106 Grade B pipe, while the four replacement tubes were specified to be 316 stainless steel.

Performance of other parts in same or similar service

The waste heat boiler system had been retrofitted with new larger superheaters. The prior superheater U-bends had also cracked, but the cracking had reportedly been caused by the bending operation and propagated in service. The newly installed tubes started leaking within six weeks of installation. However, these failures involved a localized wasting of the inside surface and grooving that started at this wasting and continued around the bend in the direction of flow.

Selection of specimens

Initially, three SA-106 Grade B tube samples were received for analysis. The U-bends were labelled T2, T3, and T4. Subsequently, failed replacement tubes, made of 316 stainless, were received for analysis. These tube samples were labelled 1T and 4T.

Visual Examination of General Physical Features

U-bend T2 was a replacement bend made out of two 90° elbows. U-bends T3 and T4 were original to the boiler. U-bend T3 contained a patch weld over a past failure. Both tubes 1T and 4T had been longitudinally cut before arrival and tube 4T contained a hole.

Testing Procedure and Results

Surface examination

The U-bend samples are shown as received for analysis in Fig. 1. Gouging was observed at the bottom of the superheater bend sample T2, but not on the top side. Figure 2 shows the U-bend sample T2 after it had been glass-bead-blasted to remove internal deposits. The gouging is apparent. Figure 3 shows a close-up of the hole in tube 4T.

Metallography

Transverse ring sections were cut from both ends of tubes T2 and T4. Each section was ground, polished, etched, and examined on a metallurgical microscope for evaluation of the microstructure and inner and outer diameter surface conditions. The typical microstructures of U-bend samples T2 and T4 are shown in Fig. 4 and 5, respectively. The microstructure consists of pearlite in a ferrite matrix, typical of carbon steel tubing. The tubes showed no evidence of excessive thermal exposure from operation. The internal surface of the U-bend sample T2 contained a region of gouging. Figure 6 shows a cross-section through the gouging.

The longitudinal section containing the hole in tube 4T was also ground, polished, etched, and examined. The typical microstructure consisted of austenite grains, demonstrating annealing twins and flow lines, along with carbides and small stringer inclusions. Branching transgranular cracks were observed along the internal surface of the tube with the cracks extending to the external surface. Figure 7 shows the cracking, which is characteristic of stress-corrosion cracking (SCC).

Chemical analysis/identification

Internal deposits were collected from U-bend

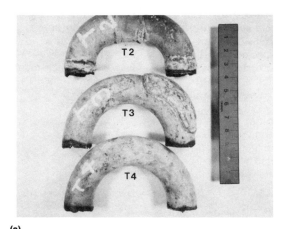

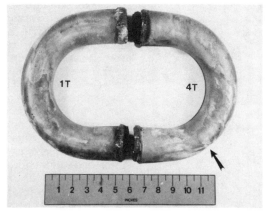

(a)   (b)

Fig. 1 U-bend samples T2, T3, and T4 are shown as received for analysis in (a), while samples 1T and 4T are shown in (b). Arrow in (b) points to location of a hole in U-bend 4T

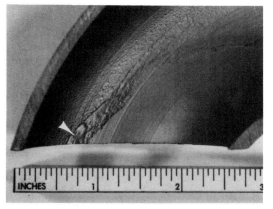

Fig. 2 The gouging (arrow) observed after the heavy internal deposit was removed by bead blasting from U-bend T2

Fig. 3 Close-up of the hole in tube 4T. Approximately 0.95×

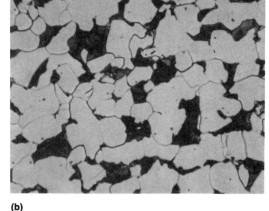

(a)   (b)

Fig. 4 Microstructure of ring sections removed from each 90° elbow that formed the U-bend sample in T2. The microstructure consists of pearlite in a ferrite matrix. The difference in grain size and pearlite distribution is a result of manufacturing. Nital etch. (a) 608×. (b) 608×

T2 and analyzed by energy dispersive X-ray spectroscopy (EDS). Compounds in the deposits were identified using powder X-ray diffraction (XRD) techniques. The EDS results of the inner diameter deposit analysis are shown in Fig. 8. The XRD re-sults showed that the deposits consisted primarily of iron oxide/magnetite (Fe_3O_4). In addition to the iron found in magnetite, EDS detected minor to trace amounts of sodium, silicon, sulfur, chlorine, and potassium. Magnetite is the normal protective oxide layer that forms on boiler tube surfaces. The remaining elements are not commonly found in superheater tubes with the possible exception of silicon. The presence of these deposits indicate carryover from the steam drum has occurred.

Figure 9 shows the results of an *in situ* EDS analysis of the base of the gouge shown in Fig. 6. The results show the presence of sodium, aluminum, silicon, phosphorus, sulfur, potassium, calcium, titanium, and iron. Sodium in the form of sodium hydroxide caused the gouging, and the attack was probably aggravated by the presence of silicon and sulfur (silicates and sulfur).

To determine the corrodent causing the cracking, internal deposits of tube 4T were analyzed using EDS. Figure 10 shows the EDS results. The deposit consisted of sodium, aluminum, silicon, phosphorus, sulfur, potassium, calcium, chromium, and iron, which indicated that carryover from the steam drum has occurred.

Mechanical properties

Hardness measurements averaged 80.5 and 95 HRB for U-bends T2 and T4, respectively. These hardnesses are higher than normal for carbon steel and are probably the result of cold work from

tube bending during manufacturing. Also, the difference in hardness between the samples is probably the result of different manufacturing procedures for the two 90° elbows from which U-bend T2 was made, as compared to the more standard method for U-bend T4.

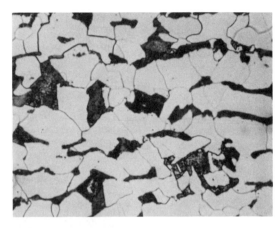

Fig. 5 Microstructure of U-bend sample T4. The microstructure consists of pearlite in a ferrite matrix. Nital etch. 608×

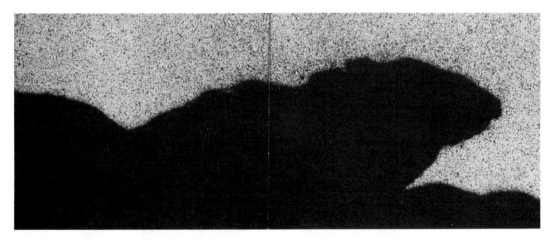

Fig. 6 Cross-section of the gouge observed in U-bend sample T2. Nital etch. 40.5×

Discussion

The carbon steel superheater tubes had experienced metal wastage in the form of caustic corrosion/gouging. For caustic gouging to occur, a caustic agent, usually in the form of sodium hydroxide (NaOH), must be present. Sodium was detected in the internal deposits as well as in the base of the gouge in U-bend sample T2, which indicates that caustic was present. Silicon and sulfur were also observed in the internal deposits and at the base of the

one gouge, and most likely aggravated the caustic attack.

Stainless steel tube 4T failed by caustic-induced SCC. Tube 1T also experienced SCC. Sodium was observed in the internal deposits of Tube 4T. Sodium, probably in the form of sodium hydroxide, caused the cracking, as it is known to stress crack austenitic stainless steels.

Conclusion and Recommendations

Most probable cause

The cause of the U-bend failures was caustic gouging for the carbon steel U-bends and caustic-induced stress-corrosion cracking for the stainless steel U-

bends. The root cause of the failures is carryover.

Remedial action

To alleviate caustic gouging and stress cracking, the carryover problem should be identified and eliminated. Carryover concentrates dissolved and/or suspended solids in the superheater tubing and possibly the turbine. If the carryover problem is eliminated and the superheater tubing is kept dry, the gouging that had previously occurred will stop. Carryover can be prevented by: keeping solids concentration in the boiler water at low levels; avoiding high water levels in the drum; avoiding overloading the boiler; avoiding sudden load changes in the boiler; avoiding contaminated condensate; and using antifoam agents (Ref 1). Until the carryover has been identified and eliminated, SCC will continue in the stainless steel tubes. Replacing the U-bends with another alloy, such as Inconel 600 or 601, may be a practical option until the carryover problem is solved.

The four stainless U-bends were replaced with Inconel 600 U-bends. At the same time, the boiler water treatment was changed to reduce the amount of carryover. Later, an in-line steam separator was installed to further purify the saturated steam going to the superheater. The superheaters are still in operation without a failure since these changes were made in 1989.

Acknowledgment

The assistance of the late Doug E. McDuffie, Jr. in developing some of the background information is gratefully acknowledged.

Reference

1. S.M. Elonka and A.L. Kohan, *Boiler Plant Questions and Answers*, McGraw-Hill Book Company, New York, NY, 1984, p 517.

Fig. 7 Typical cracking observed in U-bend 4T. The cracking is emanating from the internal surface. Electrolytic oxalic acid etch. 81×

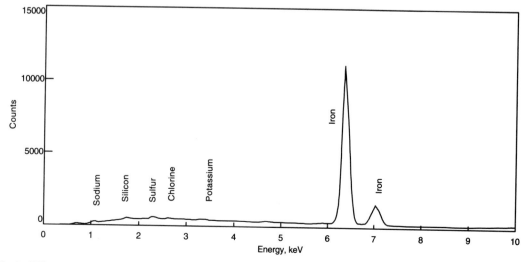

Fig. 8 EDS elemental analysis results for the internal deposits from the superheater U-bend sample T2

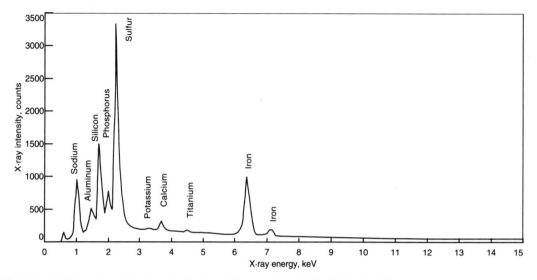

Fig. 9 *In situ* EDS spectrum of the deposits at the base of the gouge shown in Fig. 6 before bead blasting

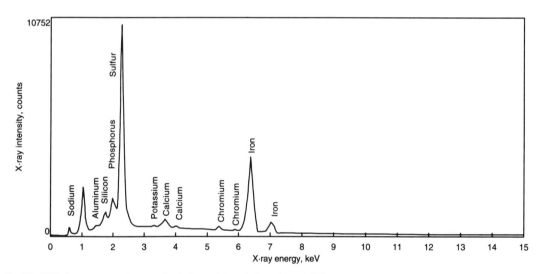

Fig. 10 EDS elemental analysis results for the internal deposits from the superheater U-bend sample 4T

Caustic-Induced Stress-Corrosion Cracking of a Flue Gas Expansion Joint

Sarah Jane Hahn, Radian Corporation, Austin, Texas

A type 430Ti stainless steel flue gas expansion joint cracked because of caustic-induced stress-corrosion cracking. Energy-dispersive X-ray spectroscopy analysis of the fracture surface deposits revealed the presence of sodium and potassium—caustics in hydroxide form. Primary fracture surfaces were all similar in appearance, and a primary crack origin could not be identified. A secondary crack brought to fracture in the laboratory showed brittle, cleavage features rather than classic, tensile overload features. This suggested that the material was embrittled.

Key Words

Ferritic stainless steels, Corrosion	Expansion joints, Corrosion	Stress-corrosion cracking
Alkalies	Flue gases, Environment	Brittle fracture
Caustic embrittlement		

Alloys Ferritic stainless steels—430Ti

Visual Examination of General Physical Features

The flue gas expansion joint sample was examined using a low-power binocular microscope, and it was photographed in the as-received condition. Figure 1 shows the expansion joint and a secondary crack observed in the sample as received for analysis. The primary fracture surfaces (the edges of the sample) were all similar in appearance, and a primary crack origin could not be identified. Figure 2 shows a secondary crack after it was fractured open in the laboratory. This crack was fractured open for comparison with the primary fracture surfaces. (The secondary crack should not have been exposed to possible handling damage and/or contamination after the primary failure.)

(a)

(b)

Fig. 1 (a) Expansion joint section as received for analysis. (b) Secondary crack (arrows) before it was fractured open in the laboratory

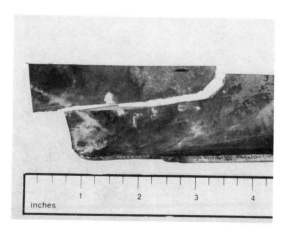

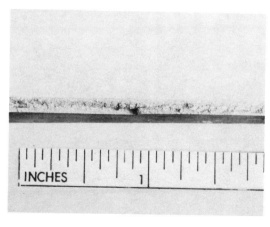

Fig. 2 Secondary crack after it was fractured open in the laboratory

Testing Procedure and Results

Surface examination

Scanning Electron Microscopy Fractography. Both primary and secondary cracks were examined using a scanning electron microscope (SEM) to document fracture features. The fracture surface deposits were analyzed by energy dispersive X-ray spectroscopy (EDS). Figure 3 shows typical SEM photographs of the crack surfaces. The primary, secondary, and lab fracture surfaces exhibited the same fracture features.

Metallography

Microstructural Analysis. Sections were cut from the expansion joint and prepared for metallographic examination by grinding, polishing, and etching. The prepared sections were examined using a metallurgical microscope to assess microstructure and internal and external surface conditions. Figure 4 shows the microstructure of the

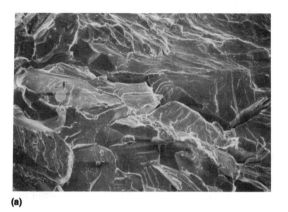

(a)

(b)

Fig. 3 SEM photographs typical of the fracture surfaces. (a) Crack fracture surface. Nital etchant, 270.6×. (b) Lab fracture surface. Nital etchant, 630×.

(a)

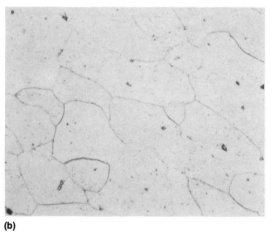

(b)

Fig. 4 Typical microstructure of the base metal. (a) Fry's reagent, 77ö. (b) Fry's reagent, 616×

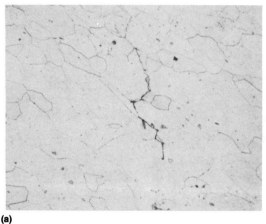

(a)

(b)

Fig. 5 (a) Mixed-mode cracking in cross section. Fry's reagent, 308×. (b) Fracture surface in cross section. Fry's reagent, 308×

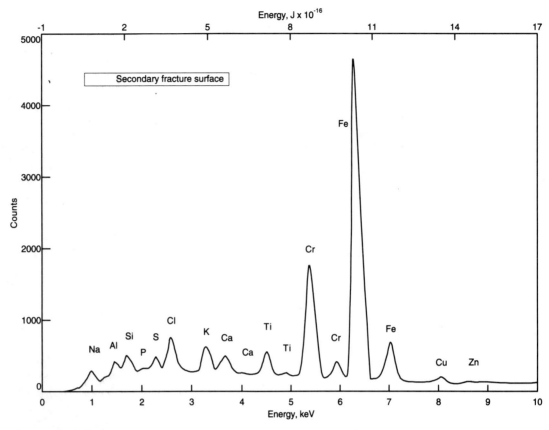

Fig. 6 Typical EDS spectrum for the deposits on the expansion joint fracture surfaces. LT = 100 ·s.

base metal, which was typical of an annealed ferritic stainless steel. Figure 5 shows the mixed-mode (transgranular and intergranular) cracking and the secondary fracture surface in cross section.

Chemical analysis/identification
Material and Weld. The base metal was analyzed by quantitative chemical analysis techniques. The chemical composition of the base metal met the requirements for type 430Ti stainless steel.

Corrosion or Wear Deposits. The EDS results for the primary and secondary fracture surfaces were the same. The results for the secondary fracture surface deposits are shown in Fig. 6. This analysis revealed sodium, aluminum, potassium, calcium, sulfur, and chlorine in addition to base metal elements.

Discussion	The brittle, cleavage features of the lab fracture (rather than classic, tensile overload features) suggest that the material was embrittled. Caustic	embrittlement is likely because the crack surface deposits contained sodium and potassium, which are caustics in hydroxide form.
Conclusions and Recommendations	The mixed mode cracking, with some branching, is characteristic of caustic-induced stress-corrosion cracking in stainless steels. The cracking in the ex-	pansion joint is therefore likely due to stress corrosion, which was caused by the caustic embrittlement.

Corrosion Failure of a Chemical Process Piping Cross-Tee Assembly

Peter F. Ellis II, Radian Corporation, Austin, Texas

A carbon steel piping cross-tee assembly which conveyed hydrogen sulfide (H₂S) process gas at 150 to 275 °C (300 to 585 °F) with a maximum allowable operating pressure of 3 MPa (450 psig) ruptured at the toe of one of the welds at the cross after several years of service. The failure was initially thought to be the result of thermal fatigue, and the internal surfaces exhibited the "elephant hide" pattern characteristic of thermal fatigue. However, metallographic failure analysis found that this pattern was the result of corrosion rather than thermal fatigue. Corrosion caused failure at this location because the weld was abnormally thin as fabricated. Thus, failure resulted from inadequate deposition of weld metal and subsequent wall thinning from internal corrosion. It was recommended that the cross-tee be replaced with a like component, with more careful attention to weld quality.

Key Words			
	Carbon steels, Corrosion Corrosion	Welded joints, Corrosion Hydrogen sulfide, Environment	Weld metal

Alloys			
	Carbon steels—A53 Grade B	Carbon steels—A105 Grade WPB	Carbon steels—A234 Grade WPB

Background

Application

This cross-tee was located in piping which conveyed concentrated H₂S gas at 150 to 275 °C (300 to 525 °F) with a maximum allowable operating pressure (MAOP) of 3 MPa (450 psig). The H₂S gas was not dry. Three legs of the cross-tee were connected by elbows to long, straight vertical runs of pipe. The fourth leg was a blind stub. The cross-tee with its elbows is shown in Fig. 1. The cross-tee was subject to both frequent thermal and pressure cycling as a result of system operation.

Circumstances leading up to failure

The component had operated in the manner described above for several years when the cross-tee ruptured at the toe of one of the welds. There were no known abnormal process occurrences immediately prior to the rupture. No other related components in the system had suffered failure.

Pertinent specifications

The cross-tee was created by welding two 2-inch pipes* 180° apart to a run of 3-in. pipe. One 2-in. leg was welded directly to the 3-in. pipe, while the other was joined through a reduction socket fitting. The design of the cross-tee specified ASTM A-53 Grade B Schedule 80 steel pipe and ASTM A-105 and ASTM A-234 Grade WPB forged steel fittings.

Performance of other parts in same or similar service

Other parts conforming to the above materials specifications have performed in this process

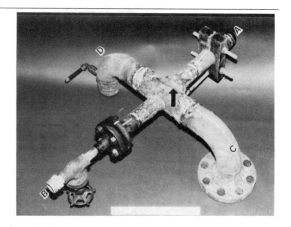

Fig. 1 The cross-tee assembly as received, showing the underside. The letters A through D identify the legs of the cross assembly. Arrow indicates the location of the rupture.

stream without a significant history of problems.

Selection of specimens

The entire failed cross-tee was submitted for examination.

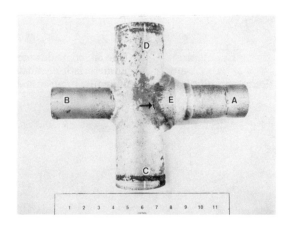

Fig. 2 Construction of the cross-tee, showing the 2-in. Schedule 80 pipe (A), joined to the 3-in. Schedule 80 pipe (C-D) by a reduction socket (E). The remaining arm of the cross (B) was a flanged nipple welded to the 3-in. pipe. The rupture is at the toe of the weld between the 3-in. pipe and the reduction socket (arrow).

Visual Examination of General Physical Features

Figure 1 shows the failed cross-tee assembly. The C-D run of the cross-tee was 3-in. pipe, while the A and B legs were 2-in. pipe. Leg B was joined directly to the C-D run by a weld, while Leg A was joined to the C-D run through a forged reduction socket (see Fig. 2).

The arrow in Fig. 1 shows the location of the rupture at the toe of the weld joining the reduction

---

* The designation 2–in. and 3–in. are nominal pipe sizes and do not relate precisely to either interior or exterior dimensions; therefore, no metric equivalents are offered.

socket to the 3-in. C-D run. The weld bead showed quite noticeable external lack of weld bead buildup at this location. The weld was visibly starved of filler metal at this location.

The internal surfaces of the piping were found to be coated with columnar crystalline deposits up to approximately 6 mm (0.25 in.) thick, deposited in concentric layers or rings.

Testing Procedure and Results

Surface examination

The cross-tee was sectioned through the plane of the legs. When the internal deposits were removed, the surface was found to be covered with a reticulated pattern suggestive of thermal fatigue (Fig. 3).

Close visual examination of the rupture showed that the lips of the rupture were extremely thin with no obvious fracture faces.

A total of nine wall thickness measurements were made on the 2-in. pipe (remote from the cross-tee welds, using a ball micrometer, while 18 comparable measurements were made on the 3-in. pipe. The approximate locations of these measurements are shown in Fig. 4. The observed values were compared with both the nominal wall thicknesses for the pipe specifications, and with the ASTM A-53 minimum allowable wall thicknesses. The average wall thickness of the 2-in. pipe was 3.4 ± 0.178 mm (0.135 ± 0.007 in.) at the 95 percent confidence level, while the nominal wall thickness for Schedule 80 50-mm (2-in.) pipe was 5.54 mm (0.218 in.), and the minimum allowable thickness under ASTM A-53 was 4.85 mm (0.191 in.). The average remaining wall thickness of the 3-in. pipe was 4.95 ± 0.305 mm (0.195 ± 0.012 in.), while the nominal wall thickness was 7.62 mm (0.300 in.) and the minimum allowable thickness under ASTM A-53 was 6.655 mm (0.262 in.).

Thus, the average metal loss for the 2-in. pipe was 2.108 ± 0.178 mm (0.083 ± 0.007 in.) based on

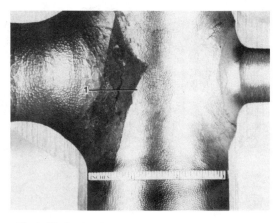

Fig. 3 The interior of the cross-tee showing the location of Weld Section 1 through the rupture at the toe of the weld joining the 3-in. pipe and the reduction socket. Note the reticulated pattern on the interior surfaces of the cross-tee, suggestive of thermal fatigue.

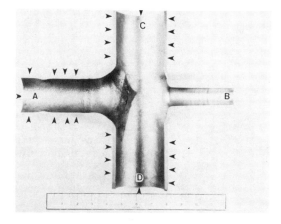

Fig. 4 Plane section of the cross-tee assembly showing the approximate locations of the wall thickness measurements

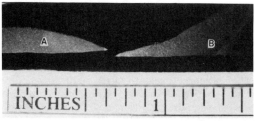

(a)

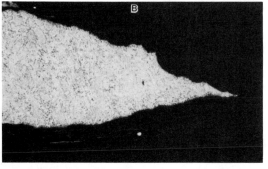

(b)

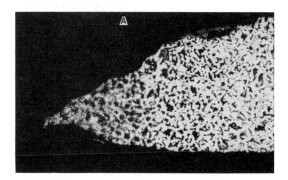

(c)

Fig. 5 The weld between the 3-in. pipe and the reduction socket at the rupture. The lower images are reversed relative to the upper image. The OD surfaces face upward in each image.

the nominal wall thickness, or 1.42 ± 0.178 mm (0.056 ± 0.007 in.) based on the minimum allowable wall thickness. For the 3-in. pipe, the average metal loss was 2.667 ± 0.305 mm (0.105 ± 0.012 in.) based on the nominal wall thickness or 1.70 ± 0.305 mm (0.067 ± 0.012 in.) based on the minimum allowable wall thickness. A composite analysis of all of the thickness data yielded an average metal loss of 2.49 ± 0.203 mm (0.098 ± 0.008 in.) based on nominal wall thicknesses or 1.626 ± 0.203 mm (0.064 ± 0.008 in.) based on minimum allowable wall thicknesses.

Metallography

Polished and etched cross sections remote from the welds showed that the pipe and forged fitting microstructures consisted of pearlite in a ferrite matrix, typical of low to medium carbon steels.

A metallographic section was prepared through the weld between the 3-in. pipe and the reduction fitting at location 1 in Fig. 3, intersecting the rupture. Figure 5 shows the microstructure of Weld Section 1. Lip B consists entirely of weld metal. Lip A consists of altered pearlite in a ferrite matrix (heat-affected zone), transitioning into the normal pearlite in ferrite matrix of the carbon steel. There was no discernible weld metal at the tip of Lip A.

From Fig. 5, it is apparent that there was no distinct crack or fracture at the toe of the weld, but rather that internal thinning had simply progressed until the toe of the weld was penetrated. Because of the extreme thinness of the lips of the rupture, fractographic examination was not possible.

Figure 6 shows a comparable cross-section through the same weld approximately 180° away from the center of the rupture. This figure demonstrates clearly that there was no preferential corrosion of the weld metal or heat affected zones.

Figure 7, a photomicrograph of the microstructure at the root of one of the reticular grooves in the 3-in. pipe, shows clearly that these grooves resulted from corrosion rather than thermal fatigue. There was no evidence of any wear or erosion morphology.

Chemical analysis/identification

Chemical analysis showed that the 2-in. and 3-in. pipe satisfied the composition requirements of ASTM A-53 Grade B pipe, while the reduction socket satisfied the requirements of both ASTM A-105 and ASTM A-243 Grade WPB forged fittings.

Deposit material from the interior of the pipe was analyzed by energy dispersive X-ray spectroscopy (EDS) and X-ray diffraction (XRD).

The deposits which coated the entire interior surface of the piping consisted almost entirely of iron and sulfur in the form of two iron sulfides, pyrite (FeS_2), and pyrohotite ($Fe_{(1-x)}S$). Both are common corrosion products of steel in a hydrogen sulfide environment. Elemental sulfur was not detected.

Mechanical properties

The Rockwell B hardness of the 3-in. pipe was 73 ± 2, while the hardness of the reduction fitting was Rockwell B 82 ± 2. The hardness of the weld metal remote from the rupture was Rockwell B 85

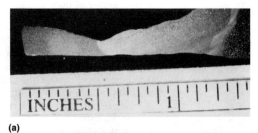

(a)

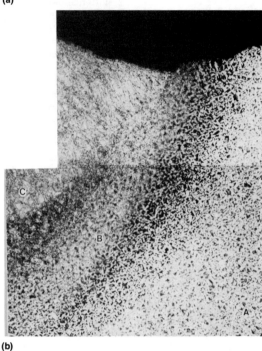

(b)

Fig. 6 The weld between the 3-in. pipe and the reduction socket 180° away from the rupture. The lower image is reversed relative to the upper image. The OD surface faces upward in each image.

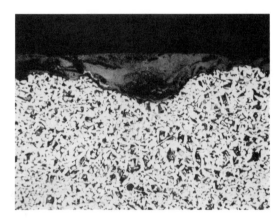

Fig. 7 Microstructure at the root of one of the reticulated grooves on the interior of the pipe.

± 2.

Based on the ASTM A-370 approximate correlation between Rockwell B hardness and tensile strength, the tensile strength of the 3-in. pipe was on the order of 434 to 455 MPa (63 to 66 ksi), while the tensile strength of the reduction fitting was on the order of 503 to 538 MPa (73 to 78 ksi). Within the uncertainty of the conversion, these values agree with the ASTM A-53 requirement of a minimum tensile strength of 414 MPa (60 ksi) for Grade B steel. These hardnesses are also far below the NACE Rockwell C 22 maximum hardness limi-

tation for steel pipe in H_2S service. Together, these data show that the components satisfied the mechanical property requirements of their respective specifications and were suitable under NACE specifications for H_2S service.

Discussion

The exterior surfaces of the cross-tee showed no evidence of significant corrosion beyond slight atmospheric rusting. Visual examination of the exterior surfaces of the welds showed clearly that the weld at the location of the rupture was much thinner than elsewhere around its circumference. The appearance was consistent with failure to complete one or more weld passes at this location on the toe of the weld.

Both the visual inspection and metallographic cross sections showed clearly that appreciable corrosion had occurred over the entire interior surface of the cross-tee assembly and as much of the associated piping as was available for examination. This corrosion was essentially uniform except for the very shallow reticulated surface pattern. There was no evidence of preferential corrosion of any of the welds or heat affected zones.

Based upon numerous measurements of the thicknesses of the 2- and 3-in. piping remote from the welds, the interior surfaces of the cross-tee and associated piping had lost a minimum of approximately 1.588 mm (0.0625 in.) due to corrosion, and the actual amount was more likely in excess of 3.175 mm (0.125 in.).

The wastage from the interior surfaces of the cross assembly and piping resulted from corrosion by the H_2S gas rather than erosion. This conclusion is supported by two observations. First, the entire interior surface was coated with a relatively thick, multilayered, and somewhat friable columnar iron sulfide corrosion product deposit. If the metal loss were the result of erosive wear, the metal surfaces at the wasted locations would have been devoid of deposits. Second, the micro-mor-

phology of the interior surfaces of the cross assembly is completely inconsistent with erosion.

Once the cross-tee was placed in service, corrosion thinned the walls of the assembly until the remaining thickness at the toe of the weld between the 3-in. pipe, and the reduction fitting could not contain the pressurized gas, resulting in a rupture. While cyclic pressure or thermal stresses may have had some secondary contributory role in the rupture mechanism, there was no evidence of a distinct cracking mechanism other than stress overload as the remaining material at the toe of the weld became too thin to contain the pressurized gas.

Reticulated corrosion patterns similar to those observed in this investigation have been observed beneath columnar deposits under service conditions in which thermal fatigue could not have occurred because there was no thermal cycling. It is probable that such reticulated corrosion patterns are produced when a somewhat protective deposit fractures by thermal contraction or structural shrinkage. The resulting fissures then allow increased mass transport of corrosive species to the metal surface at the tips of the fissures. Transport of reaction products from these sites is also facilitated. Under many circumstances, the corrosion rate of steel is mass-transport (or diffusion) controlled. The slightly enhanced mass transport results in slightly more rapid metal loss at the tips of the fissures in the deposit, producing grooves at these locations. The physical structure of the deposits in this equipment included a reticulated pattern of intercolumnar fissures, leading to the reticulated corrosion pattern on the metal surface.

Conclusion and Recommendations

Most probable cause

The proximate cause of the rupture was internal corrosion thinning of the pipe and cross-tee walls until the remaining thickness at the toe of the weld between the 3-in. pipe and the reduction fitting could not sustain the internal pressure.

The root cause of the rupture was the very thinly fabricated weld bead at the toe of the weld between the 3-in. pipe and the reduction fitting. If the weld had been fabricated with the same original thickness as at the remaining points around its circumference, no failure would have occurred because the metal thickness would have provided a

more than adequate corrosion allowance.

Remedial action

It was recommended that the cross-tee be replaced with a like component, with more careful attention to weld quality.

How failure could have been prevented

The failure could have been prevented by a careful visual inspection of the weld and possible ultrasonic thickness measurement of the weld thickness at the time of fabrication.

Corrosion Failure of Stainless Steel Thermowells

D.K. Bhattacharya and Baldev Raj, Metallurgy and Materials Programme, Indira Gandhi Centre for Atomic Research, Kalpakkam, India
E.C. Lopez, Zenford Ziegler Pvt. Ltd., Melbourne, Australia
V. Seetharaman, Universal Energy Systems, Inc., Dayton, Ohio

Pressure testing of a batch of AISI type 316L stainless steel thermowells intended for use in a nuclear power plant resulted in the identification of leakage at the tips in 20% of the parts. Radiography at the tip region of representative thermowells showed linear indications along the axes. SEM examination revealed the presence of longitudinally oriented nonmetallic inclusions that were partly retained and partly dislodged. Electron-dispersive X-ray analysis indicated that the inclusions were composed of CaO. Based on the overall chemistry of the inclusion sites, the source of the CaO was determined to be slag entrapment during the steelmaking process. It was recommended that the thermowell blanks be ultrasonically tested prior to machining and that the design be modified to make internal pressurization possible.

Key Words			
Austenitic stainless steels	Leakage		Nuclear reactor components
Pinhole	Nonmetallic inclusions		Thermocouples

Alloys Austenitic stainless steel—316L

Background

A batch of AISI type 316L stainless steel thermowells (Fig. 1) was procured for a nuclear power plant to house thermocouples at 612 locations. Pressure testing revealed that about 20% of the total number of thermowells were leaking at the tips.

Applications

The thermowells are used for the insertion of thermocouples at 306 inlet and 306 exit points of heavy-water coolant that enters the core in a nuclear power plant. Using the criterion of uniform corrosion, type 316L stainless steel was chosen for the application.

Circumstances leading to failure

At the power plant site, the thermowells were internally pressurized using a flexible hose and a soap solution. At a pressure level of only 0.69 MPa (0.10 ksi), about 20% of the thermowells were found to be leaking at the tips.

Pertinent specifications

It was specified that the thermowells were to be fabricated from 20 mm (0.8 in.) diam AISI 316L stainless steel bar stock tested ultrasonically. All of the thermowells were to be able to withstand an internal pressure of 24.5 MPa (3.55 ksi) at 30 °C (85 °F); 10% were to withstand an external pressure of the same force at the same temperature. Each thermowell was to be liquid penetrant tested.

Performance of other parts in same or similar service

The 80% of the thermowells that did not leak when pressure tested have performed well in service.

Specimen selection

Specimens for scanning electron microscopy (SEM) were obtained from several of the failed thermowells by transverse and longitudinal cutting at the tip regions.

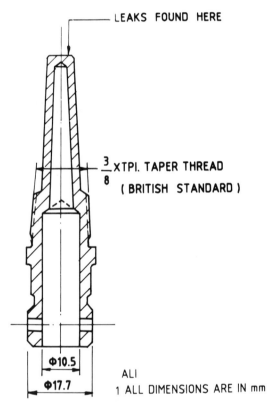

Fig. 1 Schematic of thermowell

Visual Examination of General Physical Features

No abnormalities were visible when the thermowells were examined using the naked eye.

Nondestructive evaluation

Radiography. Sample radiography at the tip regions of a few of the failed thermowells revealed linear indications along the axes.

Surface examination

Scanning Electron Microscopy/Fractography. The outside surface and the inside longitudinal surfaces near the tip of a failed thermowell were examined by SEM. Figure 2 shows through-holes present at the tip. Figure 3 shows a longitudinally oriented inclusion, partly retained and partly dislodged. Most of the areas of the internal surfaces had the appearance of a cracked wooden surface because of inclusion removal (Fig. 4).

Figure 5 shows a typical electron-dispersive X-ray analysis (EDXA) pattern from one of the inclusion sites. The presence of a large peak caused by calcium indicates that the inclusions were composed primarily of CaO. The other peaks, which are much smaller than the calcium peak, correspond to chromium and iron—from either the inclusions or the matrix. Aluminum and silicon were not present. Based on the EDXA pattern, it was concluded that the inclusions were primarily CaO, possibly containing small amounts of chromium and iron oxides. This conclusion was corroborated by the existence of pinholes at the tips of the thermowells through a thickness of about 2.5 mm (0.1 in.). It is doubtful whether the pinholes would have formed had the inclusions been a different type.

If the blanks from which the thermowells were fabricated had been ultrasonically tested, the unusual CaO inclusions would have been detected. Whether the thermowells had been pressure tested at the manufacturing site before they were dispatched to the user was not documented. It is doubtful that they were, because the thermowell design does not facilitate internal or external pressure testing to the specified pressure level.

The investigation led to the belief that during various stages of fabrication and inspection, the CaO-based inclusions came into contact with water and thereby dissolved. Although this scenario seems both simple and surprising, it is the

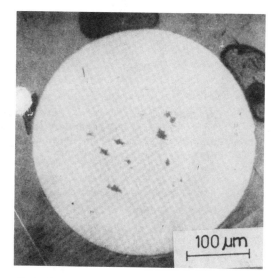

Fig. 2 SEM micrograph of thermowell tip, showing leakage paths

Fig. 3 SEM micrograph of longitudinal inclusions in the inside surface of a failed thermowell

Fig. 4 SEM micrograph of rough surface inside a failed thermowell, caused by removal of longitudinal inclusions

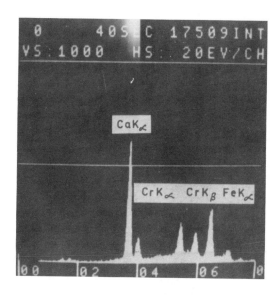

Fig. 5 EDXA pattern from type of inclusion shown in Fig. 3

only one that seems plausible. Because the occurrence of CaO-type inclusions is unusual, further discussion is warranted.

There are three main sources of calcia-bearing inclusions (Ref 1): furnace and ladle refractories, furnace and ladle slags, and deoxidants containing calcium. Refractories normally contain a large MgO content; the absence of magnesium in the inclusions indicates that the refractories did not play a role. Deoxidants containing calcium normally contain significant amounts of silicon as well; the absence of silicon in the inclusions rules out this scenario. It is thus probable that the inclusions resulted from entrapped slag of very high basicity.

Conclusion and Recommendations		
	Most probable cause	**Remedial action**
	Dislodging of the CaO-type inclusions at the tips of the thermowells probably led to the formation of pinholes.	The thermowells should be machined from ultrasonically tested blanks. In addition, the thermowell design should be modified to enable high-pressure testing (both external and internal). Although this action would not prevent failure, it would allow the detection of abnormalities.

Reference

1. R. Kiessling and N. Lange, *Nonmetallic Inclusions in Steels, Part II*, The Metals Society, London, 1978, p 82-84

Cracking in a Reducing Pipe From a Pressurized Water Reactor

Carl J. Czajkowski, Department of Nuclear Energy, Brookhaven National Laboratory, Upton, New York

Three ASME SA106 grade B carbon steel feedwater piping reducers from a pressurized water reactor showed indications of flaws near welds during ultrasonic testing. Further examination and testing indicated that the cracks resulted from a low-cycle corrosion fatigue phenomenon.

Key Words	Carbon steels	Weldments	Nuclear reactor components
	Corrosion fatigue	Pipe fittings	Pressurized water reactors
Alloys	Carbon steel—A106 grade B		

Background

During a refueling outage at a pressurized water reactor (PWR), feedwater reducers to all four steam generators were replaced. Three of the four piping reducers showed indications of flaws when examined by ultrasonic testing.

Pertinent specifications

The reducers were manufactured from ASME SA-106 grade B carbon steel.

Performance of other parts in same or similar service

Cracking of a similar nature had been previously evaluated on the feedwater nozzles of a number of PWRs.

Visual Examination of General Physical Features

Each 406 × 356 mm (16 × 14 in.) reducer was welded to a 406 mm (16 in.) steam generator feedwater nozzle on one end and a 356 mm (14 in.) schedule 60 pipe on the smaller end. One reducer was cut in half (Fig. 1) with a power saw. To prevent contamination of any cracks, cutting oil was not used. A crack approximately 125 mm (5 in.) in length was observed in the weld preparation area of the large end of the reducer. This was the only crack noted.

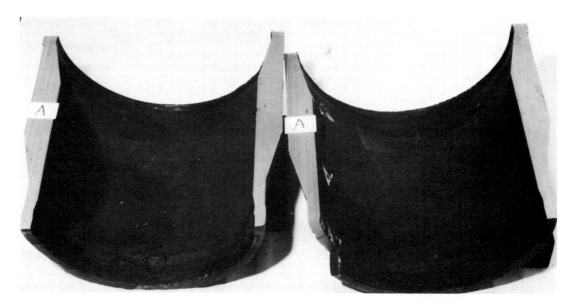

Fig. 1 Reducer section after cutting. A crack is visible in the right-hand section (upper left). There is no evidence of erosion/corrosion damage.

Testing Procedure and Results

Nondestructive evaluation

Liquid Penetrant Testing. Discussions with utility personnel suggested that the small end of the reducer might also be cracked. Because visual examination had disclosed no cracks on this side, a dye penetrant examination was performed. No cracks were revealed on the small end.

Surface examination

Visual. Two sections from the reducer were cut and opened for photographic and visual examination. First, a V-notch was cut below the crack; the cut section was then cooled to liquid nitrogen temperature and pulled apart in such a way that the fracture surface was undamaged.

Multiple initiation sites were revealed. The cracks were typically half-moon shaped and covered with a tight, black adherent oxide film. The fracture surfaces were electrolytically cleaned and then visually examined. Two fracture surfaces (Fig. 2 and 3) exhibited thumbnail-like cracks; beach marks were evident on all the fractures. The

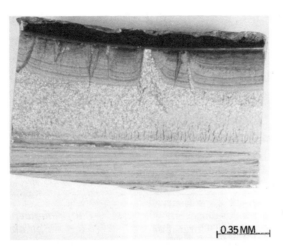

Fig. 2 Area of fracture showing beach marks on the surface

Fig. 3 Second fracture surface

Fig. 4 Low-magnification composite SEM micrograph of the first reducer crack. Beach marks and surface pitting are evident.

Fig. 5 Composite SEM fractograph of second fracture, showing multiple initiation sites and beach marks

deepest crack observed (Fig. 3) had a depth of 6 mm (0.235 in.). The second specimen had a crack with a depth of 3.8 mm (0.150 in.).

Scanning Electron Microscopy/Fractography. Two specimens were cut for scanning electron microscopy/energy-dispersive spectroscopy (SEM/EDS) evaluation prior to electrolytic cleaning and photographing. Both fracture surfaces exhibited multiple initiation sites, transgranular cracks, and beach marks (Fig. 4 and 5).

Figure 6 shows the transgranular nature of the cracking more clearly. There was evidence of pit-

ting on the inside surface of the reducer on the fracture surface (Fig. 7).

Various areas were examined using EDS. In addition to iron, trace amounts of copper, sulfur, silicon, calcium, and nickel were seen, as well as trace amounts of chlorine on two scans. No specific corrodant species were identified.

Metallography

Microstructural Analysis. Metallurgical cross sections of the reducer were cut, mounted, and polished. The specimens were etched in a 10%

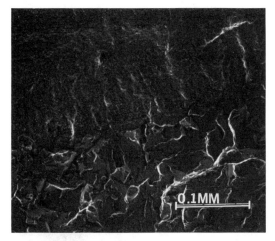

Fig. 6 Transition from transgranular to quasi-cleavage cracking (liquid nitrogen)

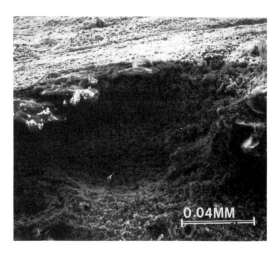

Fig. 7 SEM micrograph of pit on fracture face

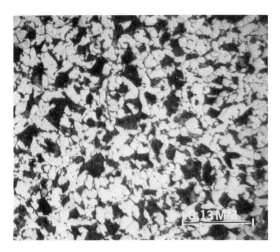

Fig. 8 Typical ferrite and pearlite microstructure of base material

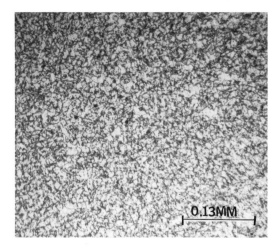

Fig. 9 Dendritic microstructure of weld metal

nital solution and then examined. The base metal, weld metal, and heat-affected zone (HAZ) microstructures (Fig. 8-10, respectively) were consistent with those of a 0.28% C welded steel.

Crack Origins/Paths. Figure 11 shows an optical micrograph of a crack prior to etching. The 4.6 mm (0.18 in.) long crack was very straight, with no apparent branching. Figure 12 depicts a section perpendicular to the main crack after etching and shows a 6 mm (0.24 in.) deep crack. This crack was also transgranular, with small areas of pitting along the crack length—indicative of a corrosion process at work in tandem with the crack propagation driving force. Higher-magnification photomicrography showed an active, fairly sharp crack tip, with very little branching.

Chemical analysis/identification

Material. Chemical analysis of the reducer material verified that the carbon steel met SA 106 grade B material requirements. The composition was 0.28% C, 1.05% Mn, 0.023% P, 0.035% S, and 0.31% Si.

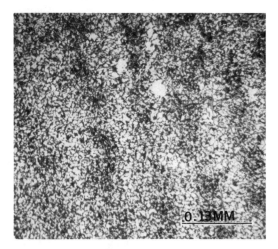

Fig. 10 Fine pearlite microstructure of HAZ

Mechanical properties

Hardness. Microhardness (Knoop) measurements were performed on the base metal, weld,

Fig. 11 Transgranular crack, 4.6 mm (0.180 in.) long. Unetched

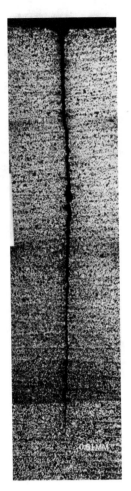

Fig. 12 Transgranular crack, 6 mm (0.24 in.) long. 10% nital etch

and HAZ of the large end of the reducer. Average values of 184, 211, and 199 HK, respectively, were obtained—typical for SA-106 grade B material.

Tensile Properties. Tensile testing yielded the following results: yield strength, 367 MPa (53.2 ksi); ultimate tensile strength, 610 MPa (88.4 ksi); elongation, 29.0%; reduction of area, 59.0%. These values also conformed to SA 106 grade B material requirements.

Impact Toughness. The Charpy impact properties of the reducer at –18 °C (0 °F) are shown in Table 1.

Table 1   Charpy impact properties

| Energy absorbed | | Shear, | Lateral expansion | |
N	ft · lbf	%	mm	in.
31	23	5	0.5	0.019
56	41	17	0.9	0.035
68	50	29	1.1	0.045

Discussion

Volume 10 of the 8th Edition of *Metals Handbook* defines beach marks as: "Macroscopic (visible) progression marks on a fracture surface that indicate successive position of the advancing crack front. The classic appearance is of irregular elliptical or semielliptical rings, radiating outward from one or more origins. Beach marks (also known as clamshell marks, tide marks, or arrest marks) are typically found on service fractures where the part is loaded randomly, intermittently, or with periodic variations in mean stress or alternating stress."

Both of the fractures examined displayed beach marks. Similar cracking had been observed previously on feedwater pipes at various nuclear power plants. The location of the large, single crack at the counterbore was similar to that observed at other PWRs. This crack and the smaller cracks exhibited crack arrests and a tight oxide film. The nature of the nozzle/reducer area of the steam generator was such that the mean stress in the pipe was occasionally superimposed by a thermal stress, which could easily have resulted in a low-cycle phenomenon. These stresses resulted from differential water temperatures (mixing) that occur during operation. Additionally, thermal stresses can result from normal plant startup and shutdown procedures. Corrosion also appeared to have contributed to crack propagation, as evidenced by pits on the fracture surface and in the cracks.

Conclusion

Most probable cause

The cracks had a large number of beach marks associated with them, indicating discontinuous crack propagation that was cyclic in nature. Corrosion appeared to play a minor but obvious role in the propagation process. The reducer cracks appeared to have resulted from a low-cycle corrosion fatigue phenomenon similar to that previously observed at other nuclear units.

Acknowledgment

This work was performed under the auspices of the U.S. Nuclear Regulatory Commission.

Cracking of Inconel 800H in a Steam Methane Reformer Furnace

Scott R. Gertler, Ashland Chemical, Dublin, Ohio

During 5.7 years of service, dye penetrant inspection of Inconel 800H pigtail connections regularly showed cracks at weld toes. Weld repairs were not able to prevent reoccurrence but often aggravated the condition. Samples containing small, but detectable, reducer-to-pigtail cracks showed intergranular cracks originating at weld toes and filled with oxidation product, which precluded determination of the cracking mechanism. All weldments exhibited high degrees of secondary precipitates, with original fabrication welds exhibiting higher apparent levels than repair welds. SEM/EDS analysis showed base metal grain boundary precipitates to be primarily chromium carbides, but some titanium carbides were also observed. Failure was believed to result from the synergism of thermally driven tube distortion, which resulted in over-stress, and from the intergranular oxidation products and intergranular carbides, which contributed to cracking. It was recommended that stresses be reduced and/or that materials and components be changed. Refinements in welding procedures and implementation of preweld/postweld heat treatments were recommended also.

Key Words

Weldments, Mechanical properties	Intergranular fracture	Tubing, Mechanical properties
Distortion	Carbide precipitation	Thermal stresses
Intergranular oxidation		

Alloys   Nickel alloys—Inconel 8004

Background

Maintenance shutdown inspections regularly showed cracks in the welded outlet pigtail connections of a steam-methane reformer furnace.

Applications

The catalytic steam reforming process uses light to medium hydrocarbons in a certain ratio to steam in order to produce different synthesis gases (i.e. methanol, hydrogen, towngas, CO-CO_2, oxoalcohols, or reduction gas). The heart of the steam reforming process is the primary reformer furnace.

The design consists of two parallel cells, each fired on both side walls and containing two staggered rows of vertical tubes. In the radiant section, tube skin temperature of about 850 °C (1562°F) is maintained by multi-sidewall radiant burners mounted in vertical tiers of horizontal rows. An induced draft fan in the exhaust stack breeching moves the combustion air.

An endothermic catalytic reforming process runs in the catalyst-filling tubes. Gas (2.07 to 2.41 MPa at 450 to 550 °C, or 300-350 psig at 842 to 1022 °F) flows from an inlet manifold system in the top of the furnace through inlet pigtails at the tops of the reformer tubes. At the bottom, tubes are welded to an assembly consisting of a reducer joined to a pipe called the bottom pigtail. The bottom pigtails are welded to multiple collection headers, each of which "tees" to a main transfer line. The top tube ends are flanged for ease of catalyst addition and removal, and the inlet pigtails are accessible and easily pinched off in case of tube leaks. The exit pigtails are in a bottom trough shielded from heat radiation, and are inaccessible while the reformer is in service. The tubes are supported at the top by concrete counterweights.

Circumstances leading up to failure

The components had seen about 50,000 h (5.7 years) of service at about 850 °C (1562 °F).

Pertinent specifications

Tubes were originally manufactured in accordance with fabricator specifications, which closely adhered to Sections IX and VIII of the ASME Boiler and Pressure Vessel Code.

Performance of other parts in same or similar service

Pigtail cracking is an industry-wide problem (Ref 1, 2).

Specimen selection

A specimen of tube-to-pigtail connection was cut from each of three tube assemblies removed due to unrelated stress rupture of the tubes. Two of the specimens appeared to have original, unaltered weldments. One had evidently been previously repaired. Two specimens had shown indications of cracking when inspected using dye penetrant.

Visual Examination of General Physical Features

A schematic of the tube/pigtail/header assembly is shown in Fig. 1. In addition to the subject cracking at the reducer-to-pigtail weld, cracking was also seen at the pigtail-to-weldolet and weldolet-to-header welds. Pigtails were often discovered to be deformed, as shown in Fig. 2.

Testing Procedures and Results

Nondestructive evaluation

Dye Penetrant Testing. All system components were abrasively blasted with ground walnut shells prior to inspection and selection of specimens. The intent was to provide a clean test surface without inducing any deformation which might affect detectability. Solvent-based visible dye penetrant was utilized.

Metallography

Microstructural Analysis. In general, precipitates were observed in significantly varying distributions and morphologies. Both pipe and weld heat-affected zone (HAZ) microstructures showed extensive grain boundary precipitation, in addition to stringers and fine matrix precipitates

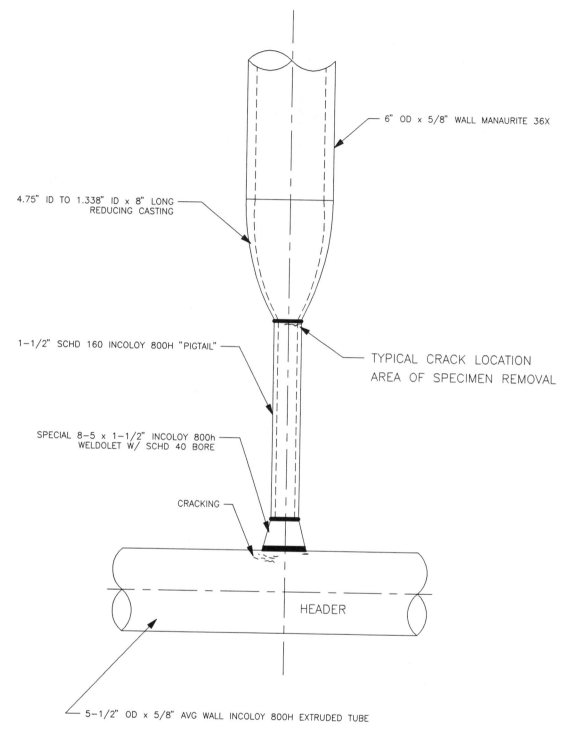

6" OD x 5/8" WALL MANAURITE 36X

4.75" ID TO 1.338" ID x 8" LONG
REDUCING CASTING

TYPICAL CRACK LOCATION
AREA OF SPECIMEN REMOVAL

1-1/2" SCHD 160 INCOLOY 800H "PIGTAIL"

SPECIAL 8-5 x 1-1/2" INCOLOY 800h
WELDOLET W/ SCHD 40 BORE

CRACKING

HEADER

5-1/2" OD x 5/8" AVG WALL INCOLOY 800H EXTRUDED TUBE

Fig. 1 Schematic of pigtail connection

(Fig. 3). Grain boundary carbide morphology was noted as both globular and "film-like." Matrix precipitates were often seen as long carbide stringers, a result of the pipe forming process.

The exteriors of all Inconel 800H pipe samples showed intergranular oxidation, presumably due to the depletion of protective chromium at grain boundaries (Fig. 4). Intergranular oxidation was noted to penetrate 0.38 to 1.27 mm (15 to 50 mils) from the surface. These could be translated to

rates of 0.064 to 0.203 mm (2.5 to 8 mils) per year. These rates are consistent with those of 800HT exposed to a furnace environment (Ref 3).

Crack Origins/Paths. Samples containing small, but detectable reducer-to-pigtail cracks showed intergranular cracks originating at weld toes and filled with oxidation product. Determination of the cracking mechanism was precluded by the oxidation damage (Fig. 5 and 6).

Weldments. All weldments exhibited high de-

Fig. 2 Close-up of actual pigtail showing deformation due to applied moment

Fig. 3 SEM photograph showing grain boundary and matrix precipitates

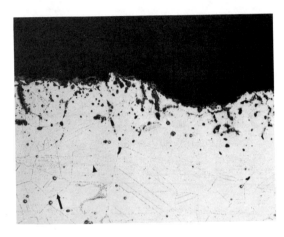

Fig. 4 Photomicrograph showing intergranular oxidation, grain boundary precipitates, and matrix stringers. 63×

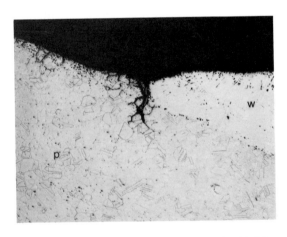

Fig. 5 Photomicrograph showing crack at toe of weld. Note voids along fusion zone and intergranular corrosion within base metal and weld metal. 15.75×

grees of secondary precipitates with original fabrication welds exhibiting higher apparent levels than repair welds. The original weld(s) exhibited relatively deeper penetration by intergranular corrosion. This difference was attributed to the orientation and morphology of the columnar weld grains, which in this case offered a path of lesser resistance. Root welds showed more equiaxed microstructures as compared to the primarily columnar microstructure of cover passes.

Chemical analysis/identification

Material and Weld. The compositions of the three submitted pigtail samples met the specifications for Inconel 800H.

Original tube-to-pigtail weld specified Inconel 82 for GTAW (TIG) root pass and Inco-Weld A electrode for SMAW cover passes. Considering iron dilution effects, it appeared the analyzed weldments met the specifications of these filler metals.

SEM/EDS analysis showed base metal grain boundary precipitates to be primarily chromium carbides, but some titanium carbides were also ob-

served. Most apparent stringer compounds were titanium carbides (and titanium cyanonitrides), but chromium carbides were also observed. SEM/EDS analysis of an original weldment showed a secondary phase in the cover pass rich in nickel, niobium, and silicon. This phase varied in terms of the relative niobium content. Niobium is also an effective carbide stabilizer.

Corrosion Deposits. Grain boundary oxidation products were identified through EDS as consisting mostly of chromium with smaller levels of aluminum, silicon, titanium, manganese, iron, and nickel.

Discussion

Metallurgical changes can occur as a result of high-temperature exposure. Such changes can strongly influence performance characteristics. Conditions such as intergranular oxidation and creep voiding are detrimental and irreversible. Oxidized grain boundaries are believed contributory to crack formation and growth and it has been suggested that carbides may participate in fatigue cracking processes (Ref 4). Carbides are necessary for high-temperature properties; however, in the improper form or morphology, their presence can be detrimental. Constitutional liquation, the separation of constituents out of a mixture due to the application of heat, occurs near the HAZ or partially melted regions, and can result in cracking. Brittle grain-boundary precipitates (chromium and titanium carbides) and oxidation products are

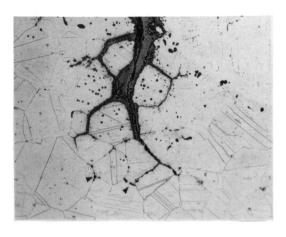

Fig. 6 Photomicrograph showing close-up of crack at toe of weld. Note precipitates within weld metal. 63×

known contributors to post-weld cracking of Fe-Ni-base superalloys. Such precipitates can be put back into solution through solution-anneal heat treatment.

Conclusion and Recommendations

Most probable cause

The damage mechanism is assumed to be synergistic. Thermally-driven tube distortion is the biggest contributor, as it imposes cyclic, moment stresses which are highest at the subject connection. Welding and high-temperature operation result in: increased creep and intergranular oxidation rates; precipitation of brittle chromium and titanium carbides at grain boundaries; introduction of high residual stresses; changes in mechanical properties; and stress risers at weld defects and weld transi-

tions. All of these factors contribute to accelerated deterioration and reduced repair weldability.

How failure could have been prevented

It was recommended that the moment stresses be reduced and/or materials and components be changed. Refinements in welding procedures and implementation of preweld/postweld heat treatments were also recommended.

References

1. Helmut Thielsch and Florence M. Cone, "Failures, Integrity and Repairs of Petrochemical Plant Reformers".
2. Marketing Brochure, Fired Heater Technology Corporation, Liberty Corner, NJ.
3. "Incoloy Alloys 800 and 800HT," Inco Alloy International, Inc., 1986, p 17.
4. "Superalloys," *Metals Handbook, Desk Edition*, American Society for Metals, Metals Park, OH, 1985, p 16-13.

Dealloying in Aluminum Bronze Components

Carl J. Czajkowski, Department of Nuclear Energy, Brookhaven National Laboratory, Upton, New York

Various aluminum bronze valves and fittings on the essential cooling water system at a nuclear plant were found to be leaking. The leakage was limited to small-bore socket-welded components. Four specimens were examined: three castings (an ASME SB-148 CA 952 elbow from a small-bore fitting and two ASME SB-148 CA 954 valve bodies) and an entire valve assembly. The leaks were found to be in the socket-weld crevice area and had resulted from dealloying. It was recommended that the weld joint geometry be modified.

Key Words			
Aluminum bronzes	Hydraulic valves		Welded joints
Nuclear reactor components	Cooling systems		Dealloying
Control valves	Cooling water		Leakage

Alloys Aluminum bronzes—SB-148 CA 952; SB-148 CA 954

Background

Leaks were found in 90 out of 782 aluminum bronze valves and fittings in an essential cooling water (ECW) system at a nuclear power plant. The leakage was limited to small-bore socket-welded components.

Applications

The ECW system provides cooling water to various safety-related systems during normal plant operation/shutdowns and both during and after postulated design basis accidents (DBA). The design pressure of the system is 825 kPa (120 psig), with normal operating pressures of about 275 kPa (40 psig).

Pertinent specifications

Four specimens were evaluated, three of which were castings: an elbow (ASME SB-148 CA 952) from a small-bore fitting and two valve bodies (ASME SB-148 CA 954), one of which did not leak.

Visual Examination of General Physical Features

The first specimen examined (Fig. 1) was an elbow fitting section that appeared to have been cut from a 25 mm (1 in.) diam elbow. There were no visible cracks on either the weld or the outside surface of the fitting. Specimens 2 and 3 were valve bodies, one of which had visible dealloying in the socket/crevice area (Fig. 2).

Specimen 4 was an entire valve assembly with an elbow, two pieces of welded pipe, and a threaded fitting. An area of staining (probable leakage) was noted on the elbow (Fig. 3). To determine whether cracks were present near the area of probable leakage, the valve assembly was leak tested by closing the valve and then pressurizing the assembly through the open end.

Pressurization was accomplished using a bottle of nitrogen gas and a gas regulator. The assembly was pressurized to 2.8 MPa (400 psig) (design pressure of an ESW system is 825 kPa, or 120 psig), and the outside surface of the assembly was brushed with a leak-detection solution. No leakage was noted.

As a second check, the valve assembly was immersed in room-temperature water and again pressurized to 2.8 MPa (400 psig). No evidence of leakage (bubble formation) was observed.

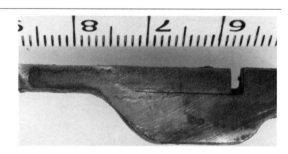

Fig. 1 As-received sample cut from 25 mm (1 in.) diam elbow fitting

Fig. 2 Dealuminized area from valve body

Testing Procedure and Results

Metallography

Microstructural Analysis. After leak testing, the valve assembly was cut on both sides of the elbow fitting. The elbow was then sectioned longitudinally in the stained, potentially leaking area. The other three as-received specimens were also sectioned, and all four specimens were mounted in epoxy, ground, and metallurgically polished. Prior to microscopic examination, the specimens were swabbed with a solution composed of 50 cm^3 (1.7 fl oz) NH$_4$OH, 50 cm^3 (1.7 fl oz) H$_2$O, and 30 cm^3 (1 fl oz) H$_2$O$_2$ (3% solution) for approximately 5 s, and then water rinsed.

The resulting microstructure in the pipe material (Fig. 4) was typical of wrought aluminum bronze material (α phase). This microstructure was found in all of the specimens examined. The socket-weld material was dendritic in appearance (Fig. 5), and no evidence of cracking was found in any of the sections examined.

The cast material outside the dealloyed area (Fig. 6) was composed of an $\alpha + \beta$ phase; small iron rosettes were evident in the structure. The dealloyed areas (Fig. 7) had all of the previously mentioned phases, as well as a γ_2 phase. This structure

Fig. 3 Staining (probable leakage area) on fitting of valve assembly

Fig. 4 Microstructure of pipe material, composed of α phase. 332×

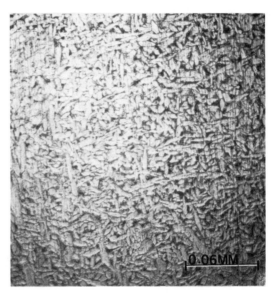

Fig. 5 Dendritic microstructure of weld metal. 332×

Fig. 6 Microstructure ($\alpha + \beta$) of cast material, showing iron rosettes. 332×

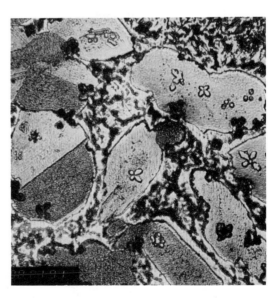

Fig. 7 Appearance of γ_2 phase in dealloyed area (arrows). 332×

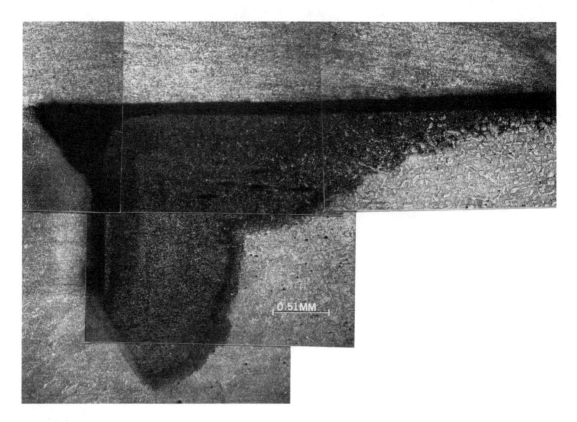

Fig. 8 Dealloying in a valve section. 31.5×

Table 1 Results of EDS scans (semiquantitative analysis)

| | Composition, % | | | | | |
| | EDS scan | | | Mill test report | | |
Specimen	Cu	Al	Fe	Cu	Al	Fe
Elbow						
Pipe	94.68	2.79	2.52	...	...	...
Dealloyed area	98.95	0.60	0.45	...	...	...
Casting	93.07	3.77	3.16	86.77	9.07	3.42
Valve No. 1						
Pipe	94.63	2.70	2.66	...	...	...
Dealloyed area	99.01	0.59	0.40	...	...	...
Casting	93.55	4.58	1.87	85.41	10.44	4.04
Valve No. 2						
Pipe	94.83	3.02	2.14	...	...	...
Dealloyed area	98.15	0.97	0.88	...	...	...
Casting	93.16	5.53	1.31	85.41	10.44	4.04
Valve No. 3						
Pipe	95.79	2.50	1.71	...	...	...
Dealloyed area	93.61	3.72	2.67	...	...	...
Casting	92.89	4.69	2.42	...	...	...

is typical of alloys dealuminized in salt water.

Lower-magnification micrographs (Fig. 8 and 9) showed varying degrees of dealloying (dealuminization). In all cases, no cracks were evident and the dealloying was confined to the crevice area of the socket-weld joint.

Chemical analysis/identification

Material and Weld. A qualitative chemical analysis was performed on the pipe material, dealloyed area, and casting material using energy-dispersive spectroscopy (EDS). This analytical technique is capable of performing elemental analysis of microvolumes, typically on the order of a few cu-

bic micrometers in bulk samples, and considerably less in thinner sections. Analysis of X-rays emitted from a sample is accomplished using crystal spectrometers, which use energy-dispersive techniques that permit analysis by discriminating among X-ray energies.

The feature of electron-beam microanalysis that best describes this technique is its mass sensitivity. For example, it is often possible to detect less than 10^{-16} g of an element present in a specific microvolume of a sample. The minimum detectable quantity of a given element, or its detectability limit, varies with many factors and, in most cases, is less than 10^{-16} g/microvolume. It should

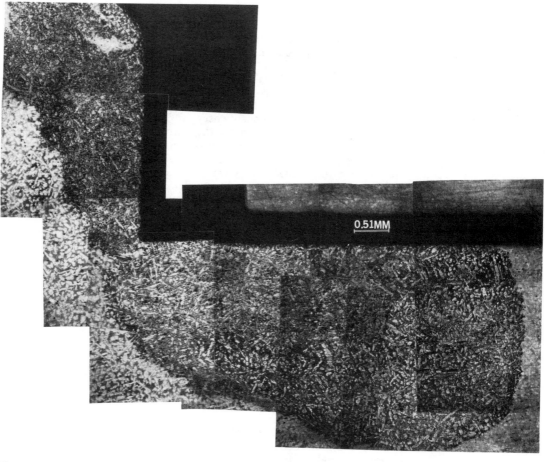

0.51MM

Fig. 9 Cross section of valve with well-defined area of dealloying. Etched

be noted that EDS will only discern elements with atomic numbers greater than 11.

Each of the four specimens were evaluated by EDS. Three scans were performed on each specimen: on the pipe material, in the area of dealloying, and on the casting material.

Figure 10 graphically depicts typical EDS scans. With the exception of the fitting from valve 3, each of the scans shows a significant lessening of the aluminum peak in the dealloyed area. Although there is a tendency for the EDS scans to be low (because of detection limitations), the qualitative evidence of the dealloying is clear. The probable cause of the anomaly in the scans on valve 3 is the very selective nature of this type of attack and its nonuniformity about the periphery of the specimen.

The data in Table 1 numerically interpret the scans and are compared, where possible, to the mill test reports provided by the utility. Because EDS is a semiquantitative measurement that relies on spot sampling, location of incident electron beam, and other factors, the information presented should be viewed from a qualitative, rather than quantitative, perspective.

Although the fittings and valve bodies were welded assemblies (gas-tungsten arc welded socket welds), the procedures used in their installation required the use of a welding flux. This flux was incorporated into the weld by applying a mixture of flux and isopropanol onto the filler wire used to weld the joint.

Because dealloying was becoming a distinct causative factor in the failure analysis, the flux was chemically analyzed for chlorides, fluorides, and total halogens (possible dealloying contributors). Chemical analysis revealed less than 10 ppm chloride, 1 ppm fluoride, and 1 ppm total of other halogens.

Mechanical properties

Hardness. Microhardness (Knoop) testing (200 g load) yielded the following average values for the elbow and two valve bodies, respectively: 135, 148, and 163 HK (casting material); 85, 79, and 59 HK (dealloyed areas). Obviously, the microhardness values in the dealloyed (dealuminized) regions were significantly less than those in the casting material (by an approximate factor of 2 to 2.5).

Discussion   Volume 10 of the 8th Edition of *Metals Handbook* defines dealloying as: "The selective corrosion of one or more components of a solid solution alloy." The National Association of Corrosion Engineers (NACE) has termed the dealloying phenomenon associated with aluminum-copper alloy systems as "dealuminization." NACE further emphasizes that although color changes may be associated with dealloying, no cracks, pits, dimensional changes, grooving, or obvious metal loss characteristics are visible. The affected material does become brittle, porous, and lighter, and loses its original mechanical properties (that is, tensile strength).

The selective attack of aluminum bronze alloys by seawater has been previously observed. This type of dealloying has been linked to the distribution and quantity of an aluminum-rich γ_2 phase. It is also recognized that, although γ_2 is the most susceptible phase, β will also dealloy when subjected to the proper environmental conditions. The formation of this phase appears to vary with composition and heat treatment. Previously examined occurrences also have shown the dealloying of these materials to be associated with crevices.

Conclusion and Recommendations

Most probable cause

The appearance of γ_2 phase in the structures examined, coupled with the lower hardnesses in dealloyed areas and the absence of general metal loss characteristics, led to the conclusion that the leakage associated with these components was caused by the selective leaching of aluminum—that is, dealuminization. This phenomenon is associated with the crevice formed in the socket-weld joints that were examined.

Remedial action

It was recommended that the socket-weld joint geometry be modified to eliminate the crevice.

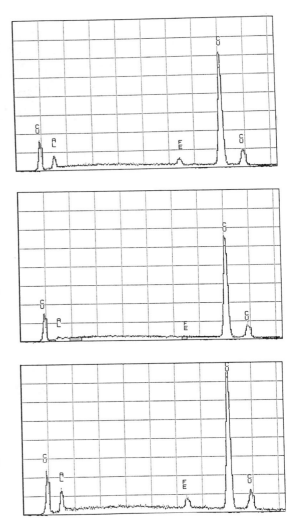

Fig. 10 EDS scans of various areas on elbow specimen. (a) Pipe material. (b) Dealloyed area. (c) Casting material

Acknowledgment

This work was performed under the auspices of the U.S. Nuclear Regulatory Commission.

Failure of Carbon Steel Superheater Tubes

Joseph P. Ribble, Betz Laboratories, Inc., The Woodlands, Texas

Two superheater tubes from a 6.2 MPa (900 psig) boiler failed in service because of creep rupture. One tube was carbon steel and the other was carbon steel welded to ASTM A213 Grade T22 (2.25Cr-1.0Mo) tubing. The failure in the welded tube occurred in the carbon steel section. Portions of the superheater were retubed five years previously with Grade T22 material. The failures indicated that tubes were exposed to long-term overheating conditions. While the carbon steel tube did not experience temperatures above the lower transformation temperature 727 °C (1340 °F), the welded tube did experience a temperature peak in excess of 727 °C (1340 °F). The long-term overheating conditions could have been the result of excessive heat flux and / or inadequate steam flow. In addition, the entire superheater bank should have been upgraded to Grade T22 material at the time of retubing.

Key Words			
Low-carbon steels, Mechanical properties Creep rupture		Chromium-molybdenum steels, Mechanical properties Overheating	Boiler tubes Superheaters

Alloys		
Low-carbon steels—A192		Chromium-molybdenum steels—A213 Grade T22

Background

Two tubes from the superheater section of a two-drum boiler failed while in service. The tubes were identified as being the "old tube" (tube 1) and the "new tube" (tube 2). Tube 1 was an original equipment tube and was in service for 19 years. Tube 2 was in service for approximately five years.

Applications

The superheater tubes had different nominal outer diameters and nominal wall thicknesses. Tubes 1 and 2 measured 44 mm (1.75 in.) and 50.8 mm (2.00 in.) in diameter, respectively. The wall thicknesses of tube 1 and the failed end of tube 2 measured 4.5 mm (0.176 in.) and 3.8 mm (0.150 in.) thick, respectively. The wall thickness at the end of tube 2 opposite the failure measured 6.0 mm (0.235 in.). These tubes were steam bearing tubes for a 6.2 MPa (900 psig) boiler.

Pertinent specifications

Tube 1 and the failed end of tube 2 were fabricated from low-carbon steel consistent with ASTM A192 specifications. The non-failed end of tube 2 was fabricated from ASTM A213 Grade T22 seamless ferritic alloy steel superheater tubing.

Performance of other parts in the same or similar service

Several other superheater tubes in a boiler of identical design and manufacturer, at the same plant, had failed two years prior to this failure by the same apparent mechanism. Portions of the superheater in the present boiler were retubed five years ago. The two submitted tubes were in parallel in the superheater section at the time of failure.

Selection of specimens

Two selections approximately 406 mm (16 in.) long were submitted for laboratory examination.

Visual Examination and General Physical Features

The external surface of the superheater tubes is shown in Fig. 1. The failures in tube 1 consisted of two ruptures in the middle of localized bulges. The two failure sites are indicated as "a" and "b" in Fig. 1. The ruptures were oriented approximately 90° apart with respect to the tube circumference, and were separated by approximately 70 mm (2.75 in.). The tube 2 sample consisted of two tubes welded together with the failure entirely within the carbon steel portion. The tube 2 failure was a thin-lipped rupture displaying a high degree of deformation. The low-alloy portion of tube 2 had no failures associated with it.

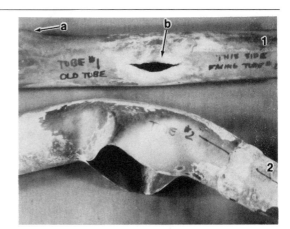

Fig. 1 The submitted superheater tube sections. 0.23×

Testing Procedure and Results

Surface examination

The tube 1 outer diameter surface exhibited two ruptures parallel to its longitudinal axis. Wall thinning to 0.76 mm (0.030 in.) and secondary creep cracks were observed at both tube 1 failure sites (Fig. 2 and 3). Secondary creep cracks and wall thinning to 0.51 mm (0.020 in.) were observed at the tube 2 failure lip (Fig. 4).

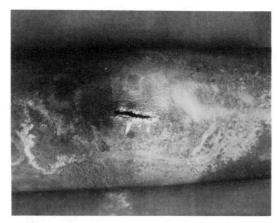

Fig. 2 Tube 2 failure site a (see Fig. 1). Note the heavy oxidation and longitudinal secondary creep cracks. 0.60×

Fig. 3 Tube 1 failure site b (see Fig. 1). Note the heavy oxidation and longitudinal secondary creep cracks. 0.60×

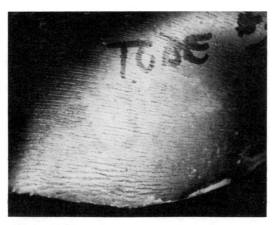

Fig. 4 The failure edge of tube 2. Note the secondary creep cracks. 1.06×

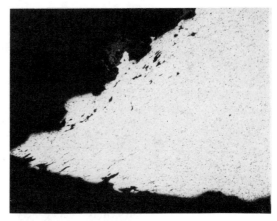

Fig. 5 Photomicrograph of the b failure lip in tube 1. Note the creep voids along the fracture surface. Nital etch, 38×

Fig. 6 Photomicrograph of the mid-wall microstructure observed in the tube 1 b failure area shown in Fig. 5. Note the spheroidization and large graphite nodule. Nital etch, 285×

Fig. 7 Photomicrograph showing the mid-wall microstructure opposite the failure of Fig. 5 and 6. Note the spheroidization. Nital etch, 285×

The internal surface of tube 1 was covered with a thin layer of yellow-brown deposits over a tightly adherent black scale. The internal surface of tube 2 was covered with a thin layer of tightly adherent black magnetite scale.

Metallography

Representative sections from select areas of both tubes were prepared for metallographic examination. Creep voids were observed along the fracture lip b of tube 1 (Fig. 5). The microstructure observed at the mid-wall location of the tube 1 b failure and opposite failure a was in a very advanced stage of spheroidization (Fig. 6 and 7). In addition, isolated graphite nodules were also observed in the microstructure.

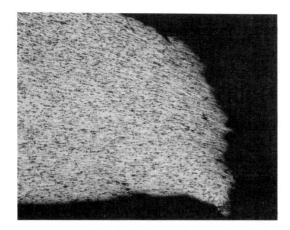

Fig. 8 Photomicrograph of the microstructure taken at the failure lip of tube 2. Note the creep voids and elongated grains. Nital etch, 57×

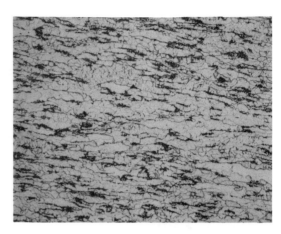

Fig. 9 Photomicrograph of the mid-wall microstructure at tube 2 failure lip. Note the reformed pearlite in the ferrite matrix. Nital etch, 285×

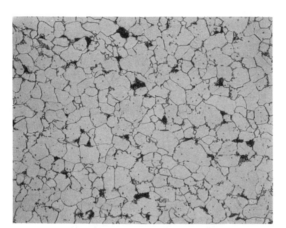

Fig. 10 Photomicrograph of the mid-wall microstructure opposite the tube 2 failure. Note the spheroidal carbides and pearlite islands. Nital etch, 285×

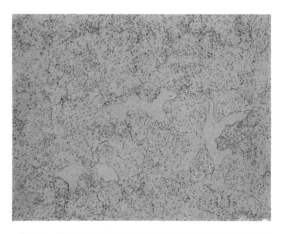

Fig. 11 Photomicrograph of the tube 2 mid-wall microstructure on the failure side, in the non-failed portion opposite the weld. Nital etch, 285×

Elongated grains were observed at the failure lip of tube 2 (Fig. 8). The microstructure at the mid-wall location of the tube 2 failure consisted of ferrite and reformed pearlite (Fig. 9). The presence of transformation products (reformed pearlite) at the failure lip indicated that this portion of tube 2 experienced temperatures above its lower transformation temperature, 727 °C (1340 °F), at the time of failure. Opposite the tube 2 failure, the mid-wall microstructure consisted of spheroidal carbides and occasional islands of pearlite in a ferrite matrix (Fig. 10). The mid-wall microstructure observed in the low-alloy portion of tube 2 that did not fail consisted of a fine dispersion of spheroidal carbides (Fig. 11).

Chemical analysis/identification

Material analysis. Portions of tube 1 and 2 from either side of the weld joint were tested for chemical analysis using optical emission spectroscopy (OES) to determine the alloy composition. The results (Table 1) confirmed that tube 1 and the portion of tube 2 containing the failure were fabricated from low-carbon steel consistent with ASTM A192 tubing specification. The similarity of the chemical composition of the failed tube sections

Table 1 Alloy analyses of failed tube sections

	Composition, %		
Element	Tube 1	Tube 2	ASTM A192 specification
C	0.11	0.12	0.06 to 0.18
Mn	0.49	0.46	0.27 to 0.63
P	0.006	<0.005	0.035 max
S	0.026	0.25	0.035 max
Si	0.02	0.01	0.25 max
Cr	0.03	0.03	ns
Ni	0.07	0.07	ns
Mo	0.02	0.02	ns
Cu	0.14	0.11	ns
Fe	rem	rem	rem

ns, not specified

suggest they were from the same heat of steel. Results of the alloy analysis on the non-failed portion of tube 2 (Table 2) indicated that it was fabricated from ASTM A213 Grade T22 tubing.

Internal deposit analysis. Scanning electron microscope-energy dispersive spectroscopy (SEM-EDS) was used to determine the elemental composition of the internal deposits on both tubes.

Table 2 Alloy analysis of non-failed section of tube 2

| Element | Composition, % | |
	Non-failed section	ASTM A213 Grade T22 specification
C	0.11	0.15 max
Mn	0.49	0.30 to 0.60
P	0.016	0.30 max
S	0.017	0.30 max
Si	0.27	0.50 max
Cr	2.30	1.90 to 2.60
Ni	0.22	ns
Mo	1.02	0.87 to 1.13
Cu	0.09	ns
Fe	rem	rem

ns, not specified

Table 3 Analyses of the internal deposits

| Element | Composition, % | |
	Tube 1	Tube 2
Fe	47.1	83.5
Ca	22.2	5.5
Si	16.6	4.5
Na	9.8	4.6
Cl	0.7	...
S	1.0	<0.5
Al	1.7	...
Mn	0.9	...
Cr	...	<0.5
Mg	...	1.6

The results (Table 3) revealed significant quantities of calcium, silicon and sodium species in tube 1. The same species were also present in tube 2 but in smaller quantities. This indicated that some carryover of boiler water was taking place in the superheater section.

Discussion

Heavy oxidation and longitudinal cracking parallel to the failures in both tubes indicated a creep rupture mechanism. Chemical analysis revealed that tube 1 and the failed section of tube 2 were fabricated from plain carbon steel. The non-failed side of tube 2 was fabricated from ASTM A213 Grade T22 low-alloy steel tubing. In steam environments, the generally accepted oxidation limit for plain carbon steel is 454 °C (850 °F), while for Grade T22 it is approximately 580 °C (1075 °F). A spheroidized microstructure with graphite nodules indicated that tube 1 was operating at a temperature significantly above its oxidation limit and below its lower transformation temperature of 727 °C (1340 °F) for an extended period of time.

Transformation products in the failed region of tube 2 indicated that the failure region experienced a peak temperature in excess of 727 °C (1340 °F). However, secondary creep cracks around the failure and creep voids along the rupture lip indicated that it experienced long-term overheating effects prior to rupture. The presence of a transformed microstructure in the carbon steel portion of tube 2 made an assessment of the degree of long-term overheating impossible. The superior oxidation resistance and microstructural stability of Grade T22 tubing, via its chromium and molybdenum additions, resulted in the virtual absence of significant scale and fine dispersion of spheroidal carbides in the microstructure.

Conclusions and Recommendations

Most probable cause

The tube 1 failure was caused by creep mechanism via long-term overheating. Microstructural analysis indicated that the tube 1 peak metal temperatures were greater than 454 °C (850 °F) but less than 727 °C (1340 °F). Secondary creep cracking indicates long-term overheating conditions were present prior to the failure of tube 2. However, microstructural analysis indicated that tube 2 experienced a brief high-temperature excursion greater than 727 °C (1340 °F) at the time of failure. The long-term overheating of tubes 1 and 2 was likely caused by an excessive heat flux and/or insufficient steam distribution. The high-temperature event coinciding with the violent rupture of tube 2 most likely resulted from a sudden loss in steam distribution through this tube circuit. Although SEM-EDS analysis identified some carryover deposits on the internal surfaces, only a small amount of transported deposit was present. Internal deposits were not a primary cause of the tube failures.

Remedial action

A review of steam flow in the superheater should be conducted to ensure adequate steam distribution in the circuit. The entire superheater should be upgraded to ASTM A213 Grade T22 tubing.

How failure could have been prevented

Alloy analyses of the submitted tube samples indicates the overheating failures occurred in low-carbon steel sections. The portion of tube 2 that did not contain the failure tested as being consistent with ASTM A213 Grade T22 boiler tubing specification. The similarity of the chemical composition of the failed sections suggests these sections were installed at the same time. When the superheater was retubed, the entire superheater assembly should have been upgraded to Grade T22 material.

Failure of a High-Pressure Steam Pipe

Carmine D'Antonio, Department of Metallurgy and Materials Science, Polytechnic University, Brooklyn, New York

A high-pressure steam pipe specified to be P22 low-alloy steel failed after 25 years of service. Located at the end of the steam line, the pipe reportedly received no steam flow during normal service. Visual examination of the failed pipe section revealed a window fracture that appeared brittle in nature. Specimens from the fracture area and from an area well away from the fracture were examined metallographically and chemically analyzed. Results indicated that the pipe had failed by hydrogen damage that resulted in brittle fracture. Chemical analysis indicated that the pipe material was 1020 carbon steel, not P22. The misapplication of pipe material was considered to be a contributing factor. Position of the pipe within the system caused the localized damage.

Key Words	Carbon steels	Steam pipe	Hydrogen embrittlement
	Brittle fracture		
Alloys	Carbon steel—1020		

Background

A high-pressure steam pipe failed after 25 years of service.

Applications

The pipe (25 mm, or 1 in., NPS [National Pipe Standard] by schedule 160) was located at the end of a 15 MPa (2100 psi), 565 °C (1050 °F) steam line. The pipe reportedly received no steam flow during normal service. The only time steam flowed through the length of pipe was during startup or shutdown procedures. A sealer compound (composition unknown) was periodically pumped into the steam to seal leaky valve stems in the system.

Pertinent specifications

P22 low-alloy steel was specified for this application.

Specimen selection

Metallographic samples were cut from the fracture area and from an area well away from the fracture site. Drillings were taken for wet chemical analysis.

Visual Examination of General Physical Features

Figure 1 shows the as-received pipe section. Failure occurred when a segment of pipe 38 mm (1½ in.) long by 13 mm (½ in.) wide blew out of the pipe wall. This type of failure is known as a window fracture. Figure 2 is a close-up view of the window. Note that the fracture was brittle in nature, with very little bulging of the pipe wall adjacent to the window. The entire fracture surface was covered with scale, which precluded scanning electron microscopic examination of the fracture.

In addition to the scale on the fracture surface, the inside of the pipe was covered with a thick layer of scale that was confined to the vicinity of the fracture. Accordingly, the pipe was sectioned to expose the inside surface at the fracture. A layer of scale, more than 13 mm (½ in.) thick in some areas with underlying severe corrosion, was found. Figure 3 shows a longitudinal section at the fracture area (the section containing the window is at the top). Thick scale and the loss in wall thickness are evident. These two features were only found in this area of the pipe.

Fig. 1 Section of pipe containing the blowout window fracture

Testing Procedure and Results

Metallography

Microstructural Analysis. Figures 4 and 5 are micrographs taken on a transverse section of the pipe that show the fracture surface in cross section. The inside surface of the pipe is shown in Fig. 4 (with the fracture along the top). Oxide scale on the fracture was apparent, as was a lack of plastic deformation adjacent to the fracture.

Of greater importance, however, was the presence of scale-filled fissures within the pipe wall. These fissures were most abundant near the inside surface and gradually reduced in number toward the outside surface. It was also evident that the fracture progressed from fissure to fissure. Figures 6 and 7 show typical fissure arrays at the inside surface.

Fig. 2 Close-up view of window

Fig. 3 Cut longitudinal section of pipe fracture area, with section containing the window (top)

Table 1 Chemical composition of the pipe material

| Element | Composition, % | | |
	Pipe material	Low-P22 alloy steel(a)	AISI 1020 carbon steel(a)
Carbon	0.20	0.15 (max)	0.18-0.23
Manganese	0.68	0.30-0.60	0.30-0.60
Phosphorus	0.024	0.030 (max)	0.040 (max)
Sulfur	0.031	0.030 (max)	0.050 (max)
Silicon	0.16	0.50 (max)	NR
Chromium	0.05	1.90-2.60	NR
Nickel	0.01	NR	NR
Molybdenum	0.01	0.87-1.13	NR
Vanadium	0.04	NR	NR

(a) NR, no requirement

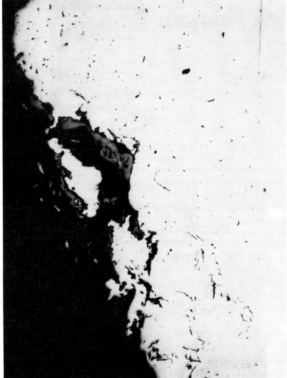

Fig. 4 Transverse section of inside surface of the pipe. Unetched. 59.25×

Table 2 Chemical composition of scale

Element	Composition, %
Iron oxides	
Fe$_3$O$_4$	70
Fe$_2$O$_3$	30
Vanadium	<0.01
Molybdenum	<0.01
Magnesium	0.005

Figure 8 shows the general microstructure of the pipe, which consisted of ferrite and pearlite in a Widmanstätten array. Figure 9 shows the microstructure near the inside surface in the fracture area. It was apparent that oxide-filled fissures occurred only in pearlite areas.

Chemical analysis/identification

The results of wet chemical analysis are presented in Table 1. The pipe material was AISI 1020 carbon steel, not P22 low-alloy steel as specified. AISI 1020 steel is not as resistant to high-temperature steam as P22; consequently, this must be considered a contributing factor in the failure of the pipe.

Analysis of the scale (Table 2) showed it to be primarily Fe$_3$O$_4$ plus Fe$_2$O$_3$. No other elements were found, except in trace amounts. It is thus unlikely that the sealing compound introduced into the system contributed to the failure.

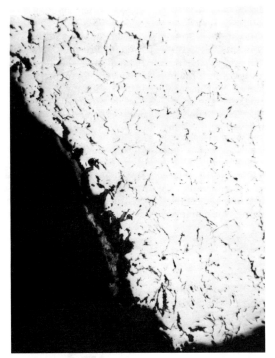

Fig. 5 Transverse section of outside surface of the pipe. Unetched. 68.25×

Fig. 6 Typical fissure arrays on the inside surface. Unetched. 335×

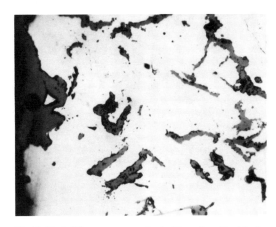

Fig. 7 Typical fissure arrays on the inside surface. Unetched. 335×

Discussion

The type of failure exhibited in this case history is typical of hydrogen damage. The features that indicate hydrogen damage are a window fracture, little or no bulging of the pipe wall, the presence of a thick internal scale consisting of Fe_3O_4 and other oxides, and the presence of fissures in the carbide (pearlite) zones of the microstructure.

The thick internal deposit is the most damaging factor. There is usually an space between the scale and the pipe metal where hydroxyl or hydrogen ion concentrations are increased, causing active corrosion and the production of hydrogen. The local temperature of the metal under these deposits also increases, which enhances further reaction and metal dissolution. These reasons underlie the fact that metal loss was found only under the thick oxide scale.

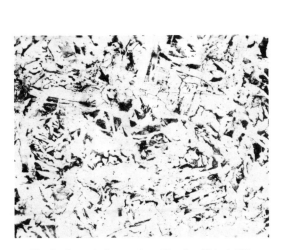

Fig. 8 General microstructure of the pipe. Etched. 134×

Fig. 9 Microstructure near the inside surface. Etched. 134×

Hydrogen produced during the corrosion process easily diffuses into the steel as nascent hydrogen, resulting in localized decarbonization. A product of the reaction of the hydrogen ions with iron carbide is methane, which collects and causes an increase in pressure and internal fissures. This is clearly what occurred in this instance, as the only affected zones in the microstructure were the pearlitic areas. The fissures represented a brittle network and resulted in a blowout-type failure because of internal operating pressures.

The fact that the failure occurred in a "dead" section of the system indicates that, within this area, conditions were such that an adherent scale formed and was not removed by steam flow. Once the scale formed, the process became autocatalytic, resulting in severe metal loss because of corrosion and infiltration of nascent hydrogen, which caused embrittlement.

Conclusion

The pipe failed by hydrogen damage that resulted in brittle fracture. Because the pipe was made of AISI 1020 carbon steel and not P22 low-alloy steel, this must be considered a contributing cause of failure, inasmuch as P22 would not be expected to deteriorate as rapidly as 1020 under high-pressure steam conditions. Position of the pipe within the system caused the localized damage.

Failure of a Pipe Slip-on Flange

Mohan D. Chaudhari, Columbus Metallurgical Services, Inc., Columbus, Ohio

A cracked 356 mm (14 in.) diam slip-on flange (Ni-Cr-Mo-V steel) was submitted for failure analysis. Reported results and observations indicated that the flange was not an integral forging or a casting, as specified. It had been fabricated by welding and machining a ring insert within a flange with a larger internal diameter. The flange cracked because the welds between the flange and the insert were inadequate to withstand the bolting pressures. A warning was issued to end users of the flanges, which are being inspected nondestructively for conformance to specifications.

Key Words   Nickel-chromium-molybdenum steels   Cracking fracturing   Pipe flanges
Welded joints

Background   A 356 mm (14 in.) diam slip-on flange cracked.

Pertinent specifications

The specification called for an integral forging or a casting. Instead, the slip-on flange had been fabricated by welding and machining a ring insert within a flange with a larger internal diameter.

Specimen selection

The cracked slip-on flange was received in one piece. It had two diametrically opposite cracks located on the pipe and gasket sides.

Testing Procedure and Results

Nondestructive evaluation

The flange was ultrasonically tested from the outside diameter. The indications were rather puzzling, because the reflections were consistently from a cylindrical surface about 95 mm (3.75 in.) from the OD. Dry magnetic particle examination confirmed that the cracks extended more or less continuously in a circular path. Nondestructive evaluation was concluded with dye penetrant tests, as shown in Fig. 1 to 4. It was quite evident that the flange was not a single-piece component. A large inside diameter flange was reduced to a 356 mm (14 in.) ID unit by welding a 19 mm (0.75 in.) thick ring insert. The welded faces had been machined.

Metallography

Figure 5 shows the radial macrosection of the flange. Note that the insert ring is welded to the main flange. When another similar piece was further sectioned to retrieve specimens for mounting and polishing, the insert separated from the main flange. The general microstructure (ferritic-pearlitic) is shown in Fig. 6.

Fig. 1   Gasket side of the flange

Fig. 2   Closeup view of the gasket side, showing the crack indication

Fig. 3   Pipe side of the flange

Table 1 Results of chemical analysis

| Material | Composition, % | | | | | | | | | | | |
	C	Mn	P	S	Si	Cu	Sn	Ni	Cr	Mo	Al	V
Main flange steel	0.23	0.62	0.024	0.038	0.24	0.27	0.016	0.094	0.079	0.026	0.006	0.003
Welded insert steel	0.26	1.02	0.017	0.039	0.28	0.063	0.003	0.050	0.046	0.032	0.008	0.000
ANSI/ASTM A 105(a)	0.35	0.60 to 1.05	0.040	0.050	0.35		...	...	...	...	...	...

(a) This specification is listed for reference only.

Fig. 4 Closeup view of the pipe side, showing the crack indication

Chemical analysis/identification

Table 1 presents results of chemical analysis by optical emission spectrometry.

Mechanical properties

The hardness of the main flange was measured at three sites, with readings of 69, 70, and 72 HRB, for an average hardness of 70 HRB. Three readings of the insert measured 76, 78, and 80 HRB, for an average of 78 HRB. No further work was deemed necessary.

Fig. 5 Macrograph of a radial section etched in 50% HCl for 15 min

Conclusion

It was concluded that the flange failed at a weak welded joint. The flange did not conform to the ANSI specification, which called for an integral component.

The defective flange was one of a large shipment of similar components imported by the customer. A warning was issued to the end users of the flanges, which were then inspected nondestructively for conformance to specifications.

Fig. 6 General microstructure in the main flange (lower half), insert (upper half), and weld metal (left edge). Nital etch. 15.75×

Failure of a Reactor Tube

J. Ciulik, Mechanical and Materials Engineering Department, Radian Corporation, Austin, Texas

A low-carbon steel (St35.8) tube in a phthalic anhydride reactor system failed. Visual and stereomicroscopic examination of fracture surfaces revealed heavy oxide/deposits on the outer surface of the tube, tube wall thinning in the area of the fracture, and discolorations and oxides/deposits on the inner surface. Cross sections from the fracture surface were metallographically examined, and the deposits were analyzed. It was determined that the tube had thinned from the inner surface because of a localized overheating condition (probably resulting from a runaway chemical reaction within the tube) and then fractured, which allowed molten salt to flow into the tube.

Key Words

Carbon steels
Tubing
Chemical processing industry

Oxidation
Chemical reactors

Fused salts, environment
Overheating

Alloys

Low-carbon steel—St35.8

Background

A low-carbon steel (St35.8) tube in a chemical reactor system failed.

Circumstances leading to failure

An explosion occurred in a phthalic anhydride reactor system that had been producing phthalic anhydride from naphthalene and orthoxylene for 19 years. It was reported that one of the 10,000 reactor tubes had failed and that salt from the recirculating heat transfer bath had flowed into the failed tube and eventually collected in the gas cooler head section. The subsequent explosion resulted in no appreciable damage to the reactor, but severely damaged related equipment.

In the reactor, an air oxidation of a mixture of naphthalene and orthoxylene occurs to form phthalic anhydride, with titanium oxide/vanadium pentoxide material acting as a catalyst. A recirculating salt bath composed of a mixture of 53% KNO_3, 40% $NaNO_2$, and 7% $NaNO_3$ cools the reactor tubes. The operating temperature of the salt bath typically ranges from 365 to 370 °C (690 to 700 °F).

The tubes were reportedly manufactured from Werkstoff St35.8 II steel. Each reactor tube was filled with hollow silicon carbide cylinders (catalyst carriers) coated with a mixture of 96% titanium oxide and 4% vanadium pentoxide. The tube temperature reportedly varied from 300 °C (570 °F) at the upper tube sheet to, typically, 450 °C (840 °F) at the hottest region of the tube, where the majority of the reaction occurred.

The reactor tubes had been recharged with catalyst 5 months prior to the explosion. Recatalyzation is typically done every 3 to 5 years. Prior to the explosion, the reactor had been shut down for routine cleaning. After startup, the reactor operated for several days until the explosion occurred.

Specimen selection

The fractured tube was submitted for analysis to determine the cause of failure by the following:

- Examination of the fracture surfaces for evidence of tube melting
- Examination of the microstructure at the fractured region and comparison with that of the adjacent tube in a similar location
- Analysis of the inner and outer oxides/deposits at the fractured region using energy-dispersive spectroscopy (EDS)
- Examination of and analysis of a deposit sample removed from the material reportedly entrapped within the tube in the fractured region to determine the presence of melted or fractured tube material

Visual Examination of General Physical Features

Visual examination of the failed ends of the tube (Fig. 1) revealed heavy oxides/deposits on the outer surface of the tube within 19 mm (3/4 in.) of the fracture surfaces (Fig. 2). The fracture surfaces exhibited a very thin tube wall with a large buildup of oxides/deposits on the inner surface (Fig. 3).

Longitudinal splitting of the fractured tube revealed localized tube wall thinning in the fractured region. The wall thicknesses at the fracture surfaces were "paper thin" (0.13 mm, or 0.005 in.). The tube thinning occurred from the inner surface and was observed for a distance of approximately 19 mm (3/4 in.) from the fracture surface, tapering to the nominal wall thickness of 2.5 mm (0.1 in.). The inner surface also exhibited discolorations and oxides/deposits within 19 mm (3/4 in.) of the fracture surfaces.

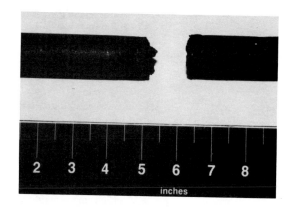

Fig. 1 As-received fractured tube section

Fig. 2   One fractured end of the failed tube

Fig. 3   Fracture surface of the tube section. ~1.7×

Testing Procedure and Results

Surface examination

Visual and stereomicroscopic examination of the fracture surfaces revealed no evidence of melting. The very thin fracture surfaces appeared to be mechanical fractures.

Metallography

Microstructural Analysis. One of the tube halves was metallographically cross sectioned to determine the microstructure at the fracture surface. Examination of the cross section revealed extreme tube wall thinning at the fracture surface and thick oxides/deposits on the inner surface (Fig. 4). The microstructure of the tube at the fracture surface was dramatically different from the typical tube microstructure (Fig. 5).

At the fracture surface, the microstructure consisted of tempered martensite and ferrite (Fig. 6), indicating that this area of the tube had been heated to a high enough temperature to fully austenitize, and then had been rapidly quenched to form martensite. No evidence of melting was observed. The microstructure changed gradually with distance from the fracture surface—from tempered martensite (Fig. 6) to the typical tube microstructure of equiaxed ferrite with small amounts of pearlite (Fig. 7) within 19 mm (3/4 in.) along the tube length.

At the fracture surface, the oxide/deposit thickness was 0.38 mm (0.015 in.) in several locations on the outer surface and approximately 2.5 mm (0.1 in.) on the inner surface. Much of the outer surface oxide had spalled. The nominal oxide thickness was 20 μm (0.8 mil) on the outer surface and 66 μm (2.6 mil) on the inner surface several inches from the fracture surface.

A microhardness profile was performed across the fracture surface microstructure into the typical tube microstructure (Fig. 8). The microhardness at the fracture surface was measured as 324 HV, which corresponds to 33 HRC. The microhardness of the typical tube microstructure, 50 mm (2 in.) from the fracture surface, was 116 HV (65.5 HRB).

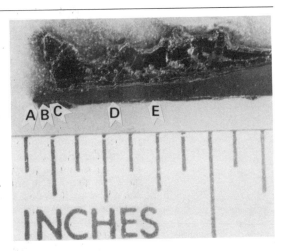

Fig. 4   Cross-sectioned tube fracture surface. Arrows indicate regions of microstructure shown in Fig. 5 and 6.

Fig. 5   Microstructure at fracture surface. Arrow A indicates region of tempered martensite. Arrow B indicates mixture of tempered martensite and bainite. Nital etch. 64×

Chemical analysis/identification

Chemical analysis of a tube sample showed that it met the requirements of Werkstoff St35.8 II (Table 1).

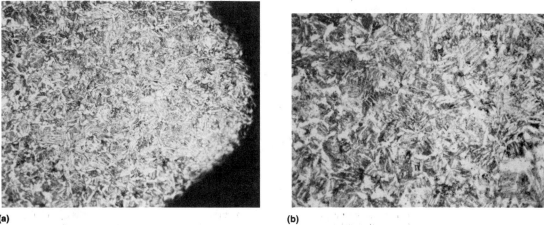

(a)   **(b)**

Fig. 6   Microstructure of region A in Fig. 5, consisting of tempered martensite and ferrite. Nital etch. (a) 268×. (b) 610×

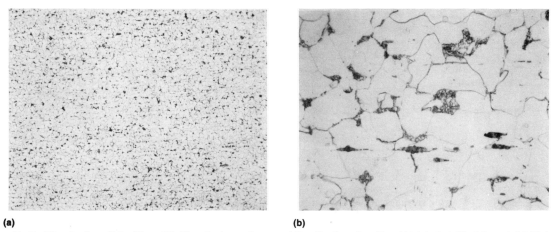

(a)   **(b)**

Fig. 7   Microstructure of tube, 50 mm (2 in.) from fracture surface, consisting of ferrite and pearlite, which is typical of the tube material. Nital etch. (a) 61×. (b) 610×

Table 1   Results of chemical analysis

Material	Composition, %								
	C	Mn	P	S	Si	Ni	Cr	Mo	Fe
Tube	0.10	0.56	0.009	0.024	0.15	0.05	0.05	< 0.05	bal
Werkstoff St35.8 II	0.17 (max)	0.40-0.80	0.040 (max)	0.040 (max)	0.10-0.35				bal

EDS Analysis of Inner and Outer Deposits. The cross-sectioned fracture surface was examined in a scanning electron microscope (SEM), and the inner and outer surface oxides/deposits were analyzed using EDS. In the SEM, the thick oxide/deposit on the inner surface near the fracture surface appeared to contain many constituents (Fig. 9). These areas, as well as the oxide/deposit on the outer surface, were each analyzed (Fig. 10-13). Most of the deposit contained primarily iron (areas 1 and 5 to 9), indicating that the majority of the material was iron oxide (magnetite). The darker-appearing material in Fig. 9 primarily contained either silicon (areas 3 and 10), indicating that it was SiC from the catalyst carriers, or sodium and potassium (area 2), indicating that it was salt from the salt bath. Other regions contained mixtures of iron, silicon, potassium, and titanium, indicating that magnetite, SiC carrier material, salt

from the salt bath, and catalyst material were present in the oxide/deposit.

X-Ray Diffraction (XRD) Analysis of Deposit Sample. Visual examination of the deposit sample removed from material reportedly trapped within the tube in the fracture location revealed large particles that appeared to be SiC catalyst carrier material, as well as numerous smaller particles that appeared to be magnetite particles. No evidence of melted material was observed. Most of the deposit appeared to be fractured magnetite (iron oxide).

X-ray diffraction of the deposit sample determined that magnetite (Fe_3O_4), silicon carbide (SiC), and rutile (TiO_2) were present. The SiC and TiO_2 were presumably from the catalyst and catalyst carrier material. Several peaks in the XRD spectrum could not be identified. No evidence of metallic iron was found in the deposit.

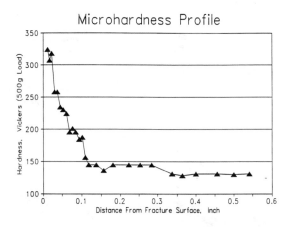

Microhardness Profile

Fig. 8 Microhardness profile from fracture surface to unaffected tube material

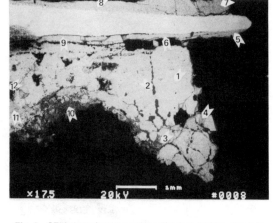

Fig. 9 SEM view of cross-sectioned fracture surface shown in Fig. 4. Arrows indicate areas analyzed. 10.9×

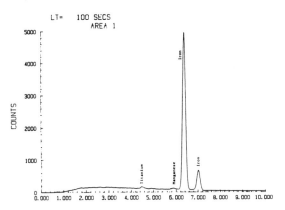

(a)

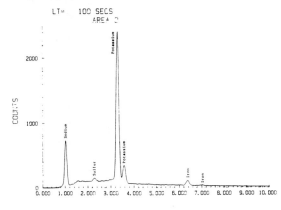

(b)

Fig. 10 EDS spectra of region 1 (a) and region 2 (b) shown in Fig. 9

(a)

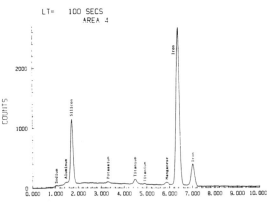

(b)

Fig. 11 EDS spectra of region 3 (a) and region 4 (b) shown in Fig. 9

Discussion

Examination of the fractured ends of the tube revealed that tube thinning occurred from the inner surface and that little thinning occurred from the outer surface. Examination of the entire length of the tube did not reveal other areas of inner surface discolorations or tube wall thinning.

The microstructure of the tube at the fracture surface was tempered martensite, which indicated that the tube temperature had exceeded 860 °C (1580 °F) in the failed region, and then had been rapidly quenched to a temperature below approxi-

mately 480 °C (895 °F). These temperatures are based on the temperature needed to austenitize the tube alloy (0.10% C steel) and the quenching temperature necessary to form martensite. Because the salt bath temperature was reported to be approximately 370 °C (700 °F), it is apparent that molten salt acted as the quenching medium.

The microstructure changed gradually from the fracture surface (martensite) to the typical microstructure (ferrite with pearlite) within approximately 19 mm (3/4 in.) of the tube length, indicating

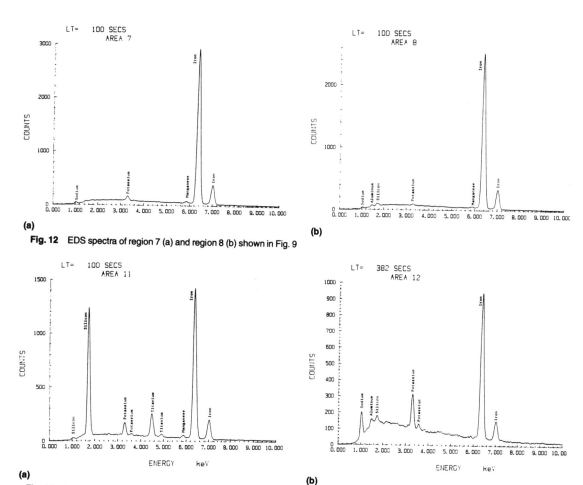

Fig. 12 EDS spectra of region 7 (a) and region 8 (b) shown in Fig. 9

Fig. 13 EDS spectra of region 11 (a) and region 12 (b) shown in Fig. 9

that the tube had been extremely overheated, fractured, and then rapidly quenched (probably by the salt bath), which extracted heat from the fracture surface area.

Because no evidence of melted metal was found at the fracture surface or in the deposit sample removed from the fractured region of the tube, it appears that the tube had locally thinned from the inside surface until it could no longer support the internal pressure and fractured. The formation of a thick oxide on the outer surface of the tube in the failed region caused the tube temperature to locally increase, which accelerated the oxidation rate and eventually led to catastrophic failure.

The oxide/deposit material found on the inner surface of the tube contained primarily magnetite, silicon carbide, and titanium oxide catalyst material. This indicates that the catalyst carriers at the failed region had become entrapped in the growing oxide on the inner surface. In addition, salt from the salt bath was found in the oxide/deposit, indicating that salt flowing into the tube had quenched the tube and been incorporated/entrapped in the oxide.

No evidence of melted or fractured tube material was found in the oxide/deposit in the heat-affected region of the cross-sectioned tube or in the submitted deposit sample, based on XRD analysis. These results indicate that the tube had not reached a temperature high enough to melt the tube alloy (~1530 °C, or 2785 °F).

Conclusion

It appears that the tube thinned from the inner surface because of a localized overheating condition and then fractured, whereupon molten salt flowed into the tube. It seems likely that a runaway chemical reaction within the tube occurred to cause internal overheating. Based on examination of the tube microstructure, the tube temperature reached at least 860 °C (1580 °F), the tube fractured, and then the fracture surface was rapidly quenched to a temperature below approximately 480 °C (895 °F) (probably by molten salt from the salt bath). No evidence of melted tube material was found on the fractured tube halves or in the deposit material that had been removed from within the tube at the fracture location, indicating that the tube temperature did not exceed the melting point of the steel.

Failure of an Aluminum Brass Condenser Tube

H.S. Khatak and J.B. Gnanamoorthy, Metallurgy Division, Indira Gandhi Centre for Atomic Research, Kalpakkam, India

Leaks developed at random locations in aluminum brass condenser tubes within the first year of operation of a steam condenser in a nuclear power plant. One failed tube underwent scanning electron microscopy surface examination and optical microscope metallography. It was determined that the tube failed from crevice corrosion under seawater deposits that had formed on the inner surface. Mechanical cleaning of the condenser tubes every 6 months and installation of intake screens of smaller mesh size were recommended.

Key Words   Aluminum bronzes, corrosion   Crevice corrosion   Condenser tubes, corrosion
Steam condensers   Nuclear reactor components, corrosion   Seawater, environment

Background

Leaks developed at random locations in aluminum brass condenser tubes in a steam condenser within the first year of operation.

Applications

The condenser was used to condense steam from a nuclear power plant. Seawater was used for cooling.

Circumstances leading to failure

The first leak was detected after less than 1 year of operation. A total of 59 tubes had failed in 6 years.

Pertinent specifications

Seawater flows inside the aluminum brass condenser tubes with a velocity of about 1.5 m/s (5 ft/s). Traveling water screens of 10×10 mm (0.4×0.4 in.) mesh are used on the suction side of the condenser water pump to avoid the entry of shells and other debris into the condenser tubes. Continuous low-dosage chlorination and intermittent $FeSO_4$ dosing have been adopted, as well as occasional booster doses of chlorination. Regular back washing, in addition to mechanical wire-brush cleaning, is performed during annual plant shutdown to flush stagnant seawater from the tubes, which sag. The maximum operating temperature is 40 °C (100 °F).

Specimen selection

A failed tube was sent for analysis. Failure locations were determined visually.

Visual Examination of General Physical Features

The various failed condenser tubes had leaked at random locations. In the tube under examination, an irregularly shaped hole approximately 3 mm (0.12 in.) in diameter and two smaller holes were observed. The outer surface (steam side) of the tube appeared to be uniformly black. The inner surface exhibited a uniform brown corrosion product. There was no indication of preferential deposition.

Dimensional measurements at different locations showed the wall thickness to vary between 1.2 and 1.3 mm (0.047 and 0.051 in.). Loss of thickness was found only near the puncture. The tube was cut axially, and examination of the inner surface revealed severe localized corrosion attack up to an approximate diameter of 10 mm (0.4 in.) around the puncture. No other corrosion attack was visible, nor did the corroded surface exhibit any deposit. Corrosion product or debris might have been washed away after the leak. Crevice-type corrosion attack at the inner surface and the shape of the puncture on both the inner and outer sides indicated that corrosion could have initiated from inside the tube (the seawater side).

Testing Procedure and Results

Surface examination

Scanning Electron Microscopy. The regions around the hole, both inside and outside, were examined using a scanning electron microscope (SEM) (Fig. 1). The corroded inside surface showed typical localized dissolution.

Metallography

Optical microscopic examination of the polished and etched surface of a tube sample taken near the punctured region revealed a normal microstructure.

Chemical analysis/identification

Metal chips taken from a location away from the failed region were analyzed. The chemical composition was found to be 77.8% Cu, 19.8% Zn, 2.2% Al, 0.02% Sn, and 225 ppm As.

Discussion

Chemical analysis and microstructural examination did not reveal any abnormalities in the material. The absence of preferential deposit and lack of corrosion at surfaces other than the punctured region suggested that proper water chemistry and flow conditions were maintained. From the topography/morphology of the region around the puncture and the localized nature of attack, it is obvious that failure occurred because of dissolution under deposits. This type of crevice attack is caused by an oxygen concentration cell that forms between an oxygen-deficient area under the deposit and the rest of the surface of the tube. Many such localized failures (pitting/crevice) have been reported in

Fig. 1 SEM micrographs. **(a)** Outside surface. **(b)** Inside surface

copper-base alloys.

The deposit in the present case was not observed because of the operating conditions of the condenser, but deposits (seashells, debris, biological growth, and the like) are carried by seawater. The 10 × 10 mm (0.4 × 0.4 in.) mesh water screen used on the suction side of the pump was coarse enough to allow debris to pass through.

Because the presence of scales, shell fragments, or corrosion products stimulates pitting and, in particular, crevice corrosion, one obvious

method to reduce or even eliminate susceptibility to attack is to keep the tube clean. Biofouling can be controlled by modern intake screening techniques, periodic chlorination of the cooling water, or periodic thermal shock. Intake screens are now available with mesh sizes as fine as 0.5 mm (0.02 in.). The thermal shock method involves recirculating the cooling water to allow the temperature to increase to 50 °C (120 °F), a condition that ensures the death of most organisms.

Conclusion and Recommendations	**Most probable cause**	**How failure could have been prevented**
	The failure of the tube occurred because of crevice corrosion under the deposit.	The use of an intake screen of smaller mesh size, frequent mechanical cleaning, and occasional thermal shocks to kill organisms could have prevented the failure.

Failure of an Embrittled Tube

S.A. Bradley, UOP, Des Plaines, Illinois

The causes of cracking of an as-drawn 90-10 cupronickel tube during mechanical working were investigated to determine the source of embrittlement. Embrittlement was sporadic, but when present was typically noted after the first process anneal. Microstructural and chemical analyses were performed on an embrittled section and on a section from a different lot that did not crack during forming. The failed section showed an intergranular fracture path. Examination of the fracture surfaces revealed the presence of tellurium at the grain boundaries. The source of the tellurium was thought to be contamination occurring in the casting process that became concentrated in the recycled skimmings. It was recommended that future material specifications for skimmings and for externally obtained scrap copper include a trace analysis for tellurium.

Key Words			
	Cupronickel	Intergranular fracture	Tubes
	Segregations	Embrittlement	Tellurium, impurities
Alloys	Cupronickel		

Background

An as-drawn 90-10 cupronickel tube readily cracked when mechanically worked. No unusual processing conditions during manufacture of the tube could be identified. To complicate matters, many of the process variables were changed during the height of the problem in an attempt to eliminate the tube embrittlement, including the alkaline dip, pickling operation, furnace atmosphere, drawing lubricant, and temperature control during hot drawing and annealing. Furthermore, tube embrittlement was not always observed, but when present was typically noted after the first process anneal. Flattening the tube demonstrated the embrittlement (Fig. 1).

(a)

(b)

Fig. 1 Expected formability of tube (a) and observed cracking in embrittled tube after flattening (b)

Testing Procedure and Results

Metallography

Microstructural Analysis. Metallographic analysis was performed on the embrittled tube and on a tube section from a different lot of material that did not crack during forming. A failed section had a fracture path that was definitely intergranular. Initial metallographic examination with etchants typically used for copper-base alloys revealed little difference between the two tubes. When an etchant with a small amount of silver nitrate added to the nitric acid was used, the grain boundaries of the embrittled tube were attacked (Fig. 2), suggesting that an intergranular constituent was causing the embrittlement.

Crack Origins/Paths. A fracture surface was produced and examined in the scanning electron microscope (SEM); a typical brittle failure was noted. With energy-dispersive analysis, only copper, nickel, iron, and manganese were detected, and all constituents were present in the expected proportions. The metallographic cross section shown in Fig. 2 was also studied in the SEM. Again, no unusual elements were detected in the grain boundaries.

The embrittled tube was then analyzed using an Auger electron scanning microprobe. A notched sample was fractured under ultrahigh vacuum. Tellurium and carbon, along with oxygen, copper, and nickel, were detected on much of the intergranular fracture surface. Scatter maps, showing the location of tellurium and carbon (Fig. 3), demonstrated tellurium inclusions at many grain boundaries. Carbon was usually, but not always, present with the tellurium. The tellurium quickly disappeared during depth profiling; thus, the inclusions were probably quite thin. Subsequent multiple trace chemical analyses by inductively coupled plasma spectroscopy that was specifically targeted for tellurium did detect a concentration level of 15 to 25 ppm in the embrittled tube. Tellurium could not be detected in the nonembrittled tube; the detectability limit for tellurium is estimated to be 1 ppm.

Chemical analysis/identification

Emission spectroscopic analysis determined that the tube was within chemical specification. Elements known to embrittle copper-nickel alloys, such as bismuth, were not initially found.

Mechanical properties

Hardness. The microhardnesses of the two tube sections were virtually identical.

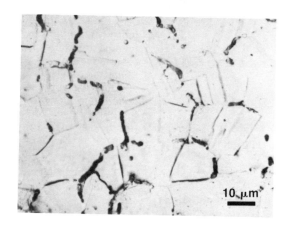

Fig. 2 Intergranular etching of embrittled tube

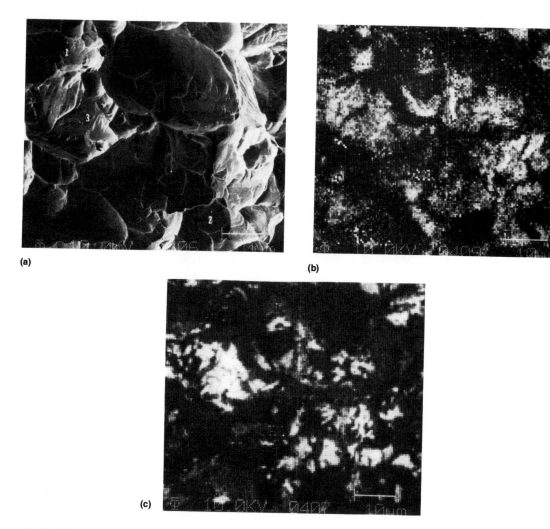

Fig. 3 (a) Fracture surface. (b) Scatter map for tellurium. (c) Scatter map for carbon

Discussion

Many polycrystalline metals can become brittle and fail along grain boundaries when a low stress is applied. The brittleness has been attributed to intergranular weakness caused by the precipitation or segregation of impurities to grain boundaries. For example, Auger electron spectros-copy of intergranular fracture surfaces has shown that antimony segregation to grain boundaries can embrittle iron and that the presence of nickel can enhance the antimony segregation (Ref 1). Copper, even though a ductile metal when pure, is no exception. Embrittlement studies of copper have been limited to the effects of antimony (Ref 2), bismuth (Ref 3), and oxygen (Ref 4). The embrittlement is caused by the reduction of grain-boundary cohesion associated with the segregation of solute atoms to grain boundaries.

Observations of tellurium segregation to grain boundaries and of the tellurium being present as a few atomic layers on the intergranular fracture indicated a similarity to bismuth-induced embrittle-ment of copper (Ref 5). Furthermore, as with bismuth, tellurium has virtually no solubility in either copper or nickel. Thus, by analogy, tellurium was concluded to be the probable embrittling agent. Although carbon was detected in some locations on the fracture surface, it probably did not contribute to the embrittlement. This conclusion is based on carbon having some solubility in both copper and nickel.

Conclusion and Recommendations

The embrittling agent for the failed cupronickel tube was most likely tellurium (source unknown). The tellurium probably entered the casting process as a contaminant. The general practice is to recycle the skimmings, which entrap impurities in the slag. The brittle slag is typically removed by a crushing operation. The tellurium impurities were probably sufficiently concentrated in the metallic portion of the skimming to embrittle subsequently cast tubes. The tellurium level was quite low in the tube and, unless specifically sought, was difficult to detect. Examination of skimmings that were not associated with these embrittled tubes did not reveal any detectable quantities of tellurium. This was not unexpected, because the problem was sporadic. The source of the carbon was unknown, but may have been associated with the tellurium. It was recommended that future material specifications for skimmings and for externally obtained scrap copper include a trace analysis for tellurium.

References

1. L.L. Briant, *Acta Metall.*, Vol 35, 1987, p 149
2. H.R. Tipler and D. McLean, *Met. Sci. J.*, Vol 4, 1970, p 103
3. J.D. Russell and A.T. Winter, *Scr. Metall.*, Vol 19, 1985, p 575
4. T.G. Nieh and W.D. Nix, *Metall. Trans. A*, Vol 12, 1981, p 893
5. A. Joshi and D.F. Stein, *J. Inst. Met.*, Vol 99, 1971, p 178

Galvanic Corrosion Failure of Austenitic Stainless Steel Pipe Flange Assemblies

Edward C. Lochanski, Metalmax, Thiensville, Wisconsin

Catastrophic pitting corrosion occurred in type 304L stainless steel pipe flange assemblies in an industrial food processor. During regular service the pumped medium was pureed vegetables. In situ maintenance procedures included cleaning of the assemblies with a sodium hypochlorite solution. It was determined that the assemblies failed due to an austenite-martensite galvanic couple activated by a chlorine bearing electrolyte. The martensitic areas resulted from a transformation during cold-forming operations. Solution annealing after forming, revision of the design of the pipe flange assemblies to eliminate the forming operation, and removal of the source of chlorine were recommended.

Key Words	Austenitic stainless steels, corrosion Food processing	Galvanic corrosion Pitting (corrosion)	Pipe flanges, corrosion
Alloys	Austenitic stainless steel—304L		

Background

Catastrophic pitting corrosion occurred in type 304L stainless steel pipe flange assemblies.

Applications

The pipe flange assemblies were used in an industrial food processor. During regular service, the pumped medium was pureed vegetables. During regular *in situ* maintenance, the inside and outside surfaces of the pipe flange assemblies were cleaned with a sodium hypochlorite solution, the active ingredient in common bleach. The sodium hypochlorite solution was then removed by a series of rinse and neutralization operations.

Circumstances leading to failure

After cost-driven change in raw stock vendors, new stock pipe flange assemblies began to corrode at the outer pipe bends within 2 days of installation. A simultaneous darkening of the processed food product suggested that corrosion was also occurring at the pipe inner bends. No changes in the regular service schedules or the regular maintenance schedules were reported.

Pertinent specifications

The pipe was specified to be manufactured from a formed section of seamless AISI 304L stainless steel tube welded to two machined sections of AISI 304L stainless steel flanges. Except for the weld zones, the pipe and flange microstructures were specified to contain only austenite. The weld zone microstructures were specified to contain 3 to 5% ferrite in a matrix of austenite.

The pipe flange assemblies were required to be passivated by the vendor per ASTM A380, "Cleaning and Descaling Stainless Steel Parts, Equipment and Systems." The pipe assemblies were required to be checked for proper passivation by in-house quality-control personnel per ASTM A380, "The Copper Sulfate Test" (section 7.2.5.3). No hardness or tensile specifications were cited.

Specimen selection

Several pipe flange assemblies were submitted for analysis—both unused and used, and both old and new.

Visual Examination of General Physical Features

The used old stock pipe flange assemblies exhibited smooth topographies, with no evidence of oxide discoloration at the weld zones. The used new stock pipe flange assemblies exhibited catastrophic pitting corrosion at the outer and inner walls of the pipe. The corrosion damage was most apparent at the outer and inner bends. All other surfaces exhibited smooth topographies, with no evidence of oxide discoloration at the weld zones.

Testing Procedure and Results

Nondestructive evaluation

The copper sulfate test is a hypersensitive test for iron metal (such as embedded steel machining particles) and for iron oxide (such as welding scale). The presence of iron metal and iron oxide is highlighted by copper plating. Neither the unused old nor the unused new stock pipe flange assemblies showed any indications of deposited copper—evidence that both types had passed.

Metallography

Microstructural Analysis. Figure 1 shows a typical unetched transverse, metallographic cross section of a used old stock pipe, revealing extremely small, round duplex inclusions. The corresponding etched view (Fig. 2) illustrates a fully austenitic matrix, except in the weld zone, which contained approximately 3 to 5% ferrite in an austenitic matrix (Fig. 3). Thus, the used old stock pipe met the specified AISI 304L microstructural requirements.

Figure 4 shows that a typical unetched transverse metallographic cross section of a used new stock pipe also displayed extremely small, round duplex inclusions. The corresponding etched views illustrate a fully austenitic matrix in uncorroded areas (Fig. 5) and a part austenitic and part martensitic matrix in corroded areas (Fig. 6). The weld zone contained approximately 3 to 5% ferrite in an austenitic matrix (Fig. 7). Thus, the used new stock pipe met the specified AISI 304L microstructural requirements, except at bend areas.

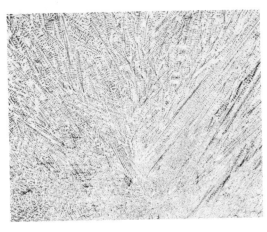

Fig. 1 Typical unetched transverse metallographic cross section of a used old stock pipe, displaying extremely small, round duplex inclusions. 63×

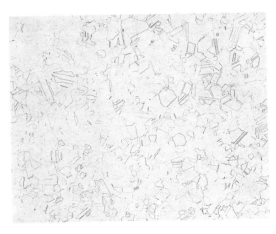

Fig. 2 Etched view corresponding to Fig. 1 of a typical area away from the weld zone, illustrating a fully austenitic matrix. 63×

Fig. 3 Etched view corresponding to Fig. 1 of a typical area in the weld zone, showing approximately 3 to 5% ferrite in an austenitic matrix. 63×

Fig. 4 Typical unetched transverse metallographic cross section of a used new stock pipe, displaying extremely small, round duplex inclusions. 63×

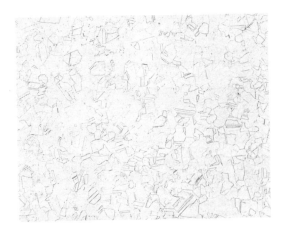

Fig. 5 Etched view corresponding to Fig. 4 of a fully austenitic matrix in uncorroded areas. 63×

Fig. 6 Etched view corresponding to Fig. 4 of a part austenitic and part martensitic matrix in corroded areas. 63×

Chemical analysis/identification

Material. Optical emission spectroscopy and combustion (carbon and sulfur) results verified that the used old stock pipe and the used new stock pipe met the specified AISI 304L chemical composition requirements (Table 1).

Corrosion Deposits. Qualitative energy-dispersive spectroscopy (EDS) analysis of the typical clean pipe and the typical pipe corrosion products found in the unetched transverse metallographic

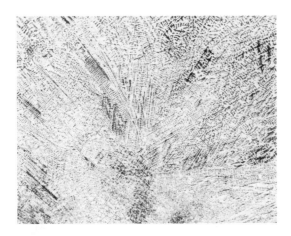

Fig. 7 Etched view corresponding to Fig. 4 of a typical area in the weld zone, showing approximately 3-5% ferrite in an austenitic matrix. 63×

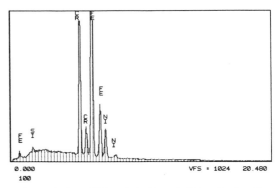

Fig. 8 Qualitative EDS analysis of the typical clean pipe (found in the unetched transverse metallographic cross section of the used new stock pipe), showing the base metal elemental composition

Table 1 Results of optical emission spectroscopy and combustion (carbon and sulfur)

Element	New stock pipe	Old stock pipe	AISI 304L requirements(a)
Chromium	18.21	18.55	18.0-20.0
Nickel	11.08	10.84	8.0-12.0
Molybdenum	0.22,0.16	NR	
Carbon	0.014	0.022	0.03 (max)
Manganese	1.29	1.58	2.00 (max)
Silicon	0.44	0.55	0.75 (max)
Phosphorus	0.016	0.023	0.045 (max)
Sulfur	0.009	0.017	0.030 (max)
Nitrogen	< 0.009	< 0.002	0.10 (max)
Titanium	< 0.01	< 0.01	NR
Niobium	< 0.01	< 0.01	NR
Copper	0.02	0.01	NR

(a) Nr, no requirement

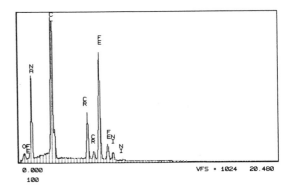

Fig. 9 Qualitative EDS analysis of the typical pipe corrosion products (found in the unetched transverse metallographic cross section of the used new stock pipe), identifying chlorine as the primary corrosive agent

cross section of the used new stock pipe assembly clearly identified chlorine as the primary corrosive agent (Fig. 8 and 9). EDS results are shown for all elements above atomic number 4. The vertical scales represent X-ray counts per unit time, and the horizontal scales represent X-ray energy in thousands of electron volts.

Discussion

Metallography and chemical analysis confirmed that the used new stock austenitic stainless steel pipe flange assemblies failed because of an austenite-martensite galvanic couple activated by a chlorine bearing electrolyte. The undesirable martensitic areas were clearly the result of an austenite-to-martensite transformation produced during the pipe cold-forming operations. The old stock pipes were solution annealed after forming, which eliminated any martensitic areas at the pipe outer bends. It was discovered that the new stock pipes were not solution annealed after forming. The new vendor had incorrectly ascertained that their new-generation forming equipment would not cold work the pipe outer bends enough to form martensitic areas. The austenite-ferrite microstructures of the weld zones did not exhibit microstructurally induced galvanic corrosion because of a unique joint design. The weld zones were completely encapsulated in plastic, molded to provide a gasket surface.

The undesirable chlorine residues were the result of contamination. After a methodical check of chlorine sources (trichloroethane degreasers, hydrochloric acid pickling solutions, sodium chloride water softener salts, bleaches, etc.), the source of the chlorine contamination was traced to the sodium hypochlorite cleaning solution. Therefore, it was reasoned that the maintenance procedure was not correctly followed.

After days of on-site observation, nothing out of the ordinary in the service or maintenance of the food processor could be cited. The pipes remained free of corrosion at the outer bends. The day after the on-site observation team left, the same pipes began to show severe corrosion at the outer bends. The food processor operator claimed that all of the daily procedures were identical to those followed during the observation period. However, a security video camera inadvertently filmed the operator skipping the maintenance rinse and neutralization processes.

Conclusion and Recommendations

Most probable cause

The most probable cause of the new stock pipe flange assembly failures was an austenite-martensite galvanic couple at the new stock pipe outer bends. The old stock pipe flange assemblies exhibited no austenite-martensite areas at the pipe outer bends; that is, the outer bends were fully austenitic. A strong contributing factor to the failure was chlorine contamination.

Remedial action

Several recommendations were offered. The pipe should be solution annealed after the forming operation. A solution anneal reverts any martensite back into austenite, eliminating the potential for an austenite-martensite galvanic couple. A fully austenitic condition usually ensures the highest possible degree of corrosion resistance.

The pipe flange assembly design should be revised so that the pipe forming operation is eliminated. The pipe bend area material could be replaced with a molded inert plastic such as Teflon. The source of chlorine-bearing electrolyte should be removed by use of a chlorine-free cleaner that is aggressive to the pumped food product, but not the pipe.

Graphitization-Related Failure of a Low-Alloy Steel Superheater Tube

David J. Kotwica, Betz Laboratories, Inc., Woodlands, Texas

A carbon-molybdenum (ASTM A209 Grade T1) steel superheater tube section in an 8.6 MPa (1250 psig) boiler cracked because of long-term overheating damage that resulted from prolonged exposure to metal temperatures between 482 °C (900 °F) and 551 °C (1025 °F). The outer diameter of the tube exhibited a crack (fissure) oriented approximately 45° to the longitudinal axis and 3.8 cm (1.5 in.) long. The inner diameter surface showed a fissure in the same location and orientation. Microstructure at the failure near the outer diameter surface exhibited evidence of creep cracking and creep void formation at the fissure. A nearly continuous band of graphite nodules was observed on the surface of the fissure. In addition to the graphite band formation, the microstructure near the failure exhibited carbide spheroidization from long-term overheating in all the tube regions examined. It was concluded that preferential nucleations of graphite nodules in a series of bands weakened the steel locally, producing preferred fracture paths. Formation of these graphite bands probably expedited the creep failure of the tube. Future failures may be avoided by using low-alloy steels with chromium additions such as ASTM A213 Grade T11 or T22, which are resistant to graphitization damage.

Key Words	Molybdenum steels, Mechanical properties Graphitization, Heating effects	Boiler tubes, Mechanical properties Creep (materials), Microstructural effects	Superheaters Overheating
Alloys	Molybdenum steels—A209 Grade T1		

Background

A low-alloy steel superheater tube failed in a two drum Sterling design boiler in service at a chemical plant. At the time of failure, the superheater tube had been in service for approximately 24 years.

Application

The superheater tube as-submitted had a nominal outer diameter of 63.5 mm (2.5 in.) and a nominal wall thickness of 4.4 mm (0.172 in.). It was intended as a steam bearing tube for an 8.6 MPa (1250 psig) boiler.

Pertinent specifications

The superheater tube had been fabricated from annealed ASTM A209 Grade T1 seamless carbon-molybdenum alloy steel.

Performance of other parts in same or similar service

A tube in the waterwall bank had failed in service due to long-term overheating at the same time. The exact cause of the overheating of the waterwall tube (e.g., excessive heat flux, internal deposits, and/or insufficient circulation in this section) was not specifically detailed.

Selection of specimens

A 61-cm (24-in.) long section of the superheater tube that failed was examined in the laboratory.

Visual Examination and General Physical Features

The external surface of the superheater tube is shown in Fig. 1. The failure consisted of a single fissure (Fig. 2), with no visible evidence of bulging or secondary cracking. No evidence of significant ash build-up was observed on the outer diameter surface. However, an iron oxide scale layer approximately 0.2 mm (0.008 in.) thick was observed under a very thin layer of brownish deposit on the outer diameter surface.

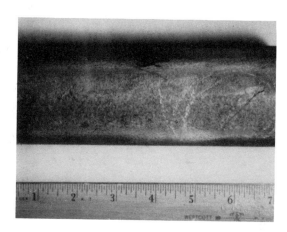

Fig. 1 The external surface of the submitted superheater tube. 0.46×

Fig. 2 The fissure oriented at 45° on the outer diameter surface. 0.76×

Testing Procedure and Results

Surface examination

The outer diameter surface of the tube exhibited a crack (fissure) oriented approximately 45° to the longitudinal axis and 3.8 cm (1.5 in.) long. No apparent wall thinning or bulging accompanied the fissure. The tube was split longitudinally with a band saw to facilitate examination of the inner diameter surface. The inner diameter surface exhibited a fissure in the same location and orientation as that observed on the outer diameter surface (Fig. 3). Therefore, the fissure had propagated through the entire wall thickness. Also shown in Fig. 3 are secondary fissures oriented similarly. The entire internal surface was covered with a uniform adherent dark gray iron oxide scale (most likely magnetite) 1.5 mm (0.006 in.) to 0.2 mm (0.008 in.) thick.

Metallography

A microstructural study of the failure and adjacent inner diameter fissures was performed by cross-sectioning the tube perpendicular to the fissures. The microstructure at the failure near the outer diameter surface exhibited evidence of creep cracking and creep void formation at the fissure (Fig. 4). Upon close examination, an almost continuous band of graphite nodules on the surface of the fissure was observed. Inspection revealed slight curvatures of the fissure surface accentuating the individual nodule positions.

A second array of graphite nodules was observed adjacent to the fissure and exhibiting the same orientation in the microstructure (Fig. 5). This band of nodules corresponded to a secondary fissure on the inner diameter surface (see Fig. 3). Examination of the secondary band of graphite nodules revealed small cracks linking the individual graphite nodules (Fig. 6). In addition to the graphite band formation, the microstructure near the failure exhibited carbide spheroidization from long-term overheating in all the regions examined. Figure 7 shows the typical mid-wall microstructure adjacent to the failed region.

Chemical analysis/identification

Material. A portion of the superheater tube was chemically analyzed by optical emission spectroscopy (OES) to confirm the alloy composition. The results, given in Table 1, indicate that the tube met the compositional requirements for ASTM A209 Grade T1 superheater tubing.

Internal Surface Coating. Scanning electron microscope/energy dispersive spectroscopy (SEM/EDS) was used to determine if any contaminants were present in the inner diameter surface magnetite scale. A specimen was prepared by scraping the scale off and mixing to produce a homogeneous sampling. The results indicated that

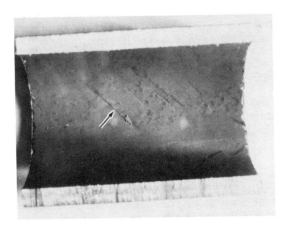

Fig. 3 The inner diameter surface of the superheater tube, at the failure. The arrow indicates the failure site. 0.68×

Fig. 4 Photomicrograph showing the crack morphology observed at the failure adjacent to the outer diameter surface. Unetched. 38×

Fig. 5 Photomicrograph showing secondary band of graphite nodules adjacent to the outer diameter surface. Nital etch. 29×

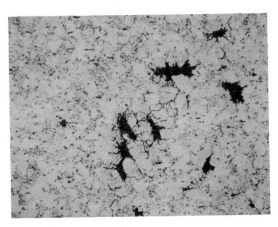

Fig. 6 Photomicrograph showing the linkage of individual graphite nodules by a network of cracks. Nital etch. 152×

Table 1 Alloy analysis by wt%

Element	Tube sample	ASTM A209 Grade T1 specification
C	0.12	0.10 to 0.20
Mn	0.48	0.30 to 0.80
P	0.005	0.045 max
S	0.016	0.045 max
Si	0.21	0.10 to 0.50
Mo	0.47	0.44 to 0.65
Ni	0.07	ns
Cr	0.10	ns
Cu	0.17	ns
Fe	Rem	Rem

ns, not specified

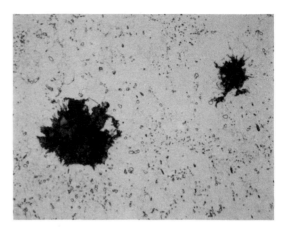

Fig. 7 Photomicrograph showing a close-up of two graphite nodules and carbide spheroidization typical of the mid-wall microstructure of the tube. Nital etch. 285×

the inner diameter scale consisted entirely of iron oxides. No impurities were detected.

Discussion

The failure of the Grade T1 superheater tube can be attributed to severe long-term overheating and the subsequent formation of bands of graphite nodules. Graphitization is a microstructural change that sometimes occurs in carbon and 0.5% molybdenum low-alloy steels subjected to moderately excessive metal temperatures 482 to 551 °C (900 to 1025 °F) for long periods of time (Ref 1). Because the carbon in this steel as-manufactured is usually present in a metastable state as carbides, thermal decomposition of the cementite (Fe_3C) platelets in the pearlite grains can result in a mixture of carbon (graphite) and ferrite. A competing form of pearlite decomposition (carbide spheroidization) will also occur at these temperatures, which was found in the microstructure. In superheater tubing, usually the formation of spheroidized carbides is predominant when metal temperatures exceed 551 °C (1025 °F).

The band of preferred graphite nodule formation at the failure likely occurred due to preferential nucleation sites. The pattern of the nodules (orientation 45° to the longitudinal axis) suggests that they formed at sites associated with internal strain bands (possibly slip planes that formed in the steel from the original manufacturing of the tube) in the steel (Ref 2, 3). Thus, the term "slip-plane graphitization" has been coined. Nucleation then proceeds at sites of higher dislocation densities and/or stacking faults in a preferential manner. In conjunction with the slip-band graphitization, a general weakening of the mechanical properties of the tube and the onset of creep, which was observed at the outer diameter surface of the failure, caused the final failure of the tube.

Conclusion and Recommendations

Most probable cause

The failure in the Grade T1 superheater tube was precipitated by exposure to higher-than-design metal temperatures (long-term overheating at metal temperatures 551 °C, or 1025 °F). The most likely causes of overheating in superheater tubing are inadequate steam distribution and/or excessive heat flux. Microstructural analysis of the failed region indicated the formation of graphite nodule arrays (slip-band graphitization) at 45° to the longitudinal axis. Since graphite nodules have little tensile strength, the slip-band graphitization severely weakened the steel along these paths. The failure of the tube occurred as creep cracking and creep void development occurred along the tube surfaces at the site of localized deterioration of mechanical properties (a graphite nodule array).

Remedial action

In addition to identifying and eliminating the excessive metal temperatures experienced in this region of the superheater bank, avoiding failures of this nature may include consideration of upgrading the superheater tubing from A209 T1 to a chromium-containing grade (e.g., ASTM A213 Grade T11 or T22). Both of these grades contain enough chromium to be immune to carbide graphitization. Additionally, T11 and T22 offer enhanced creep/oxidation resistance properties.

References

1. *Metals Handbook, Vol 11: Failure Analysis and Prevention*, 9th ed., ASM, Metals Park, OH, 1986, p 603-614.
2. R.W. Emerson and M. Morrow, *Trans. ASME*, Vol 68, 1946, p 605.
3. R.D. Port, "Non-Weld-Related Graphitization Failures," *CORROSION/89*, Paper No. 248, NACE Corrosion Conference, New Orleans, 1989.

Leaks in Copper Tubing From Cooling Coils of a Large Air-Conditioning Unit

J.O. Edwards and R.I. Hamilton, Metals Technology Laboratories, CANMET, Ottawa, Canada

Copper tubes from the cooler assemblies of a large air-conditioning unit exhibited leakage upon installation of the unit. Sections from two leaking tubes and one nonleaking tube were subjected to pressure testing and microscopic examination. The cause of leaking was determined to be pitting corrosion. Extensive pitting was found on the insides of all sections examined, with deep and numerous pits in leaking areas. Circumstantial evidence indicated that antifreeze solution left in the tubes from the manufacturing operation was the most likely cause of the pitting.

Key Words			
Copper, corrosion	Air-conditioning equipment	Condenser tubes, corrosion
Leakage	Pitting (corrosion)	

Background

Copper tubes from the cooler assemblies of a large air-conditioning unit exhibited leakage upon installation of the unit.

Specimen selection

Two 46 cm (18 in.) long tubes from newly installed chilled water coils were sent to a laboratory for testing. Both had been pressure tested—one had exhibited leaks, but the other did not. A shorter 23 cm (9 in.) section from a leaking tube was also supplied.

Visual Examination of General Physical Features

The exteriors of the tubes were clean and bright and showed no obvious leaks. All had been heavily deformed by the fin winding machine.

Testing Procedure and Results

Nondestructive evaluation

To more positively identify the source of leakage, both long tubes were pressurized with nitrogen at 276 kPa (40 psi) and brushed with soap solution. The leaking tube exhibited bubble generation over a considerable area at one end, and the source of these numerous leaks was identified for subsequent microexamination. Careful examination of the exterior showed minute brown spots as the source of the leaks. The other tube did not leak.

Surface examination

Macroexamination. Sections of the tube were split longitudinally. Of the two leaking tubes, the short tube exhibited a yellow, rusty interior, with some "water line" effects that indicated top and bottom orientation. The longer tube that leaked displayed a relatively smooth, blue-black interior deposit. The tube that did not leak had a rusty interior similar to the first tube. At higher magnification, all interior deposits exhibited small, bright crystalline facets. Sections from all the tubes were dipped in 10% H_2SO_4 solution to remove the interior deposits, after which they were examined at 20×. The short tube that leaked showed extensive pitting over the entire interior surface. Many of these pits were obviously deep, and some presumably penetrated the tube, although this was not apparent from the outside. There was an overall mottled "raindrop" effect on the interior, and although pits were often located at the center of the drop, this may have been a chance occurrence. The typical appearance of the pits is shown in Fig. 1. Similar pitting, but much less severe, occurred in an area adjacent to established leaks in the long tube,

Fig. 1 Interior of cleaned, short length from leaking tube, showing extensive pitting and "raindrop" staining. 5.2×

but only a trace of such pitting was visible in the tube that did not leak.

Scanning Electron Microscopy. Samples from the three tubes were submitted for examination by scanning electron microscopy (SEM) to determine the nature of the interior corrosion deposit. The predominant elements were shown to be copper and iron, with smaller quantities of phosphorus, sulfur, and calcium. A section from a pitted tube was flattened and the scale picked off; in this case, significant amounts of chlorine were found in the vicinity of the pit.

Cracks were also noted at the edges of the pits after flattening, and small surface cracks were also visible in some areas away from the pits, indicating embrittlement. Figure 2 shows such cracking at the edge of the pit. The crystalline nature of the corrosion deposit at the bottom of the figure is also apparent.

Fig. 2 SEM micrograph of pit with corrosion deposit partially removed. Note cracking at edge of pit and crystalline nature of remaining corrosion deposit. 388.8×

Fig. 3 Shallow and moderate pits on the interior of the long length of leaking tube. 52×

Fig. 4 Complete penetration of the tube by a pit. Note deformation of exterior (top) of tube by fin winding and relatively thick corrosion deposit on the interior (bottom) of the tube. 52×

Fig. 5 Branching pit. Corrosion deposit on interior is typical and is probably some form of cuprous oxide. 52×

Metallography

Sections were taken from opposite sides of both long tubes and examined after polishing. These sections showed marked pitting attack over the entire interior surface of the leaking tube. Shallow pits appeared as indentations filled with corrosion deposits, probably essentially cuprous oxide, with progressively greater damage with the deeper pits. Examples are shown in Fig. 3 to 5, taken from the area in which the leaks originated.

In the nonleaking tube, several superficial pits were present. This tube also had a much thinner corrosion deposit over the interior of the tube compared with the tube that leaked.

The microstructure was typical of phosphorus-deoxidized copper in the annealed or lightly cold-worked condition, with an average grain diameter of 0.045 mm (0.002 in.).

Discussion

Figures 1 to 5 show the very extensive pitting damage observed on two of the three tubes examined and the superficial damage of the same type visible on the third, nonleaking sample. Such pitting is often caused by solutions that are only mildly corrosive and are the result of local differences in potential caused by local concentrations of ions. Hence, pitting is favored by loose deposits of dirt and scale and by stagnant conditions. Water with high concentrations of oxygen and carbon dioxide is a common pitting agent for copper. In any event, it is apparent that, for pitting to occur, the surface must be wet. In general, pitting usually consists of a small number of deep pits or a large number of shallow pits. In the present case, the large number of pits combined with the depth of the pits is remarkable.

In some circumstances, pitting can be encouraged by galvanic cells. In the present instance, however, where the copper tubing is connected to iron headers, the iron would corrode preferen-

tially.

Through the investigation, a large majority of cooler assemblies were reported to leak on testing, and the leakage reportedly affected numerous tubes, being randomly located with regard to top-bottom and front-back of the assemblies, which are mounted horizontally—for example, 20 tubes high and 5 tubes deep.

The fact that the coils leaked immediately upon installation indicates that damage occurred during manufacture or storage. The extensive leaking corresponded with the extensive pitting damage found. The water left in the coils after installation 2 months prior to testing could have caused pitting, but it seems most unlikely that it would have occurred to the extent observed.

A sticker found on the coils indicated that the units had been shipped "wet," and although it is assumed that the antifreeze solution used was inhibited, such inhibitors are not stable indefinitely. Breakdown of inhibition, usually by oxidation processes, can result in the formation of ammonia or weak organic acids, both of which are corrosive to copper. Additionally, the coils would be full of air at the time they were sealed so that there would be an excess of available oxygen compared with the relatively small amount of antifreeze left in the tube. This oxygen enrichment would itself promote corrosion. (The modern car radiator designed with a catching and recycling system for the coolant is able to exclude oxygen, thus reducing corrosive attack.)

Filling the tubes with water for pressure testing would, unfortunately, tend to flush out and dilute any soluble corrosion product. It thus seems impossible to determine the exact origin of corrosion at this stage. The presence of chlorine in the area of the pit found by SEM is an indication of contamination of the antifreeze solution with chlorides, although most tap water is also chlorine treated and contains traces of chlorides.

Conclusion and Recommendations

Most probable cause

Although not conclusive, the circumstantial evidence pointed to the residual antifreeze solution in the closed tube assemblies as the cause of pitting corrosion during the storage period.

Remedial action

The widespread infestation, the extent and depth of pitting on the leaking tubes, the presence of pitting on the nonleaking tubes, and the difficulty of detecting pits made it impossible to repair or replace the tubes on a piecemeal basis.

With regard to retubing, it was recommended that the units be cleaned and dried before storage or, alternatively, flushed with fresh inhibitor solutions known to provide adequate protection for the storage period and conditions anticipated. Attack would probably have been much less severe had the units been completely filled with inhibited antifreeze solution before storage.

Mercury Liquid Embrittlement Failure of 5083-O Aluminum Alloy Piping

Jerome J. English, Occidental Chemical Corporation, Alvin, Texas and David J. Duquette, Rensselaer Polytechnic Institute, Troy, New York

The failure mode of through-wall cracking of a butt weld in a 5083-O aluminum alloy piping system in an ethylene plant was identified as mercury liquid metal embrittlement. As a result of this finding, 226 of the more than 400 butt welds in the system were ultrasonically inspected for cracking. One additional weld was found that had been degraded by mercury. A welding team experienced in repairing mercury-contaminated piping was recruited to make the repairs. Corrective action included the installation of a sulfur-impregnated charcoal mercury-removal bed and replacement of the aluminum equipment that was in operation prior to the installation of the mercury-removal bed.

Key Words

Aluminum-base alloys	Chemical processing equipment	Ethylene
Pitting (corrosion)	Butt welds	Piping
Mercury	Intergranular fracture	Liquid metal embrittlement
Hydrocarbons	Repair welding	

Alloys Aluminum-base alloy—5083

Background

A through-wall crack occurred in a butt weld of a 5083-O aluminum alloy piping system in an ethylene plant. The failure mode was identified as mercury liquid metal embrittlement, and 226 of the more than 400 aluminum alloy butt welds in the system were ultrasonically inspected for cracking.

Applications

Aluminum alloy piping, such as alloy 5083-O, and brazed alloy 3003 aluminum plate-fin heat exchangers are commonly used in hydrocarbon and natural gas processing plants because of their excellent cryogenic and heat-transfer properties, and because of the economics of construction. Process temperatures in ethylene manufacturing plants are as cold as –170 °C (–270 °F), thus making aluminum alloys the preferred materials of construction. Normally, the gaseous and liquid hydrocarbons in the aluminum equipment are not corrosive. The process streams are dry, and the organic chemicals do not react with the aluminum alloys.

Circumstances leading to failure

The ethylene plant in which the piping failure occurred was designed to process distilled feedstocks that are expected to have low levels of mercury. The plant operated for 5 years using these distilled feedstocks. In the fifth year of operation, the plant began processing an unrefined feedstock. Shortly thereafter, mercury was found in an in-line type 304 stainless steel pipe strainer just upstream from the location where the subsequent piping failure occurred, namely, the inlet piping to an aluminum plate-fin heat exchanger. Analysis of the various barge shipments of new feedstock showed the presence of up to 40 ppb Hg by weight.

The piping and heat exchangers contained inside a 3.7 m (12 ft) diam by 23 m (76 ft) high cold box were constructed of aluminum alloys. The connecting piping outside the cold box was constructed of type 304 stainless steel. The inlet piping to a brazed aluminum plate-fin exchanger within the cold box was constructed of 5083-O aluminum alloy and operated at –72 °C (–98 °F),

which is below the melting point of mercury (–39 °C, or –38 °F). Mercury is not known to embrittle aluminum while it is in the solid state. Therefore, the history of the plant was examined to determine when the area of failure exceeded the melting point of mercury. It was found that the failed piping was heated above the melting point of mercury for a few hours in 1985 and 1986, as well as during a plant shutdown in 1987. An effort had been made by the manufacturing plant to keep the equipment refrigerated below –40 °C (–40 °F) during the 1987 plant shutdown. However, this effort was not successful, and a leak was detected in the piping within 12 h after the piping exceeded the –40 °C (–40 °F) temperature.

Pertinent specifications

Aluminum alloy 5083-O was the specified seamless pipe composition, and the weld filler metal was 5183. The ASTM designation of the alloy piping is not known. Either the gas tungsten arc or gas metal arc welding process was used to make the butt weld where the failure occurred.

Specimen selection

The through-wall fracture was located in a 200 mm (8 in.) diam butt weld between a straight section of piping and a 90° elbow. Power hacksaw cuts were made in the straight section of piping 75 to 100 mm (3 to 4 in.) from the butt weld to remove the pipe section for examination. No water or cutting lubricant was used. The weld containing the through-wall crack was designated W-3.

During the ultrasonic inspection of 226 aluminum alloy piping welds associated with the cold box equipment, a second weld was found in a 200 mm (8 in.) diam pipe section that had branched cracking indications. No cracklike indications were visible on the exterior surface of this weld. This particular area of the ethylene manufacturing process normally operated at a temperature above the melting point of mercury. A 0.6 m (2 ft) long pipe length containing this weld was removed for examination. The weld was designated W-19.

Visual Examination of General Physical Features

Figure 1 shows the 200 mm (8 in.) diam elbow section containing failed weld W-3. Figure 2 shows an interior view of the elbow. Note the white residue on the flow-impingement surface of the elbow.

Figure 3 shows the branched cracking that was readily visible in the weld cap of the 200 mm (8 in.) diam butt weld. A gray stain formed on the exterior surface of the piping where the mercury had escaped from the cracks and wetted the aluminum. This through-wall crack extended along approximately 35 to 40% of the circumference of the weld.

Figure 4 shows weld W-19, in which branched cracking was identified by ultrasonic inspection. As noted earlier, no cracks were visible on the exterior surface of this weld. A high-speed cutter was used to excavate the weld to determine whether mercury was present. The freshly cut surface turned a gray color, and fine white oxide whiskers began to grow from the cracks exposed by the excavation. A backing ring had been used in the fabrication of this weld.

One of the visual characteristics of mercury degradation is the reactivity of aluminum when the mercury-contaminated surface is first exposed to moist, ambient air. The aluminum turns a gray color, and a voluminous white gamma alumina is observed growing from the freshly exposed aluminum surface.

Testing Procedure and Results

Radiography. Radiographic examination of the welds was not an effective nondestructive method for identifying mercury-induced cracking. Experiments conducted by an inspection service showed that the tight cracks produced by mercury liquid metal embrittlement did not appear when radiography was performed with an iridium radioisotopic source (Ref 1). Iridium is the radioactive source used for field radiography.

A complication in the radiographic interpretation occurred because of the gas tungsten arc welding process used to make many of the welds. Deficiencies in the welding procedure resulted in tungsten inclusions in the welds as well as tungsten residue on welding backup rings from arc strikes. These tungsten-rich areas had the same high-density characteristics as mercury on the X-ray film and were first misinterpreted as mercury contamination. When mercury was suspected as having been detected by radiography, these pipe weld areas were heated with a gas torch and radiographed again. If mercury was present in the suspect area, the mercury would have been redistributed by the heat. Many false leads, but no mercury regions were detected when radiography was used to assess the other welds in the equipment. Radiographic examination was helpful when used in combination with ultrasonic examination to try to distinguish original fabrication defects from mercury-induced degradation.

Ultrasonic examination was effective in locating planar weld defects. Of the more than 400 butt welds present in the aluminum piping systems, 226 were ultrasonically inspected. Significant planar defects were found in 89 of these welds. This high number of defects was attributed to the fact that the welds were not radiographically examined during the original construction of

Fig. 1 Pipe section and the through-wall crack in weld W-3

Fig. 2 Interior view of the elbow and the white residue in the flow-impingement area

Fig. 3 Weld cap at the through-wall crack location, showing branched cracking. The weld cap had a gray stain where mercury had contacted the surface.

Fig. 4 Weld W-19 after it was excavated to detect the presence of mercury

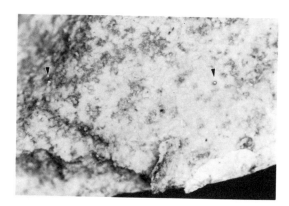

Fig. 5 Macrograph of fracture surface at weld W-3 before removal of the white oxide deposit. Note the mercury droplets. 10.26×

Fig. 6 SEM micrograph of the through-wall crack. 14.25×

the piping systems.

Of the 89 welds containing planar defects, one also showed branch cracking in the weld and in the base metal next to the weld. No visible cracks were present on the exterior of the piping at this weld location. This is the weld (W-19) that was excavated in the field and found to contain mercury-induced cracks. A proprietary ultrasonic procedure similar to the Electric Power Research Institute intergranular crack detection procedures was used to detect the branch cracking.

Surface examination

Visual. The through-wall crack was sectioned out of the pipe/elbow weld (W-3) and opened by applying a bending force to the pipe side of the weld. The crack opened easily, exposing a fracture surface covered with a white, powdery deposit. Small, reflective, metallic-appearing droplets were observed throughout the white deposit. These droplets are indicated by arrows in Fig. 5 and were identified by energy-dispersive spectroscopy (EDS) as mercury.

Scanning Electron Microscopy/Fractography. A small section of the elbow side of the weld fracture was removed for examination by scanning electron microscopy (SEM). The white deposit, aluminum oxide, obscured most of the details of the fracture surface. This deposit was removed by rinsing the surface with concentrated nitric acid, which is known not to affect the fracture appearance (Ref 2).

Figure 6 shows the general appearance of the through-wall crack. The crack originated on the inside (hydrocarbon side) of the pipe (to the right of the arrow in Fig. 6). This area was actually the weld root-pass surface in the inside of the pipe. Figure 7 shows the intergranular path of the cracking in the early propagation region. A number of pits were visible on the grain facets. The nature and extent of the pitting indicated a corrosion reaction of the mercury with the aluminum.

Figures 8 and 9 show the fracture surface of the through-wall crack at increasing distances from the inside surface. In each case, the fracture appearance was typical of mercury-induced cracking of aluminum (Ref 2). The cracking was primarily intergranular, penetrating the entire wall of the pipe. No shear lip or ductile ligament could be

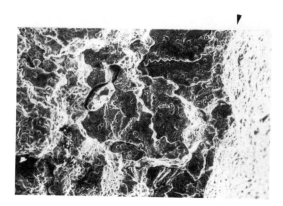

Fig. 7 SEM micrograph showing intergranular cracking near the inside surface of the pipe weld. 67×

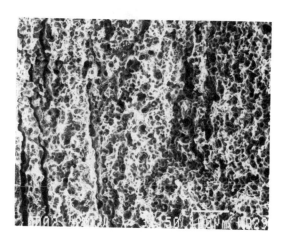

Fig. 8 SEM micrograph showing the through-wall fracture near the interior surface of the pipe. Note pitting. 93×

identified near the outer surface of the pipe. The pitting corrosion was heaviest near the inner surface of the fracture, then decreased as the fracture progressed toward the exterior surface (the outside diameter) of the pipe. This suggested that the crack did not progress as a singular event from the inside to the outside diameter.

Scanning electron microscopy was not performed on weld W-19, which was excavated by high-speed cutting in the field. The appearance of this weld is described further in the next section.

Crack Origins/Paths. Two metallographic

Fig. 9 SEM micrograph showing the through-wall crack near the exterior surface of the pipe. Note the less severe pitting compared with Fig. 8. 93×

Fig. 10 Metallographic cross section at the through-wall crack location

Fig. 11 Cross section of weld W-3 near the termination of the through-wall crack. Arrows indicate exfoliation cracks. 3.42×

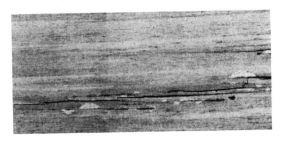

Fig. 12 Micrograph showing the exfoliation region of the pipe. 14.25×

sections were prepared from weld W-3, which contained the through-wall crack. One section was a full cross section of the through-wall crack (Fig. 10). The crack was primarily intergranular and extended from the toe of the weld root pass (inner surface) to the weld cap. Secondary cracks branched off the primary crack into the weld metal. A large number of delamination or exfoliation cracks were present in this section, propagating into the base metal (pipe) perpendicular to the weld axis. One of these exfoliation cracks is identified in Fig. 10 by an arrow. Shortly after polishing and etching of the primary, secondary, and exfoliation cracks, all the cracks exuded the white residue characteristic of an Al-Hg-O_2-H_2O interaction.

The second metallographic section was removed near the end of the through-wall crack. In this location, the crack had not propagated completely through the weld metal (Fig. 11). Several exfoliation cracks were present near the inside surface of the pipe. Figure 12 is a higher-magnification view of this region of exfoliation. Figure 13 shows the crack tip of one of the cracks, the intergranular path, and the considerable amount of branching.

Metallographic examination of the base metal next to weld W-19, which had been excavated in the field, revealed exfoliation cracking. This cracking was perpendicular to the weld axis and parallel to the pipe surface. The condition extended about 38 mm (1.5 in.) from the weld fusion line. Many of these cracks contained a light-colored phase (Fig. 14), later identified as a hard, chromium-manganese-rich intermetallic compound. Figure 15 shows this segregation.

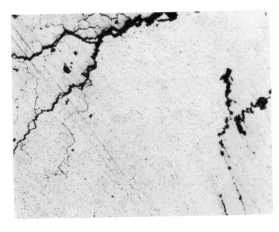

Fig. 13 Micrograph of a crack tip, showing branching and an intergranular mode of fracture. 30.5×

Chemical analysis/identification

Material and Weld. Results of chemical analyses of the pipe, elbow, and weld metals, as determined by x-ray fluorescence, are given in Table 1. Both the pipe and elbow were within the specifications for aluminum alloy 5083. The weld metal composition corresponded to the specification for 5183 filler metal.

Corrosion Deposits. White deposits from the inside surface of the pipe/elbow section were dissolved in acid and analyzed by atomic absorption. The composition of the white deposit was as follows: 26.4% Al, 1.84% Hg, and 226 ppm Zn. This white residue was identified as aluminum oxide, which resulted from the reaction of an aluminum-mercury solid solution with atmospheric oxygen and/or water vapor. Small droplets of mercury were dispersed throughout this residue. Figure 16 shows metallic droplets on the interior surface of the pipe/elbow area. A mercury X-ray image of this area is shown in Fig. 17.

EDS analysis was performed to identify the light-colored phase shown in Fig. 15. This material was found in the pipe cross section next to the field-excavated weld. The EDS spectrum of these light areas confirmed the presence of chromium and manganese in amounts greater than normally present in the 5083 aluminum alloy.

Scanning Auger microscopy was used to examine the fracture surface of the through-wall crack. Wide-energy-range spectra were taken at various points on the surface. All spectra were similar, showing a surface that was essentially pure aluminum oxide, with small amounts of carbon and magnesium. Depth-profile studies of the fracture surface, using ion etching, revealed a uniform composition as a function of depth, with no additional elements detected as the etching progressed. A specific search for mercury was conducted, both during etching and after removing about 1 μm from the surface; none was detected.

Mechanical properties

Hardness. Vickers hardness values (200 g load) for the base metal and weld metal in the area of the through-wall crack were as follows: elbow base metal, 79 HV; pipe base metal, 85 HV; weld metal (root pass), 83 HV. Hardness values for the phases shown in Fig. 15 were as follows: gray-colored matrix, 92 HV (35 HRB); light-colored chromium-manganese-rich phase, 411 HV (42 HRC). It was suspected that the high-hardness light-colored phase was an aluminide intermetallic compound.

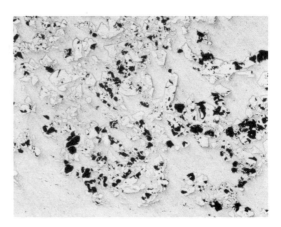

Fig. 14 Micrograph of light-colored phase found in the base metal (pipe) next to weld W-19. 30.5×

Fig. 15 SEM micrograph showing the polished transverse section containing the segregation in the base metal next to weld W-19. 310×

Table 1 Results of chemical analysis

Material	Composition, %							
	Mg	**Mn**	**Cr**	**Zn**	**Si**	**Fe**	**Cu**	**Al**
Pipe	4.59	0.80	0.24	0.11	0.13	0.16	0.05	bal
Elbow	4.12	0.75	0.07	0.01	0.06	0.13	...	bal
5083 Specification(a)	4.0-4.9	0.4-1.0	0.05-0.25	0.25	0.40	0.40	0.10	bal
Weld metal	4.45	0.73	0.12	0.04	0.09	0.20	0.02	bal
5183 specification(a)(b)	4.3-5.2	0.5-1.0	0.05-0.25	0.25	0.40	0.40	0.10	bal

(a) Single values are maximums. Source: Ref 3. (b) Filler metal for gas metal arc welding

Table 2 Results of tensile testing

Specimen No.	Origin	Environment	Strain rate	Ultimate tensile strength		Elongation in 25 mm (1 in.), %	Reduction of area, %	Failure location(c)
				MPa	**ksi**			
W-3-1	Production weld W-3	Air	50%/min	308.2	44.7	21.9	51.6	WM/HAZ
W-3-2	Production weld W-3	Air	50%/min	302.6	43.9	21.9	44.1	WM/HAZ
D-3	Laboratory weld D	Hg	0.5%/min	146.8	21.3	3.0	8.9	HAZ/BM
D-1	Laboratory weld D	Hg	50%/min	146.1	21.2	0	8.0	WM
C-1	Laboratory weld C	Hg	0.5%/min	169.6	24.6	3.0	8.5	WM
C-3	Laboratory weld C	Hg(a)	0.5%/min	218.5	31.7	(b)	6.0	BM

(a) Oxide coating on specimen not broken with diamond scribe prior to testing. (b) Failure occurred outside gage length. (c) WM, weld metal; HAZ, heat-affected zone; BM, base metal

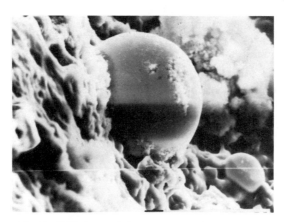

Fig. 16 SEM micrograph showing deposits on the interior surface of the weld

Fig. 17 Mercury X-ray image of the deposit on the inside surface of the pipe

Tensile Properties. Tensile tests were performed at 25 °C (78 °F). Tensile specimens were machined from the elbow/pipe weld (W-3) opposite the through-wall crack. Two new welds were made in the laboratory from 5083 piping for additional tests. The filler metal was 5183. These new welds were designated C and D. Tensile specimens were machined along the pipe axis to include the weld and base metal. Results of tensile testing are given in Table 2. Table 3 lists typical tensile properties for wrought aluminum alloy 5083 and as-welded 5083 with 5183 filler metal. The tensile properties of the two specimens obtained from the elbow/pipe weld W-3 met or exceeded the typical results for welded 5083.

Two specimens from each of the new laboratory welds were tested in liquid mercury at 25 °C (78 °F). After immersion in the mercury bath and prior to testing, the oxide coating on three of the tensile specimens was broken with a diamond scribe to facilitate wetting of the aluminum by the mercury. Table 2 shows that the mercury environment resulted in severe embrittlement of the welded aluminum alloy, as measured by the significant reduction in elongation and reduction of area. The oxide coating on one of the laboratory weld specimens was left intact. This specimen also embrittled in mercury, but had a higher ultimate tensile strength than the scribed specimens. This behav-

Table 3 Typical tensile properties of wrought and as-welded aluminum alloy 5083

Material	Average		Average elongation in 50 mm (2 in.),%
	MPa	ksi	
Wrought 5083-O(a)	289.5	42.0	22.0
As-welded 5083 with 5183 filler metal(b)	296.5	43.0	16.0

(a) Source: Ref 3. (b) Source: Ref 4

ior suggested that plastic deformation of the aluminum substrate will cause rupture of the protective oxide and result in wetting of the aluminum.

The tensile specimens tested in mercury had different fracture paths (Table 2). Furthermore, specimen D-3 contained a considerable amount of exfoliation cracking in the base metal (pipe) that initiated at the tensile fracture surface and propagated in a direction parallel to the tensile axis. A metallographic section prepared from the fracture surface of this specimen indicated that the crack initiated in the weld heat-affected zone (HAZ), with the crack propagating in a mixed intergranular-transgranular mode. After a short distance, the crack traveled into the base metal, propagating in a transgranular mode with little or no associated plastic deformation.

Discussion

Because mercury had been detected in the process just upstream from the piping failure, mercury liquid metal embrittlement was the suspected mode of failure. The distinctive visual effects on the weld cap of the through-wall crack, such as the branch cracking and the gray staining, supported this hypothesis. Fortunately, this mode of failure was suspected when the through-wall fracture was found. Safety precautions were taken to avoid the inhalation of mercury fumes during removal of the samples.

Intergranular cracking was the preferred mode of failure in the weldment and was also reflected in the exfoliation type of attack that ran parallel to the pipe walls. This type of behavior is similar to that observed in sensitized alloys and suggests a rapid reaction between the liquid

mercury and the grain boundaries of the alloy. The specific nature of this interaction is not presently known. However, the interaction apparently involves only a few atomic layers of the fracture surface, as no mercury was observed in the secondary cracks to a magnification of 1000×, nor could mercury be identified on fractured aluminum surfaces by surface analytical methods, such as Auger spectroscopy.

The presence of mercury in the secondary cracks, on fracture surfaces, and on free surfaces was unequivocal, however. Exposure of these cracks or surfaces to aqueous environments, including humid air, produced voluminous flocculent oxides containing spheroids of mercury.

Severe pitting on the fracture surface of the through-wall crack at weld W-3 indicated a corrosion reaction of mercury droplets with the aluminum. This suggested that the mercury was in contact with the fractured surface for an extended period of time. This pitting corrosion on the fractured surface was more severe near the inside diameter of the fracture, which also suggested that the crack did not progress as a single event from the inside to the outside diameter. The history of weld W-3 supported the suggestion that the crack did not propagate as a single event. Weld W-3 was normally at a temperature below the melting point of mercury and was exposed to liquid mercury for only brief periods before the failure occurred.

Mercury degradation of weld W-19, detected during ultrasonic inspection, appeared to have been aggravated by segregation of the hard chromium-manganese-rich inclusions. These inclusions could have acted as crack initiation sites. Also, the particles appeared to be more reactive to mercury than the matrix. Another factor that contributed to the degradation of weld W-19 was its backing ring. This crevicelike geometry helped to collect mercury next to the root of the weld.

The tensile tests performed in liquid mercury on the laboratory-welded specimens demonstrated that the 5083-O aluminum alloy welded with 5183 filler metal was highly susceptible to mercury liquid metal embrittlement. The fracture modes in these specimens simulated the cracking found in the plant failure, including exfoliation in the base metal.

Detection of the mercury liquid metal embrittlement by nondestructive methods was difficult. Ultrasonic examination was effective in detecting mercury degradation when branch cracking and the exfoliation type of attack were present. However, because of the large number of fabrication-related weld defects, there was no assurance that all mercury-degraded welds had been removed from the production equipment. Replacement of the mercury-contaminated aluminum equipment thus became a high priority.

The presence of mercury contamination can produce very porous welds. Repair welding of the pipe segments that were removed required special techniques by experienced welders employed by a company familiar with making repairs on mercury-contaminated equipment. The welders wore full facemasks with bottle-supplied air in order to avoid breathing the poisonous mercury vapor during welding.

Conclusion and Recommendations

Most probable cause

It was concluded that the through-wall failure in the pipe weld resulted from mercury liquid metal embrittlement. The most probable cause of the embrittlement was the use of unrefined feedstock in the ethylene plant. This feedstock was found to contain mercury at levels as high as 40 ppb. In a plant operation where 2 billion pounds of feedstock are processed per year, mercury impurity levels in the parts per billion range present a serious threat to the integrity of the aluminum equipment.

Remedial action

Prior to restarting the plant, all of the aluminum alloy piping was pneumatically tested with dry nitrogen and was checked for leaks. A mercury-removal system consisting of a sulfur-impregnated activated-charcoal bed was subsequently installed upstream of the aluminum equipment. The plant operated for 1 year in this manner until a new aluminum alloy cold box could be constructed and installed. This installation replaced the cold box that had been in service prior to the installation of the mercury-removal system.

Weld quality control during the construction of the new cold box consisted of 100% radiographic examination of all butt welds. All radiographs were filed for future reference. The use of weld backup rings was minimized in the construction. When backup rings were used, the upstream side of the ring was seal welded to minimize mercury entrapment.

How failure could have been prevented

A complete chemical analysis of trace impurities in the new feedstock for the ethylene plant could have prevented the mercury-related failure of the aluminum equipment. Upon detection of the mercury in the feedstock, a mercury-removal system could have been installed upstream of the aluminum equipment before the feedstock was processed.

References

1. R. Davies, private communication, Det Norske Veritas Industry, Houston, 1987
2. J. Kapp, "Crack Growth in Mercury Embrittled Aluminum Alloys Under Cyclic and Static Loading Conditions," Ph.D. thesis, Rensselaer Polytechnic Institute, 1982
3. *Metals Handbook Desk Edition,* American Society for Metals, 1985
4. *Welding Handbook,* 7th ed., Vol 4, American Welding Society, 1982

Short-Term Failure of Carbon Steel Boiler Tubes

Clifford C. Bigelow, Packer Engineering, Inc., Naperville, Illinois

Two identical "D" tube package boilers, installed at separate plants, experienced a number of tube ruptures after relatively short operating times. The tubes, which are joined by membranes, experienced localized bulging and circumferential cracking along the fireside crown as a result of overheating and thermal fatigue. It was recommended that recent alterations to the steam-drum baffling be remodified to improve circulation in the boiler and prevent further overheating. Several thermocouples were attached to tubes in problem areas of the boiler to monitor the effects of the steam-drum modifications on tube wall temperatures.

Key Words

High temperature	Carbon steels, Mechanical properties	Boiler tubes,
Cracking (fracturing)	Thermal fatigue	mechanical properties

Background

Two identical "D" tube package boilers experienced premature tube failures after short operating times. The tubes were joined by membranes, and many of the failures occurred in the roof area within 0.61 m (2 feet) of the steam-drum inlets.

Applications

The package boilers were used in two similar plants to generate process steam where they operate at relatively constant load.

Circumstances leading to failure

One of the boilers, designated boiler 1, had been in operation for approximately 14 months, while the other, boiler 2, had been in service for only three weeks prior to the tube failures. It was reported that boiler 1 previously experienced problems with water carryover from the steam drum. To alleviate the problem, the baffling in the steam drum was modified. Prior to these modifications, there were no detected problems with tube failures in the boiler. The baffle modifications eliminated the carryover problem. However, the tube failures occurred shortly afterwards. Similar modifica-

tions to the steam drum were performed on boiler 2 before it was put in service, but the boiler then experienced tube failures after only three weeks in operation.

Pertinent specifications

The boilers are natural gas fired with a capacity of 117,934 kg (260,000 lb) of steam per hour at a design pressure of 1.5 MPa (225 psig) and an outlet temperature of 193 °C (380 °F). The boiler tubes were specified as 6.35-cm (2.5-in.) outer diameter carbon steel tubing with 0.368-cm (0.145-in.) wall thickness. The exact grade of the steel was not reported.

Selection of specimens

Failed sections of two tubes were removed from the rear roof area of boiler 1 where a total of eight tubes failed, and numerous other sections were bulged and bowed. After visual examination of the failed tubes, metallographic and hardness test samples were taken from the fireside at and away from the bulging, and from along the coldside at representative areas of the tube.

Visual Examination

The tubes were joined by membranes and identified as tube 3 and tube 4 (Fig. 1). The outer diameter scale along the fireside was moderate in thickness and appeared to be a combination of fuel ash and oxide scale. The waterside walls were coated with thin, sooty, black deposits which appeared to be an iron oxide. Both tubes exhibited localized, periodic bulging with numerous circumferential, parallel cracks

along the fireside (Fig. 2 and 3). Associated with the fireside cracking were numerous secondary fissures and craze cracking of the outside scale, indicative of tube wall stretching (Fig. 3). The cross-section of the tubes were obviously enlarged and oval-shaped along the fireside, while retaining a basically semicircular shape on the coldside (Fig. 4).

Fig. 1. Fireside of tubes removed from the roof area of the boiler 1 showing numerous circumferential cracks and fissures. This damage was typical in both boilers.

Fig. 2. Profile of tube 4 fireside showing the localized nature of the bulging. 0.04×

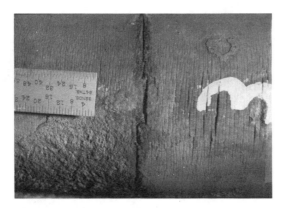

Fig. 3. Closer view of circumferential cracks and fissures on tube 3. Note the craze cracking in the scale indicative of relatively rapid tube wall stretching.

Fig. 4. Cross-section through tube 4. Note the pronounced ovality of the fireside compared to the semicircular cross section of the coldside.

Procedure and Results

To evaluate the damage to the boiler tubes, dimensional measurements were taken to quantify the distortion, while hardness measurements and metallographic examination were performed to evaluate the thermal affects to the tubes.

The outer diameter of the tubes measured across the membranes ranged from 6.30 to 6.50 cm (2.48 to 2.56 in.), while the diameter across the fireside crown of the tubes measured as large as 7.44 cm (2.93 in.). Considering that most, if not all, of the tube swelling occurred on the fireside of the tubes, these differences in measurement represent an approximate 30% growth in the radius. Tube wall thickness measurements taken along the coldside averaged 0.356 cm (0.140 in.), consistent with the specified thickness, while along the fireside, tube wall thickness measured as low as 0.178 cm (0.070 in.) near the fissures.

Metallography

Optical metallography on the coldside of tube 4 revealed a ferritic structure with small islands of lamellar pearlite, typical of the as-manufactured condition of a carbon steel boiler tube (Fig. 5). Along the fireside of tube 4 at one of the tube wall fissures, the structure was ferrite with globular carbides congregated primarily at the grain boundaries. The fissure lip showed evidence of grain deformation and subgrain formation, but no evidence of creep voids (Fig. 6). Subgrains are evidence of dynamic recrystallization, and indicates the metal was stretched relatively rapidly while at high temperature. At areas on the fireside of the tubes away from the cracking, the microstructure was also completely spheroidized.

Metallographic cross sections through secondary fissures exhibited wedge-shaped crack openings with relatively sharp, oxide-filled crack tips. The oxide filling the fissures often was cracked open as well (Fig. 7 and 8).

Mechanical properties

Hardness readings taken at the heat-affected areas along the fissure lip on the fireside of tube 4 measured 72 HRB, while at an area near the membrane (90° location) the hardness averaged 84 HRB. The lower hardness readings along the crown are consistent with the spheroidized carbide microstructure and confirms the absence of high-temperature transformation phases.

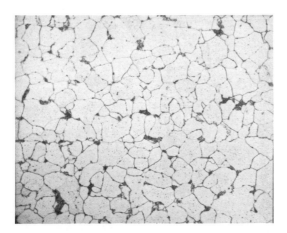

Fig. 5. Typical microstructure along the coldside of tubes 3 and 4 showing small colonies of lamellar pearlite in a ferrite matrix. 2% nital etch. 244×

Fig. 6. Cross-section through a fissure lip from tube 4. Note the preponderance of subgrains in the ferrite matrix due to distortion and recrystallization. The carbides have completely spheroidized with no evidence of lamellar pearlite. 2% nital etch. 244×

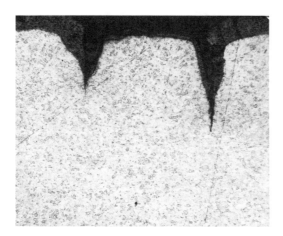

Fig. 7. Cross-section through circumferential outer diameter fissures from tube 4. Note the cracked oxide which partially fills the fissures. 2% nital etch. 61×

Fig. 8. Higher magnification of an outer diameter fissure. Note the oxide filling the wedge-shaped fissure and the tight, oxide-filled crack tip. 2% nital etch. 316×

Discussion

The local bulging along the fireside of these boiler tubes indicates the tubes have suffered from overheating, while the circumferential fissures and microstructural features of oxide-filled crack openings and subgrain formation were the result of thermal fatigue. The overheating and periodic bulging along the fireside of the tubes probably resulted from poor water circulation in the tubes, which allowed the tube wall temperature to rise and the diameter to swell. Eventually, steam blanketing or film boiling probably occurred along the fireside, further decreasing heat transfer and creating local hot spots. As the steam blanket breaks down, the tube wall was intermittently wetted and cooled. This repeating condition results in wide temperature swings and stresses in the tube wall, eventually causing thermal fatigue cracks.

The localized nature of the cracking and bulging of the tubes was likely enhanced by the constraint induced by the tube membranes which restrict the tubes from freely expanding and contracting.

The original lamellar pearlite carbides along the fireside of the tube had completely spheroidized *in situ* after very short operating times in both boilers. This condition suggests the boiler tube walls experienced sustained temperatures of 538 °C (1000 °F) or higher (well above the outlet temperature of 193 °C, or 380 °F). However, the absence of high-temperature transformation phases in the microstructure, such as bainite or martensite, indicates tube wall temperatures were below 732 °C (1350 °F) at the time of the rupture.

Considering the circumstances of the tube failures occurring shortly after the steam-drum modifications were made, and the evidence of overheating in the tubes, it was suspected that the steam-drum modifications created a restricted circulation problem in the boiler and allowed certain tubes to overheat.

Conclusion and Recommendations

The failure of the boiler tubes was a result of overheating and localized bulging, but the cracking mechanism was due to thermal fatigue. This condition resulted from poor circulation in the boiler, allowing for excessive tube wall temperatures and steam blanketing conditions on the waterside. Because the failures occurred after the steam drum was modified, the alterations were suspected and identified as the most probable cause of the failures.

Remedial action

The boilers were inspected to identify and replace all severely damaged tubes. The steam-drum baffling was re-worked to improve circulation, and several thermocouples were placed on tube walls, especially at areas where failures occurred, to monitor the effect of steam-drum baffle modifications on the tube wall temperatures.

How failure could have been prevented

The tube failures could have been avoided with improved water circulation through the circuit which, in this case, involved modifications to the baffling in the steam drum. Thermocouples attached to tubes could have been used to monitor tube metal temperatures and detect overheating situations before failures occurred.

References

1. D.N. French, *Metallurgical Failures in Fossil Fired Boilers*, John Wiley and Sons, New York, 1983.

2. R.E. Reed-Hill, *Physical Metallurgy Principles*, PWS-Kent, Boston, 1973.

Steam Superheater 800H Tube Failure

Edward J. Franz, Ashland Chemical, Inc., Dublin, Ohio

An SB407 alloy 800H tube failed at a 100° bend shortly after startup of a new steam superheater. Three bends failed and one bend remote from the failure area was examined. Visual examination showed that the fracture started on the outside surface along the inside radius of the bend and propagated in a brittle, intergranular fashion. Chemical analysis revealed that lead contamination was a significant factor in the failure and phosphorus may have contributed. The localized nature of the cracks and minimum secondary cracking suggested a distinct, synergistic effect of applied tensile stress with the contamination. Stress analysis found that stress alone was not enough to cause failure; however, the operating stresses in the 100° bends were higher than at most other locations in the superheater. Reduced creep ductility may be another possible cause of failure. Remedial actions included reducing the tube temperature, replacing the Schedule 40 100° bends with Schedule 80 pipe, and solution annealing the pipe after bending.

Key Words

Ferrous alloys, Mechanical properties
Superheaters
Lead (metal), Impurities
Ductility

Superalloys, Mechanical properties
Brittle fracture
Contaminants

Boiler tubes, Mechanical properties
Intergranular fracture
Creep (materials)

Alloys

Nickel alloys—SB407 Alloy 800H

Background

A replacement steam superheater failed shortly after start-up.

Applications

The replacement was the same design as the original with minor changes. The tube spacer design was changed, flange connections were added to the header ends, and two tube rows were changed to SB 407 alloy 800H from alloy 800. The superheater (Fig. 1) includes inlet and outlet headers at the top, with tubes suspended from these headers. The assembly is one of several exchangers suspended in an exhaust flue downstream of a methanol reformer. The internal coil design pressure is 8.96 MPa (1300 psig) at 649 °C (1200 °F) and flue temperatures are about 988 °C (1810 °F).

Circumstances leading to failure

The superheater operated for twenty-seven days when a tube failed, completely separating at a 100° bend near the outlet header. Operating personnel suspected a tube leak for five days before complete tube separation forced a shutdown. The failed bend was replaced and the unit returned to service. Operation continued for another twenty-seven days when a second tube failed. This bend was also replaced and the unit returned to service again. After fifteen days, a third leak developed. Operations shut down the unit before total tube failure. This failed tube bend was the replacement bend from the first failure. The three failures were located near the header ends and in the first tube row where the flue gases are hottest.

Pertinent specifications

The first two rows of tubes on the upstream side of the flue duct were bare 38.1-mm (1.5-in.) diameter Schedule 40 SB 407 alloy 800H pipe. The remaining rows of downstream tubes are of a lower alloy. The superheater was built to ASME Section 1 specifications. All the bends were cold bent without post-bend heat treatment. The ASME Code does not require post-bend heat treatment.

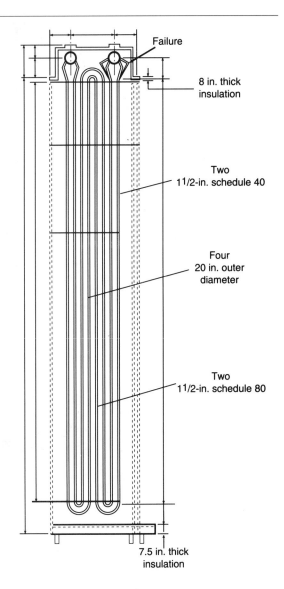

Fig. 1 Elevation side view of the steam superheater with an arrow pointing to the location of tube failures

Performance of the previous superheater

The previous superheater operated for 100,000 hours. The records show a failure in the alloy 800 at the top near an end. It is unknown if the failure was in a 100° bend. Leaks did occur on occasion at the connections to the header. After 80,000 hours of service, analysis showed a stress-rupture failure in alloy 800 near the bottom in a straight pipe section. The tubes exhibited distortion with scale and deposits on the exterior surfaces. Operations re-ported lower flue gas temperatures for this unit compared to the replacement unit.

Selection of specimens

Four tube bends, including the three bends that failed and one bend remote from the failure area, were selected for examination. Also, deposits from upstream areas of the reformer system and from surfaces of the superheater were collected.

Visual Examination of General Physical Features

All three failed tubes exhibited primary through-the-wall cracks starting on the inside radius of the bends (Fig. 2). Minimal secondary cracks, no tube wall thinning, and no deposits on the tube interior were observed.

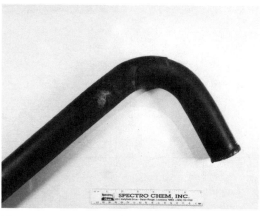

(a)

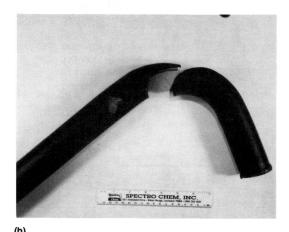

(b)

Fig. 2 Side view of the second tube failure in both photos is similar to the other two failures. The arrow points to the crack initiation site.

Testing Procedures and Results

Non-destructive evaluation

Radiography of the first failed bend revealed only the visible primary crack. Radiography of the unfailed bend showed no defects.

Liquid Penetrant Examination of the other 100° bends in the first row showed no defects.

Ultrasonic Examination. Thickness measured by ultrasonics at the outside radius of the 100° bends was 3.3 to 3.9 mm (0.13 to 0.15 in.). The nominal thickness for Schedule 40 pipe is 3.7 mm (0.146 in.) before bending.

Surface examination

Visual. All three failures showed a single crack located in the 100° bend. Examination showed no tube bulges, no deformation, and no wall thinning. Marks on the interior from the bending operation were observed; however, from later observations and analysis, these marks probably did not contribute to the failures. The third tube failed like the earlier two failures (Fig. 3). A vendor formed the third bend by the sand method which did not impart internal forming marks. The comparative bend also had internal forming marks. Using this unfailed bend, investigators stressed it to failure at room temperature (Ref 1). The fracture mode was ductile.

Macrofractography. Analysis located the fracture origin on the inside radius of the bend

Fig. 3 Side view of the third tube failure showing that the fracture started on the inside of the bend

(Fig. 4). The fracture propagated in a brittle, intergranular fashion. It started on the inside radius and then moved along the side of the tube until the tube separated in a ductile manner on the outside radius.

Scanning Electron Microscopy. The fracture surface (Fig. 5) shows an intergranular fracture mode with corrosion product.

Metallography

Microstructural Analysis. At the fracture origin, the intergranular fracture (Fig. 6) revealed minimal branching. In addition, a uniform oxide measuring 0.0038 to 0.0076 mm (0.15 to 0.30 mils) thick was present along the fracture. The third tube failure had a secondary intergranular crack about 12.8 mm (0.5 in.) from the primary fracture, which started at the inner radius and on the outside surface. This secondary crack penetrated the tube wall 23%, and like the primary crack, exhibited minimal branching with a gray-colored oxide along the grain boundaries (Fig. 7).

The grain size of ASTM 4 to 5 met the specification.

Chemical analysis/identification

Material. Base-metal analysis showed that the chemistry conformed to SB 407 (UNS N08810). Table 1 lists the analysis. Investigators used a plasma emission spectrometer and a carbon/sulfur combustion-type analyzer (Ref 1). The chemistry for the third failure, although not reported here, also conformed to specification. This failed bend replaced the first failure. This replacement bend was from a different lot of material. The supplier material test certification was used to report the chemistry for the Schedule 80 replacement bends.

Coatings and Surface Layer. Unlike the first and second tube failure, operations shut down the unit as soon as they suspected a third failure. A

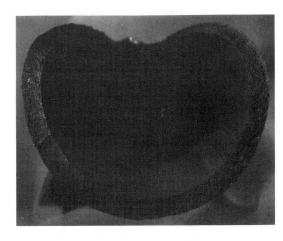

Fig. 4 Plan view of the second tube failure illustrating a brittle fracture

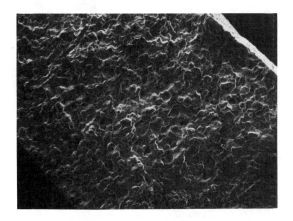

(b)

Fig. 5 SEM fractographs of the inside radius of the bend. (a) A distinct intergranular fracture surface for the entire tube thickness at the inside radius of the bend. (b) The oxide-type corrosion product masking the intergranular facets of the fracture.

(a)

Fig. 6 An as-polished section across the primary fracture through the second tube failure. 250×. (a) The inside diameter. (b) The outside diameter (fracture origin). A uniform oxide type corrosion product is visible along the fracture surface.

Table 1 Chemical composition (wt%) for the 100° bends

Element	First failure	Second failure	Final Schedule 80 Replacement	SB 407 (UNS N08810) specification
C	0.079	0.082	0.08	0.05 to 0.10
S	0.004	0.003	<0.001	0.015 max
Ni	31.3	30.0	31.18	30 to 35
Cr	20.4	19.7	19.58	19 to 23
Mn	0.88	0.95	0.92	1.5 max
Si	0.17	0.085	0.21	1.0 max
Cu	0.4	0.47	0.50	0.75 max
Al	0.43	0.42	0.49	0.15 to 0.60
Ti	0.48	0.54	0.52	0.15 to 0.60
Nb	0.09	0.21	...	...
Mo	0.29	0.30	...	...
Co	0.23	0.23	...	...
P	<0.01	<0.01	...	...
Fe	Rem	Rem	46.52	39.5 min

yellow-green to beige deposit covered the upper area of the superheater. This area is recessed out of the main flue and appears to act as a dead flow area. The deposit was not apparent in the first and second tube failures. Most deposits in the first and second failure may have been washed away by escaping steam. X-ray fluorescence spectra of the dust from the tube surface, showed lead, silicon, aluminum, and calcium as the predominant contaminants, with lesser concentrations of iron, chromium, nickel, copper, and titanium.

Corrosion Deposits. The fracture surface was enriched in iron (Ref 1). Sporadic areas of iron-nickel and chromium-iron occurred throughout the fracture. Detailed chemical analysis of the second and third fracture surface using SEM/EDS and XRF revealed the elemental contaminants lead, phosphorus, chlorine, aluminum, silicon, and calcium. The corrosion product in the secondary crack of the third tube failure contained a band of metal between the oxide-appearing material. Lead was identified in this secondary crack. Escaping steam did not blow across this crack. Chemical analysis of the gray-colored oxide revealed chromium enrichment compared to the iron and nickel relative to the base-metal composition. The band of metallic material between the gray-colored oxide contained high concentration of iron and nickel and low chromium as compared to the base metal of the tube. Grain boundary precipitates, i.e. carbides, were distinguished through the tube next to the fracture.

Chemical analysis of the initial tube fracture surface was checked for lead. This fracture had steam blowing over it for an extended time, which probably washed away some evidence. X-ray spectra from two different locations of the fracture at the inside radius of the bend provided poorly defined lead peaks, indicating a lead concentration of 5 to 10 ppm .

During a planned shutdown three months after the third tube failure, deposits were gathered from several locations upstream of the superheater. From these samples, the source of lead was identified as a large bank of air preheat coils. The plant preheats incoming burner combustion air to the reformer. The coils, installed ten years earlier, now exhibit severe corrosion on the fins. Analysis of the fin surface showed a coating of 100% lead. A check

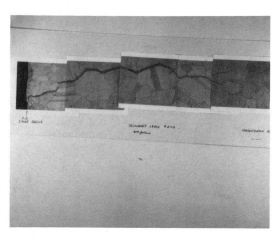

Fig. 7 The secondary crack through the third superheater tube failure. Note the distinct intergranular nature of the crack with development of an oxide type corrosion product along the grain boundaries. 400×

of the vendor drawing revealed a notation that the fin material was lead-coated steel.

Mechanical properties

Hardness. The maximum hardness in the bend measured 27 HRC. Away from the bend, hardness measured 85 HRB. The comparative unfailed tube showed similar hardness. Hardness of the Schedule 80 replacement bends after their post-bend heat treatment was similar to the base metal before bending.

Stress analysis

Analytical. A finite element model of a single tube row was developed to evaluate the structural and thermal stresses. The analysis included the effects of dead weight, temperature, pressure, aerodynamic drag, and maldistribution of steam flow. Since the steam flow enters the header in the middle, it is conceivable that the end tubes receive less flow and the center tubes more flow. The failures occurred near the ends of the header. The calculated stresses did not exceed ASME Code values.

Conclusion and Recommendations

Most probable cause

The tubes failed by a brittle intergranular cracking mechanism. Lead contamination was a significant factor and phosphorus may have contributed. Suppliers report that the high-nickel alloys are prone to embrittlement by lead and phosphorus, among other contaminants (Ref 2). The highly localized nature of the cracks and minimal secondary cracking suggests a distinct synergistic effect of applied tensile stress with the contamination. Because the cracking mechanism is not present in all tubes, the applied stress on selected tubes is a controlling factor. The stress analysis found that stress by itself was not enough to cause failure. However, the operating stresses in the 100° bends are higher than at most other locations in the superheater.

Reduced creep ductility is another possible cause of failure. Others report that precipitates at the grain boundaries can cause the reduction (Ref 3). Carbon content, manufacturing procedures such as cold forming, and the heat treatment influence the formation and location of precipitates. Also, stress enhances precipitation. For these failed bends, the evidence shows parameters that favor a reduction of creep ductility.

Remedial action

After the third tube failure, the 100° bends were wrapped in insulation. Insulating a small part of the total exchanger did not measurably affect the overall heat transfer. The localized insulation reduced the tube temperature at the bend and may have served as a barrier to contaminants. No failures occurred from December 9, 1990 through the planned shutdown on March 10, 1991. During the planned shutdown, the 100° bends were replaced with new 800H Schedule 80 pipe. After cold bending, they were solution annealed. The bends were reinsulated. No tubes have failed since startup in April 1991.

How failure could have been prevented

Several actions could have been taken to improve reliability. These include a reduction in stress, a more favorable microstructure, a lower operating temperature, and eliminating or providing a barrier to contaminants. Heat treatment after cold bending can reduce residual stresses and provide a more favorable microstructure. A thicker pipe wall reduces stresses from external loads.

References

1. G.G. Paulson and J.P. Jendrzejewski, "Reports: Superheater Tubes," Nov. 1990 and Jan. 30, 1991.
2. "Fabricating," Inco Alloys International, Publication No. IAI-21, 2nd ed., 1987.
3. A.A. Tavassoli and G. Colombe, "Mechanical and Microstructural Properties of Alloy 800," *Metall. Trans. A*, Vol 9, Sept. 1978.

Stress-Corrosion Cracking in a Downcomer Expansion Joint

Ralph D. Bowman, Consulting Metallurgical Services, Inc., Marietta, Georgia

A type 321 stainless steel downcomer expansion joint that handled process gases was found to be leaking approximately 2 to 3 weeks after installation. The expansion joint was the second such coupling placed in the plant after failure of the original bellows. The failed joint was disassembled and examined to determine the cause of failure. Energy-dispersive X-ray analysis revealed significant peaks for chlorine and phosphorus, indicating failure by chloride stress-corrosion cracking (SCC). Cracks in the liner and bellows exhibited a branched pattern also typical of SCC. Cracks through the inner liner initiated on the outer surface of the liner and propagated inward, whereas cracks in the bellows originated on the inner surface and propagated outward. Stress-corrosion cracking of the assembly was caused by chloride contaminants trapped inside the bellows following hydrostatic testing. Checking the test fluid for chloride and removing all fluids after hydrostatic testing were recommended to prevent further failure.

Key Words	Expansion joints, corrosion	Stress-corrosion cracking	Chlorides, environment

Background

A type 321 stainless steel downcomer expansion joint (Fig. 1) was found to be leaking approximately 2 to 3 weeks after installation.

Applications

The expansion joint handled process gases at 1790 kPa (260 psi) pressure and temperatures of 810 to 850 °C (1490 to 1560 °F).

Circumstances leading to failure

The expansion joint was the second such coupling placed in the plant after failure of the original bellows.

Performance of other parts in same or similar service

An independent analysis of the failure of the original expansion joint revealed that the coupling had failed from stress-corrosion cracking (SCC) of the bellows.

Specimen selection

The as-received joint was disassembled (Fig. 2) and sectioned (Fig. 3 and 4) to remove the bellows and inner liner. The outer surface of the bellows was examined for cracks using a stereoscopic microscope and dye penetrant. Upon disassembly, two areas of corrosion were found and were submitted for energy-dispersive X-ray analysis (EDXA). The cracked portions were removed for microscopic examination.

Fig. 1   As-received failed expansion joint

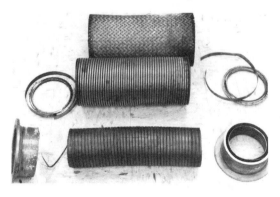

Fig. 2   Disassembled expansion joint. Note outer braided sleeve, bellows, and inner guide tube.

Fig. 3   Sectioned tube at flange-to-bellows joint (bellows at left)

Visual Examination of General Physical Features

After the assembly was sectioned, no cracks were visible or detected in the bellows. However, the outer surface of the bellows contained corrosion in the form of red rust and a white powdery by-product (Fig. 5). Removal of the inner liner revealed a cracked portion (Fig. 6).

Testing Procedure and Results

Nondestructive evaluation

Liquid Penetrant Testing. The bellows was dye penetrant inspected for possible cracks using the fluorescent dye method. Thorough inspection did not reveal any through cracks in the main bellows assembly.

Surface examination

Macrofractography. Stereoscopic examination of the outer surface of the bellows revealed no cracks, but did locate several dents on the tops of the convolutes. The dents represented possible stress-concentration sites.

Sectioned and mounted areas from the cracked portion of the inner liner and a portion of the

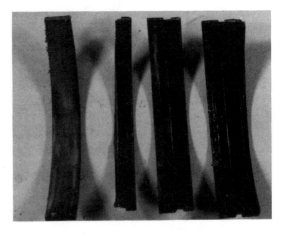

Fig. 4 Sectioned bellows for microscopic examination

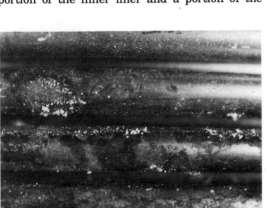

Fig. 5 Corrosion particles on surface of bellows (white powder)

Fig. 6 Inner surface of bellows, showing circumferential crack

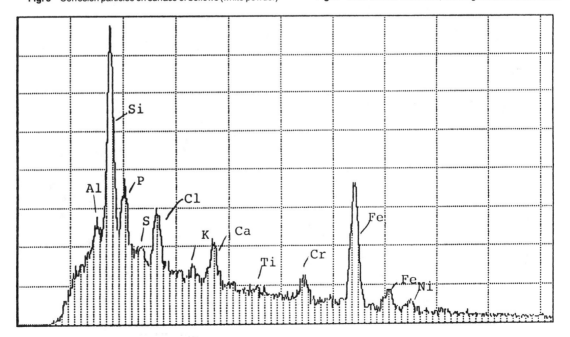

Fig. 7 EDXA chart of contaminants on bellows surface, showing significant peaks for chlorine and phosphorus

sealed weld of the outer bellows were examined microscopically. The cracks were examined after etching and were determined to be branch cracks, a pattern typical of SCC. The crack initiated on the outer surfaces of the liner and propagated inward.

Microscopic examination of the sectioned bellows and braiding seal welds showed misalignment of the bellows with respect to the pipe. This resulted in the application of a weld bead on top of the convolution, thus introducing welding and restraint loading into a highly stressed area of the bellows. Further examination of the fractured bellows revealed stress-corrosion branch cracking that originated on the inner surface of the bellows and propagated outward.

Metallography

Crack Origins/Paths. The branch cracking through the inner liner initiated on the outer surface and propagated inward. The cracking in the bellows, however, originated on the inner surface and propagated outward.

Chemical analysis/identification

Corrosion or Wear Deposits. EDXA analysis of the corrosion by-product from the outer bellows surface revealed significant peaks for chlorine and phosphorus (Fig. 7). The strong chlorine peak indicated that the bellows failure was caused by chloride SCC.

Discussion

Microscopic examination of specimens revealed that SCC of the assembly had initiated between the outer bellows and the inner liner and had propagated in both directions. EDXA analysis confirmed the assumption of SCC. The significant presence of chlorine would have caused corrosive attack and helped initiate SCC.

Conclusion and Recommendations

Most probable cause

The failure was caused by SCC. The chloride that was present became trapped between the braid and bellows after the bellows fractured and contaminant leakage occurred. The contaminants were trapped inside the bellows in a concentration high enough to initiate the crack.

The design of the expansion joint allowed hydrostatic test water or other fluids to become trapped between the bellows and the liner. This fluid remained in the assembly until boiled out when processing temperatures became sufficiently high. Hydrostatic test water is the most probable source of the contaminants.

Welding of the pipe, bellows, and braiding also resulted in a high stress concentration in the bellows due to residual stresses and restraint problems in a critical area. The dents found on the outer surfaces probably caused stress concentrations in the bellows as well.

Remedial action

Better manufacturing controls should be implemented. Steps to remove any fluids from the assembly after hydrostatic testing should be taken in conjunction with checking the test water for chloride content.

Better care in the use of the assembly to prevent dents and stress concentrations would reduce the stress that is capable of forming cracks as well. Prevention of misalignment of the bellows with respect to the welded pipe might prohibit welding and restraint loading of an already highly stressed area.

Stress-Corrosion Cracking in a Stainless Steel Emergency Injection Pipe in a Nuclear Reactor

P. Muraleedharan and J.B. Gnanamoorthy, Metallurgy Division, Indira Gandhi Centre for Atomic Research, Kalpakkam, India

A section of type 304 stainless steel pipe from a standby system used for emergency injection of cooling water to a nuclear reactor failed during precommissioning. Leaking occurred in only one spot. Liquid penetrant testing revealed a narrow circumferential crack. Metallographic examination of the cracked area indicated stress-corrosion cracking, which had originated at rusted areas that had formed on longitudinal scratch marks on the outer surface of the pipe. The material was free from sensitization, and there was no significant amount of cold work. It was recommended that the stainless steel be kept rust free.

Key Words			
	Austenitic stainless steels, corrosion	Stress-corrosion cracking	Cooling systems
	Pipe, corrosion	Rusting	Water cooling
	Nuclear reactor components	Marine environments	Nuclear reactors

Alloys   Austenitic stainless steel—304L

Background

A section of 250 mm (10 in.) diam type 304 stainless steel pipe failed during precommissioning.

Applications

The pipe was part of the standby system for emergency injection of cooling water to a nuclear reactor in case of a failure in the primary heat transport system.

Circumstances leading to failure

Pressure testing revealed a leak during precommissioning of the system. The exact time interval between the installation of the pipe inside the reactor and the observation of leak was not known.

Specimen selection

Leaking occurred in only one spot. The defective portion of the pipe was drilled out, and the pipe was repaired. The drilled-out piece was submitted for analysis.

Visual Examination of General Physical Features

Visual examination of the pipe in the leaking region showed a number of deep longitudinal scratches. The pipe was rusted in this area.

Testing Procedure and Results

Nondestructive evaluation

Liquid penetrant testing of the leaked portion revealed a narrow circumferential crack.

Metallography

Microstructural Analysis. The drilled-out sample was mounted in a mold of quick-setting resin. Both the inside and outside surfaces of the pipe were polished and etched. The etched structure indicated that the material had not been sensitized.

Crack Origins/Paths. Optical microscopic examination of the defective portion of the pipe revealed a crack on the outer surface running in the circumferential direction of the pipe (Fig. 1). No crack was visible on the inner surface. When the specimen was examined in the microscope after polishing, cracks were visible on both surfaces. These cracks were highly branched and resembled transgranular stress-corrosion cracking (SCC) (Fig. 2 and 3). The cracks were tighter at the inner

Fig. 1 Optical micrograph of the outer surface of the as-received material showing a crack and perpendicular scratch marks

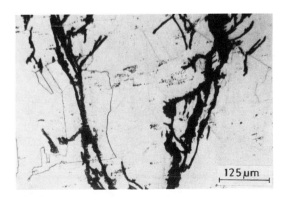

Fig. 2 Branched transgranular stress-corrosion cracks at the outer surface of the material. Polished and etched

surface than at the outer surface, indicating that initiation probably occurred at the outer surface.

Mechanical properties

Hardness. The hardness values measured on both the outer and inner surfaces were similar to that of the solution-annealed material.

Discussion

Optical microscopic examination of the leaking portion of the stainless steel pipe established that the leak was caused by SCC. The crack originated on the outer surface of the pipe and penetrated through the wall thickness. Longitudinal scratch marks were observed near the crack origin. The material was rusted in the scratched portion.

The pipe material was in good metallurgical condition. It was free from sensitization, and there was no significant amount of cold work.

All stainless steel components are pickled and passivated as a last step in the manufacturing process. This ensures removal of all surface impurities, especially embedded iron, and also provides a uniform passive film for good corrosion resistance. However, it is essential that stainless steel components be kept in a rust-free condition. Whenever rusting is noticed, the component must be immediately cleaned and passivated.

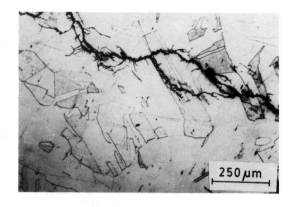

Fig. 3 Transgranular stress-corrosion cracks at the inner surface of the material. Polished and etched

In the present case, abrasion and rusting of the stainless steel pipe occurred during installation or storage. The pipe was exposed to a coastal atmosphere, and the rust had been present for a long period. The rust contained enough ferric ions to raise the electrochemical potential of the steel locally. Absorption of chloride ions and moisture from the coastal atmosphere and residual stresses present in the material caused SCC.

Conclusion and Recommendations

Most probable cause

Rusting of the stainless steel surface because of loss of passivity by abrasion was the primary cause of failure. The rust provided the necessary environment for SCC.

Remedial action

Rusting of stainless steel surfaces is very harmful. Removal of rust, by abrasion with emery, pickling, and repassivation can prevent failure.

Stress-Corrosion Cracking of Stainless Steel Superheater Tubing

Mel J. Esmacher, Betz Laboratories, Inc., The Woodlands, Texas

Several 304H stainless steel superheater tubes fractured in stressed areas within hours of a severe caustic upset in the boiler feedwater system. Tests performed on a longitudinal weld joint, which connected two adjacent tubes in the tertiary superheater bank, confirmed caustic-induced stress-corrosion cracking, promoted by the presence of residual welding stresses. Improved maintenance of check valves and routine inspection of critical monitoring systems (conductivity alarms, sodium analyzers, etc.) were recommended to help avoid future occurrences of severe boiler feedwater contamination. Additional recommendations were to eliminate these short longitudinal weld joints by using a bracket assembly joint between the tubes, use a post-weld heat treatment to relieve residual welding stress, or select a more stress-corrosion cracking resistant alloy for this particular application.

Key Words

Austenitic stainless steels, Corrosion
Superheaters
Caustic cracking

Boiler tubes, Corrosion
Welded joints

Stress-corrosion cracking
Residual stress

Alloys

Austenitic stainless steels—304H

Background

Stainless steel superheater tubes fractured in stressed areas within a few hours after caustic regenerant leaked into the boiler feedwater system.

Applications

In typical paper mill recovery boilers, stainless steel superheater banks are often installed because of their superior fireside corrosion resistance, excellent high-temperature strength, and relatively good weldability with chromium/molybdenum low-alloy steel tubing in the system. This particular tertiary superheating bank operates at 10.5 MPa (1500 psi), with a steam temperature range of 510 °C to 538 °C (950 °F to 1000 °F).

Circumstances leading to failure

The superheating tubing had been in service for eight years without any previous cracking or failure problems. Plant personnel reported that a sudden leakage of caustic into the boiler feedwater had occurred and that it had initially gone undetected. It was subsequently discovered that a faulty check valve in the dilution water line (used in the caustic regeneration of an anion unit in the demineralizer system) had caused the upset. At the time of the upset, a sudden spike in the boiler water pH (from 9.5 to 11.8) was recorded. There was no warning of excessive sodium content in the steam because the sodium analyzer was not operative. Within hours after the first indication of the caustic upset, a large differential between the feedwater flow and the steam flow (indicating massive leaks in the boiler system) was detected. At this point, the entire unit was forced off-line via an Emergency Shutdown Procedure (ESP).

Pertinent specifications

The tubes were manufactured in accordance with ASTM A213/304H specifications. The tube outer diameter was indicated to be 51 mm (2.0 in.), with a nominal wall thickness of 5 mm (0.200 in.).

Performance of other parts in same or similar service

Subsequent inspection of the superheater banks showed no leaks in the primary or secondary banks, which were fabricated from 2.25Cr-1.0Mo low-alloy steel. The tube ruptures were confined to the 304H sections of the tertiary bank, principally at the bottom bends and at other areas of potentially high residual stress. Analysis of the water retained in an unfailed bend after the emergency shutdown had a pH of 10.6, a sodium content of 200 ppb, and a conductivity of 225 μohm. All of these chemistry parameters significantly exceeded nominal conditions that should be present in the steam.

Visual Examination of General Physical Features

Examination of the tubing removed from the superheater bank confirmed an extensive crack network stemming from a longitudinal fillet weld, or a "stitch" weld, which connected two adjacent sections. Linking-up of these multiple crack fronts had caused a "window" to be completely blown out of a section identified as Tube 5 (Fig. 1). Inspection of the side directly opposite the "window" rupture also showed branched cracking stemming from a stitch-weld region (Fig. 2). The tube surfaces were coated with a thin layer of dark brown oxide. No excessive build-up of deposits was discerned.

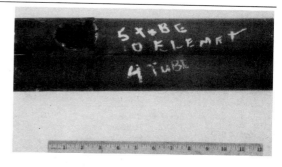

Fig. 1 Ruptured superheater tubing

Testing Procedures and Results

Surface examination

The fracture surfaces appeared markedly brittle. There were no signs of significant corrosion pitting or general corrosive attack. The cracks were multidirectional and branched.

Metallography

Microstructural Analysis. Cross-sections were removed from the fractured regions and prepared for metallographic study. The microstructure exhibited generally equiaxed grains typical of as-fabricated austenitic stainless steel tubing. The cracks were finely branched and predominantly transgranular (Fig. 3 and 4). The cracking was confirmed to have originated on the internal surface.

Chemical analysis/identification

Material. A portion of the tubing was chemically analyzed by optical emission spectroscopy (OES) to determine the alloy composition. Table 1 indicates that the tube was in compliance with the chemical requirements of ASTM A213 specifications for Grade 304H austenitic stainless steel.

Scanning electron microscope-energy dispersive spectroscopy (SEM-EDS) was used to determine the elemental composition along a fracture surface. A portion of Tube 5, adjacent to the "window" fracture, was broken open to expose a fresh fracture surface. The semi-quantitative results are listed in Table 2. The results confirmed a significant sodium concentration on the fracture surface (6% Na). This finding is consistent with what would be expected with a sudden carryover of caustic (NaOH) into the steam system.

Mechanical properties

Hardness. The hardness of the stainless steel tubing ranged from 75 to 77 HRB. Thus, the material was in compliance with ASTM A213 specifications for Grade 304H stainless steel, which stipulates that the metal hardness should not exceed 90 HRB.

Fig. 2 Cracks extending from the "stitch" weld, on the back side of the rupture site. 0.46×

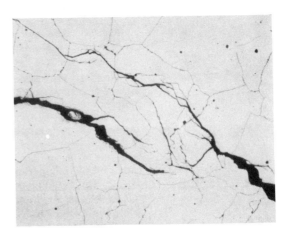

Fig. 3 Branched transgranular SCC of 304H stainless steel by caustic. 10% oxalic acid electrolytic etch. 152×

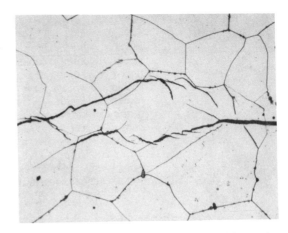

Fig. 4 Fine details of the cracking damage in the stainless steel. 10% oxalic acid, electrolytic etch. 304×

Table 1 Chemical analysis of the stainless steel

Element	Composition, wt% Tube metal	ASTM A213/Grade 304H
Carbon	0.06	0.04 to 0.10
Manganese	1.54	2.00 max
Phosphorus	0.025	0.040 max
Sulfur	0.021	0.030 max
Silicon	0.58	0.75 max
Nickel	10.10	8.00 to 11.00
Chromium	18.30	18.00 to 20.00
Molybdenum	0.30	...
Copper	0.28	...
Iron	Bal	Bal

Table 2 SEM-EDS analysis of a fracture surface

Element	Composition, wt%
Iron	67.5
Chromium	17.0
Sodium	6.0
Nickel	6.0
Silicon	2.0
Calcium	1.0
Aluminum	<0.5
Sulfur	<0.5

Discussion

Inspection of the tubing from a ruptured region confirmed that the material was in a typical condition for as-fabricated 304H-SS tubing stock. The extensive development of branched stress-corrosion cracks occurred in the "stitch" weld used to join two adjacent tube sections (where residual welding stresses would be anticipated to be high).

Stress-corrosion cracking (SCC) caused by concentrated caustic in stainless and low-alloy steels is often observed to be intergranular in nature.

However, the morphology of caustic-induced SCC in austenitic stainless steel alloys can sometimes be transgranular, indistinguishable from that caused by chlorides (Ref 1). The lack of cracking in welded and other stressed areas of the low-alloy steel portions of the superheater system suggests that austenitic stainless steels are extremely sensitive to rapid transgranular crack growth when exposed to caustic at high temperatures.

Conclusions and Recommendations

Most probable cause

The brittle rupture at the welded portion of the 304H-SS superheater tubing was most likely the result of stress-corrosion cracking caused by caustic carryover in the steam.

Remedial action

Under normal operating conditions, the possibility of caustic SCC of stressed areas of the 304H-SS superheater bank is remote. However, high residual stresses, resulting from short longitudinal weldments to "stitch" together adjoining tubes, increase the risk of environmentally induced crack-

ing if chemistry upsets occur in the steam system. Check valves and critical monitoring systems (conductivity alarms, sodium analyzers, etc.) must be maintained to avoid accidental chemical upsets that could rapidly result in brittle environmental cracking conditions. If post-weld heat treating of these attachment welds is impractical, use of mechanical bracketing to connect adjoining sections is a viable alternative. Another option is the installation of replacement tubing with a more SCC-resistant material, such as Incoloy 825.

Reference

1. C.P. Dillon and D.R. McIntyre, *MTI Publication No. 15: Guidelines for Preventing Stress Corrosion Cracking in the Chemical Process Industries*, Materials Technology Institute of the Chemical Process Industries, Columbus, OH, 1985, p. 41.

Stress-Corrosion Cracking of a Swaged Stainless Steel Reheater Pendent Tube

Anthony C. Studer, Radian Corporation, Austin, Texas
Robert S. Dickens, The Hartford Steam Boiler Inspection and Insurance Company, Pittsburgh, Pennsylvania

A cold-formed Grade TP 304 stainless steel swaged region of a reheater tube in service for about 8000 hours cracked because of sulfur-induced stress-corrosion cracking (SCC). Cracking initiated from the external surface and a high sulfur content was detected in the outer diameter and crack deposits. Comparison of the microstructure and hardness of the swaged region and unswaged Grade TP 304 stainless steel tube metal indicated that the swaged section was not annealed to reduce the effects of cold working. The high hardness created during swaging increased the stainless steel's susceptibility to sulfur-induced SCC. Because SCC requires water to be present, cracking most likely occurred during downtime or startups. To prevent future failures, the boiler should be kept dry during downtime to avoid formation of sulfur acids, and the swaged sections of the tubes should be heat treated after swaging to reduce or eliminate strain hardening of the metal.

Key Words	Austenitic stainless steels, Corrosion Sulfur	Boiler tubes, Corrosion Cold working (swaging)	Stress-corrosion cracking Strain hardening
Alloys	Austenitic stainless steels—304 Grade TP		

Background

Applications

The reheater tube is part of a single drum boiler that generates 421 400 kg/h (927 000 lb/h) of steam.

Circumstances leading to failure

The swaged sections are cold formed, and the manufacturer does not heat treat the sections unless they are above a specified level of hardness. The tube was in service for about 8000 hours.

Pertinent specifications

The reheater tube sample contained a swaged section which reduced the tube dimensions from a 7.0 cm (2.75 in.) outer diameter with a 0.46 cm (0.180 in.) wall thickness, to a 5.1 cm (2.0 in.) outer diameter with a 0.41 cm (0.160 in.) wall thickness. The swaged tube section was welded to a 5.1 cm (2.0 in.) outer diameter with a 0.41 cm (0.160 in.) minimum wall thickness tube. Both the swaged tube and the 5.1 cm (2.0 in.) tubing were reported to be ASTM A 249, Grade TP 304 stainless steel.

Testing Procedure and Results

Surface examination

Visual. Figure 1 (top) shows the tube section as received for analysis. Figure 1 (bottom) shows a circumferential crack observed in the swaged section of the reheater tube.

Scanning Electron Microscopy Fractography. The crack shown in Fig. 1 was broken open to reveal the fracture surface. Figure 2 shows the typical appearance of the fracture surface. The rock candy appearance of the fracture surface indicates the mode of fracture is intergranular separation.

Metallography

Microstructural Analysis. The typical microstructure observed in the swaged and unswaged sections of the tube are shown in Fig. 3 and 4, respectively. Both microstructures consist of austenite grains with carbides at the grain boundaries. Both sections of the tube are in the early stages of sensitization as indicated by the carbide precipitation at the grain boundaries and along slip lines in the work-hardened swaged metal. The only significant difference between the microstructures are the slip lines observed within the grains of the swaged structure. The banding is a result of the swaging operation.

Crack Path. The typical appearance of the crack, extending from the external surface of the tube inward, is shown in Fig. 5. The crack has penetrated approximately 95% of the wall thickness. The mode of cracking is intergranular separation with a small degree of side branching.

Chemical analysis/identification

Tube Material. Table 1 presents the chemical composition of the tube metal, along with the com-

(a)

(b)

Fig. 1 The reheater tube sample as received for analysis. The bottom photograph shows a close-up of a circumferential crack.

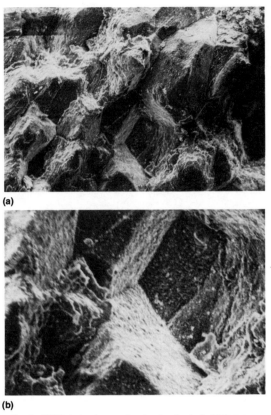

(a)

(b)

Fig. 2 SEM photomicrographs showing the typical intergranular appearance of the fracture surface of the crack. Top, 124×. Bottom, 310×

(a)

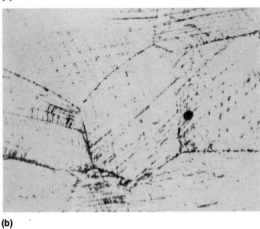

(b)

Fig. 3 Microstructure of the swaged section of the tube. The microstructure consists of austenite grains with carbides along the grain boundaries and slip lines within the grains. Oxalic acid electrolytic etch. Top, 62×, Bottom, 496×

(a)   (b)

Fig. 4 Slightly sensitized microstructure of the unswaged section of the tube, consisting of austenite grains with carbides along the grain boundaries. Oxalic acid electrolytic etch. Top, 62×. Bottom, 496×

positional requirements for ASTM A 249, Grade TP 304 stainless steel. The tube metal met the chemical requirements for 304 stainless steel.

Corrosion Deposits. Figure 6 gives the deposit analysis results from the fireside deposits. The external deposits contained major amounts of silicon, sulfur, iron, and potassium. EDS also detected minor to trace amounts of aluminum, calcium, arsenic, phosphorus, titanium, chromium zinc, and nickel. Figure 7 shows the *in situ* EDS,

analysis results from the crack deposits. The deposit composition was similar to the external deposit.

Mechanical properties

Hardness measured in the swaged section averaged 35.6 HRC, compared to an average value of 78.3 HRB for the unswaged material. Note that the hardness values are given on different scales, Rockwell C and B. A hardness value of 22 HRC is approximately equivalent to 100 HRB.

Fig. 5 The primary crack and secondary side branching extending inward from the external surface. Top is outer diameter, bottom is inner diameter. Oxalic acid electrolytic etch. Approximately 24.6×

Discussion

Metallographic and fractographic analysis showed the primary mode of cracking to be intergranular separation. A small amount of secondary branching was observed. The crack morphology was characteristic of sulfur-induced stress-corrosion cracking (SCC).

Intergranular SCC is the preferential corrosive attack of grain boundaries in a stressed material. Factors which promote stress-corrosion cracking are sustained stresses, a corrosive agent in the presence of water, and a susceptible material. Intergranular cracking of alloy steels with a hardness of 22 HRC or greater can be caused by sulfur-in-

duced SCC. As the hardness of the material increases, so does its susceptibility to SCC. The swaged section of the tube contains residual stress from the imposed cold working. The threshold stress for SCC in type 304 stainless steel decreases dramatically for cold reductions greater than about 25%.

The EDS analysis of the fracture surface deposits revealed the presence of sulfur. Sulfur was also detected at the crack terminus. Sulfur, in the form of sulfur acids such as sulfurous acid of polythionic acid, is known to attack sensitized stainless steel intergranularly. When this intergranular attack is

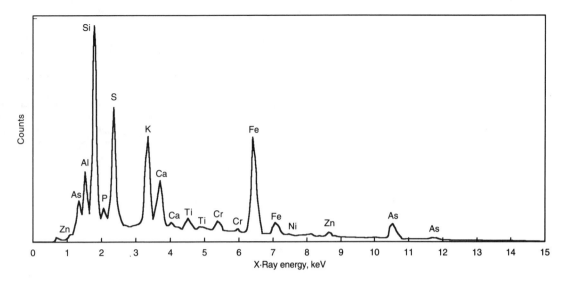

Fig. 6 EDS elemental analysis from the bulk external deposits removed from the reheater tube

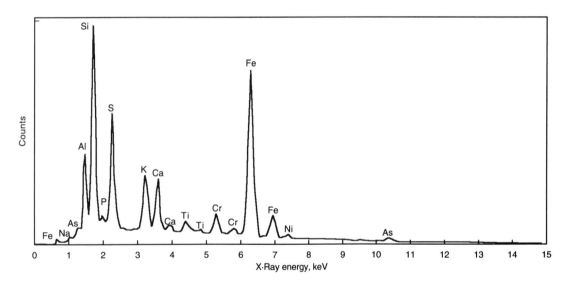

Fig. 7 *In situ* EDS elemental analysis from the crack deposits

combined with tensile stresses, either static or residual, intergranular SCC occurs. The requirement of an aqueous medium suggests that the cracking occurred during downtime or startups.

Unstabilized grades of stainless steels become sensitized when heated or cooled through the temperature range of about 427 to 900 °C (800 to 1650 °F). Sensitization causes carbon to migrate to the grain boundaries and combine with the chromium in the alloy, forming chromium carbides. The precipitation of these chromium carbides depletes the grain boundary areas of chromium, lowering their corrosion resistance. The sensitization of the stainless steel combined with the high hardness of the swaged section made the reheat tube susceptible to sulfur-induced SCC.

Table 1 Chemical composition (wt%) of the reheater tube

Element	Tube metal	ASTM A249, Grade TP 304 stainless steel
C	0.06	0.08 max
Mn	1.51	2.00 max
P	0.024	0.040 max
S	0.018	0.030 max
Si	0.50	0.75 max
Ni	9.58	8.00 to 11.00
Cr	18.38	18.00 to 20.00
Mo	0.20	ns
Fe	rem	rem

ns, not specified

Conclusion and Recommendations

Most probable cause

The reheater tube failure was due to sulfur induced stress-corrosion cracking. Most likely, moisture reacted with sulfur in the external deposits forming sulfur acids, such as sulfurous acid or polythionic acid. The sulfur acid(s), combined with the increased metal hardness and residual stresses from the swaging operation, caused the SCC failure.

Remedial action

All the swaged sections of reheater tubing should be inspected for cracks. Any sections containing cracks should be replaced with swaged sections that have been properly annealed. Swaged sections that do not exhibit cracking should either be properly heat treated to reduce the cold work/hardness or replaced with annealed swaged sections.

How failure could have been prevented

If the original swaged sections had been annealed, they would not have been susceptible to sulfur-induced stress-corrosion cracking.

Thermal Fatigue Failure of Alloy UNS NO8800 Steam Superheating Tubes

Moavinul Islam, CC Technologies, Columbus, Ohio

Alloy UNS NO8800 (Alloy 800) tubes of the steam superheating coils of two hydrocracker charge heaters in a refinery failed prematurely in service. Failure analysis of the tubes indicated that the failures could be attributed to thermal fatigue as a result of temperature fluctuations as well as restriction to movement. Fatigue cracks initiated intergranularly from both the flue gas and steam sides. Enhanced general and grain boundary oxidation coupled with age hardening of the alloy led to the formation of incipient intergranular cracks that acted as sites for the initiation of the fatigue cracks.

Key Words	Thermal fagitue	Steam superheating coils	Hydrocracker charge heater
Alloys	UNS NO8800 (Alloy 800)		

Background

Several tubes of the steam superheating (SSH) coils of two hydrocracker (HCR) charge heater units in a refinery failed prematurely in service. The tubes were made of Alloy UNS NO8800 (Alloy 800). Unit 1 suffered five successive tube failures over a period of about three months within two years after it was commissioned. All failures were in the finned tubes located near the intermediate support towards the inlet side and were similar in nature, having circumferential cracks. Two of the tubes were completely fractured into two pieces. It was also noticed that the fins at the intermediate support were damaged, indicating restriction in movement. Unit 1 was recommissioned after retubing with the same alloy having double the original wall thickness. Slight modifications were also made to the tube supports.

One of the finned tubes in Unit 2 failed in an identical manner three years after start-up. In addition, hydro-testing of the unit during a shutdown shortly afterwards revealed leaks in another finned tube that was located between the intermediate support and the outlet.

The tubes had a thickness of 2.75 mm (0.108 in.) and a diameter of 88.9 mm (3.50 in.). Some of the tubes were circumferentially finned on the outside with type 304 stainless steel while others were bare. The fins (thickness 1.25 mm or 0.0492 in., width 12.8 mm or 0.504 in.) were spaced approximately 7 mm (0.28 in.) apart and were joined to the tube surface by electric resistance welding. The SSH coils had superheated steam inside and hot flue gas on the outside. The design temperature of the unit was 767 °C (1412 °F), metal temperature, but the normal operating temperature was 325 °C (616 °F). The design pressure was 1.42 MPa (206 psig). Figure 1 shows the arrangement of the SSH coil in the HCR units and the locations of the failed tubes.

Selection of specimens

Failure analysis was carried out on a tube section from Unit 1, approximately 200 mm (8 in.) long (which was completely fractured in two), and a tube section from Unit 2, about 350 mm (14 in.) long which had 40 fins.

Visual Examination of General Physical Features

The examined tube section of Unit 1 was completely fractured into two pieces and had 26 welded fins. Some of the fins showed a lack of bonding. Visually, the position of the fracture appeared to coincide with the base of one of the welded fins. No secondary cracks were visible. Figure 2(a) shows a photograph of the failed tube with the matching fracture faces with one of the fins re-

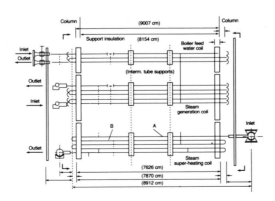

Fig. 1 Schematic arrangement of the steam superheating coil in the hydrocracker charge heater units showing locations of tube failures in (a) Unit 1 and (b) Unit 2

Fig. 2 Photographs showing: (a) fractured finned tube from Unit 1 (arrow indicates location of area marked "X" in photograph); and (b) fracture surface of the failed tube

moved. Figure 2(b) is a photograph of the fracture surface. The arrow in Fig. 2(a) corresponds to the area marked "X" in Fig. 2(b).

The tube from Unit 2 was cut in half and a section 350 mm (14 in.) long with 40 fins was examined. Some of the fins showed a lack of bonding. No cracks were readily visible on the outer side. Four circumferential cracks 2 to 6 mm (0.08 to 0.24 in.) in length were visible on the inside surface, two of which are shown in Fig. 3(a). The cracks were all located in the 9 o'clock position in the direction of

the steam flow. Examination of the outer surface after removal of the fins revealed that the four cracks observed on the inner surface had completely penetrated through the tube wall. Sixteen other circumferential cracks of various dimensions were also found on the outer surface. Fig. 3(b) shows some of these cracks. It was noticed that all the cracks were located adjacent and to the right side of the welded zones. None of the cracks originated in areas where there was lack of welding.

Testing Procedure and Results

SEM examination

The fracture surface of the tube in Unit 1 was examined with the scanning electron microscope (SEM). The fracture surface was not treated for oxide removal prior to the SEM examination. Figure 6(a) shows a low magnification view of the area marked "X" in Fig. 2, while Fig. 6(b) shows a higher magnification micrograph from the same area. Fatigue striations are clearly visible in Fig. 6(b). The relatively wide spacing of the striations suggests the occurrence of low cycle fatigue. The direction of striations indicates that the fracture path is from the steam side to the flue gas side. However, the direction of other fatigue striations found in another area of the fracture surface suggests a fracture path in an opposite direction. The fracture surface was mostly transgranular in nature, except for some in-

tergranular cracking near the steam side.

One of the cracks shown in Fig. 3 (the tube of Unit 2) was carefully broken open to expose the

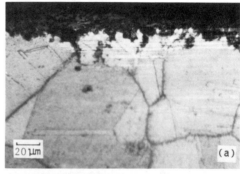

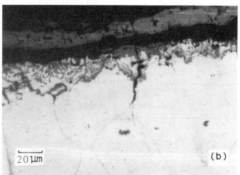

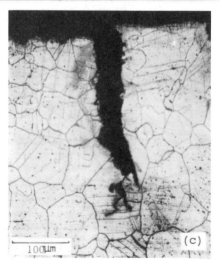

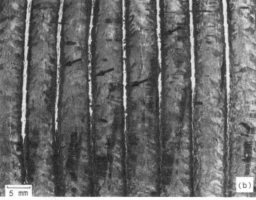

Fig. 3 Photographs showing: (a) two of the four visible cracks observed on the inner surface of the tube of Unit 2; and (b) the outer surface of the tube after removal of the fins. Arrows indicate locations of cracks adjacent to the welds.

Fig. 4 Micrographs of cross sections from the tube of Unit 1 showing: (a) oxidation of the flue gas side with intergranular and transgranular penetration; (b) intergranular oxidation on the steam side; and (c) transgranular crack originating from the gas

fracture surface for examination with the SEM. Some intergranular and transgranular fracture features were seen in addition to fatigue striations. Again, as in the case of Unit 1, the fatigue cracks were found to propagate from both the steam side and the gas side.

Metallography

Cross-sectional and longitudinal sections from the Unit 1 tube were prepared from near the fracture area and from an area slightly away from the location of fracture. Microstructural analysis indicated that the grain size of the alloy was not uniform and varied between ASTM grain sizes No. 2 and 3. Intergranular and intragranular carbides were observed in the microstructure of the material.

Examination of the metallographic cross sections revealed the presence of intergranular fissuring in addition to oxidation with intergranular and transgranular penetration on both sides of the tube (Fig. 4a, 4b). The intergranular fissuring and oxidation were more pronounced on the inner surface (steam side) than on the outer surface (flue gas side). A metallographic cross section through another crack revealed that the crack was transgranular without branching (Fig. 4c) and had progressed about one-fourth of the wall thickness.

Examination of metallographic cross sections of the Unit 2 tube showed that cracks started near the base of the weld and were transgranular in nature without crack branching (Fig. 5a). As in the case of the tube in Unit 1, intergranular fissuring was observed on both the steam and the flue gas sides and again was more prominent on the steam side (Fig. 5b, 5c). However, unlike the tube of Unit 1, there was no oxidation layer in the tube of Unit 2, and the grain boundary oxidation was much less than in the tube of Unit 1. The grain size of the Unit 2 tube was found to be between ASTM No. 4 and 5.

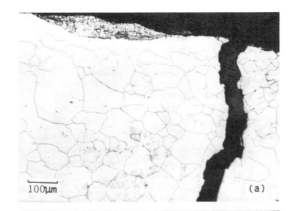

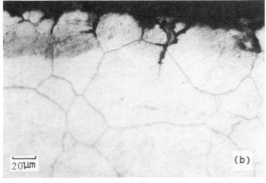

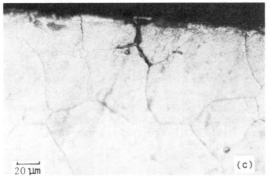

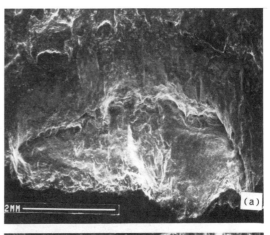

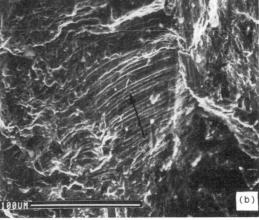

Fig. 5 Micrographs of cross section from the tube of Unit 2 showing: (a) transgranular crack originating from near the base of a weld; (b) incipient intergranular and transgranular fissuring on the steam side; and (c) intergranular fissuring on the flue gas side

Fig. 6 SEM micrographs from the tube from Unit 1 showing: (a) low magnification view of the area marked "X" on the fracture surface shown in Fig. 2(b); and (b) fatigue striations observed in the same area in Fig. 6(a) marked "Y." Photo is oriented in same direction as Fig. 2(b). Direction of crack propagation is from the steam side (bottom of micrographs) to the flue gas side of the

Table 1 Tensile and hardness properties of UNS NO8800

Data	Yield strength, MPa	Ultimate tensile strength, MPa	Elongation, %	Hardness, HRB	Grain size ASTM No.
Laboratory	337	670	34	85	2-3
Manufacturer	238	556	55	69	4-5
ASTM B 407-88	172 (min)	450 (min)	30 (min)	95 (max)	5

Table 2 Chemical composition of UNS NO8800

Element	Composition, %, heat 1	Composition, %, heat 2	Composition, %, ASTM B 407-88
Ni	31.02	31.35	30.0-35.0
Cr	20.20	20.39	19.0-23.0
Fe	46.00	45.00	39.5 min
C	0.06	0.06	0.05-0.10
Mn	0.72	0.89	1.50 max
Si	0.46	0.49	1.0 max
Cu	0.14	0.14	0.75 max
Ti	0.36	0.34	0.15-0.6
Al	0.36	0.26	0.15-0.6
S	0.005	0.002	0.015 max

Mechanical properties

Rockwell B hardness tests were carried out on both the inside and the outside surfaces of the Unit 1 tube. Tensile tests were conducted on longitudinal specimens machined from the tube material. None of these specimens had surface cracks at or near the welded areas. Table 1 shows the tensile and hardness results in addition to the values provided by the manufacturer and the ASTM specifications. The values given in the table show that the tube material has yield strength, ultimate tensile strength, and grain size higher than those specified by the manufacturer or the ASTM specification B 407-88 (ASME SB-163). Therefore, these results suggest that the material had age-hardened during service.

Because of the extensive cracking of the tube in Unit 2, tensile tests could not be performed. However, hardness tests were done and the value was found to be 82 on the Rockwell B scale.

Discussion

The metallurgy of Alloy UNS NO8800 (Alloy 800) is rather complex. Minor constituents, in particular Ti, Al, and C, together with the annealing heat treatment, play a critical role in determining its service performance characteristics. It has been established that the maximum Ti+Al level should be 0.6% and the Ti to C ratio should be a minimum of 12 to maximize the stress-corrosion cracking resistance, the creep rupture strength, and rupture ductility properties (Ref 1, 2). Variations of the above limits create instability in the microstructure.

Table 2 gives the chemical composition of the two heats of UNS NO8800 used in the HCR units as provided by the manufacturer, in addition to ASTM B 407-88 specifications. It can be seen that the Ti+Al content of the failed alloy is between 0.6 to 0.72% and the Ti/C ratio is around 6. Hence some microstructural instability may be expected during service. However, the ASTM specification does not specify a Ti and Al level for Alloy 800, but specifies that the level should be 0.85 to 1.20 in the higher temperature grades. The specification also does not specify a Ti/C ratio limit.

The service performance of UNS 8800 in various applications is generally good. It is not surprising, therefore, that not many instances of failures have been reported for this alloy. The one most relevant to the present failure case is that reported by Lippold (Ref 3), which describes the creep-fatigue failure of Alloy 800 in superheated steam.

In order to identify the most probable cause of failure of the SSH coils, it is necessary to consider the findings of the investigation in parallel with the operating conditions of the HCR. According to information gathered from the refinery, it was found that Unit 1 was, on occasion, operated at the design temperature of 767 °C (1412 °F), and the SSH coil was also run dry for some time. The normal operating temperature should be 325 °C (616 °F). ASTM B 407-88 specifies 595 °C (1100 °F) as the maximum service temperature for Alloy UNS NO8800. The time-temperature-precipitation diagram for Alloy 800 presented in Lippold's paper (Ref 3) indicates that in the temperature range above 540 °C (1004 °F), carbides start to precipitate within the first week (~100 h) of service if there is carbon available in solution. Tubes that operate in the temperature range below 480 °C (896 °F) should not form carbides in several years. High temperatures would also induce age hardening and grain coarsening. The total effect would be to deplete the chromium content of the alloy, particularly at the grain boundaries and the surface.

The tube material of Unit 1 suffered from oxidation and intergranular attack from both the steam side and the gas side (Fig. 3a, b). The microstructure was quite coarse (ASTM grain sizes No. 2 and 3), and there was carbide precipitation at the grain boundaries. All these features are consistent with high-temperature operating conditions. Moreover, the presence of striations on the fracture surface (Fig. 4b) clearly indicates that fatigue was involved, and that cracking was from both sides.

The occurrence of fatigue failure in the tube of Unit 1 is not readily explainable because in general the fatigue life of Alloy 800 is quite good (Ref 3). However, the situation appears logical when one considers the information available in the literature on the low cycle fatigue of Alloy 800. The fatigue life of tubular samples was found (Ref 4) to be only 20% of that of solid bar samples when tested in static steam at 650 °C (1202 °F) with a total strain range of 0.5%. A more dramatic decrease in fatigue life was observed when a tensile and/or compressive hold period was incorporated into the fatigue test (Ref 4, 5). Nearly a 50-fold decrease in the fatigue life was observed because of specimen geometry (solid versus tubular) and fatigue loading conditions (hold versus no hold time) (Ref 3). Tensile or compressive hold time during fatigue can result in creep damage to the material during the hold period. As a result, the material damage is termed creep-fatigue behavior rather than low cycle fatigue.

It is well known that steam equipment may fail by fatigue if mechanical service stresses are fluctuating or vibrating, or if thermal cycles or thermal gradients impose sufficiently high peak stresses. Some form of mechanical constraint and temperature fluctuations gives rise to thermal fatigue (Ref 6). Constraint may be external (for example, imposed by rigid mountings) or it may be internal, in which case it is set up by a temperature gradient within the part. It has also been established that a reduction in fatigue life takes place due to the occurrence of intergranular cracking and due to the effect of oxidation (Ref 7). Grain boundary oxidation may produce a notch effect that may limit the fatigue life. In the present case, creep damage was not observed in the alloy as apparent from the absence of any creep voids or intergranular fissures within the alloy matrix.

It therefore appears that thermal fatigue was responsible for the failure of the tubes in Unit 1. Operation of the unit at high temperatures resulted in aging, grain coarsening, and carbide precipitation, all of which led to enhanced oxidation of the alloy and grain boundary attack, thus reducing the fatigue life of the alloy. Moreover, the presence of grain boundary oxidation acted as stress raisers that, together with hold times during fatigue, led to the premature failure of the tubes of Unit 1.

Unit 2 was found to have been operated under tighter control (i.e., the unit was not run dry and there were no excursions to high temperature). The observed microstructure (finer grain size, less carbide precipitation) and the less severe oxidation of the alloy in comparison to that of Unit 1 appears to be consistent with the operating conditions of Unit 2. The metallographic and fractographic evidence indicate that the failure of the tube of Unit 2 was similar to that of Unit 1. However, because of the longer time taken to form incipient intergranular fissures, the fatigue initiation time was longer in Unit 2, and hence the service life was longer (approximately 3 years as compared to 2 years for Unit 1).

In the case of Unit 2, the fatigue cracks were mostly from the flue gas side. A few cracks were found propagating from the steam side. Crack initiation and propagation from the outer side of the tube is quite logical because the highest tensile stresses are concentrated on the outer diameter of the tube. It can be hypothesized that when the fins are properly welded to the tube, complex stress patterns are set up in the weld zone due to thermal gradients as well as geometric restrictions leading to more rapid cracking.

Conclusion

The failure of the Alloy 800 tubes in the HCR units was due to thermal fatigue. The source of fatigue was possibly due to temperature fluctuations in addition to restriction of movement of the coils.

The initiation and propagation of fatigue cracks were mostly from the outside (flue gas side). However, some cracks were found to propagate from the steam side.

In the case of Unit 1, uncontrolled operating conditions (running dry and excursions to high temperatures) resulted in age hardening of the material and enhanced general and grain boundary oxidation with incipient intergranular cracks, leading to early initiation of fatigue cracks. In the case of Unit 2, with better controlled operating conditions, the grain boundary oxidation and incipient intergranular cracking were less severe, which manifested in a longer fatigue life.

References

1. J. Orr and J.B. Marriot, Proceedings of the CEC Conference on Alloy 800, Brussels, Sept., 1976

2. J. Orr, Proc. Petten International Conference on Alloy 800, North Holland Publishing, 1978, p 25

3. J.C. Lippold, *Mater. Performance*, Vol 24 (No. 10), 1985, p 16

4. W.B. Jones and J.A. Van Den Avyle, "Fatigue and Creep-Fatigue Testing of Steam Filled Tubular Alloy 800 Specimens," Report SAND. 82-0856, Sandia National Laboratories, May, 1982

5. H. Teranishi and A.J. McEvily, Proc. Petten International Conference on Alloy 800, North Holland Publishing, 1978, p 125

6. *Failure Analysis and Prevention, Metals Handbook*, 8th ed., Vol 10, American Society for Metals, 1975, p 122

7. C. Levaillant and A. Pineau, Low-Cycle Fatigue and Life Prediction—ASTM STP 770, 1982, p 169

Brittle Fracture Explosive Failure of a Pressurized Railroad Tank Car

Samuel J. Brown, Quest Engineering Development Corp., Humble, Texas

A 127 m^3 (4,480 ft^3) pressurized railroad tank car burst catastrophically. The railroad tank was approximately 18.26 m (719 in.) long (from 2:1 elliptical heads), 305 m (120.25 in.) in outer diameter and 15.875 mm (0.625 in.) thick. The chemical and material properties of the tank were to comply with AAR M-128 Grade B. As a result of the explosive failure of the tank car, fragments were ejected from the central region of the tank car between the support trucks from ground zero to a maximum of approximately 195 m (640 ft). The mode of failure was a brittle fracture originating at a pre-existing lamination and crack in the tank wall adjacent to the tank nozzle. The mechanism of failure was overpressurization of the railroad tank car caused by a chemical reaction of the butadiene contents. The interrelationship of the mode, mechanism, and consequences of failure is reviewed to reconstruct the sequence of events that led up to the breach of the railroad tank car. Means to prevent similar reoccurrences are discussed.

Key Words

Tank cars
Brittle fracture, Pressure effects
Explosions

Pressure vessels,
Mechanical properties

High strength low alloy steels,
Mechanical properties

Background

The breaching or explosion of a tank car occurred at a storage facility in which butadiene is transferred from large storage tanks to railroad tank cars, a routine process for the approximately 45454-kg (100,000-lb) railroad cars with approximately a 126800 l (33,500 gal) capacity (127 m^3, or 4,480 ft^3). The butadiene is maintained in the tanks at about –5.6 °C (22 °F). The dimensions of the railroad tank car are shown in Fig. 1. The railroad car had a rated safety relief valve setting of 1.93 MPa (280 psig). The tank specifications were to meet the chemical and mechanical requirements of AAR M-128 Grade B.

Applications

The transportation of fluid products via railroad tank cars provides an important link in the transportation of raw and finished goods for industry, government and the consumer. Because of the large quantities of pressurized fluids carried by railroad tank cars, these costly and dangerous failures must be prevented.

Circumstances leading up to the failure

Prior to this incident, the storage facility had routinely filled railroad tank cars with butadiene. The facility had the capability to receive several trains of cars, to fill a number of cars simultaneously, and to flare through vent-line connections at numerous locations. The flare was provided with a

nitrogen purge system which flowed toward the flare.

Normally, the storage facility would fill tank cars that had the same service or had been cleaned and inerted. Approximately three months prior to the failure of the railroad tank car, it had been maintenanced, repaired, and hydrotested, but not inerted, at an area repair yard. It arrived at and entered the storage facility due to an error and misidentification. It was queued with several other tank cars at the facility for loading with approximately 126800 l (33,500 gal) of butadiene. During 1½ h, the –5.6 °C (22 °F) butadiene was pumped from the storage facility into several connected tank cars, including the subject car. The subject tank car was the last to fill, since it was noticed that pressure had built up in the tank car during the pumping process. The pressure in the railroad tank car was approximately 0.79 to 0.83 MPa (115 to 120 psig). To relieve the built-up pressure, a vent line was attached to the railroad tank car to discharge to the facility flare stack, which in turn would burn the discharged vapors. Through coordination with a number of facility personnel, the vent line from the tank car was opened to the facility flare stack. Following the opening of the vent line from the tank car, the flare caught and displayed a nice flame which burned for a few seconds. This was followed by two "pops" at the flare and then the flame went out. Meanwhile, a "pop"

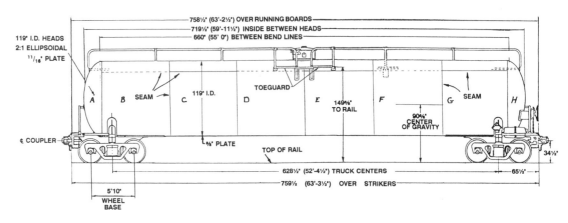

Fig. 1 Dimensional drawing of railroad tank car

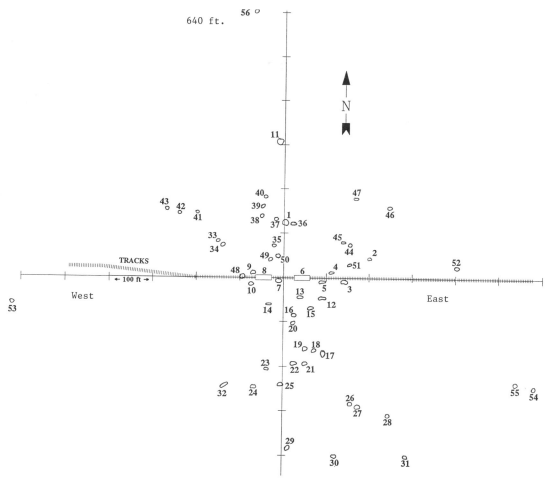

Fig. 2 Fragment distribution of railroad tank car after rupture

was heard at the railroad tank car and a thin pencil-like flame emerged from the safety relief valve. The flame column reportedly rose higher until it reached over 15.2 m (50 ft) above the tank car, at which point the railroad tank car burst or exploded violently, ejecting fragments from the central part of the tank car, from ground zero to approximately 195 m (640 ft) away; see Fig. 2. Approximately 60 major fragments as large as 5227 kg (11,500 lb) were ejected from the tank car. At the time of tank breach, a vapor space at the top of the car occupied approximately 20% of the tank volume (approximately 24.9 m^3 or 880 ft^3).

Pertinent specifications

The specification drawings for the railroad tank car requires that the pressure vessel tank material properties comply with specification requirements AAR M-128 Grade B.

The tank, according to specification drawings, was to be built, tested, and stamped per Class ICC 112A340-W specification. The nominal thickness cited in the specification drawings is 1.587 cm (0.625 in.) and the heads are 1.746 cm (0.687 in.). The safety relief valve is reported to have been set at 1.93 MPa (280 psi).

Performance of other parts in same or similar service

Butadiene is commonly shipped in railroad

tank cars at –4.4 °C (24 °F) as a liquid under its own vapor pressure.

Selection of specimen

Fifty-six fragment samples of the railroad tank car were located, numbered on a grid (see Fig. 2), and assembled at a central storage site. The fragments were measured and photographed. After a field examination of the fracture surfaces of the fragments, seven samples were selected as representative of various sections of the tank car:

- Fragment 3, shown in Fig. 3, contains the east end of the tank car head and first shell plate (see Fig. 1, segments H and G).

- Fragment 7, shown in Fig. 4, consists of the west head and part of the first plate in the cylindrical shell (see Fig. 1, segments A and B).

- Fragment 10 was chosen as a sample from the first shell, from the west end of the tank car, because of the appearance of the fracture mode changing from brittle to ductile propagation.

- Fragment 11 was chosen as a sample from a shell plate near the middle of the car and the bottom of the tank (see Fig. 1, segment D).

- Fragment 12 was used as a sample because of its proximity to the east end of the tank car head and the transition of the fracture mode

Fig. 3 Fragment 3, vessel segments G and H

Fig. 4 Fragment 7, vessel segments A and B

at a weld (see Fig. 1, segment G).

- Fragment 17, shown in Fig. 5, exhibited characteristics of a fracture origin and delamination, and inclusions were visible. Fragment 17 came from the top-south side of the tank car adjacent to the nozzle at the top center of the car. Refer to Fig. 1, segment E.

- Fragment 27 showed areas of delamination at the fracture surface. Its origin appears to have been at the top-middle-south side of the tank car. See Fig. 1, segment E.

Visual Examination of General Physical Features

As shown in Fig. 1, the pressure vessel portion of the railroad tank car consists of a cylindrical shell, 16.6 m (655.5 in.) long, composed of six cylindrical rings or sections with a length of 2.8 m (109.25 in.) between their circumferential seams. The ends of the vessels are closed with 2:1 elliptical heads. The six 2.8-m (109.25-in.) long segments that compose the length of the tank car cylinder have longitudinal welds at an elevation of approximately 127 m (50 in.) above the center line of the tank at the top and alternated for each successive ring on either side of the manway at the center-top of the railroad tank car. The manway is a 78-cm (31-in.) outer diameter, 46-cm (18-in.) inner diameter, 5-cm (2.0-in.) min thick cylindrical nozzle that intersects the railroad tank at a right angle at the top of the tank and midway between the vessel ends.

The tank car is supported by two trucks, one on either end. The support load between the tank and the railroad support trucks is distributed by tank slabbing and saddle plates under each vessel elliptical head and first tank plate ring.

The size or mass distribution of fragments consisted of eight to ten large fragments, with the remaining fragments decreasing in size. The large fragments consisted of the following:

- 2 railroad truck (wheel) assemblies—east and west
- 2 ends or heads with portions of their adjacent cylindrical shell

- 4 to 6 large pieces of the 4 central 2.8-m (109.25-in.) ring sections, segments C, D, E, and F in Fig. 1, on either side of the manway

A field examination of the failure surfaces revealed characteristic brittle fracture chevrons, brittle fracture transition into ductile tearing, delamination along the mid-plate thickness plane and a fracture origin on Fragment 17 (from the top center of the tank) as well as the presence of inclusions in the Fragment 17 plate material.

The characteristics of the fragmentation of the vessel and the fracture morphology appeared to be typical of burst tests performed and reported over the last five decades. For example, in the case of very long cylindrical pipelines, the fracture runs axially and initiates at a defect (the fracture may run hundreds of feet). In the case of a pipeline with circumferential stiffening rings or endcaps, the axial fracture transitions to a circumferential tear at the circumferential stiffening rings or endcaps. In the case of a cylindrical vessel with endcaps and an intersecting nozzle in the middle, the nozzle is typically the source of the axial breach initiation. The number of fragments is dictated by fracture toughness or ductility of the material and the number of attachments or stiffening members. Examples are provided in Ref 1. The Pressure Vessel Research Council (Welding Research Council) contains a considerable body of literature in their bulletins on this subject. An early text on this phenomena was written by Nadai (Ref 2).

Testing Procedures and Results

Surface examination

Visual. The failure or initiation site occurred at the center top of the tank car, in the shell, adjacent to the nozzle. The brittle fracture failure paths radiate from the initiation site in both longitudinal and cir-

cumferential patterns. The brittle fracture paths are observed to transition to ductile tearing which is particularly evident near the two end caps.

Figure 6 provides a close-up of the origin of the failure in Fragment 17 (see also Fig. 5). Figure 6

Fig. 5 Fragment 17, vessel segment E

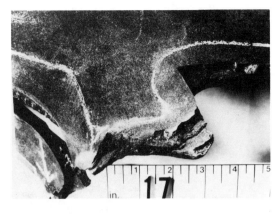

Fig. 6 Photomacrograph of shell midwall laminations and fracture site in Fragment 17

Fig. 7 Photomacrograph of fracture origin and midwall lamination cracks

Fig. 8 Photomacrograph of Fragment 3 showing brittle fracture to ductile transition

shows the laminations and delamination that occurred in the shell adjacent to the manway. Note the plate has mid-wall separation parallel to the plate surface (see Fig. 6 and 7). Some delamination one quarter of the plate thickness from both surfaces was observed.

Figure 7 is a magnification of the fracture origin shown in Fig. 5. Note that high-temperature oxidation is evident on the interior fracture face in some areas.

Figure 8 shows the transition of a failure path from brittle fracture to ductile failure as the crack propagation transitions from the cylindrical shell to the vessel end. See Fig. 3 for the location of this enlargement.

Figure 9 provides a close-up of Fig. 5, which is in the cylindrical segment close to the west vessel head. Figure 9 shows the transition from ductile to brittle failure.

The fracture surfaces shown in Fig. 6 through 9 were cleaned by ENDOX (electrolytic analysis process that uses an active anion of cyanide to remove oxidation).

Metallography

Microstructural Analysis. Figure 10 is a photomicrograph of a longitudinal cross-section through the fracture face of Fragment 17 (See Fig. 6). The microstructure is a fine-grained banded

Fig. 9 Photomacrograph of Fragment 10 showing ductile to brittle transition

ferrite and pearlite, with regions of a mixed transformation microstructure. This material was in the normalized condition. However, it is so contaminated that the heat treatment was not completed. The material is sufficiently contaminated to be a contributing source of the laminations prior to the breech of the railroad tank car, and to cause delamination during the failure process.

For comparison, a photomicrograph of a longitudinal cross-section through the fracture face of

Fig. 10 Photomicrograph of longitudinal microstructure through fracture origin, Fragment 17. 3% nital. 31.5×

Fig. 11 Photomicrograph of longitudinal microstructure through fracture face of Fragment 3. 3% nital. 31.5×

Table 1   Chemical analysis of tank fragments

Fragment	Composition, wt%									
	C	Mn	P	S	Si	Ni	Cr	Cu	V	Al
17, 27	0.24	1.28	0.023	0.018	0.24	0.16	0.08	0.03	0.04	0.02
10, 11	0.23	1.25	0.020	0.016	0.23	0.22	0.08	0.04	0.04	0.02
3-shell, 12	0.28	1.40	0.016	0.017	0.27	0.12	0.14	0.35	0.01	0.009
3-head	0.23	1.20	0.011	0.012	0.23	0.20	0.15	0.27	0.01	0.06
7-head	0.24	1.23	0.015	0.023	0.21	0.21	0.05	0.03	0.04	0.02
AAR M-128	0.28 max	0.92 to 1.62	0.035 max	0.040 max	0.13 to 0.45	0.25 max	0.25 max	...	0.08 max	0.02 min

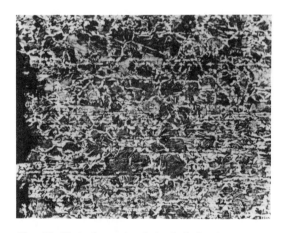

Fig. 12 Photomicrograph of longitudinal microstructure through fracture face of Fragment 7. 3% nital. 31.5×

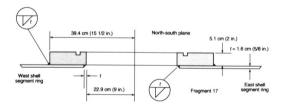

Fig. 13 Nozzle/plate lap weld to two center shell segment rings D and E

Fragment 3 in the shell is presented in Fig. 11. This shell segment adjacent to the east head is relatively free from nonmetallic inclusions when compared to Fig. 10. The microstructure is coarse-grained, banded, nonhomogeneous ferrite and pearlite.

As a final comparison, a photomicrograph of a longitudinal cross-section through the fracture face of Fragment 7 (west head) is shown in Fig. 12. The mixed microstructure is ferrite, pearlite and bainite. This is not a normalized microstructure, and it is probably in the as-formed condition.

Crack Origins/Paths. In the as-fabricated condition, the AAR M-128 Grade B shell panel containing Fragment 17 contributed to crack initiation and propagation because of the presence of bands of nonmetallic inclusions in concert with localized high stress nozzle-shell juncture. During normal operation, this condition resulted in delamination in the presence of the highly stressed region at the attachment of the nozzle to the tank shell: outer diameter fillet weld south and north direction, and inner diameter fillet weld east and west direction. During fabrication and installation of the nozzle, welding most likely promoted crack sites in the lamination. In the absence of contaminated material with laminations and cracks, the nozzle-to-vessel juncture would be the origin of crack growth and failure during a burst pressure test of the vessel.

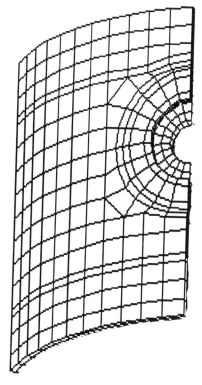

Fig. 14 Finite element model of nozzle-to-shell segment rings D and E

Table 2 Charpy V-notch impact properties for tank fragments

Fragment	Charpy V-notch impact strength, N · m (ft · lbf)	
	−45.6 °C (−50 °F)	−2.8 °C (27 °F)
17-shell, 27	40.7 (30)	58.3 (43)
10-shell	36.6 (27)	54.2 (40)
3-shell, 12	4.1 (3)	23.0 (17)
3-head (east)	24.4 (18)	42.0 (31)
7-head (west)	4.1 (3)	25.8 (19)

Chemical analysis and identification

Table 1 provides a representative chemical analysis of five samples from the various locations represented by the seven fragments selected in this analysis. Fragments 3-shell and 12 come from the same shell segment and do not meet the chemical specifications of AAR M-128 Grade B. Fragments 3-shell and 12 were below the requirements for aluminum which may account for these material specimens being coarse grained.

Mechanical properties

Hardness tests were performed on ten samples that were used for chemical testing. Average values ranged from 88 to 97 HRB.

Tensile Properties. The average tensile strength ranged from 586 to 723 MPa (85 to 105 ksi) which compares to the AAR M-128 Grade B requirements of 558 to 696 MPa (81 to 101 ksi).

Impact Properties. Charpy V-notch impact specimens were tested at −45.6 °C (−50 °F) and −2.8 °C (+27 °F). Table 2 gives impact values. The AAR M-128 Grade B impact requirements are 20.3 N · m (15 ft · lbf) at −45.6 °C (−50 °F). Fragment 3-shell exhibited low Charpy values. This specimen also showed below-specification aluminum content. Fragment 7, which had acceptable chemical properties, exhibited low impact values which may be attributed to its large grain structure; the microstructure indicates that the head was not normalized.

Stress analysis

The prediction of stresses in cylindrical vessels containing circular and rectangular attachments has been investigated for many years and is cur-

rently under investigation. The localized stresses around an attachment are typically intensified in certain locations. The investigations summarized in Ref 3 reflect the quest of designers to contour the nozzle and provide fabrication techniques to minimize attachment stresses.

This investigation considered classical stress analysis methods from Bijlaard in the 1950s to Brooks (Ref 4) in the 1990s for the Pressure Vessel Research Council. Considering the minimum tensile strength in the tank is 586 MPa (85 ksi), then the burst pressure of the cylindrical shell segment away from the nozzle should be approximately 5.5 MPa (800 psi). Utilizing the classical solution, the localized stress at the shell to nozzle connection at the north and south outer diameter (assuming a monoblock attachment and that plane sections remain plane), reach an ultimate stress at about 2.8 MPa (405 psi). The classical solutions provide useful data for estimating critical crack sizes. In our case, the attachment of the shell segment to nozzle ring is not a monoblock casting, forging, or full penetration butt weld. Figure 13 illustrates the fillet weld lap joint attachment of the tank car nozzle to the two ring shell segments of the tank car shell.

Figure 14 illustrates a finite element computer simulation of the tank car shell segments and the nozzle pad. The finite element method permits a more detailed picture of the actual stresses with concentration and defect effects in the nozzle, welds, and shell as a result of pressure, moments, forces, and temperature loadings (steady state or dynamic). The effects of fatigue, K_I, and variable shear modulus G due to lamination and delamination are evaluated by the FEM.

Figure 15 shows the Von Mises stress contour from internal pressure. Notice the high-stress region along the outer circumference of the nozzle to shell weld. The FEM model shown in Fig. 14 provides a comparison between the overlap fillet weld arrangement (Fig. 13) versus a monoblock design with respect to stresses in the fillet welds from pressure, forces, moments, and temperature or thermal interaction. FEM results show the susceptibility of the fillet weld arrangement to failure from 1.93 to 2.76 MPa (280 to 400 psi) as a result of the presence of laminations and inclusions in the shell to which the nozzle is fillet-welded. The results are discussed in more detail in Ref 5. FEM results indicate that the pressure tank car failed at a pressure between 1.96 and 2.76 MPa (285 and 400 psi).

Simulation tests

Since the exact pressure of the vessel burst

was not recorded at the time of the incident, a number of theoretical simulations were performed to obtain independent estimates of the burst pressure. One method was a stress analysis as described above. Another theoretical method was based on the fragment distribution in Fig. 2 and energy calculations of the pressurized contents. Since at least 20% of the tank car (approximately 24.9 m³, or 880 ft³) was compressed gas, a considerable amount of energy to drive the fragments would be available. References 6 and 7 list references for performing the energy, missile velocity and missile range calculations (as well as an overview of blast and fragmentation). The blast

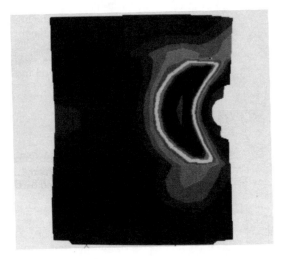

Fig. 15 FEM stress contours with maximum stress of 689 MPa (100 ksi). Bands denote decreasing increments of 48 MPa (7 ksi). Load case pressure, 2.76 MPa (400 psi)

Fig. 16 1/10-scale model of tank (not exact)

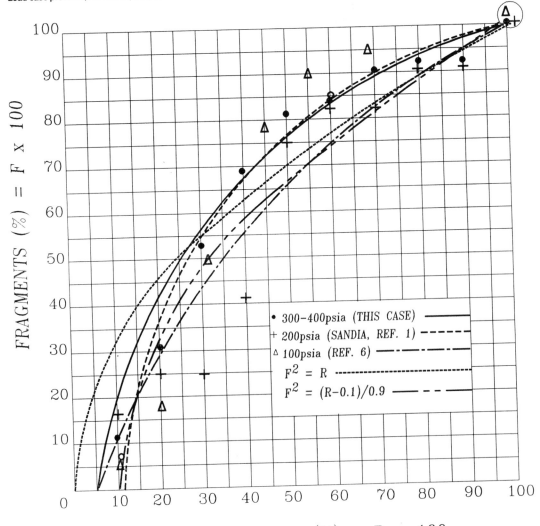

FRAGMENTS (%) = F x 100

RANGE OF FRAGMENTS (%) = R x 100

- 300–400psia (THIS CASE) ———
+ 200psia (SANDIA, REF. 1) – – – –
△ 100psia (REF. 6) —·—·—
$F^2 = R$ ·············
$F^2 = (R-0.1)/0.9$ — — —

Fig. 17 Comparison of fragment range vs fragment number to other low-pressure large vessel bursts

and fragmentation simulations performed in this analysis confirm that the fragmentation range and dispersal pattern illustrated in Fig. 2 could be achieved if the gas in the tank car was pressurized between 1.93 to 2.76 MPa (280 to 400 psi). Reference 5 shows that the pressure produced energy levels in the tank car equivalent to between 27.3 to 54.5 kg (60 to 120 lb) of TNT.

An experimental burst test of a 1/10 scale vessel (Fig. 16) with a circular plug was shown to burst below the shell theoretical burst pressure ($P = (\sigma_{ult} t/R$, where P is pressure, σ_{ult} is ultimate tensile strength, t is thickness, and R is radius). The crack initiated in the outer diameter heat-affected zone (HAZ) south of the weld. A number of published burst tests correlate significant reduction of the theoretical burst pressure in cylinders with weld attachments at the high stress junction. Figure 17 compares the fragment distribution of the railroad car versus two large-volume low-pressure vessel bursts. Pressures of 689 and 1379 kPa (100 and 200 psia) achieved maximum fragment ranges of 274 and 405 m (900 and 1330 ft), respectively.

Discussion

Tests were performed to determine the mode, the mechanism, and the casual factors of failure. Samples from the failed AAR M-128 Grade B railroad tank car shell had good tensile properties longitudinal to the shell but poor transverse properties, as evidenced by lamination and inclusions in some of the shell segments. Some shell segments exhibited good Charpy impact values while others showed extremely low values. Most samples exhibited chemical properties in compliance with specifications; however, some were below specification. Structural elements such as the tank car nozzle attached to the surface of a plate with lamination by means of a fillet weld will fail at considerably reduced pressures than monoblock nozzles with a blended or tapered transition to the shell thickness for full-penetration butt welding.

Examination of the failure site indicates that failure occurred along the lamination and introduced delamination along these weak planes. The computer simulation of stress and deformation of the lap joint arrangement between the vessel shell and nozzle demonstrated that the fillet weld induces high localized stresses between the fillet weld and the outer plies or fiber of the shell with laminations. The analysis also shows that the unwelded space or crack between the two fillet welds acts as a initiation site for crack growth and/or brittle fracture. The stress analysis indicated that failure of the tank car most likely occurred after the safety relief valve opened at 1.93 MPa (280 psi) but before it reached 2.76 MPa (400 psi).

The theoretical computer simulations utilizing the fragment distribution, the recorded volume of compressed gas, and the eyewitness testimony regarding the events correlates with the estimated pressure of between 1.93 and 2.76 MPa (280 and 400 psi) calculated by the theoretical stress analysis.

The eyewitness description of events confirm that the explosive failure of the pressure vessel was a result of the pressure increase from the loading pressure of 0.83 and 1.93 MPa (120 to 280 psi) when the safety relief valve opened as a result of combustion of the butadiene with the oxidizer (the air contained in the empty tank car before filling). The combustion process of heating and increasing the pressure in the tank was relatively slow as evidenced by the observed slow rise of the flame from the safety relief valve (the speed was slow relative to a detonation process). As the pressure increased above 1.93 MPa (280 psi), a longitudinal crack (circumferential to the nozzle and parallel to the weld) extended until it intersected a lamination, which caused the shell to separate from the inner nozzle weld sufficiently to cause the initial crack to become critical due to the increased stress, and then to run. This occurred in milliseconds. The pressure weld ejected the fragments principally perpendicular to the axis of the railroad car vessel. However, the effects of the thick girth or circumferential weld seams caused a greater number of fragments than would have probably occurred if the tank were fabricated from a single cylinder. The thick circumferential seams and outer insulation casing on the tank probably affected the fragment distribution.

Conclusion and Recommendations

Most probable cause

Causality is usually assigned to one or more of the following factors: design, material selection, material defects, fabrication, operation and use, maintenance, and inspection. In this case, all of these factors contributed to the failure of the railroad tank car in varying degrees. The explosion or breach of the pressure vessel tank car below its burst pressure was primarily a result of poor material properties and defects in the AAR M-128 Grade B railroad tank shell, in concert with the type of design and fabrication used for the tank shell to nozzle surface fillet welds.

It may be argued that the railroad tank car could have contained the combustion process if it had not prematurely burst, or if the safety relief valve had been properly sized. A counter-argument could be made that an oxidizer such as air should not have been present in the railroad tank car prior to the loading of the butadiene. Clearly, the inadvertent and unintentional movement of the tank car (also filled with air) into the butadiene loading facility was not foreseen. But it set the stage for the mechanism for increasing the tank pressure which during loading would normally not have exceeded 827 kPa (120 psi) and during transport would not have exceeded the vapor pressure of butadiene (considerably below 827 kPa, or 120 psi).

The most likely source of ignition was static electricity caused when the vent line from the tank car was opened. However, other mechanisms for initiating the combustion process were possible, but less likely.

Remedial action

It was recommended that the method of attachment of the nozzle to the shell be modified to incorporate a nozzle ring/plate with a tapered transition region around the perimeter of the nozzle which permits a full-penetration butt weld to the shell. This fabrication is preferred as a means of reducing failures associated with weld placement at high stress shell thickness transition regions. This method of fabrication is less susceptible to failure associated with weld or shell base metal imperfections.

A more rigorous material inspection should be performed on all shell segment samples to ensure proper mechanical properties, grain formation, and to monitor and minimize the size, frequency, and location of inclusions and defects, particularly in highly stressed regions.

Inerting of tank cars prior to filling with butadiene should be incorporated as a standard practice at the facility as well as a checking of previous use and current contents.

Acknowledgments

The participation and assistance of Dr. Ed Bravenec of Anderson and Associates/ISI and Troy Brown of QED/ISI in developing the data presented in the case and reported here is gratefully acknowledged.

References

1. S.J. Brown, *The Product Liability Handbook: Prevention, Risk, Consequences, and Forensics of Product Failure*, Van Nostrand Reinhold, New York, NY, 1991, Chapter 10.
2. A. Nadi, *Theory of Flow and Fracture of Solids*, McGraw Hill, New York, NY, 1950.
3. S.E. Moore, W.L. Greenstreet, and J.L. Mershon, "The Design of Nozzles and Openings in Pressure Vessels," in *A Decade of Progress—1982*, ASME, New York, NY, 1982.
4. G.N. Brooks, "Spring Constraints for Rectangular Attachments to Cylindrical Shells," in *ASME-PVP Vol 194: Analysis of Pressure Vessel and Heat Exchanger Components*, ASME, New York, NY, 1990 (See also "On Eliminating a Parameter in Welding Research Bulletin #198," in *ASME-PVP Vol 188*, 1990.
5. S.J. Brown, *Handbook of Explosions and Hazardous Release*, Butterworth Publishing Co. (to be released).
6. S.J. Brown, "Energy Release Protection for Pressurized Systems, Part I: Review of Studies into Blast and Fragmentation," *Appl. Mech. Rev.*, Vol 38, No. 12, 1985.
7. S.J. Brown, "Energy Release Protection for Pressurized Systems, Part II: Review of Studies into Terminal Ballistics/Impact," *Appl. Mech. Rev.*, 1986.

Failure of Welded Helium Tanks

H.S. Khatak, V. Seetharaman, and J.B. Gnanamoorthy, Metallurgy Programme, Indira Gandhi Centre for Atomic Research, Kalpakkam, India

Two tanks made of AISI type 304 stainless steel exhibited cracking in the heat-affected zone (HAZ) of the weld that joined the dished end and the shell. The dished ends had been produced by cold deformation. Hardness measurement and simulation tests showed that the deformation was equivalent to a 30% reduction in thickness. Residual stresses were measured at about 135 MPa (20 ksi). The HAZ was found to be sensitized. The tanks had been stored in a coastal atmosphere for about 4 years before installation. The failure was attributed to intergranular stress-corrosion cracking in a sensitized HAZ due to chloride from the environment. Use of low-carbon type AISI 304L was recommended. Minimization of fit-up stresses and covering with polyethylene sheets during storage were also suggested.

Key Words			
Austenitic stainless steels, corrosion	Welded joints		Welding parameters
Storage tanks, corrosion	Heat-affected zone		Nuclear reactor components
Stress-corrosion cracking	Sensitizing		Marine environments

Alloys	Austenitic stainless steels—304, 304L

Background

Two tanks made of AISI type 304 stainless steel exhibited cracking in the heat-affected zone (HAZ) of the weld that joined the dished end and the shell. The dished ends had been produced by cold deformation.

Application
The tanks had been fabricated for the storage of helium gas used in the core of a pressurized heavy-water reactor.

Circumstances leading to failure
After fabrication, the tanks were stored unprotected in a coastal atmosphere for about 4 years.

Performance of other parts in same or similar service
Many such stainless steel weldments that were not sensitized had not exhibited failures in a coastal environment for much longer durations.

Specimen selection
After visual and low-magnification microscopic examinations, the locations of the samples were established. The types of samples selected and their locations are shown in Fig. 1.

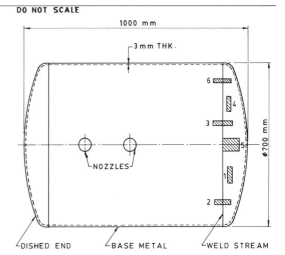

Fig. 1 Schematic of the helium storage tank. The areas where samples were taken are indicated: 1, intergranular corrosion testing; 2, hardness testing; 3, microscopy; 4, fractography; 5, residual stress measurement; 6, chemical analysis

Visual Examination of General Physical Features

Visual examination of the first tank revealed red corrosion products on the surface where the dished ends were joined to the main body. These products, in the form of lines, propagated toward the dished end from the weld seam. No corrosion product was observed on the weld metal itself.

Pressurizing the tank at 69 kPa (10 psi) and testing with soap solution revealed leaks at areas where the corrosion products were seen. The leaks were found all along the circumference. A few leaks were also detected in the base metal near the weld joining a nozzle to the shell. Reddish brown corrosion products were also seen on the tank surface at a location approximately 30 mm (1.2 in.) from one of the nozzle welds. A slight bulge (a few millimeters in height) was seen at the knuckle of the dished end.

In the case of the second tank, no leak was detected in the dished end. Leaks were seen on the tank surface near the weld deposit of the nozzle only. No leak was detected in the weld seam. In addition, at one point, approximately 30 mm (1.2 in.) from the weld deposit, there was a pit from which two cracks originated.

These tanks were fabricated by cold-forming operations without any final stress-relieving treatment. From the appearance of bulges at the dished ends, it was inferred that severe hammering could have been resorted to in the fit-up operations before welding. Thus, complex patterns of residual stresses could be expected in the joints because of the combined influence of cold working, fit-up, and welding.

Testing Procedure and Results

Surface examination

Scanning Electron Microscopy/Fractography. The specimens for fractographic studies were obtained by pulling in tension, to the point of fracture, the samples containing cracks. The induced fracture surfaces thus obtained were examined in a scanning electron microscope (SEM); typical micrographs are shown in Fig. 2. The fracture path near the fusion line was transgranular (Fig. 2a), then changed to intergranular in the HAZ (Fig. 2c), and switched back to the transgranular mode in the unaffected base metal (Fig. 2d). The regions that had been fractured intentionally by tensile loading revealed dimples characteristic of ductile fracture (Fig. 2e).

Metallography

Microstructural Analysis. The microstructure in the HAZ was found to be sensitized. Regions very near the weld metal and away from the HAZ were not sensitized. This was confirmed by intergranular corrosion testing (ASTM A262, practice E).

Crack Origins/Paths. Crack morphology (size, shape, and distribution) indicated that cracks had started from the outside surface. Optical microscopy was carried out on specimens taken from the dished end as well as on specimens cut near the nozzle where pitting had been observed. After standard metallographic preparation, the

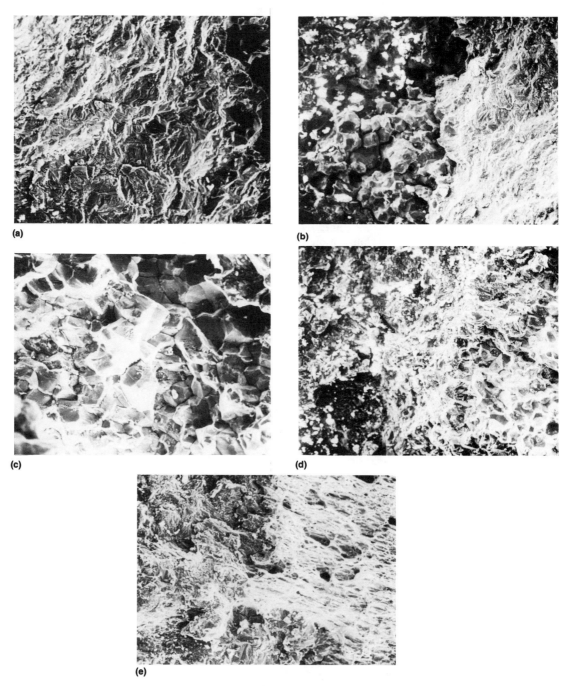

Fig. 2 SEM fractographs showing mode of fracture at different stages of crack propagation. (a) Close to weld seam. (b) First intermediate location in HAZ. (c) Second intermediate location in HAZ. (d) Third intermediate location in HAZ. (e) Dished end unaffected by heat of welding

(a)

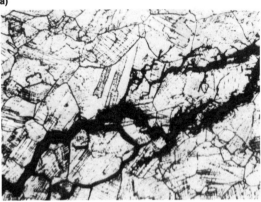

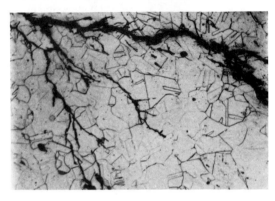

(b)

Fig. 4. Micrographs showing a pit and a crack near the nozzle. (a) Crack starting from a pit. (b) End of the crack in (a)

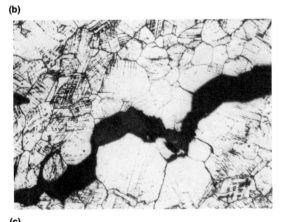

(c)

Fig. 3. Micrographs showing transition in mode of fracture. Electrolytically etched in 10% oxalic acid. (a) Transgranular cracks near the weld. (b) Transition from transgranular to intergranular mode while propagating toward the dished end. (c) Intergranular mode in HAZ

Table 1 Results of chemical analysis

Element	Composition, %	
	Shell	**Dished end**
Carbon	0.063	0.09
Sulfur	0.014	0.013
Phosphorus	0.038	0.038
Manganese	0.45	0.47
Silicon	0.53	0.60
Nickel	8.3	8.3
Chromium	16.9	17.4

samples were etched. Micrographs taken from the dished end are shown in Fig. 3. The cracks near the weld zone were transgranular, and a gradual transition to the intergranular mode of propagation occurred as the cracks propagated through the HAZ. The crack was branched, characteristic of stress-corrosion cracking (SCC). The general microstructure also showed evidence of copious precipitation of carbides along the grain boundaries and deformation bands.

Figure 4 shows micrographs of the etched specimens taken near the nozzle. The transgranular crack, with a number of branches propagating in different directions, had nucleated from a pit. It

is important to note that this region did not appear to be sensitized.

Two samples for intergranular corrosion testing were chosen from an area that was relatively free from cracks and was located close to the weld. These samples were tested in acidified Cu-CuSO₄ solution (ASTM A262, practice E). The samples were found to be heavily sensitized up to a distance of 10 mm (0.4 in.) from the edge closest to the weld. In fact, the metal close to the weld became almost powderlike after the test.

Chemical analysis/identification

Material. Samples from the dished end and the shell were chemically analyzed. The compositions are listed in Table 1. The carbon content of the dished end material was higher than that of the shell material.

Mechanical properties

Hardness testing was performed on both the inside and outside surfaces using a 20 kg load. The hardness values obtained at different locations on the samples are shown in Fig. 5. The average hardness values found on the shell, the weld deposit, and the dished end were 181, 176, and 310 HV, respectively. The values were low in the HAZ of the dished end, presumably because of partial annealing. High hardness values associated with the dished end indicated that it had been subjected to severe cold working. Such an extensive deformation could have resulted from the cold-forming operation as well as the subsequent hammering blows delivered during fit-up.

It was deemed useful to determine the exact amount of cold work present in the dished end. Accordingly, a few samples from the dished end were solution annealed at 1050 °C (1920 °F) for 1 h. The samples were then rolled to different thicknesses. The variation in hardness with the amount of cold reduction is shown in Fig. 6. A 30% reduction in thickness led to a hardness value of 310 HV. It was thus inferred that the deformation associated with the dished end was about 30%.

Stress analysis

Experimental. Residual stresses present in the dished end at approximately 30 mm (1.2 in.) from the fusion line were measured by X-ray diffraction techniques and found to be about 135 MPa (20 ksi). This value of residual stress is quite large and is sufficient for initiating SCC.

Discussion

Optical microscopic examination of microstructures and intergranular corrosion testing showed that the HAZ of the dished end was sensitized. Chemical analysis also indicated a high carbon content in the dished end. Hardness and residual stress measurements indicated high levels of residual stress. From metallographic and visual examinations, it was inferred that the cracks had initiated from the outside surface and were typical stress-corrosion cracks. The mode of fracture was intergranular in the sensitized regions and transgranular in the nonsensitized regions. These observations were corroborated by SEM examination.

The surface morphology of cracks with multiple branching also indicated the presence of typical stress-corrosion cracks.

It was concluded that the tanks failed by SCC. The SCC of stainless steel at room temperature can be caused by chloride ions. High residual stresses combined with the sensitized microstructure resulted in SCC in the presence of chloride ions from the coastal environment.

Conclusion and Recommendations

Most probable cause

The tanks failed because of chloride SCC in the sensitized HAZ of the weldment.

Remedial action

Low-carbon-content steel and a low level of cold work in the dished end could have prevented sensitization in the HAZ and therefore the failure. Use of type 304L stainless steel and dished ends in the solution-annealed condition was recommended.

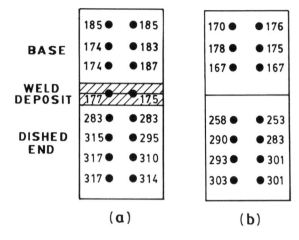

Fig. 5 Hardness profile across the weldment on the inner surface of the tank (a) and on the outer surface (b)

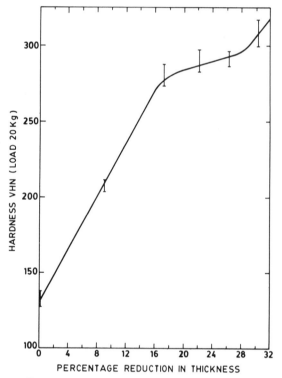

Fig. 6 Variation in hardness with cold work in a solution-annealed sample from the dished end

Failure of a Stainless Steel Tank Used for Storage of Heavy Water/Helium

R.K. Dayal, J.B. Gnanamoorthy, Indira Gandhi Centre for Atomic Research, Kalpakkam, India
N. Chandrasekharan, formerly at Madras Atomic Power Project, Kalpakkam, India

The dished ends of a heavy water/helium storage tank manufactured from 8 mm (0.3 in.) thick type 304 stainless plate leaked during hydrotesting. Repeated attempts at repair welding did not alleviate the problem. Examination of samples from one dished end revealed that the cracking was confined to the heat-affected zone (HAZ) surrounding circumferential welds and, to a lesser extent, radial welds that were part of the original construction. Most of the cracks initiated and propagated from the inside surface of the dished ends. Microstructures of the base metal, HAZ, and weld metal indicated severe sensitization in the HAZ due to high heat input during welding. An intergranular corrosion test confirmed the observations. The severe sensitization was coupled with residual stresses and exposure of the assembly to a coastal atmosphere during storage prior to installation. This combination of factors resulted in failure by stress-corrosion cracking. Implementation of a new repair procedure was recommended. Repairs were successfully made using the new procedure, and all cracks in the weld repair zones were eliminated.

Key Words   Austenitic stainless steels, corrosion   Stress-corrosion cracking   Heat-affected zone
                   Nuclear reactor components   Domes, corrosion   Repair welding
                   Weld defects   Nuclear reactors   Storage tanks, corrosion
                   Welded joints   Sensitizing

Alloys   Austenitic stainless steel—304

Background

The dished ends of a heavy water/helium storage tank manufactured from 8 mm (0.3 in.) thick type 304 stainless steel plate leaked during hydrotesting.

Circumstances leading to failure

The heavy water/helium storage tank was designed for a pressure of 0.1 MPa (14.5 psi) and a temperature of 67 °C (150 °F) to contain helium gas at a maximum pressure of 0.035 MPa (5 psi) and a temperature of 40 °C (105 °F), as well as to provide emergency storage of heavy water at atmospheric pressure and at a maximum temperature of 67 °C (150 °F). The tank (Fig. 1) had a capacity of about 132,000 L (35,000 gal) with an outside diameter of about 4.42 m (14.5 ft). The entire fabrication was made from ASTM A240-304 stainless steel. The wall thickness of the shell was 6 mm (0.2 in.) and the thickness of the dished ends was 8 mm (0.3 in). For various reasons, the tank was assembled and prepared for initial hydrotesting on site 5 years after the manufacture of the dished ends and the shell courses. The tank was located very close to the seashore.

During the hydrotest, with a static head of water of only about 1 m (3 ft), leaks were observed in the bottom dished end, and the test was abandoned. All of the dished end welds, including the heat-affected zones (HAZs), were liquid penetrant tested on both sides. Examination revealed numerous cracks confined to the HAZ areas. These were ground and weld repaired. During the weld repair and grinding operations, more cracks opened in adjacent areas; the final repair encompassed 75 sites. More cracks were detected on the inside of the dished end than on the outside.

The condition of the top dished end was also in question. Liquid penetrant examination conducted on the welds and on the HAZ of this dished end also revealed numerous small hairline cracks. The dished end was weld repaired, followed by liquid penetrant examination and spot radiographic examination. The vessel was again prepared for

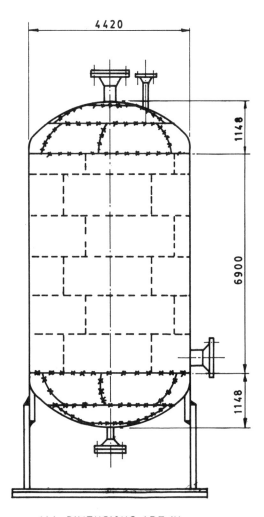

4420

1148

6900

1148

ALL DIMENSIONS ARE IN mm

Fig. 1 Schematic of heavy water/helium storage tank. Dimensions given in millimeters

Table 1 Results of chemical analysis

Elements	Composition, %		
	Base metal	HAZ	Weldment
Carbon	0.08	0.08	0.08
Chromium	18.4	18.4	18.4
Nickel	10.0	10.0	10.0
Manganese	2.0	1.5	0.5
Silicon	0.77	0.82	0.95
Sulfur	0.01	0.01	0.01
Phosphorus	0.045	0.045	0.045

hydrotesting.

The second hydrotest was conducted at 0.16 MPa (23 psi). No indications of leaking were observed. Subsequently, a pneumatic soap bubble leak test was conducted at 0.10 MPa (14.5 psi). This controlled soap bubble test revealed very minute leaks at about four locations, and the cracks were repaired.

The cracks in the tank all occurred at the dished ends. No problems were visible in the shell courses welds or the dished end-to-shell weldings. Because of their large diameters, the dished ends were made of several plates, joined by two circular welds and radial welds, as shown in Fig. 2. The cracks were confined primarily to the circumferential weld HAZs and to some extent to the radial weld HAZs. Most of the cracks initiated and propagated from the inside surface of the dished ends.

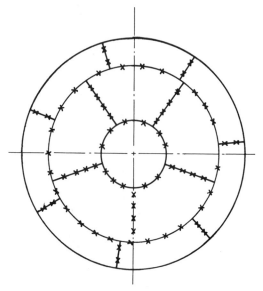

Fig. 2 Weld details of the leaking dished end

Testing Procedure and Results

Metallography

A 120 × 15 × 8 mm (4.7 × 0.6 × 0.3 in.) sample was cut out from one dished end. This sample included a weld deposit about 40 mm (1.6 in.) long at one end. The longitudinal cross section of the sample was electrolytically etched in 10% oxalic acid. Figure 3 illustrates the weld metal, and Fig. 4 shows the interface of the weld metal and the base metal. As shown in Fig. 3, a number of weld passes had been made on two closely adjacent areas, leaving islands of base metal in between the weld deposit. The microstructures of the base metal in the island, as well as that up to 10 mm (0.4 in.) from the weld metal, indicated sensitization of the material. The remaining base metal did not indicate sensitization.

Chemical analysis/identification

Material and Weld. Chemical analysis of the base metal, the HAZ, and the weld metal (Table 1) did not indicate any deviations that could have led to abnormal corrosion.

Simulation tests

The sample was subjected to intergranular corrosion testing per ASTM Standard A 262, Practice E (boiling in Cu/CuSO₄ for 24 h followed

Fig. 3 Macrograph of the etched cross section on the weld metal.

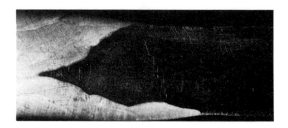

Fig. 4 Macrograph of the etched cross section on the interface of the weld metal and the base metal. 3.72×

by a U-bend test). During the bend test, the sample fractured very easily in the HAZ (Fig. 5).

Discussion

Severe sensitization had occurred in the HAZ, perhaps the result of use of an improper fabrication technique. With the joint design, fit-up problems could have caused higher metal deposition, as well as possible weld repair. In either case, excessive heat input to the already low wall thickness plate had resulted in sensitization. The dished ends had been in storage a long time before assembly and were left unattended during that period. They were possibly exposed to the open atmosphere at the fabricator's site and subsequently were kept in a coastal atmosphere before commis-

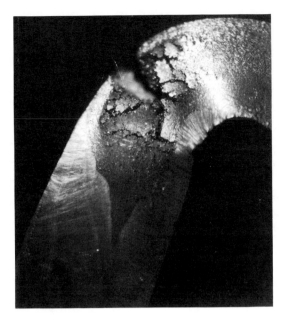

Fig. 5 Macrograph of the specimen after intergranular corrosion testing. 4.74×

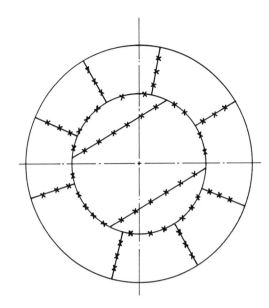

Fig. 6 Modified dished end design

sioning. Thus, the HAZ had experienced severe sensitization and residual stresses. The intergranular corrosion susceptibility could have accelerated the corrosion cracking, resulting in premature failure.

Conclusion and Recommendations

Most probable cause

A large heat input occurred during the repeated repair weld passes, leading to severe sensitization of the HAZ. The presence of fit-up and welding stresses combined with exposure to the coastal atmosphere provided all of the conditions necessary for stress-corrosion cracking.

Remedial action

To alleviate the problem of new cracks occurring during and after grinding and welding, a new repair procedure was adopted. This involved heating the repair zone to a temperature of about 315 °C (600 °F) to bring previously unnoticed potential cracks to the surface. The new procedure revealed the severity of the cracks in the HAZ, allowing the viability of weld repair to be ascertained. Repairs were successfully carried out using the new procedure, which eliminated all cracks in the weld repair zones.

How failure could have been prevented

As shown in Fig. 2, the dished end contained two circumferential joints. Although the applicable code (ASME Section VIII, Div. 1) does not prohibit such joints, it is not a recommended practice. In fact, the Indian Boiler Regulation does not permit such circular joints in dished ends. Even the British Pressure Vessel Code BS 5500-1976 advises the use of a single plate. This is understandable, because the closing circular joint will experience severe thermal restraints, resulting in a high level of residual stresses. To control and correct distortions, many fixtures must be used, which in turn introduce additional residual stresses. The problem is accentuated by fit-up problems and accompanying weld repairs, and is especially acute in the case of austenitic stainless steels, because of their higher thermal expansion and lower thermal conductivity compared with carbon steel. In the case of a similar tank in another plant, one of the critical circumferential joints was eliminated, thus contributing to its successful operation (Fig. 6).

Fatigue Failure at Fillet-Welded Nozzle Joints in a Type 316L Stainless Steel Tank

D.K. Bhattacharya, S.K. Ray, and Placid Rodriguez, Metallurgy and Materials Programme, Indira Gandhi Centre for Atomic Research, Kalpakkam, India

Upon arrival at the erection site, an AISI type 316L stainless steel tank intended for storage of fast breeder test reactor coolant (liquid sodium) exhibited cracks on its shell at two of four shell/nozzle fillet-welded joint regions. The tank had been transported from the manufacturer to the erection site by road, a distance of about 800 km (500 mi). During transport, the nozzles were kept at an angle of 45° to the vertical because of low clearance heights in road tunnels. The two damaged joints were unsupported at their ends inside the vessel, unlike the two uncracked nozzles. Surface examination showed ratchet marks at the edges of the fracture surface, indicating that loading was of the rotating bending type. Electron fractography using the two-stage replica method revealed striation marks characteristic of fatigue fracture. The striations indicated that the cracks had advanced on many "mini-fronts," also indicative of nonuniform loading such as rotating bending. It was recommended that a support be added at the inside end of the nozzles to rigidly connect with the shell. In addition to avoiding transport problems, this design modification would reduce fatigue loading that occurs in service due to vibration of the nozzles during filling and draining of the tank.

Key Words			
Austenitic stainless steels		Fatigue failure	Fast nuclear reactors
Nozzles		Crack propagation	Breeder reactors
Fillet welds		Storage tanks, design	

Alloys	Austenitic stainless steel— 316L

Background

Upon arrival at the erection site, an AISI type 316L stainless steel tank exhibited cracks on its shell at two of four shell/nozzle fillet-welded joint regions.

Applications

The tank is used for storage of liquid sodium, the coolant in a fast breeder test reactor.

Circumstances leading to failure

After fabrication, the tank was transported to the erection site by road, a distance of about 800 km (500 mi). During transport, the nozzles were kept at an angle of 45° to the vertical because of low clearance heights in road tunnels. When the tank arrived at the erection site, cracks were observed on its shell at two shell/nozzle fillet-welded joints (Fig. 1 and 2).

Pertinent specifications

The tank was fabricated in accordance with the section of the ASME Pressure Vessel Code that applies to nuclear components. Every stage of the fabrication was stringently controlled, as were the choice of parent metal and filler wires. The root pass of the fillet weld was made by the tungsten inert gas welding, followed by manual metal arc welding with basic coated electrodes. After each pass, the weldments were inspected visually and by using dye penetrants. After fabrication was completed, the tank passed a hydraulic test.

Specimen selection

Small interconnecting holes were drilled around the two nozzles to remove the cracked regions. The nozzles were then separated from the cut-out disks by machining the fillet weldments, taking care that sufficient weldment material remained with the disks (Fig. 3). The two disks contained all the radial cracks and the circumferential cracks intact.

The circumferential cracks were found to have

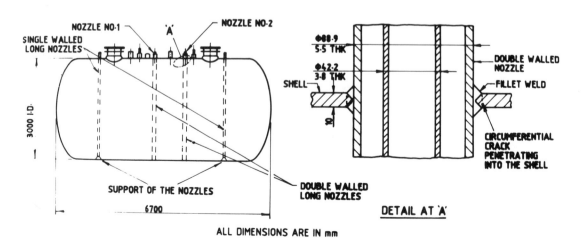

Fig. 1 Diagram of the cylindrical tank. Double-wall nozzles 1 and 2 were not supported at their ends inside the vessel. The other two nozzles, which were single walled, were supported as shown.

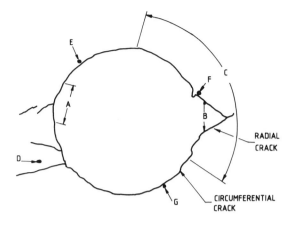

Fig. 2 Diagram of the circumferential and radial cracks observed at the fillet weld/shell boundary and in the shell of nozzle 2

started both from the outside and the inside surfaces of the shell and propagated toward the center of the shell thickness. The disks were cut radially to reveal the circumferential crack surfaces.

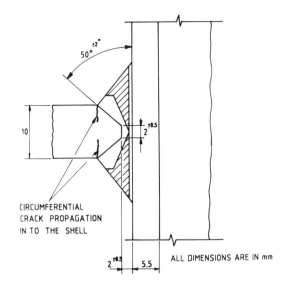

Fig. 3 Fillet weld joint between the shell and double-wall nozzles. The hatched portion in the weld was removed by lathe machining.

<table>
<tr><td>**Visual Examination of General Physical Features**</td><td></td></tr>
</table>

Both the outside and the inside surfaces of the tank exhibited severe grinding marks. Most of the radial cracks followed the course of the grinding marks, at least in the initial stage of crack propagation (Fig. 4).

Figure 4 shows the outside surface of the shell at position A of Fig. 2. Interconnecting drill marks are visible at the top, with weldment at the bottom. The lower crack (marked "l") followed the course of the grinding marks after initiation. Crack "u" was connected to crack "l" by a faint wavy crack ("w").

Testing Procedure and Results

Surface examination

Figure 5 shows a typical opened crack. Cracks propagated from the outside as well as the inside surface of the tank shell and met somewhere near the center of the shell thickness. Ratchet marks ("R") were visible at the edges of the fracture surface. These ratchet marks indicated that the loading was a rotating bending type, and crack propagation started at numerous points on the surface.

Electron Microscopy/Fractography. The fracture surfaces were examined at various locations using the two-stage replica method. A cellulose nitrate replica was obtained in the first stage. This replica was shadowed by chromium, and a thin carbon layer was deposited on its surface. Next, the cellulose nitrate was dissolved in alcohol. The carbon replicas with chromium shadowing material were examined in a transmission electron microscope (TEM).

Figure 6 shows a TEM fractograph of a typical radial crack. Striation marks characteristic of fatigue fracture were visible, caused by discontinuous or repetitive fatigue crack propagation. Each striation corresponded to the crack advance during one cycle of loading. Examination of the striations showed that the cracks had advanced on many "mini-fronts." This behavior is expected when the

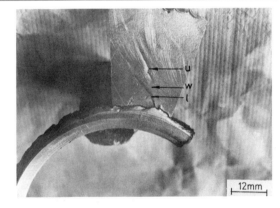

Fig. 4 Outside surface of the shell at position A of Fig. 2

Fig. 5 Surface of a portion of an opened circumferential crack. P, plasma-cut surface (created by removal of the sample by sawing); R, ratchet mark

loading is nonuniform, as in rotating bending.

Metallography

Microstructural Analysis. Figure 7 is a typical optical micrograph of a radial crack. Because of the large width of the radial cracks, it was not pos-

Fig. 6 TEM fractograph of a typical radial crack

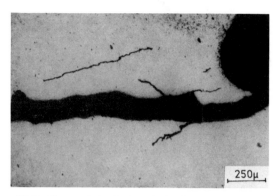

Fig. 7 Micrograph showing branched and parallel cracks associated with a wide radial crack

sible to determine whether the cracks were transgranular or intergranular. Many finer branched cracks originated from the main radial cracks (Fig. 7). These cracks were transgranular. A few fine cracks parallel to the main cracks were also observed. This type of crack formation is characteristic of fatigue crack propagation.

Crack Origin/Paths. Crack fronts originated at numerous points on both the inside and the outside surfaces of the tank shell. The cracks propagated across the shell wall because of rotating bending fatigue loading and met at the approximate midpoint of the shell thickness.

Discussion	Evidence indicated that the nozzle joints failed by rotating bending fatigue. The following features pointed to fatigue failure: • Initiation of cracks at a number of points • Ratchet marks on the fracture surfaces, characteristic of fatigue failure under rotating	bending loading • Periodic branched cracks inclined to the main cracks and occasional cracks parallel to the main crack fronts • Appearance of fatigue striations in the TEM fractographs
Conclusion and Recommendations	Severe grinding marks were also visible, and the cracks often followed these marks. Utmost care must be taken not to introduce grinding marks or other stress raisers near weld joints. Fatigue failure was consistent with the type of loading present. The nozzles, positioned at angles of 45° to the vertical during transport of the tank, acted as cantilevers. Cyclic loading resulted from the usual jerks and vibrations associated with	road journeys. Such loading could have been avoided had the two nozzles been supported at their ends inside the vessel, similar to two other nozzles that did not show failure. In addition to avoiding transport problems, this design modification would reduce fatigue loading that occurs in service due to vibration of the nozzles during filling and draining of the tank.

Failure of AISI Type 347 Stainless Steel Bellows

N.G. Muralidharan, Rakesh Kaul, K.V. Kasiviswanathan, and Baldev Raj, Indira Gandhi Centre for Atomic Research, Kalpakkam, India

A number of AISI 347 stainless steel bellows intended for use in the control rod drive mechanism of a fast breeder reactor exhibited leaks during helium leak testing before being placed in service. The bellows had been in storage for 1 year in a seacoast environment and exhibited a very small leak rate, of the order of 10^{-7} cm^3/s (6×10^{-8} in.^{3}/s). Optical metallography revealed numerous pits and cracks on the surfaces of the bellow convolutes, which had been welded to one another using an autogenous gas tungsten arc welding process. Microhardness measurements indicated that the bellows had not been adequately stress relieved. It was recommended that a complete stress-relieving treatment be applied to the formed bellows. Improvement of storage conditions to avoid direct and prolonged contact of the bellows with the humid, chloride-containing environment was also recommended.

Key Words			
Austenitic stainless steels, corrosion	Weldments, corrosion	Welded joints, corrosion
Nuclear reactor components, corrosion	Bellows, corrosion	Stress-corrosion cracking
Pitting (corrosion)	Leakage	Marine environments
Fast reactors	Breeder reactors	

Alloys	Austenitic stainless steel—347

Background

A number of AISI 347 stainless steel bellows intended for use in the control rod drive mechanism of a fast breeder reactor exhibited leaks during helium leak testing before being placed in service.

Circumstances leading to failure

The bellows had been stored in polyvinyl chloride (PVC) bags for about 1 year after procurement. Helium leak testing revealed numerous leaks. Although the leak rates were very small (i.e., of the order of 10^{-6} to 10^{-7} cm^3/s, or 6×10^{-7} to 6×10^{-8} in.^{3}/s), the bellows were deemed unacceptable for service.

Figure 1 shows one of the defective bellows. The bellow was approximately 21 cm (8 in.) long and 4 cm (1.6 in.) in diameter and was constructed from approximately 150 μm (5900 μin.) thick convolutes welded to one another using an autogenous gas tungsten arc welding (GTAW) process. Figure 2 shows a cross section of a fabricated bellow. The details of the welding process were not supplied by the manufacturer.

Specimen selection

Specimens were cut from the regions of those convolutes that exhibited leakage paths during testing. Samples were chosen so that both convolute weld regions and convolute thicknesses were included.

Testing Procedure and Results

Metallography

Microstructural Analysis. The microstructure of the bellows material was typically austenitic, with a high degree of cold work as revealed by flow lines and elongated grains (Fig. 3). Only isolated inclusions of submicron size were observed.

Fig. 1 A typical failed bellow

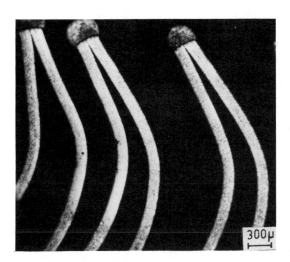

Fig. 2 Weld profile of a fabricated bellow

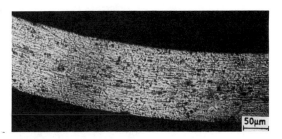

Fig. 3 Microstructure of the bellows material, showing flow lines and elongated grains

Fig. 4 Photomicrograph showing a through-crack

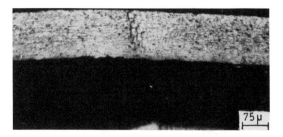

Fig. 5 Photomicrograph showing a branched crack

Fig. 6 Photomicrograph showing a crack initiating at the bottom of a pit and terminating in a large circular area. Another pit is visible on the opposite surface of the convolute.

Crack Origins/Paths. The major challenge during this failure analysis was to locate the leakage paths, which were expected to be minute pinhole defects in light of the very small measured leak rates. Only then would it be possible to investigate the microstructural details around the leaks. Polishing of the specimen chosen for optical microscopy was terminated at intervals of about 10 μm (400 μin.), and the specimen was then etched. Although this method was painstaking and tedious, several defects were located. The various features revealed by optical metallography are discussed below.

Several through-cracks across the wall thickness of a bellow convolute were observed. Figure 4 shows such a crack, which extended from one end of the convolute to the other. Figure 5 shows a branched crack in one of the convolutes.

A number of pits were observed on the surfaces of the bellow convolutes. A small pit is visible on one of the convolutes shown in Fig. 2. Figure 6 shows a fine crack originating at the bottom of a pit and terminating in a large circular area, perhaps where there had been an inclusion. A pit is also visible on the other side of the same convolute. Figure 7 shows a similar crack that starts at the bottom of a pit and extends across the wall thickness of a convolute. The nature of the pits shown in these figures is typical of the pitting attack in type 347 stainless steel caused by exposure to environments containing chloride ions.

Chemical analysis/identification

The composition of the bellows material, as determined by chemical analysis, along with the specified composition of AISI 347 stainless steel, is presented in Table 1. The composition met the chemical requirements.

Mechanical properties

Hardness. Microhardness measurements using a load of 25 gf on the cross section of the convolute metal strips near and away from the crack edge produced hardness values in the range of 385 to 395 HV. The microhardness value for type 347 stainless steel in the annealed condition is about 170 HV.

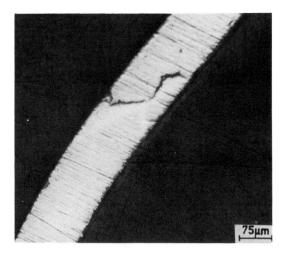

Fig. 7 Photomicrograph of a through-crack originating at the bottom of a pit and extending across the thickness of a convolute

Table 1 Results of chemical analysis

| | Composition, wt% | |
Element	Bellows	Chemical requirements for AISI 347 stainless steel
Carbon	0.076	0.08 max
Chromium	17.8	17-19
Nickel	9.4	9-13
Manganese	1.34	2.00 max
Silicon	0.85	1.00 max
Phosphorus	0.042	0.045 max
Sulfur	0.002	0.030 max
Niobium	0.44	10× C min
Iron	bal	bal

Discussion

Optical metallography revealed several pits in the regions where cracklike defects were observed. Fine cracks originated at pit bottoms and extended across the thickness of the convolutes, suggesting that pits acted as sites for crack nucleation. The high level of hardness of the parent material indicated that the bellows had not been adequately stress relieved. The possibility of the presence of locked-up residual stresses in the convolute metal strips was likely.

The failure occurred at a seacoast area near the Mediterranean with average monthly temperatures of 25 to 32 °C (77 to 90 °F). Average humidity values range from 70 to 80%, and chloride contents in the atmosphere, evaluated by the wet-candle method and expressed as milligrams of sodium chloride per square meter per day, range from 8 to 45 over the year. Therefore, even if the bellows were packed in PVC bags, the possibility of contact with chlorides was high.

The bellows were not stored in a conditioned atmosphere. Chloride ions are known to be potent corrosion enhancers, and localized adsorption of chloride ions has been reported on stainless steels. The sites of chloride ion agglomeration can act as prenuclei of pits. These prenuclei form at some sites on the metal surface where the passive oxide film is defective—for example, at interfaces between the steel matrix and nonmetallic inclusions. Pits can also nucleate at carbides, grain boundaries, and other inhomogeneities (including those that are microstructural) on the metal surface. The presence of moisture in the environment can facilitate the electrolytic path for the chloride ions. These pits act as sites for the nucleation of cracks, which may extend through the thickness of a convolute under the combined influence of residual stress and environment.

Conclusion and Recommendations

Most probable cause

Metallographic observation revealed two types of cracks: branched cracks running across the thickness of the convolutes and cracks originating at the base of pits (found on both the inside and the outside of the bellows) and extending across the thickness of the convolutes. It was concluded that the stainless steel bellows failed by stress-corrosion cracking (SCC) caused by the presence of residual stresses in the material and improper storage conditions that led to contact with chloride ions in the ambient atmosphere.

Remedial action

The likelihood of such a failure reoccurring can be reduced significantly by the use of fully stress-relieved material. SCC failures are often associated with cold-worked materials. Stress relieving should therefore be performed to obtain adequate resistance of a component to SCC. This is especially true in the case of bellows, whose prolonged performance is desired for critical applications. Techniques such as finite-element analysis and X-ray diffraction can be employed to determine the magnitude of residual stresses present in the formed bellows.

Storage conditions should be improved to avoid direct contact of chloride ions and moisture with the bellows. The packing material should include a dehumidifying agent, and the bellows should be sealed in double PVC bags.

Failure of a Martensitic Stainless Steel Ball Valve

Rakesh Kaul, N.G. Muralidharan, K.V. Kasiviswanathan, and Baldev Raj, Indira Gandhi Centre for Atomic Research, Kalpakkam, India

Repeated failures of high-pressure ball valves were reported in a chemical plant. The ball valves were made of AFNOR Z30C13 martensitic stainless steel. Initial examination of the valves showed that failure occurred in a weld at the ball/stem junction end of austenitic stainless steel sleeves that had been welded to the valve stem at both ends. Metallographic examination showed that a crack had been introduced into the weld by improper weld heat treatment. Stress concentration at the weld location resulting from an abrupt change in cross section facilitated easy propagation of the crack during operation. Proper weld heat treatment was recommended, along with avoidance of abrupt change in cross section near the weld. Dye penetrant testing at the ball stem junction before and after heat treatment was also suggested.

Key Words	Martensitic stainless steels	Valves	Cracking (fracturing)
	Weld defects	Austenitic stainless steels	Sleeves
	Dissimilar metals, welding	Welded joints	Chemical processing equipment
	Welding parameters		
Alloys	Martensitic stainless steel—Z30C13		

Background

High-pressure ball valves made from AFNOR Z30C13 martensitic stainless steel failed repeatedly while operating in a chemical plant. The ball valve stems had 27 mm (1.1 in.) long austenitic stainless steel sleeves welded to their ends.

Applications

Ball valves of this type are used to control the flow of dry synthetic gas (a 3 to 1 mixture of nitrogen and hydrogen with 4% ammonia). Figure 1 depicts the assembly of a ball valve. The valve operated at a pressure of 25 MPa (3.6 ksi) at temperatures ranging from 5 to 180 °C (40 to 355 °F).

Circumstances leading to failure

After about 8 months of operation, a ball valve failed at the junction of the ball and bottom stem. Service lives of other ball valves varied from 6 to 8 months. Figure 2, a schematic of the ball valve, shows the presence of sleeves, locations of various welds, and location of the failure. It should be noted that the ball valve drawing did not originally consider the sleeves and the welds. Rather, these were discovered during the course of the investigation described below.

Pertinent specifications

The ball valve material was reported to be Z30C13, which is basically a martensitic stainless steel with an ultimate tensile strength of 830 MPa (120 ksi) and a yield strength of 635 MPa (90 ksi). The specified chemical composition of this French steel is furnished in Table 1.

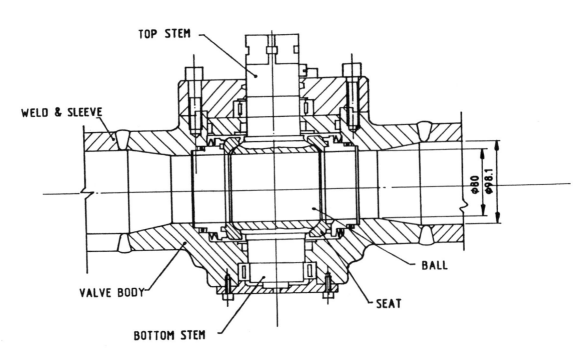

Fig. 1 Schematic of ball valve assembly

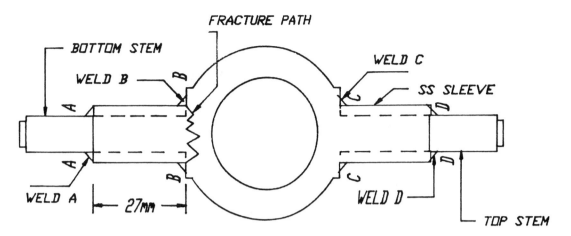

Fig. 2 Schematic of ball valve, showing failed region and locations of various welds

Visual Examination of General Physical Features

The ball valve failed at the junction of the ball and bottom stem. The location of the failure is shown in Fig. 3, which also reveals the presence of a weld on the top stem of the ball valve. The weld crown was not ground after welding.

Testing Procedure and Results

Surface examination

Visual. Figure 4 shows magnified views of the mating fracture surfaces. This typical brittle failure can be broadly divided into three zones, designated A, B, and C in Fig. 4.

Zone A appears smooth, which implies rubbing between the mating surfaces and represents a region of slow crack growth. The black spot at the edge of the fracture surface may have been caused by the seepage of oil or grease into the material through the initiated crack opening. This location may have been a crack nucleation site.

The radial lines in zone B, which point back toward the origin of the crack in zone A, are typical of brittle failure. These radial lines come to an abrupt end along a line designated XY, separating zone B from C, where the final failure took place.

Fig. 3 Failed ball valve

(a)

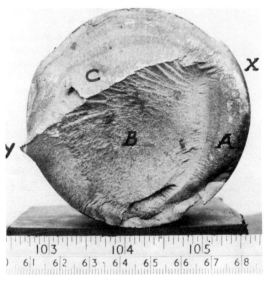

(b)

Fig. 4 Mating fracture surfaces

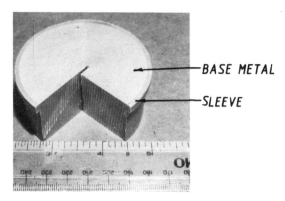

Fig. 5 Cut portion of the stem, showing presence of sleeve

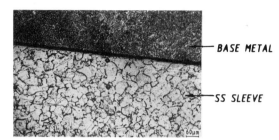

Fig. 6 Micrograph of sleeve/base metal interface

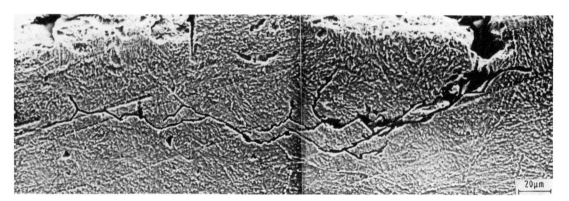

Fig. 7 Micrograph of branched crack in the base metal

Metallography

MicrostructuralAnalysis. A cut portion from the lower stem of the ball valve was mounted, polished, and etched, which revealed a typical tempered martensite structure. The cut cross section of the stem showed the presence of a 2.5 mm (0.1 in.) thick sleeve around the stem (Fig. 5). The microstructure of the interface region between the sleeve and stem is shown in Fig. 6, which reveals the tempered martensite structure for the stem region and a typical sensitized microstructure in the sleeve. The microstructure was "ditched," as described in ASTM A262-86. A sensitized microstructure in the austenitic stainless steel occurs because of the grain-boundary precipitation of chromium-rich carbides.

In situ metallography over the stem surface revealed the presence of austenitic stainless steel sleeves, approximately 27 mm (1.1 in.) long, starting from the junction of the ball and stem. The sleeves were welded to the stem at its two ends.

Metallographic examination of a cut section at the weld A region (Fig. 2) indicated a lack of fusion. Some fine cracklike indications starting from this region were also noticed.

Crack Origins/Paths. Metallographic examination of the cross-sectional surface B-B (Fig. 2) revealed a number of cracks in this region. Figure 7, a micrograph of a transgranular crack in the martensitic steel region, reveals extensive branching of the crack. A crack was also observed in the weld region (Fig. 8).

Fig. 8 Crack in the weldment

Because only a small amount of weldment area was available for metallographic examination at location B (Fig. 2), a detailed examination of the weldment could not be carried out at the location nearest to failure. A similar martensitic austenitic stainless steel weld, available on the other junction of the ball and stem at location C (Fig. 2) was used for detailed metallographic examination. The weld was checked by dye penetrant testing, which revealed some defects in the weld region.

To learn about the nature and origin of defects, this defect region was subjected to *in situ* metallography. Figure 9 shows an *in situ* micrographic mosaic of a crack starting from the weldment (following one of the weld beak contours) and extending into the base metal. This particular mosaic was taken on three different planes in order to cover the full crack length. Despite the difficulties arising from the geometrical configuration of the defect region in terms of metallographic observation, Fig. 9 reveals the contour of the entire crack. Fig-

Fig. 9 Crack starting from weldment and extending into base metal

Fig. 10 Crack and defects observed in weld region

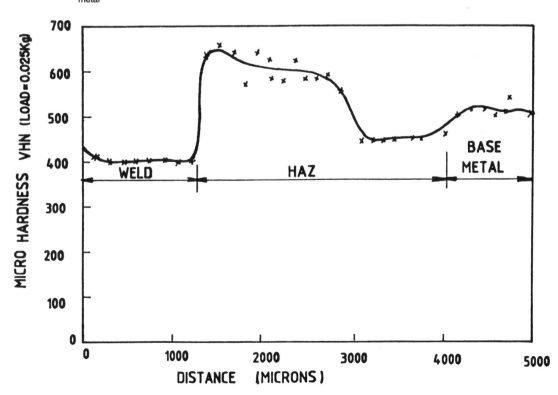

Fig. 11 Microhardness profile across weld/HAZ/base metal interface

Table 1 Results of chemical analysis

| Element | Composition, % | | |
	Base metal	Sleeve material	AFNOR Z30C13 requirements
Carbon	0.3129	0.078	0.25-0.35
Chromium	11.9	17.8	12-14
Nickel	Trace	8.7	1.00 (max)
Silicon	...	...	1.00 (max)
Manganese	...	1.12	...
Sulfur	0.005	0.006	0.03 (max)
Phosphorus	0.005	nil	0.04 (max)

ure 10 is an *in situ* micrograph of a crack joining the blowholes in the weldment. The number of blowholes in the weldment indicates poor welding quality.

Chemical analysis

Material. Results of chemical analysis of the stem of the ball and of the sleeve material are presented in Table 1. The chemical composition of the stem conformed to the specified composition, whereas the chemical composition of the sleeve material suggested that it was an austenitic stainless steel.

Mechanical properties

Hardness. Microhardness measurements (25 g load) were taken on the weldment/heat-affected zone (HAZ)/base metal interface at location B (Fig. 2). A typical microhardness profile obtained in this region is shown in Fig. 11.

Discussion

The martensitic stainless steel stems of the ball valve had 27 mm (1.1 in.) long austenitic stainless steel sleeves around them. These sleeves were welded to the stems at both ends. One of the two welds coincided with the junction of the ball and stem, which constituted a stress-concentration site because of the abrupt change in geometry.

Several cracks were noticed in welds B and C, which were located at the two junctions of the ball and stem. These cracks originated from the root of the multipass weld and propagated into the base metal of the stem of the ball.

A large variation in microhardness across the weld/HAZ/base metal indicated that proper preweld and postweld heat treatment schedules were not followed. A typical profile is shown in Fig. 13. Recommended practice for welding AISI 410 stainless steel involves preheating to at least 150 °C (300 °F) before welding and postheating to 700 °C (1290 °F), followed by cooling at a rate of 38 °C/h (68 °F/h). The presence of several blowholes in the weldment indicated use of poor welding procedures.

Improper heat treatment of the martensitic stainless steel/austenitic stainless steel weld at the ball/stem junction, combined with poor welding procedure, were largely responsible for the introduction of cracks into the weld. Once nucleated, these cracks could propagate through the material under the influence of:

- Residual stresses developed as a result of improper heat treatment or as a result of grinding of weld reinforcements
- Stresses developed as a result of misalignment of the stem, which would have been large at the welds close to the ball/stem junction because of the abrupt change in cross section

Conclusion and Recommendations

Most probable cause

Failure was attributed to the welding of an austenitic stainless steel sleeve to a martensitic stainless steel stem without proper preweld and postweld heat treatments, poor weld quality, and the presence of the weld at the stress-concentration site.

Remedial action

If possible, use of austenitic stainless steel sleeves should be replaced by chrome plating of the stem. If an austenitic stainless sleeve must be used over the stem, a proper heat treatment schedule should be followed when welding this material to martensitic stainless steel. Also, an abrupt change in cross section near the weld should be avoided.

Dye penetrant testing should be carried out on the existing ball valves at the ball/stem junctions, where the sleeves are welded. Periodic dye penetrant tests should also be performed as an in-service inspection procedure during shutdown to detect defects that may develop in the region during operation.

Failure of a Reversible Flap Valve in a Thermal Power Station

K.K. Vasu, Welding Engineering and Metallurgy Department, Larsen and Toubro Ltd., Bombay, India

A reversible four-way carbon steel flap valve in a thermal power station failed after 7 years of service. The flap had been fabricated by welding two carbon steel plates to both sides of a carbon steel forging. The valve was used for reversing the flow direction of seawater in the cooling system of a condenser. Visual examination of the flap showed crystalline fracture, indicating a brittle failure. Metallographic examination, chemical analyses, and tensile and impact testing indicated that the failure was caused by the notch sensitivity of the forging material, which resulted in low toughness. It was recommended that fully killed carbon steel with a fine-grain microstructure be used. Redesign of the flap to remove the step in the forging that acted as a notch was also recommended.

Key Words

Carbon steels
Brittle fracture
Marine environments

Forgings
Valves

Weldments
Electric power generation

Background

A reversible four-way flap valve in a thermal power station failed after 7 years of continuous operation. The valve was used for reversing the flow direction of seawater in the cooling systems of a condenser.

Pertinent specifications

The flap body of the valve was made of two separate flap pieces fastened together at the outer periphery using bolts and spacers. The flap pieces were elliptical, with an approximate size of 1927 × 1187 × 80 mm (76 × 47 × 3 in.). The bolted flap body was mounted using a dowel pin on the central shaft, which passed through a central circular groove machined along the minor axis of the elliptical flap. The flap was fabricated by welding two carbon steel plates to both sides of a carbon steel forging, followed by stress relieving at 600 °C (1110 °F) (Fig. 1). The working pressure in the seawater inlet line of the condenser was approximately 100 kPa (14.5 psi), and the rise in temperature at the inlet to the outlet was about 20 °C (40 °F) above ambient.

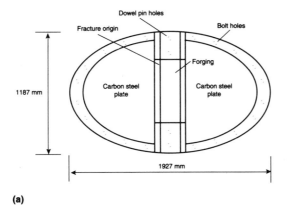

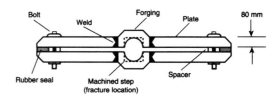

(a)   (b)

Fig. 1   Diagram of the flap of the reversible four-way valve

Visual Examination of General Physical Features

Visual examination of the flap showed crystalline fracture, indicating a brittle failure (Fig. 2). The chevron marking on the fracture surface indicated that the fracture originated at the machined step. No indication of a pre-existing flaw, local necking, or surface serrations was noticed at the cracked zone. The failure took place along the root of a step machined in the central forging (Fig. 3). Some of the bolts connecting both the flap pieces had been sheared off, and the spacers inserted between the flaps were missing. No severe corrosive attack was apparent on the flaps.

Fig. 2   Fractured pieces of the flap

Fig. 3 Fractured end of the flap

Table 1 Results of chemical analysis

Material	Composition, %				
	C	Mn	Si	S	P
Carbon steel plate	0.19	1.08	0.12	0.026	0.039
Carbon steel forging	0.10	0.50	0.10	0.036	0.042

Testing Procedure and Results

Metallography

Microstructural Analysis. Metallographic specimens were prepared from the forging and plate material of the flap body in both the as-received and normalized conditions and were etched with 2% nital. Microscopic examination of these samples revealed that the carbon steel forging specimen in the as-received condition had a very coarse grain structure with almost uniform distribution of ferrite and pearlite (Fig. 4). The grain size observed was of the order of ASTM No. 3. After normalizing at 900 °C (1650 °F), the grain size of the carbon steel forging was slightly modified, with an even distribution of ferrite and pearlite (Fig. 5). The grain size was of the order of ASTM No. 5.

The microstructure of the carbon steel plate material in the as-received condition had a uniform distribution of ferrite and pearlite (Fig. 6). The grain size was approximately ASTM No. 5. Normalizing at 900 °C (1650 °F) modified the grain size of the carbon steel plate to approximately ASTM No. 6, and the microstructure appeared more uniform (Fig. 7).

The cracked end of the carbon steel forging was also examined microscopically. The microstructure showed no evidence of plastic deformation prior to fracture (Fig. 8).

Chemical analysis/identification

Material and Weld. Chemical analysis of both the plate and the forging materials was performed. Results are presented in Table 1.

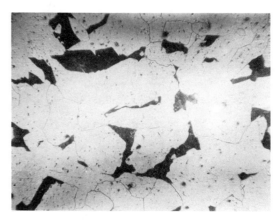

Fig. 4 Microstructure of the carbon steel forging used in the flap body in the as-received condition. Etched in 2% nital. 71×

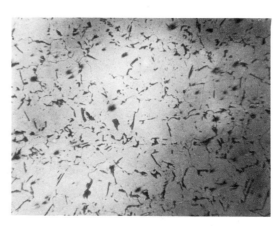

Fig. 5 Microstructure of the carbon steel forging used in the flap body after normalizing at 900 °C (1650 °F). Etched in 2% nital. 71×

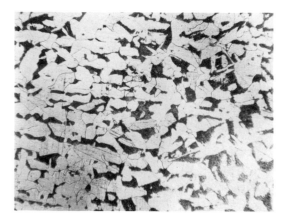

Fig. 6 Microstructure of the carbon steel plate material used in the flap body in the as-received condition. Etched in 2% nital. 71×

Mechanical properties

Tensile Properties. Tensile testing was performed on machined samples from the carbon steel plate and the forging material both in the as-received condition and after normalizing at 900 °C (1650 °F). The test results, along with typical test values for IS 226 carbon steel plate material, are

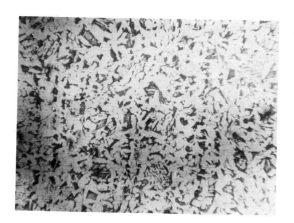

Fig. 7 Microstructure of the carbon steel plate material used in the flap body after normalizing at 900 °C (1650 °F). Etched in 2% nital.71×

Fig. 8 Microstructure of the fractured end of the forging used in the flap. Etched in 2% nital. 100×

Table 2 Tensile properties

Material	Ultimate tensile strength		Yield strength		Elongation,
	MPa	ksi	MPa	ksi	%
Carbon steel forging					
As-received condition	444.7	64.5	349.0	50.6	28.6
Normalized at 900 °C (1650 °F)	436.5	63.3	378.4	54.9	45.30
Carbon steel plate					
As-received condition	465.0	67.4	257.2	37.3	35.00
Normalized at 900 °C (1650 °F)	476.5	69.1	295.9	42.9	36.10
IS 226 plate	475.6	69.0	280.5	40.7	33.50

Table 3 Impact properties

Material	Room-temperature Charpy impact values	
	J	ft · lbf
Carbon steel forging		
As-received condition	3.9	2.9
Normalized at 900 °C (1650 °F)	9.2	6.8
Carbon steel plate		
As-received condition	15.7	11.6
Normalized at 900 °C (1650 °F)	103.9	76.6
IS 226 plate	97.7	72.0

presented in Table 2.

Impact Toughness. Impact test results for the carbon steel plate and for the forging material in the as-received condition and after normalizing at 900 °C (1650 °F), along with typical test values for IS 226 carbon steel plate material, are given in Table 3. To evaluate the effect of temperature variations in the seawater during different climatic conditions on impact properties, impact testing was carried out at +10 and +15 °C (+50 and +59 °F). The results, along with room-temperature impact values, are listed in Table 4.

Table 4 Impact properties at different temperatures

Material	Charpy impact values					
	+10 °C (+50 °F)		+15 °C (+59 °F)		+28 °C (+82.4 °F)	
	J	ft · lbf	J	ft · lbf	J	ft · lbf
Carbon steel forging						
As-received condition	3.5	2.6	3.9	2.9	4.2	3.1
Normalized at 900 °C (1650 °F)	5.2	3.8	6.2	4.6	9.1	6.7
Carbon steel plate						
As-received condition	8.8	6.5	10.8	8.0	15.7	11.6
Normalized at 900 °C (1650 °F)	82.3	60.7	99.0	73.0	103.9	76.6

Discussion

Visual examination of the fracture surface revealed transgranular cracking, indicating brittle fracture. Although the tensile properties of the forging material were found to be satisfactory, the notch sensitivity of the material (i.e., its toughness) was extremely poor, as indicated by Charpy impact testing (3.9 J, or 2.9 ft · lbf). The coarse grain structure of the forging, accompanied by inferior chemical composition with regard to manganese, resulted in poor impact values. Normalizing at 900 °C (1650 °F) slightly improved the grain size, which resulted in a marginal increase in impact properties. Normalizing the forging at a slightly higher temperature (930 °C, or 1705 °F) would probably improve mechanical properties marginally. The steel used for the manufacture of the forging was apparently not fully killed, which further contributed to poor impact strength.

The carbon steel plate material was found to have adequate tensile properties and moderate toughness, as indicated by Charpy impact testing (35.3 J, or 26 ft · lbf). Grain refinement during normalizing greatly improved the impact strength of the plate material, with a minor change in tensile properties. Temperature variations in the seawater due to climatic changes had little effect on the impact properties of either the forging or the plate material (Table 4).

Conclusion and Recommendations

Most probable cause

The basic cause of failure was overloading of the flap valve due to hydraulic shock waves in the line and use of a low-toughness forging material. The step in the forging acted as a notch.

Remedial action

To avoid such failures, steels with improved impact properties that meet a minimum Charpy impact value of 34.3 J (25.3 ft · lbf) at room temperature should be used. In general, for applications that require better impact properties, carbon steel plates that meet the specification requirements of ASTM A516, IS 2002, or IS 2026 and carbon steel forgings that meet the specification requirements of ASTM A105, A181, A226, or A350LF2 should be chosen. Fully killed and fine-grain steel in the normalized and stress-relieved condition has improved impact properties. In the case of thick forgings, quenching and tempering provide better impact properties than normalizing and stress relieving.

The flap assembly should be redesigned to avoid any notching effect on the flap body. A single-piece solid flap of a boat-type design with a thicker central part tapered toward the periphery may be an option. Operating conditions should be properly monitored to avoid such situations as the closing of the valve during the working of pumps, the separation of air due to the decreased solubility of air with an increase in coolant temperature, and air locks in the line.

Hydrogen Damage in a Waterwall Boiler Tube Section

Wendy L. Weiss, Mechanical and Materials Engineering Department, Radian Corporation, Austin, Texas

The rear wall tube section of a boiler that had been in service for approximately 38 years was removed and examined. Visual examination of the tube revealed a small bulge with a through-wall crack. Metallography showed that the microstructure of the bulged area consisted of a few partially decarburized pearlite colonies in a ferrite matrix. The microstructure remote from the bulged area consisted of pearlite in a ferrite matrix. EDS analysis of internal deposits on the tube detected a major amount of iron, plus trace amounts of other elements. The evidence indicated that the bulge and crack in the tube resulted from hydrogen damage. Examination of the remaining water circuit boiler tubing using nondestructive techniques and elimination of any heavy deposit buildup was recommended.

Key Words

Boiler tubes, corrosion
Cracking (fracturing)

Boilers, corrosion

Hydrogen damage

Background

The rear wall tube section of a boiler that had been in service for approximately 38 years was removed and examined.

Circumstances leading to failure

The boiler experienced a pH depression of the boiler water during operation approximately 8 months prior to the failure of the tube.

Performance of other parts in same or similar service

Other waterwall tubing from the same boiler had failed due to irreversible hydrogen damage.

Visual Examination of General Physical Features

The as-received tube was visually examined. A small bulge with a through-wall crack was observed.

Testing Procedure and Results

Surface examination

Visual. Figure 1 shows macrographs of the external and internal surfaces of the bulge. Note the thick deposit buildup found on the internal surface of the bulged area (Fig. 1b). Figure 2 shows a cross section through the bulge after a hydrochloric acid macroetch. The dark areas denote hydrogen damage. A large portion of the section away from the bulge suffered hydrogen damage.

Metallography

Microstructural Analysis. A cross section through the bulge area was prepared for metallographic evaluation by grinding, polishing, and etching. The prepared section was examined using a metallurgical microscope to evaluate the failure mode, microstructure, and internal and external surface conditions. The typical microstructure of the bulged area, shown in Fig. 3, consisted of a few partially decarburized pearlite colonies in a ferrite matrix with grain-boundary fissures. The microstructure remote from the bulged area consisted of pearlite in a ferrite matrix, with no signs of overheating.

Crack Origins/Paths. Figure 4 shows the through-wall crack in the bulged area. Note the severe fissuring surrounding the crack. There was a large copper deposit at the internal surface of the crack.

Chemical analysis/identification

Corrosion or Wear Deposits. Energy-dispersive X-ray spectroscopy (EDS) analysis of the internal deposits in the tube detected a major

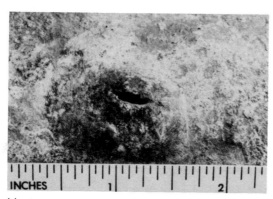

(a)

(b)

Fig. 1 Macrographs showing the external (a) and internal (b) surfaces of the bulge. Note the thick deposit on the internal surface (b). ~4.35×

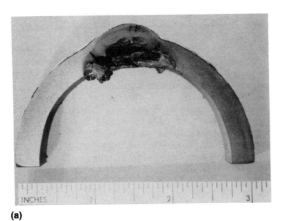
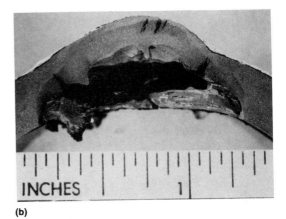

(a)

(b)

Fig. 2 Macrographs showing a cross section through the bulge after a hydrochloric macroetch. The dark areas denote hydrogen damage. A closeup view of the bulge is shown in (b).

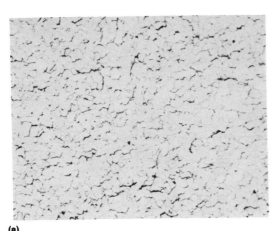
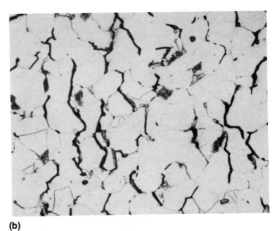

(a)

(b)

Fig. 3 Typical microstructure of the bulged area, consisting of a few partially decarburized pearlite colonies in a ferrite matrix with grain-boundary fissures. Etched in nital. (a) 76×. (b) 304×

amount of iron, plus minor to trace amounts of silicon, phosphorus, sulfur, calcium, chromium, nickel, and copper. The composition of the deposit is not unusual for a boiler using a phosphate water treatment, except for the presence of copper and nickel, which are the result of the corrosion of copper-nickel water circuit tubing.

Discussion   Metallurgical evidence indicated that the bulge and crack in the tube resulted from irreversible hydrogen damage. Hydrogen damage may occur where corrosion reactions result in the production of atomic hydrogen. Heavy deposits can act as concentration sites for oxygen and acid-forming contaminants. When a low-pH (acidic) condition is present, hydrogen is generated during rapid corrosion of the internal surface. The hydrogen atoms diffuse into the steel. Some of the diffused atomic hydrogen will combine at grain boundaries or inclusions in the metal to produce molecular hydrogen; other atomic hydrogen may react with the iron carbide (Fe_3C) in the pearlite to form methane. This results in decarburization of the steel (Fig. 3). Neither molecular hydrogen nor methane is capable of diffusing through steel, so the gases accumulate, primarily at the grain

Fig. 4 Through-wall crack in the bulged area. Note the severe fissuring surrounding the crack. The arrows point to a large copper deposit. The outside diameter is at the top, the inside diameter at the bottom. Etched in nital. 19.25×

boundaries. Eventually, gas pressures cause separation of the metal at its grain boundaries, producing discontinuous intergranular separation (fissures). As fissures accumulate, tube strength decreases until internal tube stresses exceed the tensile strength of the remaining intact tube metal. The tube metal bulged and finally cracked because of the decreasing metal strength.

Conclusion and Recommendations	*Most probable cause*	*Remedial action*
	The low pH excursion in the boiler water circuit experienced by this boiler was probably responsible for the hydrogen damage found in the rear wall tube section. Corrosive attack and hydrogen damage were aggravated by the copper present in the internal deposits due to galvanic effects.	The remaining water circuit boiler tubing should be examined by nondestructive techniques for any evidence of internal wall thinning (accelerated corrosion) or hydrogen damage (fissures). Any heavy deposit buildup in the remaining tubing should be eliminated.

References

1. Robert D. Port and Harvey M. Herro, *The Nalco Guide to Boiler Failure Analysis,* McGraw-Hill, Inc. 1991

ROTATING EQUIPMENT FAILURES

Blades, Disks, and Rotors

Impellers

Shafts

Cracking in a Steam Turbine Rotor Disk

Richard L. Colwell, Air Products and Chemicals, Allentown, Pennsylvania

An A-470 steel rotor disk was removed from the high-pressure portion of a steam turbine-powered compressor after nondestructive testing revealed cracks in the shoulder of the disk during a scheduled outage. Samples containing cracks were examined using various methods. Multiple cracks, primarily intergranular, were found on the inlet and outlet faces along prior-austenite grain boundaries. The cracks initiated at the surface and propagated inward. Multiple crack branching was observed. Many of the cracks were filled with iron oxide. X-ray photoelectron spectroscopy indicated the presence of sodium on crack surfaces, which is indicative of NaOH-induced stress-corrosion cracking. Failure was attributed to superheater problems that resulted in caustic carryover from the boiler. Two options for disk repair, installing a shrink-fit disk or applying weld buildup, were recommended. Weld repair was chosen, and the rotor was returned to service; it has performed for more than 1 year without further incident.

Key Words

Nickel-chromium-molybdenum-vanadium steels, corrosion
Compressors

Stress-corrosion cracking
Rotors
Sodium hydroxide, environment

Intergranular corrosion
Steam turbines

Alloys

Nickel-chromium-molybdenum-vanadium steel—A-470

Background

An A-470 (Grade 4) steel rotor disk was removed from the high-pressure portion of a steam turbine-powered compressor after nondestructive testing revealed cracks in the shoulder of the disk during a scheduled outage.

Applications

The disk was removed from the section anchoring the first and second rows of blades in the turbine. The turbine was designed to be powered by 3800 kPa (550 psig) steam, superheated to 330 °C (630 °F).

Circumstances leading to failure

The turbine was powered by 3800 kPa (550 psig) steam, which was superheated to 330 °C (630 °F) for the first 3½ years of its 5 year service life. During the remaining 1½ years, the boiler oper-ated without a superheater. During this period the turbine was powered by saturated steam at 3800 kPa (550 psig) and 247 °C (477 °F). As the steam performed work on the turbine, its quality degraded, thus introducing moisture, a prerequisite for stress-corrosion cracking (SCC), into the turbine system. Loss of superheat steam temperatures resulted from problems with the boiler's superheater.

During a scheduled outage, nondestructive examination revealed cracks in the shoulder of the disk. A damaged portion of it was removed by machining and submitted for analysis.

Specimen selection

Samples containing cracks were further sectioned for metallographic, fractographic, and mechanical property analyses.

Testing Procedure and Results

Surface examination

Visual examination of the turbine disk revealed multiple cracks on the inlet and outlet sides (Fig. 1). The cracks were primarily radial, and were fairly linear. The majority of cracks were located in the midradius region of the disk (Fig. 2).

Crack orientation is a function of stress state on the disk surface, which was not determined. The disk was forged and was an integral part of the rotor.

Macrofractography. One of the cracks was opened in the laboratory to facilitate examination.

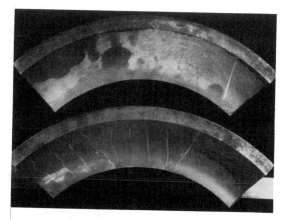

Fig. 1 Multiple cracks on the inlet and outlet sides of the sectioned turbine disk

Fig. 2 Midradius region, exhibiting the majority of cracking. 1.2×

Under a low-power stereomicroscope (30×), the fracture surface could not be classified. It was covered with a thick, adherent black scale—characteristic of caustic-induced cracking.

Scanning Electron Microscopy/Fractography. The examined crack surfaces were cleaned using an inhibited acid. Scanning electron microscopy (SEM) revealed that the crack surface morphology was entirely intergranular (Fig. 3). Intergranular cracks are characteristic of caustic-induced SCC.

Metallography

Microstructural Analysis. Metallographic cross sections of the turbine disk were prepared for examination. The resultant microstructure was bainitic, typical for quenched and tempered nickel-chromium-molybdenum-vanadium steels (Fig. 4). Grain sizes were ASTM 7 and 8 on average.

Crack Origins/Morphology. The cracks were primarily intergranular along prior-austenite grain boundaries (Fig. 5 and 6). Multiple crack branching was observed. Many of the cracks were filled with iron oxide (Fig. 7). These observations are characteristic of caustic-induced SCC.

Chemical analysis/identification

Material. Elemental analysis of the bulk alloy was performed using optical emission spectroscopy. Results, shown in Table 1, indicate that the alloy closely matched ASTM A-470 (Grade 4) specifications. The effect of the anomalous 0.1% Cu on mechanical properties was not known.

Coatings or Surface Layers. X-ray photoelectron spectroscopy (XPS) was used to identify the aggressive chemical species responsible for the cracking. Results of this analysis indicated the presence of sodium. Levels of 0.5% Na were observed on opened crack surfaces of the as-received sample. The appearance of sodium on the crack surfaces suggested the presence of caustic/sodium hydroxide (NaOH), a known promoter of SCC in high-strength steels.

Fig. 3 Micrograph showing the intergranular morphology of the crack surfaces. 630×

Fig. 4 Micrograph showing the bainitic microstructure of the turbine disk. Picral etch. 315×

Table 1 Results of elemental analysis

	Composition, %	
Element	Sample	ASTM A-470, Grade 4
Carbon	0.28	0.28 (max)
Manganese	0.53	0.20-0.60
Phosphorus	0.009	0.12 (max)
Sulfur	0.009	0.015 (max)
Silicon	0.28	0.15-0.30
Nickel	3.54	2.50 (min)
Chromium	0.39	0.75 (max)
Molybdenum	0.45	0.25 (min)
Vanadium	0.11	0.03 (min)
Copper	0.10	...
Iron	bal	bal

Fig. 5 Intergranular cracks propagating along prior-austenite grain boundaries (etch: picric acid + wetting agent). 630×

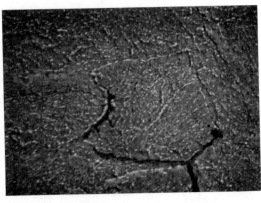

Fig. 6 Higher-magnification view of Fig. 5. 1260×

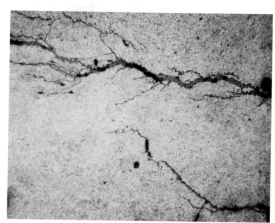

Fig. 7 Multiple branching in the stress cracks, which are filled with iron oxide. Picral etch. 126×

Mechanical properties

Tensile Properties. Tensile testing verified that the mechanical properties met the A-470 grade 4 specification. Two round tensile bars were

Table 2   Results of tensile testing

Property	Sample 1	Sample 2	Specified
Tensile strength, MPa (ksi)	784 (113.7)	805 (116.8)	720 (105) min
Yield strength, MPa (ksi)	658 (95.4)	682 (98.9)	590 (85) min
Elongation, %	22.8	19.7	17
Reduction of area, %	72.0	68.4	45

machined to different sizes; sample 1 had a gage diameter of 6.4 mm (0.250 in.), and sample 2 had a gage diameter of 4.1 mm (0.160 in.). Tensile test results are shown in Table 2.

Fracture Toughness. To qualitatively assess the rotor disk toughness, Charpy bars were machined and tested. Ten bars were machined so that their longitudinal axes were in the transverse direction. Test results indicated that the disk was not embrittled by aging. The fracture appearance transition temperature (FATT) was determined to be approximately –60 °C (–75 °F). Room-temperature impact energy was above 120 J (90 ft · lbf).

Conclusion and Recommendations

Most probable cause

The rotor disk cracked as the result of SCC. During the 1½ year period when the turbine was powered by saturated steam, caustic carryover to the turbine occurred. Exposure to caustic environments promotes SCC in carbon steels at elevated temperatures. After an incubation period, cracks initiated, then propagated. These cracks may have grown to failure in continued service.

Remedial action

There were two options for disk repair: installing a shrink-fit disk or applying weld buildup. The weld repair approach was chosen. The rotor was returned to service and has performed for more than 2½ years without further incident.

Failure Analysis of a Cracked Low-Pressure Turbine Blade

J. Ciulik, Radian Corporation, Austin, Texas

A cracked, martensitic stainless steel, low-pressure turbine blade from a 623 MW turbine generator was found to exhibit fatigue cracks during a routine turbine inspection. The blade was cracked at the first notch of the fir tree and the cracks initiated at pits induced by chloride attack. Examination of the blade microstructure at the fracture origins revealed oxide-filled pits and transgranular cracks. The oxide-filled cracks appeared to have originated at small surface pits and probably propagated in a fatigue or corrosion-fatigue fracture mode. It was recommended that the sources of the chlorides be eliminated and that the remaining blades be inspected at regular maintenance intervals for evidence of cracking.

Key Words			
Martensitic stainless steels, Mechanical properties	Turbine blades, Mechanical properties	Fatigue failure, Environmental effects
Pitting (corrosion), Environmental effects	Chlorides, Environment	Corrosion fatigue

Alloys Martensitic stainless steels—X20Cr13

Background

During a routine turbine inspection, a cracked low-pressure blade was discovered. The blade was cracked at the first notch of the fir tree.

Applications

The turbine is used to generate electricity at a major power plant.

Circumstances leading up to failure

Chloride contamination of the steam reportedly occurred a few months prior to the inspection.

Pertinent specifications

The blade was reportedly manufactured from DIN X20Cr13 stainless steel, a martensitic steel with approximately 13% chromium.

Visual Examination

Visual inspection of the blade under a stereomicroscope revealed one crack at the first notch of the fir tree (Fig. 1). The crack was visible on the leading edge side of the fir tree and in the first notch on the pressure side of the blade, extending approximately 2.5 cm (1 in.) along the blade root. Dry magnetic particle inspection and examination under a stereomicroscope did not reveal indications of other cracks.

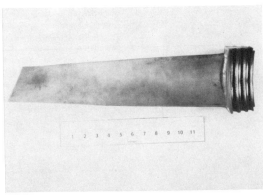
(a)

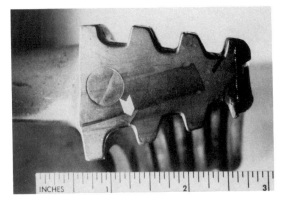
(b)

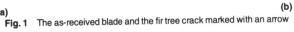

Fig. 1 The as-received blade and the fir tree crack marked with an arrow

---

Testing Procedure and Results

Surface examination

Visual. The crack shown in Fig. 1 was opened to reveal the fracture surfaces. Figure 2 shows the fracture surface and the locations of three fatigue crack origins. Beach marks were evident across the fracture surface, and ratchet lines were observed between the adjacent crack origins. From the dimensions of the fracture surface, the crack extended approximately 2.5 cm (1.0 in.) along the notch and 1.0 cm (0.4 in.) deep at the leading edge of the fir tree. The crack appeared to have originated along a machining mark in the first notch of the fir tree.

Scanning Electron Microscopy Fractography. Examination of the fracture surface in the SEM revealed a flat oxidized fracture, typical of fatigue. Pits were found at each of the three fracture origins. Two origins are shown in Fig. 3. The pits were 0.013 to 0.076 mm (0.0005 to 0.003 in.) deep. On the machined surface adjacent to the crack, many pits were observed (Fig. 4 and 5). The pits were as large as 0.10 mm (0.004 in.) in diameter. EDS analysis at the base of the pits detected large amounts of chlorine (Fig. 6), indicating that chloride attack caused the pitting.

Metallography

Microstructural Analysis. Examination of the blade microstructure at the fracture origin revealed oxide-filled pits and cracks (Fig. 7). The cracks were transgranular, typical of fatigue cracks. The oxide-filled cracks appeared to have originated at small surface pits and probably propagated in a fatigue or corrosion-fatigue fracture mode. These cracks were not detected by visual examination or by magnetic particle inspection. The microstructure of the blade consisted of tempered martensite.

Chemical analysis

A sample of the blade was chemically analyzed, and the results met the requirements of DIN X20Cr13, as shown in Table 1.

Mechanical Properties

The hardness of the blade was 20 HRC.

Fig. 2 The crack fracture surfaces. The arrows indicate three origins marked A, B, and C. Approximately 2.48×

(a)

(b)

Fig. 3 SEM fractographs of the fracture origins in Fig. 2. 310×. (a) The fracture origin labeled A. (b) The fracture origin labeled C

Table 1 Chemical analysis results

	Element, wt%								
	C	**Mn**	**P**	**S**	**Si**	**Ni**	**Cr**	**Mo**	**Fe**
Blade	0.19	0.70	0.014	0.006	0.40	<0.10	13.30	<0.05	rem
Specification DIN X20Cr13	0.16 to 0.25	1.00 max	0.030 max	0.030 max	1.00 max	1.00 max	12.00 to 14.00	NS	rem

NS, not specified

Discussion

The crack in the blade originated at three locations in the first notch in the fir tree. Corrosion pits were observed at each origin. The machined surface of the fir tree contained many small corrosion pits and the chlorine detected in the pits indicates that the corrosion occurred due to chloride attack. The oxide-filled cracks and pits revealed during the metallographic examination indicate that the initial corrosive attack and cracking of the blade were not detectable by nondestructive inspection methods.

Fig. 4 An SEM view of pitting along the fir tree notch, near the fracture origin. 21.7×

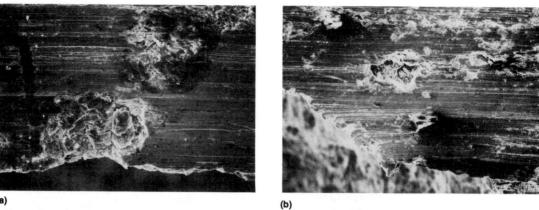

(a)   **(b)**

Fig. 5   Higher magnification SEM views of the pits labeled 1 (a) and 2 (b) in Fig. 4. (a) 310×. (b) 124×

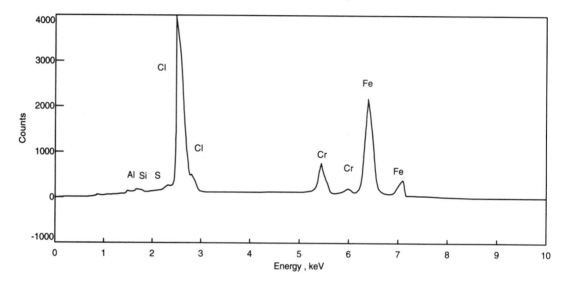

Fig. 6   An EDS spectrum from a pit shown in Fig. 5. Note the large amount of chlorine.

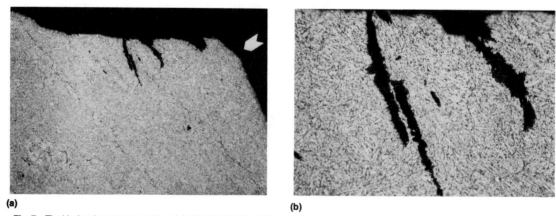

(a)   **(b)**

Fig. 7   The blade microstructure at the origin labeled A in Fig. 2. The arrow denotes the fracture surface. Note several oxide-filled pits and cracks parallel to the fracture surface. Vilella's reagent. (a) 62×. (b) 248×

Conclusion and Recommendations

The crack initiated at small pits in the first notch of the fir tree. The pits were caused by chloride attack; the crack then propagated by fatigue or corrosion-fatigue. No additional cracks were found during nondestructive examination, but several small, oxide-filled cracks approximately parallel to the fracture surface were found during the metallographic examination of the cracked region. These appeared to have initiated at surface pits.

It is recommended that the source of the chlorides be eliminated and that the remaining blades be inspected at regular maintenance intervals for evidence of cracking.

Failure of Non-Magnetic Retaining Ring in a High-Speed Generator Rotor

Thomas P. Sherlock, Structural Integrity Associates, Inc., Akron, Ohio
Michael J. Jirinec, Consolidated Edison of New York, Inc., Bronx, New York

A shrunk-fit 18Mn-5Cr steel retaining ring failed without warning during normal unit operation of a 380 MW electrical generator. The cause of the ring failure was determined to be intergranular stress-corrosion cracking (IGSCC) because of the high strength of the ring material and the presence of moist hydrogen used to cool the ring. Factors which promoted the failure were higher than normal strength levels in the ring material, lower than normal ring operating temperatures, possible moisture in the lubrication oil system, periodic poor performance of the hydrogen dryers, and a ring design which allowed water to become trapped in a relief groove. These factors caused pitting in the ring in an estimated 100 hours of operation. The ring had been inspected previously 18 months prior to the failure and no defects or pitting were found. Calculations showed that a 0.127-cm (0.050-in.) deep pit could grow to a critical size in 3000 to 4000 hours of operation. To prevent further failures, it was recommended that the ring be replaced with an 18Mn-18Cr alloy with superior resistance to IGSCC. A program of periodic inspection and replacement of other retaining rings in the system was also recommended.

Key Words	High-alloy steels, Corrosion Intergranular corrosion Rotating equipment	Retaining rings, Corrosion Pitting (corrosion)	Stress-corrosion cracking Retaining rings
Alloys	High-alloy steels—18Mn-5Cr		

Background

Applications

Retaining rings are used in electrical generators to resist the centrifugal forces of the copper windings and are among the most highly stressed components in a power plant. To reduce any electrical losses, they are often made from austenitic, non-magnetic alloys. They are shrunk onto the generator shaft after winding, and require a high yield and tensile strength to resist the high shrink and rotational stresses encountered. A schematic of a retaining ring is shown in Fig. 1.

Circumstances leading up to the failure

The ring failed during normal operation, with no warning. The ring burst into two large fragments, which were contained within the generator housing. The ring fragments did cause a hydrogen fire, which was extinguished quickly.

Pertinent specifications

ASTM A 289, which closely approximates the OEM vendor specification, is summarized in Table 1. In order to meet the specification requirements, the ring must be cold expEnded after forming. For the vintage ring in this analysis and others made at the same time, the high strengths were achieved by explosive forming.

Performance of other parts in similar service

Several thousand 18Mn-5Cr rings are in service in the United States. No previously reported

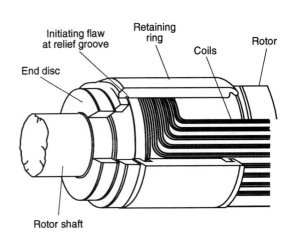

Fig. 1 Schematic of a retaining ring assembly

ring failures of this alloy are reported, though approximately a dozen failures in the past fifty years of other austenitic rings have been reported, representing approximately one failure per ten thousand machine-years of service.

Visual Examination of General Physical Features

Figure 2 shows the best preserved fracture face and a darkened area at the fracture origin. Figure 3 shows the extent of the pitting in the relief groove associated with the darkened region in Fig. 2. The chevron markings in Fig. 2 clearly point to an origin at the relief groove, and several clusters of pits in the groove had cracking originating from the pitting. The white areas on the fracture face in Fig. 2 are the results of electrical disturbances caused by the ring cutting stator windings.

Fig. 2   Fracture surface showing origin

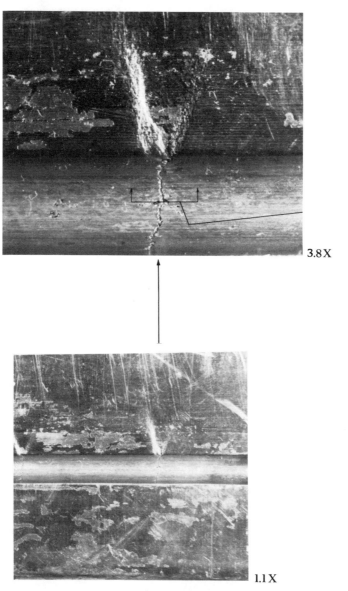

3.8 X

1.1 X

Fig. 3   Detail of pitting and cracking in the relief groove **(continued)**

Fig. 4 Closer view of fracture origin (37%)

Fig. 3 Detail of pitting and cracking in the relief groove

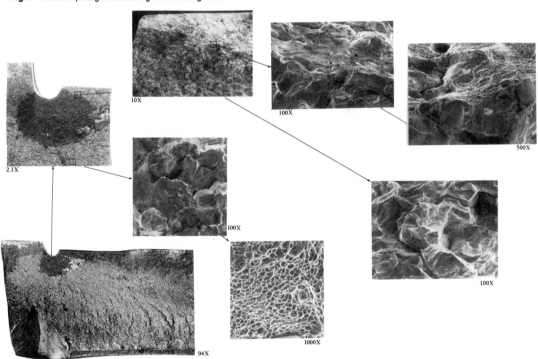

Fig. 5 Detail of the fracture features. (a) 0.32×. (b) 0.71×. (c) 34×. (d) 34×. (e) 34×. (f) 170×. (g) 34×. (h) 340×

<table>
<tr><td>**Testing Procedure and Results**</td></tr>
</table>

Testing Procedure and Results

Nondestructive evaluation

The failed ring had been inspected 18 months prior to the failure using visual, liquid penetrant and eddy current examination. No defects were found in the area of the groove.

Surface examination

Figure 4 shows a slightly enlarged photograph of the origin. Figure 5 shows the various SEM features in the darkened origin and in the transition and overload areas. The origin is entirely intergranular (IG), which changes to a quasi-IG mode with considerable dimpling on the facets, as shown in Fig. 5(g) and (h). The gradual transition from IGSCC to a ductile IG mode most probably occurred during normal unit operation.

Table 1 ASTM Specification A 289 limits

Element	Composition, wt%
C	0.4 to 0.6
Mn	16.0 to 20.0
Cr	3.5 to 6.0
Ni	2.0 max
Si	0.2 to 0.65
P	0.08 max
S	0.025 max
Fe	rem

Metallography

Figure 6 shows the nature of the pit formation, and Fig. 7 shows the IGSCC and deformation twinning associated with the cold-worked microstructure. No other picture is available concerning the pitting of the ring.

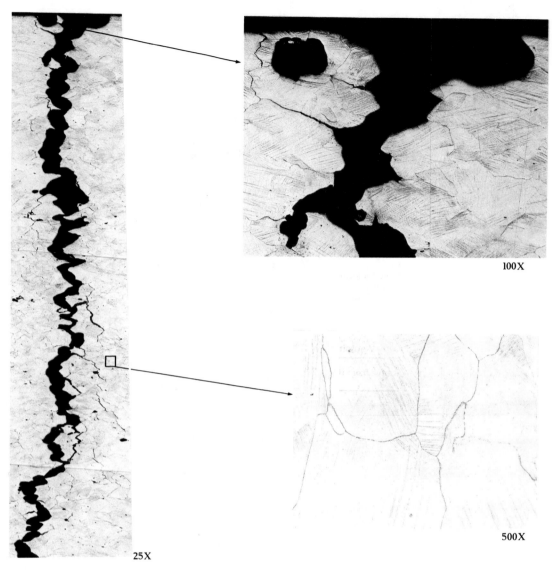

100X

500X

25X

Fig. 6 Pit formation and subsequent initiation. (a) 14.75×. (b) 59×. (c) 295×

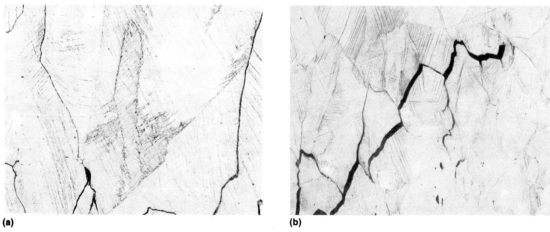

(a)　　　　　　　　　　　　　　　　　　　(b)

Fig. 7 General microstructure and crack morphology. 4% nital etch

Chemical analysis/identification

Table 2 shows the results of two analyses taken on the ring from specimens some 46 cm (18 in.) apart. In general, the material conforms to specification.

Mechanical properties

Tables 3 and 4 list results of the mechanical property and microhardness tests. These tests confirmed that the ring was at the upper end of the

Table 2 Results of two analyses taken 46 cm (18 in.) apart

Element	Composition, wt%	
	Sample 1	Sample 2
C	0.55	0.56
Mn	18.31	19.0
Cr	5.05	5.35
Ni	0.06	0.44
Si	0.36	0.38
P	...	0.04
S	0.003	0.001
V	0.06	0.044
Fe	rem	rem

Table 3 Mechanical property test results

Property	Distance from fracture plane			
	23 cm (9 in.)		46 cm (18 in.)	
	ID	OD	ID	OD
Ultimate tensile strength, MPa (ksi)	1342 (194.6)	1203 (174.5)	1349 (195.6)	1215 (176.2)
Yield strength, MPa (ksi)	1138 (165.0)	914 (132.6)	1183 (171.6)	795 (115.3)
Elongation, %	15.0	24.0	13.5	26.5
Reduction in area, %	17.0	...	21.0	33.0

Charpy V-notch properties at 42 °C (107 °F)

Distance from ID		Impact strength		Lateral expansion	
cm	in.	N · M	ft · lbf	μm	mils
0.64	0.25	44.7	33	343	13.5
2.03	0.80	40.7	30	203	8.0
3.43	1.35	46.1	34	381	15.0
4.95	1.95	65.1	48	559	22.0
6.10	2.40	75.9	56	660	26.0
7.49	2.95	124.7	92	1041	41.0

ID, inner diameter. OD, outer diameter

normal scatterband in tensile and yield strength for 18Mn-5Cr rings, resulting in lower than normal toughness.

Discussion

The most probable set of circumstances which lead to the initiation of pitting and IGSCC can be summarized briefly.

- Moisture, most likely from the lube oil system, was allowed to enter the hydrogen cooling system.
- Intermittent poor performance of the dryers failed to remove the moisture, increasing the dewpoint.
- Moisture condensed on the ring, and was centrifuged to the relief groove, where it remained entrapped for long periods.
- The relatively high strength of the ring accelerated the pitting and IGSCC.

Typically, lube oil is vacuum-treated and the retaining rings run at higher temperatures than the service environment experienced by this retaining ring, so the possibility of condensing moisture on the rings during normal operation is significantly lower. The relatively high strength contributed to low toughness of the ring material, as is shown in Fig. 8, and probably caused very short initiation

Table 4 Microhardness results

Distance from apex of relief groove		Knoop microhardness, KHN
mm	in.	
0.991	0.039	502
2.006	0.079	508
2.997	0.118	436
3.988	0.157	485
5.004	0.197	464
5.994	0.236	467
7.010	0.276	480
8.001	0.315	434
8.992	0.354	450
10.008	0.394	485
10.998	0.433	490
11.989	0.472	401
13.005	0.512	466
13.995	0.551	485
15.011	0.591	481
16.002	0.630	446

times and high crack growth rates. Figure 9 confirms that initiation times on the order of days are possible for this material when the applied stress level exceeds 413 MPa (60 ksi).

Metallurgically, the initiation and propagation followed the expected pattern for IGSCC in high-strength austenitic steels. The 18Mn-5Cr alloy is particularly susceptible because of its high

strength level, and will readily crack in aerated water. The gradual transition from pure IG fracture to a mixed IG/dimpled rupture appearance away from the initiating flaw may be due to propagation in hydrogen at very high rates (0.25 cm, or 0.1 in., or more per hour) or is simply a characteristic of this material.

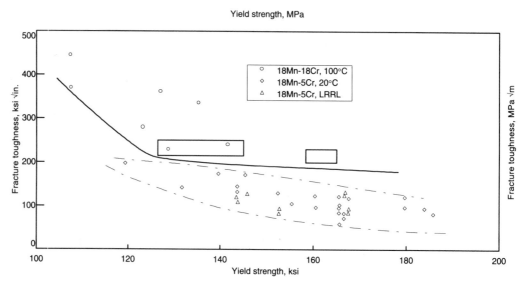

Fig. 8 Fracture toughness versus yield strength

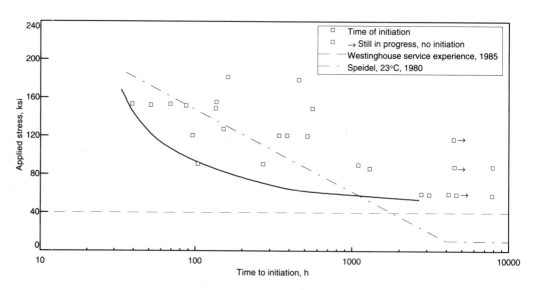

Fig. 9 Initiation time versus applied stress level at 30 °C (86 °F)

Conclusions and Recommendations	*Most probable cause*	*Remedial action*

Most probable cause

The basic cause of the failure was pitting and IGSCC due to the entrapment of moisture in the ring relief groove. Several contributing factors made this scenario possible, as discussed above.

Remedial action

The ring was replaced with an 18Mn-18Cr alloy, which has shown no susceptibility to IGSCC in laboratory tests. In addition, upgraded hydrogen dryers were purchased and more frequent monitoring of the hydrogen dewpoint has been instituted.

Failure Analysis of Turbine Blades

Antonio F. Iorio and Juan C. Crespi, Gerencia de Desarrollo, Comisión Nacional de Energía Atómica, Buenos Aires, Argentina

Two 20 MW turbines suffered damage to second-stage blades prematurely. The alloy was determined to be a precipitation-hardening nickel-base superalloy comparable to Udimet 500, Udimet 710, or René 77. Typical protective coatings were not found. Test results further showed that the fuel used was not adequate to guarantee the operating life of the blades due to excess sulfur trioxide, carbon, and sodium in the combustion gases, which caused pitting. A molten salt environmental cracking mechanism was also a factor and was enhanced by the working stresses and by the presence of silicon, vanadium, lead, and zinc. A change of fuel was recommended.

Key Words

Nickel-base alloys, corrosion
Turbine blades

Superalloys, corrosion
Gas turbines

Hot gas corrosion

Background

Two 20 MW turbines suffered damage to second-stage blades prematurely. The first failure occurred after 25,000 h of duty and the second after 24,000 h.

Pertinent specifications

Material specifications were unknown. The specified working temperature for the second stage of the turbine was on the order of 700 to 750 °C (1290 to 1380 °F), and the minimum anticipated working lifetime was 35,000 h.

Visual Examination of General Physical Features

Three blades were examined. Half of one failed blade is shown in Fig. 1. A large number of pits were observed on the pressure side zone of each blade. Liquid penetrant inspection did not reveal the presence of cracks. None of the blades exhibited a protective film coating.

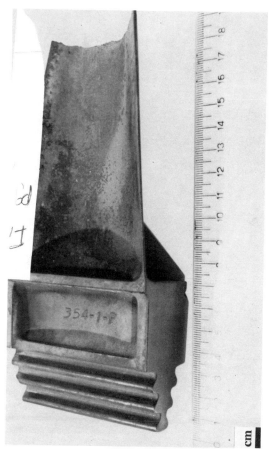

Fig. 1 Part of a failed turbine blade. (a) Zone exhibiting extensive pitting. (b) Rear side of blade in (a), showing extension of pitting area

**Testing
Procedure
and Results**

Surface examination

Macrofractography. At low magnification, the fracture zone exhibited a brittle appearance without abnormal overload indications. Figure 2 shows that crack growth was stable through one-third of the transverse section of the blade, after which final failure by overload occurred.

Scanning Electron Microscopy/Fractography. Figure 2 indicates the zones examined by scanning electron microscope (SEM) fractography. Figures 3 and 4 show patterns of stable and unstable crack growth. The stable crack growth was intercrystalline in nature, while the final fracture exhibited transcrystalline characteristics mixed with cleavage patterns.

Metallography

Microstructural Analysis. In the zone where crack initiation occurred, perpendicular and parallel sections were extracted in order to study the microstructure of the material. After polishing, the specimens were etched with Marble's reagent to reveal the distribution of carbides on grain boundaries and inside them, as well as of γ' precipitates.

Figure 5, representing a section parallel to the crack plane, shows the characteristic dendritic microstructure produced by the investment casting method, with a grain size on the order of 0.5 to 1.0 mm. On the suction side of the blade was a layer of small (50 μm) grains (Fig. 6). Figure 7 shows the grain boundaries with carbide precipitates ($M_{23}C_6$ or M_7C_6) and a homogeneous precipitation of γ' phase inside the grains. The carbides were of composite type (TiCr)C. Figure 8 shows the microstructure in a section perpendicular to the crack plane. There are no creep indications that would reveal the presence of high stresses; creep in this type of alloy begins near 650 °C (1200 °F).

Chemical analysis/identification

Material. Because of the lack of information about the chemical composition of the blade material, electron probe microanalysis was first conducted. The major elements found were nickel, chromium, and cobalt; titanium, aluminum, and molybdenum were minor alloying elements. Tungsten, niobium, and tantalum were not detected. A more accurate chemical analysis revealed the following composition: 53.7% Ni, 18 to 20% Cr, 18.2% Co, 3.7% Ti, 3.7% Mo, 2.5% Al, 0.1 to 0.2% Fe, <0.1% Nb, 0.085% C, and 0.006% S.

It was concluded that the material was a precipitation-hardening nickel-base superalloy comparable to Udimet 500, Udimet 710, or René 77. This type of alloy is not normally used for gas turbine blades.

Coatings or Surface Layers. It is common practice to apply a special coating to gas turbine blades in order to ensure that these components will meet the expected design life. There are three primary types of coatings:

Fig. 2 Macrofractograph of a failed blade showing the extension of the stable crack growth

(a)

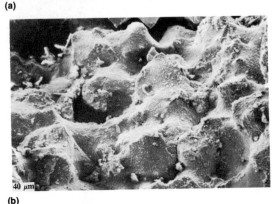

(b)

Fig. 3 Microfractographs taken in zone B of Fig. 2. The intercrystalline nature of the crack growth is evident.

- Coating systems that during diffusion produce intermetallic compounds (mainly aluminides), in this case NiAl
- Complex alloy coating systems based on NiCrAlX, where X is an active element that promotes oxide scale adhesion
- Coating systems that work as thermal barriers (metal-ceramic compounds)

In general, these coating layers can improve the hot corrosion behavior of blades, extending their lifetime by a factor of four. The testing techniques used confirmed the lack of any coating layers.

Corrosion Deposits. A specimen was taken from the stable crack growth fracture surface shown in Fig. 3 and subjected to spectrographic analysis at two different points in order to identify the presence of contaminants. Table 1 presents the

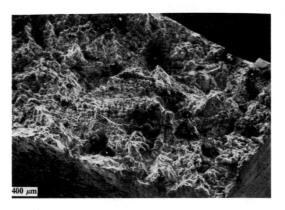

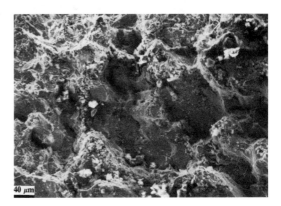

Fig. 4 Microfractographs taken in zone C of Fig. 2, showing transcrystalline characteristics mixed with some cleavage patterns

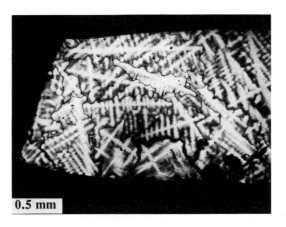

Fig. 5 Section parallel to the crack plane, showing the characteristic dendritic microstructure produced by investment casting

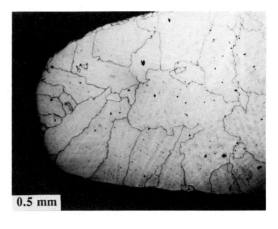

Fig. 6 Same section shown in Fig. 5, revealing the characteristic grain size and distribution

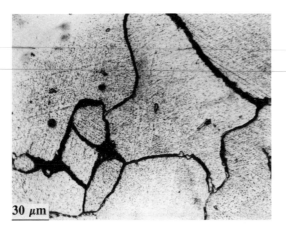

Fig. 7 Micrograph showing grain boundaries with carbide precipitates and a homogeneous precipitation of γ′ phase inside the grains

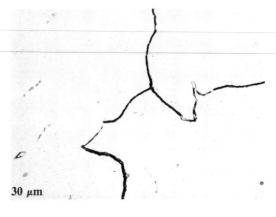

Fig. 8 Microstructure in a section perpendicular to the crack plane. Creep indications were not present.

results of the analysis; spectrums obtained by Auger electron spectroscopy and X-ray photoelectron microscopy (XPS) are shown respectively in Fig. 9 and 10.

An additional chemical analysis using a semiquantitative spectrographic method was performed on the fractured blade corrosion deposits. Lead was found present in an amount on the order of 2%.

Associated Environments. The fuel burned in the turbines was Diesel C (commonly known as Diesel Coke), which is the last fuel distilled from crude oil. It is recommended as a dissolvent agent and boiler fuel but not as turbine gas fuel because of its high degree of oxidation. Furthermore, it behaves as a highly corrosive gas due to the presence of many impurities, such as sulfur, sodium, vanadium, iron, carbon, and chlorine. The physical and chemical properties of Diesel C fuel are listed in Table 2.

Table 1 Results of electron spectroscopy at two different points on the fracture surface

Point	Depth, nm	Elements detected
1	5	Ni, O, C, Cl, S
	10	Ni, O, N, C, Cl, S
	15	Ni, O, N, C, Cl, S
	20	Ni, O, N, C, Cl, S
	30	Ni, O, N, C, Cl, S
	40	Ni, O, N, C, Cl, S
	50	Ni, O, N, C, Cl, S
2	10	Ni, O, N, C, Cl, S
	20	C, Cl, S
	30	C, Cl, S

Discussion

For blades working in the temperature range of 700 to 850 °C (1290 to 1560 °F), it is desirable that the

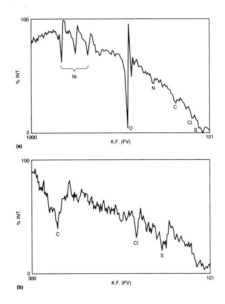

Fig. 9 Auger spectrums taken at two points on the fracture surface

Table 2 Physical and chemical properties of Diesel C fuel

Properties	Luján de Cuyo	La Plata
Density	0.825	0.850
Color, ASTM D-1500	8	8
Distillation (°C)		
1st. Drop:	165	200
10%	200	220
50%	254	280
85%	351	365
Sulphur, w/c	0.50	0.55
Cetene index	48	48
Inflammation point (°C)	60	70
Draining point (°C)	5	5
Viscosity, SSU at 37.8 °C	35	38
Ashes, w/c	Traces	Traces
Oxidation stability test	Do not pass	Do not pass
Neutralization number mg KOH/g	0.3	0.3
Conradson carbon on 10% distillation vol., w/c	0.30	0.33
Upper heat content, kcal/k	10,970	10,950
Water and sediments, vol.%	0.1	0.1

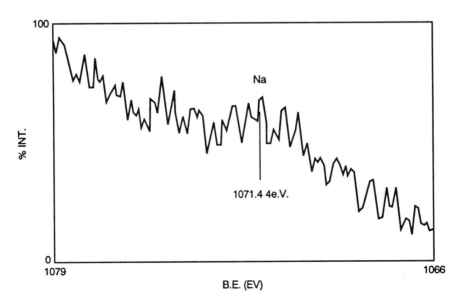

Fig. 10 XPS spectrum showing the presence of sodium

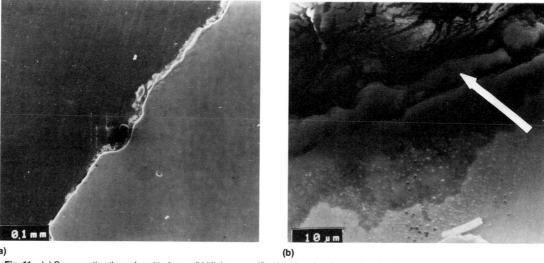

(a)   **(b)**

Fig. 11   (a) Cross section through a pitted area. (b) Higher-magnification view showing a lost surface layer present at the bottom of the pits

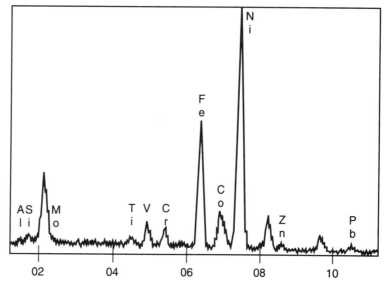

Fig. 12   Peaks of an EDAX spectrum showing the presence of silicon, vanadium, lead, and zinc

Al_2O_3 surface oxide be formed, because Cr_2O_3 can produce volatile CrO_3. Nevertheless, under such temperatures the degradation process is corrosion rather than oxidation, and Cr_2O_3 exhibits better behavior against sulfidation problems. The pits found on the three inspected blades were probably caused by this type of process, as evidenced by the high degree of SO_3 produced by the Diesel C fuel.

Actually, the process is much more complex, but can be summarized as follows. The Ni-Cr-Al system develops three oxide types:

- A surface oxide of NiO, with subscale of Cr_2O_3 and/or Al_2O_3
- A scale formation of Cr_2O_3, with a discontinuous underlying layer of Al_2O_3
- The formation of a thin continuous layer of Al_2O_3 surface scale

These oxides can be dissolved by Na_2SO_4, with the process accelerated by the presence of SO_3 and carbon. The latter elements can be caused by imperfect combustion.

As can be seen in Table 2, chlorine exists due to the presence of NaCl or HCl, which increases the corrosion rate. Although the lead content of the combustion residues on the fractured blade was high and could have given rise to a stress-corrosion cracking mechanism, the residual stresses would not be the operative ones because they would be relaxed at the service temperature. Cracking was primarily originated by service stresses that were increased by stress concentrators due to the presence of pits.

Conclusion and Recommendations

Most probable cause

The Diesel C fuel used contains excessive amounts of SO_3, carbon, and sodium, which can produce high-temperature corrosion and severe pitting. Figure 11 shows a cross section through a pitted area; a lost surface layer is present at the bottom of the pits. In Fig. 12, the peaks of a spectrum obtained by energy-dispersive X-ray analysis (EDAX) show the presence of silicon, vanadium, lead, and zinc, which may induce a molten salt environmental cracking mechanism.

Remedial action

Use of a fuel such as Gasoil is recommended in order to meet the operational lifetime of the turbine blades. Proper coating of the blades would improve their performance and perhaps permit the use of the less expensive Diesel C fuel. However, this type of remedial action has not been tried in the past, and the change of fuel is the better option.

Fatigue Cracking of Gas Turbine Diaphragms

N.A. Fleck, Dow Chemical U.S.A., Freeport, Texas

Several compressor diaphragms from five gas turbines cracked after a short time in service. The vanes were constructed of type 403 stainless steel, and welding was performed using type 309L austenitic stainless steel filler metal. The fractures originated in the weld heat-affected zones of inner and outer shrouds. A complete metallurgical analysis was conducted to determine the cause of failure. It was concluded that the diaphragms had failed by fatigue. Analysis suggests that the welds contained high residual stresses and had not been properly stress relieved. Improper welding techniques may have also contributed to the failures. Use of proper welding techniques, including appropriate prewelding and postwelding heat treatments, was recommended.

Key Words

Martensitic stainless steels
Welded joints
Gas turbine engines

Compressors
Fatigue failure

Welding parameters
Diaphragms

Alloys

Martensitic stainless steel—403

Background

Several compressor diaphragms from five gas turbines cracked after a short time in service. The vanes were made from type 403 martensitic stainless steel. Welding was performed using type 309L austenitic stainless steel filler metal.

Applications

The compressor is part of the combustion system in the gas turbine. It compresses air between the rotating blades and the stationary diaphragms and then feeds the air to the combustion system. The combustion system, after receiving the air, mixes in fuel and burns the mixture in order to raise the energy level that is then discharged to the turbine.

The stationary vanes in the compressor section are assembled into units called diaphragms (Fig. 1). The individual vanes are held in position in the diaphragm by being welded at each end into shrouds. The shrouds extend circumferentially around the inside wall of the casing, where they are supported in machined grooves. Each stage of the vane is split into two 180° shroud segments for easy access to the compressor.

Circumstances leading to failure

The gas turbines were undergoing routine maintenance and part replacement (hot gas paths) when cracks were discovered on the compressor diaphragms (Fig. 2).

Pertinent specifications

The specifications for manufacturing the compressor diaphragms were unknown. The parts were made by the original equipment manufacturer (OEM).

Specimen selection

Most of the cracks were located directly below

Fig. 1 Compressor diaphragms removed from gas turbine

Fig. 2 Diaphragm, showing fracture surface after a vane was removed and two vanes with crack indications

the shroud/blade seal weld. Several welds, along with their respective fracture faces, were removed and prepared for examination.

Visual Examination of General Physical Features

Figures 3 and 4 show the locations where the cracks occurred most frequently. Visual examination showed evidence of poor welding craftsmanship. Some of the welds appeared irregular and sloppy; others appeared to have been made without use of a filler metal.

Fig. 3 Typical crack location at outer shroud seal weld

Fig. 4 Crack locations on inner shroud after removal of seal box

Testing Procedure and Results

Nondestructive evaluation

When cracks were discovered visually, the diaphragms were removed from the turbine and inspected using the techniques described below.

Liquid Penetrant Testing. Each diaphragm was cleaned of grease and dirt. The diaphragms were set securely in a rack, with the splits facing up. Several ounces of water-washable red dye penetrant were poured inside the openings on each end. The diaphragms were completely filled with water, and all vane welds were covered; the outer sections of the diaphragms were kept dry. After a minimum of 8 h, developer was sprayed over the outer shrouds and at the vane-to-shroud intersections. Most of the time, the cracked vane-to-shroud welds could be determined visually, because the penetrant would leak out after only a few minutes. All areas displaying cracks were documented.

Ring testing was also conducted on the diaphragms, not only to determine the condition of the vane-to-shroud welds, but also to help locate cracks in the vanes themselves. Each vane was lightly tapped with a rounded metal hammer. The ringing that resulted was usually of a high to medium pitch and fairly consistent from one vane to the next. When a broken or cracked vane was tapped, a dull pitch resulted. If a definite change in sound occurred, the vane was inspected visually or using liquid penetrant.

Surface examination

Macrofractography. The shroud/blade seal welds, along with their respective fracture faces, were examined under a binocular microscope at magnifications from 10 to 50×. The results of this examination concurred with earlier observations that some welds had been made without use of a filler metal. After the fractures were opened, examination of the fracture surfaces revealed multiple ratchet and beach marks, indicative of fatigue failure. The cracks initiated on one end of the vane and propagated to the other side. In some instances, the cracks had propagated fully. The majority of the fatigue cracks initiated near the trailing edge (thinner end) of the vanes and propagated to the leading edge (Fig. 5).

Metallography

Microstructural Analysis. Microstructural examination revealed that the cracks initiated in the weld heat-affected zone (HAZ) (Fig. 6). Some sections taken from the inner shroud/vane joint revealed that the welds did not fully penetrate (Fig. 7), leaving defects beneath them. Such defects provide high stress concentrations and can lead to crack propagation. The base metal microstructures were martensitic, whereas the welds were

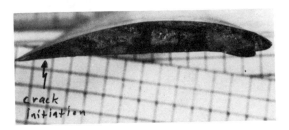

Fig. 5 Fractograph of broken vane, showing area of crack initiation

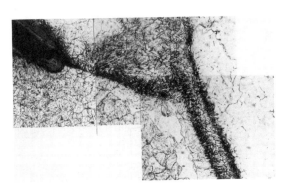

Fig. 6 Micrograph of compressor diaphragm seal weld, showing crack propagating along weld HAZ. 160×

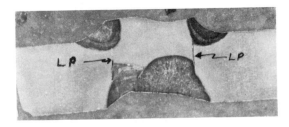

Fig. 7 Inner shroud/vane joint, showing lack of weld penetration

composed of austenite (<80%) and ferrite, which indicated that they were made using an austenitic filler metal.

Chemical analysis/identification

Material and Weld. Energy-dispersive x-ray analysis revealed that the base metal of the vanes was type 403 martensitic stainless steel. Analysis of the weld metal confirmed that the filler metal used was type 309 austenitic stainless steel.

Mechanical properties

Hardness. Microhardness readings were

Table 1 Average microhardness values

Label	Experimental welds							Actual		
	A	B	C	D	E	F	G	1	2	3
Filler material	309L	309L	309L	410	410	Inco A	410	309L	309L	309L
Stress relief	0	950	1250	1250	950	0	0	1100(a)	1100(a)	1100(a)
Weld hardness	47.5	40.0	34.5	28.5	41.5	84(b)	40.5	32.5	33.0	29
HAZ hardness	45.5	43.5	30.5	25.5	33.5	44.5	42.5	20.0	25.5	28
Base hardness	23.0	24.5	85(b)	98(b)	20.5	20.5	98(b)	88(b)	85(b)	92(b)

Note: All hardness readings are in Rockwell C units, except those noted. The base metal hardness readings were taken adjacent to the HAZ. (a) Estimated stress relief temperatures from OEM. (b) Rockwell B

taken of the HAZs and base metals of selected welds. Figure 8 shows one of the inner shroud welds and corresponding microhardness readings. As in Fig. 8, all inner shroud welds exhibited different hardness gradients, indicating inconsistency from one weld to the next. Some welds actually contained more passes than others, which had a definite effect on residual stresses. The outer shroud welds did not exhibit any sharp hardness gradients through the transition zone. The microhardness values indicated that the diaphragms had been stress relieved.

Simulation tests

Seven experimental welds were made on one of the broken vanes using different filler metals and postweld stress-relief temperatures. The same welding parameters (heat input, electrode size, etc.) were used for each experimental weld. Table 1 shows average microhardness results for the weld metal, HAZ, and base metal of the experimental welds and the actual welds. The results indicate that the welding parameters used to make the experimental welds were not the same as those used by the OEM. The weld that was made using matching filler metal (type 410 stainless steel) with a 675 °C (1250 °F) stress relief exhibited optimum properties.

Discussion   Use of an austenitic filler metal in the welding of type 403 stainless steel is not uncommon, especially in field weld applications when stress relieving or annealing is not feasible and when there is no requirement for the weld metal to have the same mechanical and corrosion properties as the base metal. Austenitic stainless steel filler metals are often used to obtain more ductile weld metal in the as-welded condition.

Martensitic stainless steels can be welded in the annealed, hardened, and hardened and tempered conditions. Regardless of the prior condition of the stainless steel, welding produces a hardened martensitic zone adjacent to the weld. As hardness increases, toughness decreases, and the zone becomes more susceptible to cracking. Preheating and control of interpass temperature are the most effective means by which to avoid cracking. Stress relieving or annealing the welds alleviates the residual stresses caused during welding. However, annealing cannot be fully effective when using an austenitic filler metal because of the difference in coefficients of thermal expansion between the

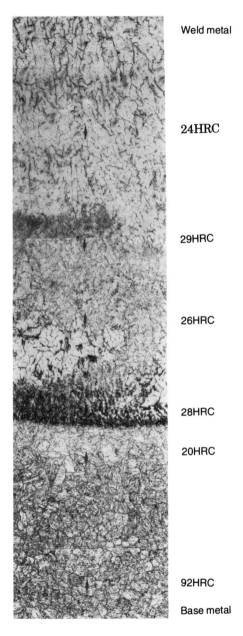

Weld metal

24HRC

29HRC

26HRC

28HRC

20HRC

92HRC

Base metal

Fig. 8 Micrograph of a diaphragm weld, showing base metal and weld metal microstructures, along with corresponding hardnesses. 122×

weld metal and base metal when the compressor is operating.

Conclusion and Recommendations

Most probable cause

The compressor diaphragm seal welds failed because of fatigue cracks that initiated in the weld HAZs on the trailing edges of the vanes. The remaining defects, caused by either no filler metal or lack of penetration, also led to fatigue cracking. The weld HAZs contained high residual stresses because of:

- Difference in coefficients of thermal expansion between the base metal (martensitic) and the weld metal (austenitic)
- Improper postweld heat treatment (that is, stress-relief temperature was too low)
- Inconsistent and poor-quality welding (lack of penetration, no filler metal, varying number of passes, sloppy beads)

Two other factors that may have contributed to cracking were the preheat temperature and the interpass temperature. However, no information was available (from the manufacturers) concerning these two parameters.

Remedial action

All inner shroud welds were ground out and rewelded to ensure 100% penetration. All other welds with crack indications were also ground out and rewelded. Type 410 stainless steel filler metal was used for all rewelding. The diaphragms (base metal) were heated to 230 °C (450 °F) prior to welding. After welding was completed, the diaphragms were stress relieved at 675 °C (1250 °F) for 2 h.

Fatigue Failure of Titanium Alloy Compressor Blades

S. Radhakrishnan, A.C. Raghuram, R.V. Krishnan, and V. Ramachandran, National Aeronautical Laboratory, Bangalore, India

The cause of low fatigue life measurements obtained during routine fatigue testing of IMI 550 titanium alloy compressor blades used in the first stage of the high-pressure compressor of an aeroengine was investigated. The origin of the fatigue cracks was associated with a spherical bead of metal sticking to the blade surface in each case. Scanning electron microscopy revealed that the cracks initiated at the point of contact of the bead with the blade surface. Energy-dispersive X-ray analysis indicated that the bead composition was the same as that of the blade. Detailed investigation revealed that fused material from the blade had been thrown onto the cold blade surface during a grinding operation to remove the targeting bosses from the forgings, thereby causing local embrittlement. It was recommended that extreme care be taken during grinding operations to prevent the hot, fused particles from striking the blade surface.

Key Words

Fatigue failure
Compressor blades
Compressors

Titanium-base alloys
Fatigue strength

Aerospace engines
Forgings

Alloys

Titanium-base alloy—IMI 550

Background

Several high-pressure compressor blades used in the first stage of an aeroengine exhibited low fatigue life during vibratory fatigue testing of new blades. Fatigue life is evaluated as the product of the amplitude and frequency (AF value). During routine fatigue testing, a few blades were found to have unusually low AF values. The reason for the low fatigue life was not immediately apparent.

Pertinent specifications

The blades were made of titanium alloy IMI 550, with a nominal composition of 4% Al, 4% Mo, 2% Sn, 0.5% Si, and the balance titanium, and were manufactured by the closed-die forging process. Targeting bosses flash buttons at the extremities of the blades serve as reference points for dimensional control and are later removed by grinding.

Visual Examination of General Physical Features

The origin of fatigue cracking in the blades exhibiting low AF values was associated with spherical beads (Fig. 1), although these surfaces were polished, examined, and found clear of any surface flaws during metrology prior to fatigue testing. The fatigue cracks had initiated in the vicinity of these beads and propagated through the blade material.

Testing Procedure and Results

Surface examination

Visual examination of the fracture surface indicated that the fracture originated at the bead-to-blade surface contact point (Fig. 2).

Scanning Electron Microscopy/Fractography. Figure 3 shows a low-magnification micrograph taken by scanning electron microscopy (SEM) of the cracked region in one of the blades. The bead had a dendritic solidification structure. Furthermore, it is clear that the fatigue crack initiated at the point of contact of the bead with the blade surface. In one of the blades, the crack was

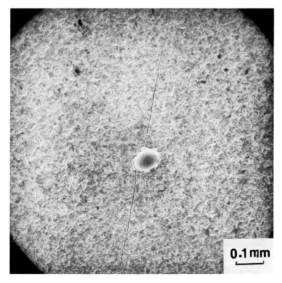

0.1 mm

Fig. 1 Fatigue crack associated with a bead on the blade surface

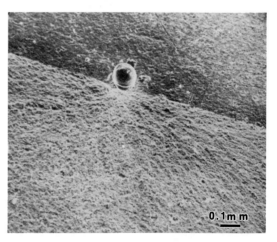

0.1 mm

Fig. 2 Fractograph showing a metallic bead at the origin of fracture

more than 19 mm (0.75 in.) long, was widest at the bead, and propagated on both sides of the bead. Other blades also contained similar beads on both the convex and concave sides of the aerofoil surface at different locations. Only those blades with beads in vulnerable locations developed cracks.

Chemical analysis/identification

Microprobe examination using the energy-dispersive analyzer of the SEM revealed that the bead had a composition similar to that of the parent blade material (Fig. 4).

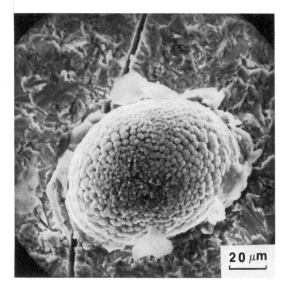

Fig. 3   Closeup view of a bead.

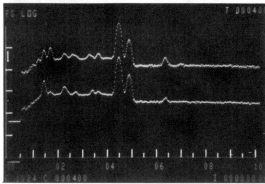

Fig. 4   Energy spectrum from the bead (top trace) and the blade material (bottom trace). Constituents present in both include titanium, aluminum, molybdenum, tin, silicon, iron, calcium, and sodium.

Discussion

The presence of beads at critical locations in the blades was directly responsible for the low fatigue life. These beads initiated fatigue cracks in the blade material at the point of contact. Elemental analysis further established that the beads were essentially of the same composition as that of the blade material. From the surface features of these beads, it is evident that they were in the fused condition at the time of impact on the blade. These particles had sufficient velocity at the time of impact to become welded to the blade surface.

To identify the particular processing operation during which particles of this nature could form, the operations preceding the vibratory fatigue testing were studied. It was learned that the targeting bosses of the forged blades were removed in the final stage by a grinding operation to facilitate mounting the blades for fatigue testing. The possibility of ground particles from the targeting bosses being thrown onto the blade surface, while still in the molten or semisolid condition, existed if grinding was severe and coolant was insufficient. This possibility was confirmed by simulation tests. Impingement of the hot particles formed during the grinding operation onto the cold surface of the blades had a localized embrittling effect, leading to initiation of fatigue cracking.

Conclusion and Recommendations

Fatigue crack initiation was associated with molten particles becoming lodged on the blade surface during grinding of the targeting bosses. This caused localized embrittling, because titanium is susceptible to oxygen absorption at high temperatures. Extreme care must be taken during grinding to prevent the molten grinding particles from falling onto the blade surface. Proper orientation of the blade during grinding and use of a suitable suction device would have prevented the particles from striking the blade surface.

Fatigue Fracture of a Fan Blade

Robert A. McCoy, Materials Engineering Department, Youngstown State University, Youngstown, Ohio

A blade from the engine cooling fan of a pickup truck fractured unexpectedly. The blade was made from type 301 stainless steel in the extra full hard tempered condition with a hardness of 47 HRC. Failure analysis indicated that the blade fractured in three modes: crack initiation, fatigue crack propagation, and final rapid fracture in a ductile manner. The fatigue crack originated near a rivet hole.

Key Words	Austenitic stainless steels	Fans	Trucks
	Automotive components	Blades	Engine components
	Fatigue failure		
Alloys	Austenitic stainless steel—301		

Background

A blade from the engine cooling fan of a pickup truck fractured unexpectedly.

Circumstances leading to failure

The fan was original. The vehicle had been in use several years before the fracture occurred.

Pertinent specifications

The blade material was specified as type 301 stainless steel, extra full hard temper. This temper is achieved by cold rolling to about a 60% reduction of thickness, resulting in a tensile strength that exceeds 1310 MPa (190 ksi). Because the austenitic structure of type 301 stainless steel is meta-

stable, this severe cold working results in the formation of a significant amount of strain-induced martensite. The presence of this martensite is confirmed by the fact that the blades are moderately magnetic. The combination of the very high degree of cold working and the formation of significant amounts of martensite greatly reduces the ductility of the steel and increases its notch sensitivity.

After the stainless steel strip is rolled to extra full hard temper, the blades are blanked and the rivet holes are pierced. The blades are then lightly rolled to impart a slight curvature. Finally, the blades are stress relieved.

Visual Examination of General Physical Features

Examination of the fan (Fig. 1) revealed that it was dirty from use. However, aside from the fractured blade, it was in good condition, with no signs of corrosion, dents, or other plastic deformation.

Testing Procedure and Results

Surface examination

Macrofractography. In order to examine the fracture surfaces at low magnification, the four rivets attaching the blade and the backer plate to the spider arm were removed. Figure 2 shows the fractured blade still attached to the backer plate by three clips.

Figure 3 shows the smaller portion of the fractured blade after separation from the backer plate. This fractured piece was ultrasonically cleaned using acetone and then examined at low magnifications using a stereomicroscope. The fatigue crack surface, which ran a length of 64 mm (2.5 in.) (Fig. 3, right), appeared very flat and normal to the plane of the blade. No beach marks or other indications of fatigue were visible. The final rapid fracture region (Fig. 3, left) appeared to follow an irregular path and was sheared at 45° to the plane of the blade.

Scanning Electron Microscopy/Fractography. Examination by scanning electron microscopy (SEM) revealed fatigue striations on the flat portion of the fracture surface (Fig. 4 and 5). From the relative curvature of these striations, it was determined that the fatigue crack originated next to one of the rivet holes (Fig. 3). The distance from the rivet hole to the fatigue crack origin was

Fractured blade

Fig. 1 Engine cooling fan, showing 0.15x location of fractured blade.

about 2 mm (0.08 in.). The 45° sheared fracture surface was dimpled (Fig. 6), indicating a ductile fracture.

Metallography

Microstructural analysis was difficult because of the highly cold worked condition of the blade. Basically, the microstructure appeared similar to that shown in Fig. 7, with highly elon-

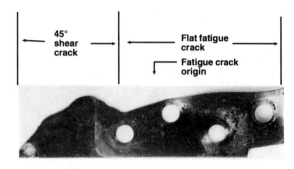

Fig. 2 Fractured blade still attached to backer plate (lower section) by three clips. 0.43x

Fig. 3 Smaller portion of fractured blade, showing fatigue crack origin near a rivet hole. 0.7×

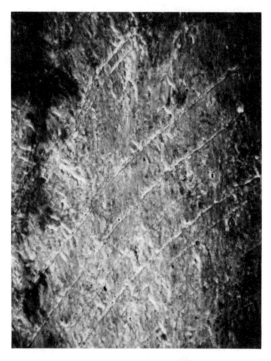

Fig. 4 Fatigue striations on fractured blade surface to left of rivet hole shown in Fig. 3. 800×

Fig. 5 Fatigue striations on fractured blade surface to the right of rivet hole shown in Fig. 3. 320×

gated and distorted grains and dark streaks, indicating strain-induced martensite.

Mechanical properties

Hardness. The approximate hardness of the blade was measured as 47 HRC. This relatively high hardness for a stainless steel suggests that its ductility was relatively low and that it was relatively notch sensitive in fatigue loading.

Discussion SEM fractography clearly showed that a fatigue crack initiated near a rivet hole and propagated in both directions for a length of 64 mm (2.5 in.). The exact crack origin area could not be identified. No surface defects were observed that could have triggered the initiation of the fatigue crack. Because this steel was so heavily cold worked, microstructural evaluation proved worthless. The development of the slow crack growth portion of the failure may have required thousands of stress cycles. Finally, the remaining cross section of the blade was no longer able to support the applied load, and rapid, ductile fracture resulted. Because the blade was only about 0.5 mm (0.02 in.) thick, ductile fracture occurred as a shear crack 45° to the blade surface.

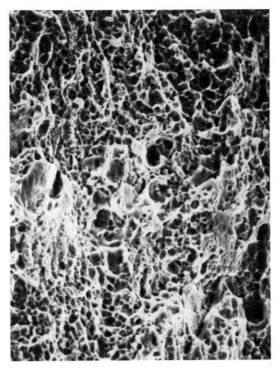

Fig. 6 Dimpled fracture surface on 45° shear crack shown in Fig. 3. 0.7×

Fig. 7 Type 301 stainless steel sheet, cold rolled to 40% reduction (full hard), showing almost complete transformation to martensite in severely deformed austenite grains. 250×. Source: Ref 1

Conclusion and Recommendations

Most probable cause

The blade fractured in three modes: crack initiation, fatigue crack propagation, and final rapid fracture. The crack initiated near a rivet hole, but it is not clear whether the hole, acting as a stress raiser, had a role in initiating the crack. The relatively high hardness of the steel may have contributed to its notch sensitivity. Once initiated, the fatigue crack propagated in both directions, producing a flat fracture normal to the plane of the blade. Finally, the fatigue crack so weakened the blade that its remaining cross section was overloaded and failed rapidly in a ductile manner, producing a 45° shear fracture.

Remedial action

Because this blade failure appeared to be an isolated case, no specific remedial action was taken.

Reference

1. *Metals Handbook*, 9th ed., Vol 9, *Metallography and Microstructures*, American Society for Metals, 1985, p 287

Fatigue Fracture of a Precipitation-Hardened Stainless Steel Actuator Rod End

Eli Levy, de Havilland Inc., Downsview, Ontario, Canada

A 17-4PH steering actuator rod end body broke during normal take-off. Results of failure analysis revealed that the wall thickness of the race was much below the design limits, thus causing the race to rest on the body's swaged edges rather than on the load carrying centerline of the body. This assembly condition generated abnormal high loads on the swaged edges, ultimately resulting in fatigue failure. To prevent a recurrence of similar failure in the future, the dimensions of the race in the spherical bearing were changed, no further failure occurred.

Key Words
Precipitation-hardening steels
Fatigue failure

Spherical bearings
Bearing races

Actuators

Alloys
Precipitation-hardening steel—17-4PH

Background

A 17-4 PH steering actuator rod end body broke during normal take-off. A rod end bearing is a three-part assembly of rod end body, race and ball. The assembly process (coining), shown in Fig. 1, is a cold-working procedure of swaging the faces of the body around the race. The applied service load in the rod end bearing is predominantly in the radial direction (see Fig. 2 and 3).

Pertinent specifications

The rod end body was manufactured of 17-4PH steel hardened to 36 to 38 HRc, whereas the race was also of 17-4PH hardened to 26 to 35 HRc. The ball was constructed of 440C steel hardened to 55 to 62 HRc. The liner was a Teflon impregnated sintered bronze matrix.

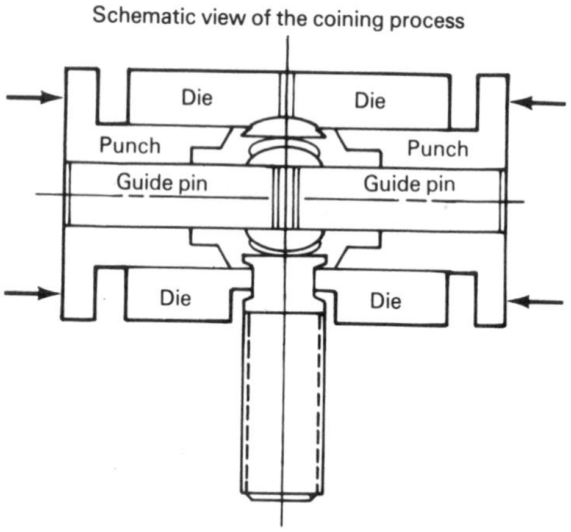

Schematic view of the coining process

Fig. 1 Schematic of the coining process

Visual Examination of General Physical Features

Rod and Body. The fracture surface was found to contain beach marks originating at the tips of the swaged edges (see Fig. 4). These marks are essentially the path along which the crack travelled. Close examination of the swaged edges disclosed circumferential secondary cracks in the inner face contour (see Fig. 4).

Race. The lined race displayed circumferential cracks along the outside periphery at its centerline. The race was then cut to remove the liner. Further examination disclosed that the race wall was only .008 in. (.20 mm) thick. The wall thickness of a race from a similar rod end bearing was .0285 in.(.72 mm, see Fig. 3). As illustrated in Fig. 3, this body/race internal setup produced abnormal high loading on the swaged edges.

Fig. 2 Broken steering actuator rod end. No environmental damage was apparent on the rod-end assembly. 0.9×

Testing Procedures and Results

Surface examination

Scanning Electron Microscopy/Fractography. The fracture surface was examined with a scanning electron microscope (SEM). The smooth "oyster shell" areas (Fig. 5a) on the fracture surface displayed faint fatigue striations throughout, some of which are shown in Fig. 5(b). The surface was also found to contain secondary cracks between some of the nonmetallic inclusions (Fig. 5c). The size and frequency of these inclusions appeared to be excessive.

Metallography

The body was evaluated for material cleanliness. The size and frequency of microinclusions in the body seemed to be ten fold compared with another rod end body.

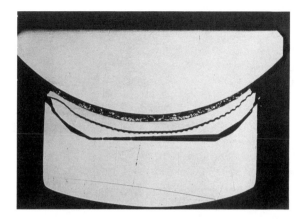

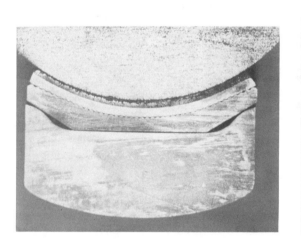

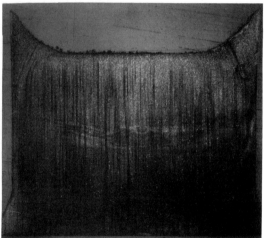

Fig. 3 (a) Macrograph of the sectioned failed rod-end assembly. The internal setup clearly demonstrates the abnormal distribution of loads. (b) Macrograph of a sectioned new rod-end assembly. The internal setup in this system minimizes the side loads encountered in (a). (c) Macrograph of the sectioned failed rod-end body. The cracks in the swaged ring should be attributed to both the high inclusion content in the material and the abnormal loads shown in (a).

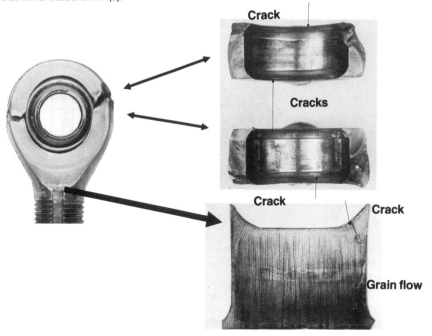

Fig. 4 Closeup view of the rod-end body. Note the fatigue marks on the fracture surfaces (top right) and the circumferential cracks in the swaged material (top and bottom).

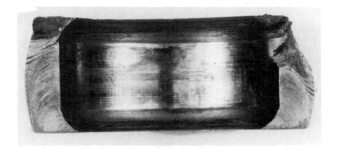

 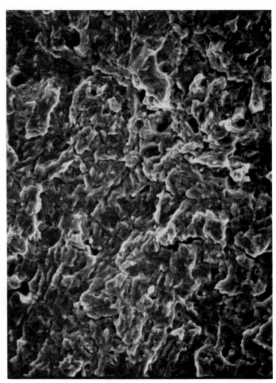

Fig. 5 (a) Macrograph of the fracture surfaces of the rod-end body shown in Fig. 4. The smooth "oystershell" character is typical of fatigue cracks. As indicated, the failure initiated at the knife-edge swaged ring and propagated inward (small arrows). (b) SEM fractograph of a field on the fracture surface (see a), showing evidence of faint fatigue striations and secondary hairline cracks. The fractograph has been rotated 92.5° relative to (a). 9400×. (c) SEM fractograph of a typical field on the fracture surface. The fractograph was overexposed to highlight the secondary cracks/inclusions network. 1128×

Conclusion and Recommendations

The steering actuator rod end body suffered fatigue failure. The wall thickness of the race in the spherical bearing was only 0.20 mm (0.008 in.) instead of 0.72 mm (0.0285). This discrepency caused the race to rest on the load bearing centerline in the body. This unusual assembly condition gener-

ated high abnormal service loads on the swaged edges ultimately resulting in fatigue failure.

To prevent a recurrence of similar failure in the future, the dimensions of the race in the spherical bearing were changed. No further failure occurred.

Metallurgical Analysis of Steam Turbine Rotor Disc

Leonard J. Hodas, Radian Corporation, Austin, Texas

Numerous cracks observed on the surface of a forged A470 Class 4 alloy steel steam turbine rotor disk from an air compressor in a nitric acid plant were found to be the result of caustic-induced stress-corrosion cracking (SCC). No material defects or anomalies were observed in the disc sample which could have contributed to crack initiation or propagation or secondary crack propagation. Chlorides detected in the fracture surface deposits were likely the primary cause for the pitting observed on the disc surfaces and within the turbine blade attachment area. It was recommended that the potential for water carryover or feedwater induction into the turbine be addressed via an engineering evaluation of the plant's water treatment procedures, steam separation equipment, and start-up procedures.

Key Words

Nickel chromium molybdenum steels, Corrosion
Turbine blades, Corrosion
Steam turbines

Rotor discs, Corrosion
Chlorides
Water chemistry

Stress-corrosion cracking
Pitting (corrosion)
Feedwater

Alloys

Nickel chromium molybdenum steels—A470 Class 4

Background

Applications

The steam turbine drives an air compressor in a nitric acid plant.

Circumstances leading to failure

The turbine is started using saturated steam, but is normally operated using superheated steam.

Pertinent specifications

The steam turbine disc was fabricated from an A470 Class 4 alloy steel forging, which is a common steam turbine disc material.

Selection of specimens

Approximately one-third of the disc was available for examination and sectioning.

Visual Examination of General Physical Features

The steam turbine rotor disc segment, which is shown as received for analysis in Fig. 1, was visually examined. Multiple cracks were observed over much of the surface of the damaged disc.

Testing Procedure and Results

Nondestructive evaluation

Magnetic Particle. The disc surfaces were examined using the dry magnetic particle technique. Multiple cracks, oriented in the radial direction on the sides of the disc, were observed. Figure 2 shows a closeup view of this damage.

Surface xamination

Visual. The turbine disc surfaces were examined using a stereomicroscope at magnifications from 7 to 67.5×. Cracking was observed on the sides of the turbine disc.

Metallography

Microstructural Analysis. A section was removed from an area of the steam turbine disc segment which did not exhibit visible damage. This sample was mounted, ground, polished, and etched for metallographic examination and photographic documentation. The photomicrographs in Fig. 3 show the typical turbine disc microstructure at two different magnifications. The microstructure consists of tempered martensite and bainite.

Crack Origins/Paths. Samples representing the typical damage were sectioned from the turbine disc segment. These samples were mounted, ground, polished, and etched for metallographic examination and photographic documentation. The photomicrographs in Fig. 4 through 6 show typical examples of the appearance of the damage.

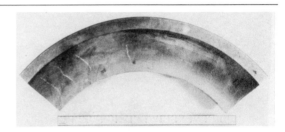

Fig. 1 The rotor disc segment is shown as received for analysis.

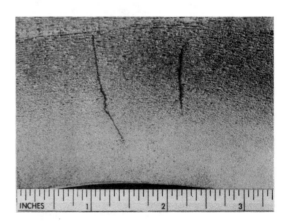

Fig. 2 This photomacrograph shows a close-up view of the typical MT indications observed on the disc side surfaces.

Multiple cracking was observed on the turbine disc sides and in the blade attachment areas. The photomicrographs in Fig. 4 show that some of the cracking has initiated from the bottoms of corrosion pits. The cracking morphology is mixed mode,

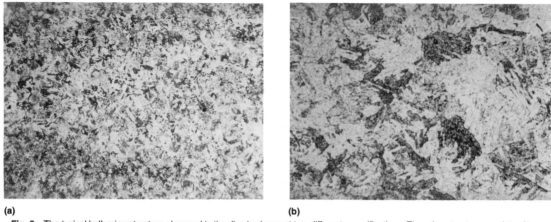

Fig. 3 The typical bulk microstructure observed in the disc is shown at two different magnifications. The microstructure consists of tempered martensite and bainite.

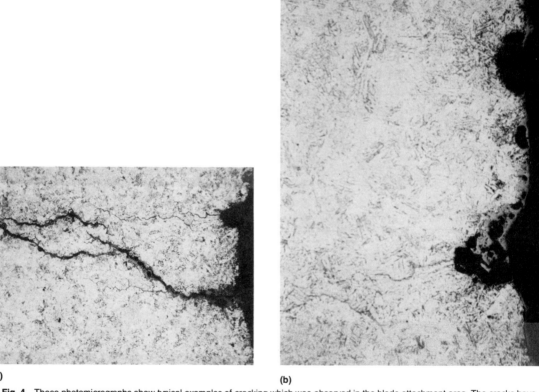

Fig. 4 These photomicrographs show typical examples of cracking which was observed in the blade attachment area. The cracks have propagated from pit bottoms.

that is, both intergranular (around grain) and transgranular (through grain). The upper photomicrograph in Fig. 6 shows two immediately adjacent cracks exhibiting different crack morphologies. The disc surfaces and primary crack paths show considerable evidence of on-going corrosion, in the form of pitting on the disc surfaces and build-up of corrosion deposits within the larger cracks.

Chemical analysis/identification

Material. A sample of the turbine disc segment was analyzed for chemical composition. The re-

sults of the compositional analysis of the steam turbine disc material are presented in Table 1. The compositional requirements for A470 Class 4 are listed in the table for comparison. The disc material meets the chemical requirements of the indicated specification.

Associated Environments. The deposits present at the termini of laboratory opened cracks were analyzed in situ using energy-dispersive X-ray spectroscopy (EDS). The typical results of an in situ EDS analysis of the crack surface deposits are presented in Fig. 7. The deposit contains major amounts of iron, silicon, calcium, potassium and

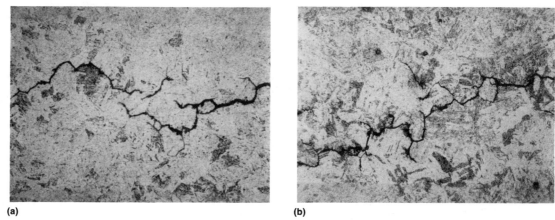

(a)   (b)

Fig. 5 These photomicrographs show typical examples of cracking which is primarily intergranular in nature.

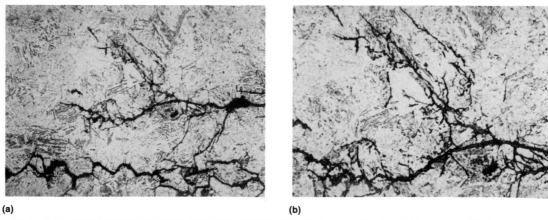

(a)   (b)

Fig. 6 The top photomicrograph shows two immediately adjacent cracks. The lower crack is primarily intergranular while the upper crack is primarily transgranular. The bottom photomicrograph shows a higher magnification view of the branched, transgranular cracking.

Table 1   Chemical composition of steam turbine rotor disc (Weight Percent)

Element	Disc	Chemical Requirements per ASTM A470 Class 4
Carbon	0.27	0.28 max
Manganese	0.51	0.20-0.60
Sulfur	0.011	0.018 max
Phosphorus	0.010	0.015 max
Silicon	0.28	0.15-0.30
Nickel	3.55	2.50 min
Chromium	0.41	0.75 max
Molybdenum	0.41	0.25 min
Vanadium	0.09	0.03 min

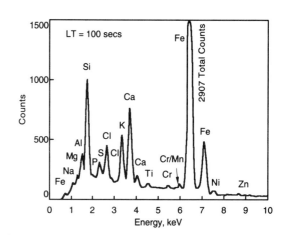

Fig. 7 The spectrum presents the results of the in-situ EDS analysis of the deposits present in the fracture terminus areas of the laboratory opened crack sample.

chlorine plus minor to trace amounts of aluminum, sulfur, sodium, magnesium, phosphorus, titanium, chromium, manganese, nickel, and zinc. The iron, nickel, manganese, and chromium are base metal constituents. The other deposit constituents likely come from feedwater carryover.

Mechanical properties

Hardness. The bulk metal hardness of the steam turbine disc was measured using the Rockwell test. The bulk metal hardness readings averaged Rockwell C (HRC) 20.2. This hardness value corresponds to a tensile strength of about 758 MPa (110 ksi) per the ASTM A370 approximate tensile strength conversion. This value meets the specification requirements.

Discussion	The multiple, radially-oriented mixed mode cracking observed in the disc is indicative of caustic-induced SCC. Both sodium and potassium were identified at the fracture terminus areas. The source of the caustic can be a result of carryover or the introduction of feedwater into the turbine or severe steam quality problems. Chlorine (as chlorides) was also detected in the fracture surface deposits. The chlorides are likely the primary cause for the pitting observed on the disc surfaces and within the turbine blade attachment area. Both pitting and SCC occur in wet environments.

The disc composition, microstructure, and hardness appeared to be normal. No material defects or anomalies were observed that could have contributed to the cracking.

Conclusions and Recommendations	**Most probable cause**

The cause of the numerous cracks observed was caustic-induced SCC.

Remedial action

An evaluation of several factors in the operation of the plant's steam system will be required to identify the source of the caustic. The operation of the steam drum/steam separators and the effectiveness of the plant water treatment program should be addressed. Additionally, the procedure of using saturated steam during start-up should be assessed with regard to the potential for inducting feedwater into the steam turbine.

Thrust Bearing Failure Leading to the Destruction of a Propeller Rotor

Tina L. Panontin, NASA Ames Research Center, Moffett Field, California

An accidental overspeed condition during wind tunnel testing resulted in the destruction of a propeller rotor. The occurrence was initially attributed to malfunction in the collective pitch control system. All fractured parts in the system were inspected. Highly suspect parts, including the pitch control thrust bearing set, head bolts, hub fork, and actuator rod end, were examined in more detail. The thrust bearing set (52100 steel) was identified as the probable source of the uncommanded pitch angle change. A complete failure analysis of the bearing indicated that failure was precipitated by excessive heating, causing cage disintegration, plastic flow of the races and balls, and eventual separation of inner and outer races. It was recommended that the bearing set be resized to accommodate the large thrust loads and that a thermocouple be added to monitor the condition of the bearing during testing.

Key Words			
	Bearing steels	Aircraft components	Thrust bearings
	Rotors	Propellers	Wind tunnels
Alloys	Bearing steel—52100		

Background

An accidental overspeed condition during wind tunnel testing resulted in the destruction of a propeller rotor.

Applications

The prop test rig (PTR) had been used for 20 years to test a number of different rotors. The configuration for the present test is shown in Fig. 1. The rotor was a right-hand rotation 7.5 m (25 ft) diam propeller with three blades. The purpose of the wind tunnel tests was to define the propulsive performance of the rotor.

A schematic of the collective pitch control system is shown in Fig. 2. The pitch of the rotor blades was changed by a stationary actuator, which moved the rotating control tube forward or backward through the thrust bearing set.

Circumstances leading to failure

The mishap occurred during a test run in which several data points had been taken and new conditions were being set up. Torque had just been reduced from 21,700 to 13,500 N·m (16,000 to 10,000 lbf · ft) so that the tunnel airspeed could be increased when the rotor torque suddenly decreased to negative 2700 N·m (2000 lbf · ft) in an uncommanded event. Because control of the rotor had been lost, the breaker to the motors driving the propeller was opened to protect the motors. However, instead of slowing, the rotor began to accelerate.

Unknown to test operators, the rotor blades had gone to a lower blade angle because of a thrust bearing failure, which allowed the tunnel airspeed to drive the rotor. Operating limits for monitored parameters on the rotor were soon reached, and the emergency stop of the tunnel was activated. By that time, the rotor had reached speeds beyond the structural limits of the blade retention straps and quickly self-destructed.

One blade tore loose and pierced the top of the test section. Because of the imbalance created, the remaining rotor and mast assembly failed and left the PTR. Most of the debris came to rest against a debris fence 90 m (300 ft) down the tunnel circuit. Damage to the wind tunnel was minimal; however, the rotor and hub were destroyed.

Fig. 1 PTR with three-bladed propeller rotor in wind tunnel test configuration

Pertinent specifications

The collective thrust bearing consisted of a pair of thin, 140 mm (5.5 in.) diam, angular contact bearings mounted back to back. The cage was made of a phenolic laminate, and the races and balls were made of 52100 heat-treated steel. Grease was used as a lubricant. The maximum permissible operation temperature for this bearing was 120 °C (250 °F). Selected to withstand thrust loads only, the bearing pair was supplied with a 0.03 mm (0.001 in.) preload. Preload was applied in the PTR assembly by squeezing the bearing pair with large retaining nuts torqued to approximately 400 N · m (300 lbf · ft). The bearing pair was installed in the rotor rig in 1969 and operated during all testing, with one documented removal and inspection in 1984.

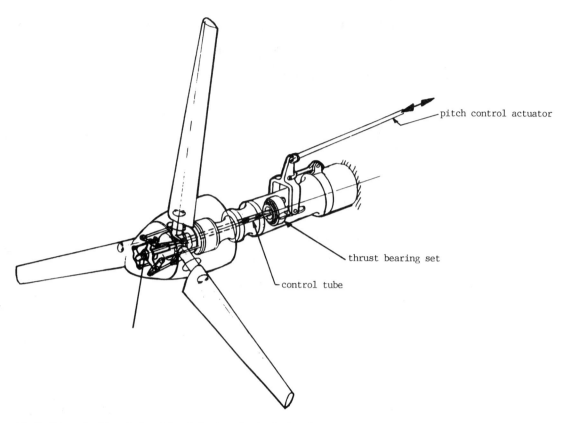

Fig. 2 Schematic of the collective pitch control system, showing the thrust bearing set, control tube, and actuator

Specimen selection

The overspeed condition reached during the mishap required an uncommanded pitch angle change, which indicated that a failure had occurred in the collective pitch control system. Accordingly, the failure analyses performed in support of the mishap investigation focused on this control system. All fractured parts in the system were inspected. Highly suspect parts, such as the pitch control thrust bearing set, head bolts, hub fork, and actuator rod end, were examined in more detail. All the parts except the thrust bearing set appeared to have failed because of overload following the rotor overspeed.

Additional evidence suggested that the thrust bearing set was the likely source of the uncommanded pitch angle change. First, it was the only one of six PTR assembly bearings that exhibited extreme heat damage and plastic deformation, even though all had experienced the overspeed. The other PTR assembly bearings showed limited evidence of overload (brinelling or fracture). Second, no documentation existed showing that the pitch control thrust bearing had been reanalyzed for the threefold increase in pitch control system loads anticipated for the present set of tests. Finally, a pitch control thrust bearing failure in the PTR assembly is a single-point failure, which allows the control tube to move with a fixed actuator position and still transfer load to the remainder of the assembly. Because of this evidence, the bearing set was subjected to a complete failure analysis.

Visual Examination of General Physical Features

To expose the thrust bearing, the rotor shaft was removed from the mast housing. It was noted that the outer race was displaced relative to the inner race by approximately 19 mm (0.75 in.). The bearing housings and the races were then cut using a high-speed grinder. As shown in Fig. 3(a) and (b), the bearings exhibited signs of extensive plastic deformation and of reaching extremely high temperatures. The bearing materials were discolored, the races had deep, axially oriented grooves, and some of the balls were flattened. Many balls were no longer in their races; others were "welded" to the races. The bearing cages had completely disintegrated.

Removal of the races from their housings showed that the races were no longer round and that the inner races were loose on their housing. As shown in Fig. 4, the inner race housing was also discolored from an apparent exposure to high temperature and was gouged and worn on the shoulder some distance from the bearing seat. Metal welded to the outer race containment nut is shown in Fig. 5.

(a)

(b)

Fig. 3 Pitch control thrust bearing races, showing signs of overheating and plastic deformation. (a) Inner races, as exposed. (b) Outer races, as exposed

Fig. 4 Pitch control thrust bearing inner race housing. Note the discoloration and wear on the shoulder. 1×

Fig. 5 Pitch control thrust bearing outer races in housing held by retaining nut. Note metal welded to the retaining nut and the ball welded to the outer race. 1.09×

Testing Procedure and Results

Surface examination

Scanning Electron Microscopy/Fractography. The bearing races were sectioned, then examined using a binocular microscope and a scanning electron microscope (SEM). Sliding wear and ball seizure (material transfer and deep wear grooves, as shown in Fig. 6) were apparent. Features indicative of rolling wear under high speed or high load (deformation tongues, surface cracking, etc.) were also evident (Fig. 7). Although these features do not suggest race fatigue as a cause, fatigue cracking and the

spalling that accompanies race fatigue may have been hidden by the extreme-temperature environment or the deformation experienced by the bearing pair.

Metallography

Microstructural Analysis. Sections of the inner and outer bearing races and balls were mounted and polished for metallographic examination. The ball paths of both races were severely damaged, as demonstrated by the smeared metal and altered microstructure shown in Fig. 8(a) and

Fig. 6 SEM micrograph of the bearing outer race, showing evidence of material transfer and deep grooves indicative of sliding wear and ball seizure. 16×

Fig. 7 SEM micrograph of bearing inner race. Note deformation tongues and surface cracking, features that suggest rolling wear caused by high speed or high load. 15.4×

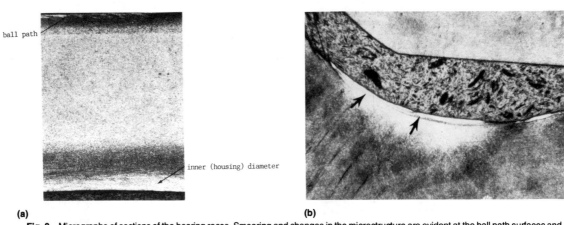

(a) **(b)**

Fig. 8 Micrographs of sections of the bearing races. Smearing and changes in the microstructure are evident at the ball path surfaces and at the inner diameter of the inner race. White layers are untempered martensite; black layers are tempered martensite. (a) Inner race. 11.5×. (b) Outer race. 65×

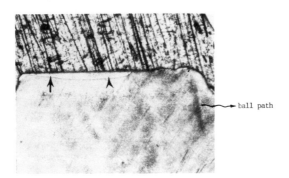

Fig. 9 Micrograph of a section of bearing outer race shoulder, showing microstructural changes similar to those described in Fig. 8. 52×

(b). The inner race also showed layers of altered microstructure on its inner diameter (mate with housing) (Fig. 8a). The outer race showed similar layers on its shoulder (Fig. 9). The balls were severely deformed and smeared, and exhibited microstructural changes similar to those of the races (Fig. 10). The white layers in Fig. 8 to 10 are untempered martensite; the black layers are tempered martensite. The core of the outer race exhibited microstructural features typical of 52100 steel (Fig. 11).

Mechanical properties

Hardness. Microhardness traverses were per-

Table 1   Results of hardness testing

| Location | Hardness, HRC | | |
	Inner race	Outer race	Ball
Ball path	50	>60	60
	49	33	
	38	49	
	40		
Core	33	52	48
	30	54	
	35	55	
	32		
Housing side	>60	>60 (shoulder)	

formed on cross sections of the inner and outer races and the balls. The results are listed in Table 1. At surfaces reaching extreme temperatures, hardness was measured at 60 HRC, which is con-

Fig. 10 Micrograph of a section of ball bearing. Severe deformation, smearing, and microstructural changes are evident. 16.75×

Fig. 11 Micrograph of a section of bearing outer race core. The microstructure is typical of heat-treated 52100 steel. 276 ×

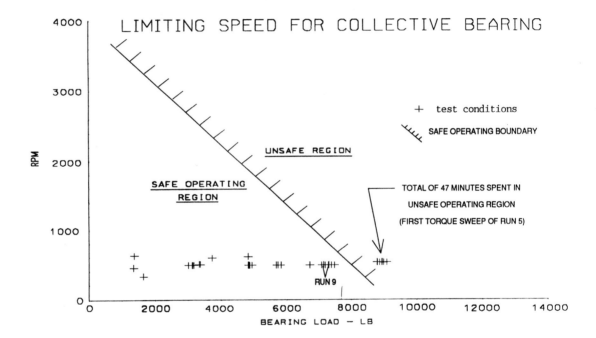

Fig. 12 Acceptable combinations of thrust load and rev/min for the pitch control thrust bearing set. Also plotted are actual load and rev/min conditions experienced by the bearing set during operation.

Table 2 Pitch control thrust bearing set: Fatigue load summary

Test dates	Thrust load range		Run time, 10^6 rev	Damage fraction	Cumulative damage fraction
	kg	lb			
1970-1975	0-1470	0-3240	0.205	0.0174	0.0174
1983	0-610	0-1350	0.270	0.0013	0.0187
1984	0-2080	0-4590	0.790	0.0916	0.1103
1985	610-1470	1350-3240	0.766	0.1081	0.2184
1987	0-2400	0-5300	0.718	0.3215	0.5399
1988	0-3380	0-7450	0.311	0.2580	0.7979
Current	0-4120	0-9080	0.331	0.4612	1.2591

sistent with the untempered martensitic structure observed. It should also be noted that the housing surface supporting the inner races also demonstrated a hardness of 60 HRC, suggesting that it too experienced extremely high temperatures. Inner race and ball core hardnesses were softer than those specified by the manufacturer, indicating

that the cores had been overtempered. This again is consistent with the appearance of the microstructure in these regions.

Stress analysis

Bearings can overheat for a number of reasons, such as lack of lubrication, excessive lubrication (leading to churning), contamination or debris, too little or too much preload, overspeed/overload, or race spalling by fatigue. Most of these causes were not supported by evidence in the rotor mishap. Based on maintenance records and the operational history of the bearing pair, the most likely cause of the overheat condition was race damage caused by fatigue or overspeed/overload.

Analytical. Fatigue damage in a bearing can cause the race to spall, which roughens the ball path and introduces debris into the bearing. This can eventually lead to smearing of the balls and to heat generation. An analysis was conducted to determine the cumulative fatigue damage that the bearing pair sustained over its operational life. The loads experienced by the bearing, the time under load, and the cumulative damage fraction are listed in Table 2. The analysis results show that, during the current phase of testing, the bearing exceeded the number of revolutions in which 10% of bearing races would begin to show fatigue damage.

The analysis results do not conclusively indicate race fatigue, because 90% of these bearings would still operate successfully with the load history of the pitch control thrust bearing set. Furthermore, no physical evidence suggesting race fatigue was observed.

Overload/overspeed conditions were also investigated for the pitch control thrust bearings. Exceeding the limiting speed for a given thrust load can cause the contact angle between the balls and the races to rise, increasing the local contact stresses and generating excessive heat. Acceptable combinations of rotational speed and thrust load for this type of bearing are shown in Fig. 12. Also plotted in Fig. 12 are the thrust load and rev/min conditions experienced by the bearing set during test runs. As shown in Fig. 12, the limiting speed of the bearings was exceeded for 47 min during run 5 and was again reached during 20 min in run 9 just prior to the mishap.

Discussion

Microstructural alterations to 52100 steel such as those observed in the pitch control thrust bearings can occur by exposure to high temperatures and/or plastic deformation. Temperatures above 760 °C (1400 °F) would have to be reached at the contact surfaces to cause the microstructure to transform to untempered martensite. Lower temperatures, such as those that would exist some distance from contact surfaces, can temper existing martensitic microstructures. According to published material data, such exposures would have to last approximately 10 to 15 min at temperatures around 540 °C (1000 °F) to soften the material to the hardness measured in the cores of the inner race and balls.

Consultations with bearing engineers confirmed that the type of damage observed on the pitch control thrust bearing set was consistent with overheating caused by operation at speeds in excess of the bearing limiting speed (see Fig. 12). They also stated that the damage probably could not have occurred during the short duration (1 to 2 s) of the rotor overspeed. All evidence gathered supported the hypothesis that damage from previous runs under overload/overspeed conditions led to rapid failure as the bearing again approached capacity just prior to the mishap. Therefore, it was concluded that the pitch control thrust bearing set failed from the overheating caused by repeated operation at or beyond bearing capacity.

Based on the evidence obtained from the bearing set and from the rotor data collected until the time of rotor separation, the sequence of events leading to the destruction of the rotor by the bearing failure can be hypothesized as follows. In the final test run, the bearing reached its limiting speed

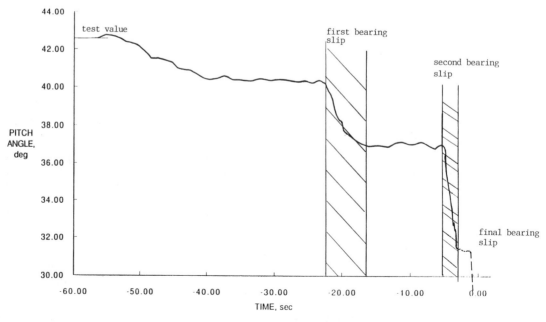

Fig. 13 Data trace showing blade pitch angle versus time until rotor separation

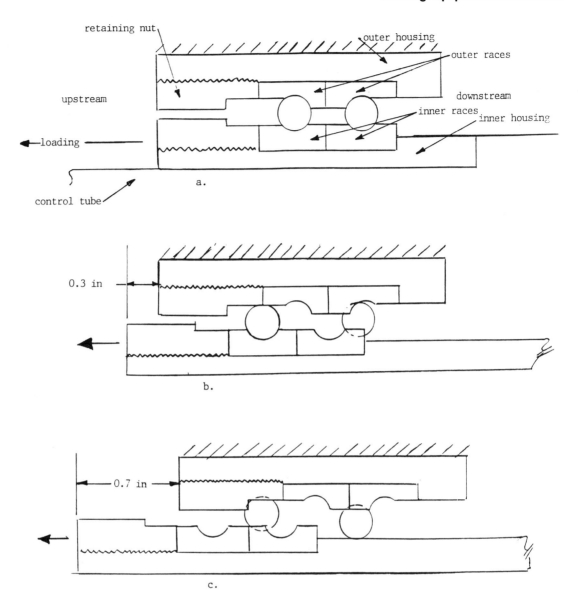

Fig. 14 Quasi-stable configurations of the pitch control thrust bearing during failure as postulated from deformation patterns observed on the races and the blade pitch data. (a) Correct bearing set position (*T* > T-23 s). (b) Bearing set position after first slip (T-19 to T-5 s). (c) Bearing set position after second slip (T-5 to T-1 s)

and operated there for about 20 min. Because of damage incurred by operation beyond its capacity during earlier testing, the bearing pair was incapable of sustaining the combination of high load and high speed. The bearing temperature probably increased rapidly to that causing cage failure (near 200 °C, or 400 °F) and race and ball material softening (near 315 °C, or 600 °F). The softened races deformed under load, and eventually the balls slipped across the races. This first slip in the pitch control system occurred about 23 s prior to the rotor separation (T-23 s), according to the data trace of blade pitch in Fig. 13. This slip was estimated from the data to be about 7.6 mm (0.3 in.) of pitch control tube travel.

After the slip, the bearing appeared to maintain a quasi-stable configuration through friction and race deformation. This configuration is depicted schematically in Fig. 14(b) and was deduced from the deformation patterns observed on the races in Fig. 15 and 16. Because of the position of

the balls after the slip, bearing rotation became increasingly difficult. It is believed that at this time the bearing seized and spun on the inner housing; this is supported by the evidence of extreme heat exposure observed on the inner diameter and housing of the inner race, as shown in Fig. 4 and 8(a). The heat generated by the seizure further elevated the temperature of the bearing race and balls, causing more softening and allowing additional race and ball deformation.

Because of this increasing deformation, the bearing balls slipped out of their quasi-stable position at T-5 s (see Fig. 13). This second slip is believed to have been about 10 mm (0.4 in.) of control tube travel. The resulting position of the balls is envisioned to be that shown in Fig. 14(c), based on deformation seen on the bearing races and housings (Fig. 17 to 19). This second quasi-stable configuration held for a few seconds, but as the load on the control tube increased with the blade pitch reductions, the inner and outer bearings separated

Fig. 15 Outer races in housing held by retaining nut. Note wear and gouges on upstream (left side) outer race shoulder and on retaining nut, suggesting that the bearing balls moved out of the ball path to travel on the shoulder and retaining nut, as hypothesized in Fig. 14(b). 1.09×

Fig. 17 Outer races in their housing held by retaining nut. The smeared material shown indicates that the bearing balls spun in the nut radius after the second slip. See Fig. 14(c). 1.11×

completely. This occurred at approximately T-1 s (Fig. 13). The control tube traveled to its full forward position (evidence that the tube impacted the tube stop was observed), and overspeed of the rotor occurred.

Fig. 16 Inner races in housing held by retaining nut. Deep formation wells on the downstream (right side) race shoulder support the hypothesized quasi-stable configuration shown in Fig. 14(b). 0.89×

Fig. 18 Inner races in their housing. Deep wells on the shoulders of the races support the hypothesized position of the bearings after the second slip shown in Fig. 14(c). 1.34×

Fig. 19 Inner races in their housing. Wear on the housing shoulder shows where the balls probably spun after the second bearing slip. See also Fig. 14(c). 1.19×

Conclusion and Recommendations

Most probable cause

The overspeed condition was caused by the progressive decrease in blade pitch allowed by the failure of the thrust bearing set in the pitch control system. The bearing set failed because of the generation of excessive heat by repeated operation at and beyond its load/speed capacity.

Remedial action

The bearing set for the PTR will be resized to withstand the actual, maximum load/speed combinations that it will experience during operation. Thermocouples will be installed to monitor the condition of the pitch control thrust bearing set (and other bearings designated as single-point failures). The fatigue life of the bearings will be closely watched, and the bearing set will be re-placed before its life is exceeded.

How failure could have been avoided

The bearing capacity should have been reexamined in light of the threefold increase in loads anticipated during the new phase of operation. Destruction of the rotor itself could have been prevented if the bearing had been recognized as a potential single-point failure and treated accordingly—for instance, by monitoring bearing temperature during testing.

Aluminum/Refrigerant Reaction Resulting in the Failure of a Centrifugal Compressor

Anthony C. Studer, Radian Corporation, Austin, Texas

An investigation of the impeller and deposit samples from a centrifugal compressor revealed that an aluminum / R-12 refrigerant reaction had occurred, causing extensive damage to the second-stage impeller and contaminating the internal compressor components. The spherical surface morphology of the impeller fragments suggested that the aluminum had melted and resolidified. The deposits were similar in composition and were identified by XRD as consisting primarily of aluminum trifluoride. In addition, EDS analysis detected major amounts of chlorine and iron. Results of a combustion test indicated that the compressor deposit was comprised of a 9.8 wt% carbon and that the condenser deposit contained 8.7 wt% carbon. It was concluded that the primary cause of failure was the rubbing of the impeller against the casting and that a self-sustaining Freon fire had occurred in the failed compressor.

Key Words		
Aluminum base alloys, Mechanical properties Reactions (chemical)	Impellers, Mechanical properties Melting	Refrigerants, Reactions (chemical) Rubbing

Background

Applications

A centrifugal refrigeration system depends upon centrifugal force to compress the refrigerant vapors. The impeller of the centrifugal compressor draws in vapor near the shaft and discharges at a high velocity at the outside edge of the impeller. The high velocity is converted into pressure. When the pressure drop is high, the compressor is built in stages.

Pertinent specifications

The compressor is part of a centrifugal refrigeration machine using R-12 refrigerant. The unit has a capacity of 319 Mg (320 tons) and was installed in 1978. The impellers are manufactured from an aluminum alloy.

Selection of specimens

The samples selected for analysis included metallic fragments and primarily black deposits removed from the compressor outlet and the condenser inlet. The metallic samples are shown as received for analysis in Fig. 1.

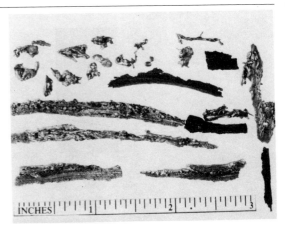

Fig. 1 Metallic pieces from the centrifugal compressor as received for analysis

Visual Examination

A section about 1.3 cm (0.5 in.) wide by 0.51 cm (0.2 in.) thick was missing from the second stage impeller. The impeller, the adjacent vane plate and the seal ring were coated by a heavy black deposit. All the steel piping, the evaporator suction pipe and the compressor volute exhibited more surface rust than is typical of hermetic systems. Regions of the casing exhibited evidence of the impeller rubbing.

Testing Procedure and Results

Surface examination

The metallic fragments were examined using a stereomicroscope and a scanning electron microscope (SEM). Figure 2 shows several of the aluminum pieces. The spherical surface morphology suggests that the aluminum has melted and resolidified. High magnification views of the spherical surface appearance are shown in Fig. 3.

Metallography

The pieces were mounted, ground, and polished using standard metallurgical techniques. The polished sections were examined on a metallurgical microscope for evaluation of microstructure. Figure 4 shows the typical microstructure observed in the aluminum pieces. The structure consists of dendrites of primary aluminum with inter-dendritic silicon particles and porosity. The porosity (shrinkage cavities) is most likely caused by solidification shrinkage.

Chemical analysis/identification

The deposit samples were analyzed for elemental constituents using energy dispersive X-ray spectroscopy (EDS). This analytical technique is used in conjunction with an SEM and can detect the presence of elements of atomic number 9 (the element fluorine) and greater. Light elements such as carbon, hydrogen, nitrogen and oxygen cannot be detected. For the elements aluminum and heavier, the EDS detection limit is in the range of 0.3 to 0.5 wt%. For the lighter elements, particularly so-

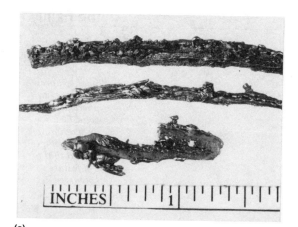

(a)　　　　　　　　　　　　　　　　　　　　　　　　**(b)**

Fig. 2　Close-up views of the aluminum pieces removed from the compressor. The beaded surface suggests a melting and resolidification of the surface layer.

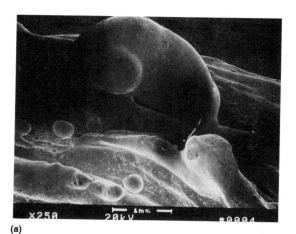
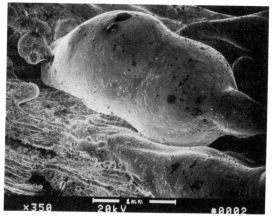

(a)　　　　　　　　　　　　　　　　　　　　　　　　**(b)**

Fig. 3　SEM photographs showing the appearance of the surface of the aluminum pieces removed from the compressor. (a) 15.75×. (b) 22.05×

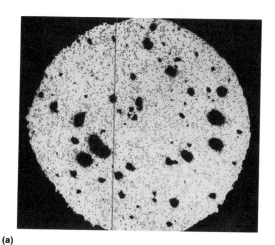
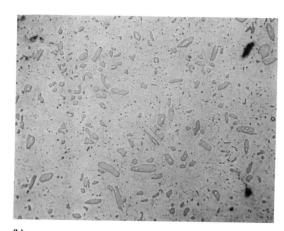

(a)　　　　　　　　　　　　　　　　　　　　　　　　**(b)**

Fig. 4　Representative microstructure observed for the aluminum pieces. The structure consists of dendrites of primary aluminum with interdendritic silicon particles. Note the presence of solidification porosity. Unetched. (a) 31.5×. (b) 252×

dium and fluorine, the detection limit is considerably higher (several percent for fluorine). A representative deposit sample was analyzed using wavelength dispersive spectroscopy (WDS) to verify the presence of fluorine. WDS can be used to detect elements of atomic number 5 (boron) and

greater and has a detection limit on the order of 100 ppm. Crystalline compounds in the deposit samples were identified using powder X-ray diffraction techniques (XRD). Since the black deposits likely contained amorphous carbon, which is not detectable by EDS and XRD analysis tech-

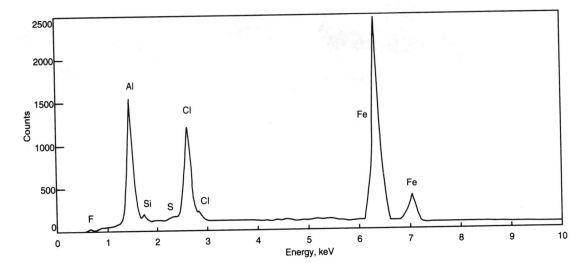

Fig. 5 EDS elemental analysis results for the compressor outlet deposit

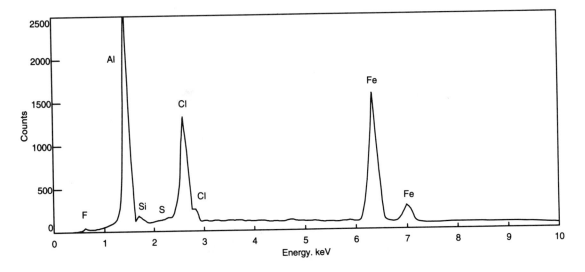

Fig. 6 EDS elemental analysis results for the condenser inlet deposit

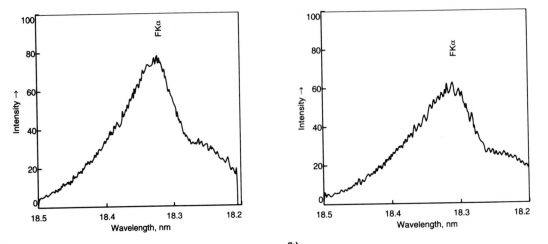

(a)　　　　　　　　　　　　　　　　　　　　　**(b)**

Fig. 7 Wavelength dispersive spectrographic analysis (WDS) over the 1.820 to 1.850 nm wavelength range of the typical deposit sample removed from the compressor outlet (a) and the condenser inlet (b). This analysis indicates that fluorine is present within both deposits.

niques, high-temperature combustion tests were performed on the deposits to determine the presence of carbon.

Figures 5 and 6 present the results of the EDS analyses for the compressor and condenser deposits, respectively. The deposits were similar in composition, and were identified by XRD as consisting primarily of aluminum trifluoride (AlF_3). In addition to the elements present in this compound, EDS analysis detected major amounts of chlorine and iron. EDS analysis also detected minor to trace amounts of silicon, sulfur and phosphorus. WDS analyses results, shown in Fig. 7, verify the presence of fluorine in the compressor and condenser deposits. The results of the combustion test indicated the compressor deposit is comprised of 9.8 wt% carbon and the condenser deposit contained 8.7 wt% carbon.

Discussion

The high-temperature reactions of halohydrocarbon refrigerants leading to the destruction of aluminum impellers and other aluminum parts in centrifugal compressors have been well documented (Ref 1). The reaction dissociates the refrigerant into carbon and hydrochloric and hydrofluoric acids, which consume aluminum. The reaction requires the presence of a film-free aluminum surface, which can result from either melting or mechanical abrasion, and sufficient heat to decompose the refrigerant. The heat required for the decomposition most likely resulted from the impeller rubbing against the casing. The reaction is highly exothermic and can be self sustaining if there is sufficient heat of reaction to melt the aluminum (approximately 650 °C, or 1200 °F). In addition to heat and light, the typical reaction products are aluminum fluorides, other metal halides, carbon and free acids. In a system circulating "pure" R-12 refrigerant, the minimum temperature at which an aluminum/refrigerant reaction could occur is about 593 °C (1100 °F), which is the decomposition temperature of the R-12. The presence of lubricating oil, possible degradation of the oil, and reactions between the oil and the refrigerant all can lower the temperature at which the aluminum-consuming reaction can occur. Temperatures as low as 149 °C (300 °F) have been reported (Ref 2).

The deposit samples contained aluminum trifluoride and carbon, which are the known products of a high-temperature aluminum/refrigerant reaction. The observation of the surface morphology of the aluminum pieces, some pieces being completely spherical, verify that a self-sustaining aluminum/refrigerant reaction had occurred. The heat generated and the free acids formed most likely caused damage to several system components.

Conclusions

The primary cause of failure was the rubbing of the impeller against the casing.

The analysis of the submitted samples verifies that a self-sustaining aluminum/refrigerant reaction (freon fire) had occurred in the failed compressor. The deposit samples consisted primarily of aluminum fluorides and carbon, known products of a high-temperature aluminum/refrigerant reaction. The highly exothermic reaction, which consumes aluminum, led to the destruction of the impeller. The aluminum fragments demonstrated melting as a result of the reaction.

References

1. B.J. Eiseman, Jr., "Reactions of Chorofluorohydrocarbons with Metals," *ASHRAE Journal*, Vol 5, No. 5, 1963, p. 63-70.

2. H.J. Borchardt, "New Findings Shed Light on Reactions of Fluorocarbons Refrigerants," *DuPont Innovation*, Vol 6, No. 2, 1975, pp 2-5.

Cracking of a Main Boiler Feed Pump Impeller

Leonard J. Hodas, Mechanical and Materials Engineering Department, Radian Corporation, Austin, Texas

An investigation was conducted to determine the cause of numerous cracks and other defects on the surface of a cast ASTM A743 grade CA-15 stainless steel main boiler feed pump impeller. The surface was examined using a stereomicroscope, and macrofractography was conducted on several cross sections removed from the impeller body. Areas that appeared to have the most severe surface damage were sectioned, fractured open, and examined using SEM. The chemistry of the impeller and an apparent repair weld were also analyzed. The examination indicated that the cracks were shrinkage voids from the original casting process. Surface repair welds had been used to fill in or cover over larger shrinkage cavities. It was recommended that more stringent visual and nondestructive examination criteria be established for the castings.

Key Words

Stainless steels	Castings	Impellers
Casting defects	Repair welds	Shrinkage
Rotary pumps	Electric power generation	

Alloys   Stainless steel—A743 grade CA-15

Background

Numerous cracks and other defects were apparent on the surface of a main boiler feed pump impeller at an electric utility.

Circumstances leading to failure

The impeller had been in operation for about 3 years when the defects were observed during a pump overhaul activity.

Pertinent specifications

The impeller was fabricated from an ASTM A743 grade CA-15 stainless steel casting.

Performance of other parts in same or similar service

The impeller examined replaced another impeller, which had been in service for about 14 years.

Specimen selection

The entire impeller element was available for examination and sectioning.

Visual Examination of General Physical Features

The damaged impeller, which is shown as received for analysis in Fig. 1, was visually examined. Apparent cracking was observed over much of its surface.

Testing Procedure and Results

Surface examination

Visual. The impeller surfaces were examined using a stereomicroscope at magnifications from 7 to 67.5×. Apparent cracking was observed on many areas of the impeller. Figures 2 and 3 show closeup views of this damage.

Macrofractography. Several cross sections were removed from the impeller body and ground, polished, and etched for macroscopic examination. Figure 4 shows typical examples of the damage observed.

Scanning Electron Microscopy/Fractography. Several areas that appeared to have the most severe surface damage were sectioned from the impeller body and fractured open in the laboratory. These surfaces were examined using a scanning electron microscope (SEM) and photographically documented (Fig. 5). The surfaces exhibited a dendritic structure, which indicated that the defects were formed during the original impeller casting process. Figure 5(a) shows a low-magnification view of the shrinkage void surface. The wavy nature of the casting dendrites and some corrosion scalloping and corrosion deposits are evident. Figure 5(b) shows an example of a dendritic casting slag found at the base of several void areas.

Weld Characteristics. Several areas on the

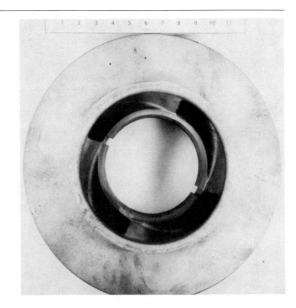

Fig. 1 As-received main boiler feed pump impeller

impeller surface appeared to exhibit surface repair welds. The arrows in Fig. 2(a) indicate impeller surface areas containing apparent repair welds.

Metallography

Microstructural Analysis. A section was removed from an area of the impeller that did not exhibit apparent damage. This sample was mounted,

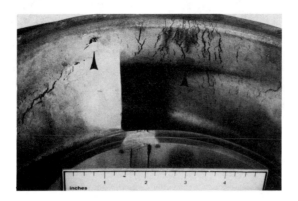

Fig. 3 Photomacrographs showing other examples of damage observed on the impeller

Fig. 2 Photomacrographs showing closeup views of typical cracking observed on the impeller surface. Arrows in (a) indicate areas containing repair welds.

ground, polished, and etched for metallographic examination and photographic documentation. Figures 6(a) and (b) show the typical impeller microstructure from an area without damage at two different magnifications. The microstructure consisted of tempered martensite, alloy carbides, and ferrite plus some nonmetallic inclusions.

Crack Origins/Paths. Samples representing typical damage and apparent weld-repaired areas were sectioned from the impeller and prepared for metallographic examination and photographic documentation. Figures 7 and 8 show typical defects, which appeared to be shrinkage voids or cavities. Figure 9 shows the appearance of a typical repair-welded area on the surface of the impeller. The weld covered an area containing an apparent casting void or shrinkage cavity.

Chemical analysis/identification

Material and Weld. A sample of the damaged impeller was analyzed to determine its chemical quantitative composition. An area that contained an apparent repair weld was also analyzed using energy-dispersive X-ray spectroscopy (EDS) under a standardless quantitative (SQ) analysis program. EDS provides semiquantitative elemental analysis of materials under SEM examination based on the characteristic energies of X-rays produced by the electron beam. EDS can normally detect elements with atomic numbers of 13 (aluminum) and above at concentrations as low as 0.3 to 0.5 wt%; fluorine, sodium, and magnesium are detectable at somewhat higher concentrations. As

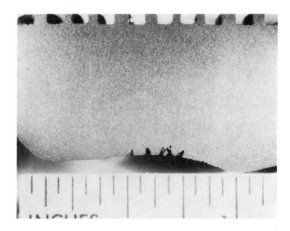

Fig. 4 Photomacrographs showing examples of typical damage observed in cross sections removed from the impeller body. The maximum defect depth is about 2.5 mm (0.10 in.).

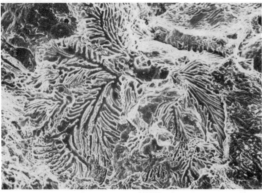

Fig. 5 SEM micrographs showing the appearance of the cross section of a shrinkage void that was fractured open in the laboratory. The void surface has a dendritic structure, which indicates that it was formed during the original impeller casting process. (a) ~18.6×. (b) ~310×

(a)

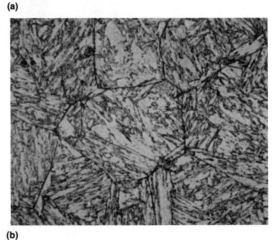

(b)

Fig. 6 Photomicrographs showing the typical microstructure of the impeller at an undamaged area. The microstructure consists of tempered martensite, alloy carbides, ferrite, and nonmetallic inclusions. Etched in Vilella's reagent. (a) 77×. (b) 616×

Table 1 Chemical composition of the damaged impeller

| | Composition, wt% | |
Element	Damaged impeller	Chemical requirements for A743 grade CA-15 stainless steel
Carbon	0.13	0.15 max
Manganese	0.91	1.00 max
Phosphorus	0.015	0.04 max
Sulfur	0.012	0.04 max
Silicon	0.72	1.50 max
Nickel	0.42	1.00 max
Chromium	11.58	11.5-14.0
Molybdenum	<0.05	0.50 max

(a)

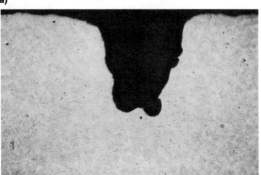

(b)

Fig. 7 Photomicrograph showing a typical shrinkage cavity in the damaged impeller. Etched in Vilella's reagent. 31×

performed in this examination, EDS cannot detect the elements hydrogen, lithium, beryllium, boron, carbon, nitrogen, or oxygen. The results of this type of analysis indicate the relative amounts of the elemental constituents.

The results of the quantitative compositional analysis of the impeller material are presented in Table 1. The compositional requirements for A743 grade CA-15 stainless steel (12% Cr) are also listed for comparison. The impeller material met the chemical requirements of the indicated specification. The results of the EDS SQ analysis of the repair weld material are presented in Table 2. The as-deposited weld filler metal composition is consistent with the nominal 12% Cr requirements.

Table 2 Composition of the repair weld, as determined by the EDS SQ method

Element	Weld	Nominal compositional requirements for a 12% Cr casting
Silicon	0.5	1.50 max
Manganese	0.9	1.00 max
Chromium	13.7	11.5–14.0
Iron	85.0	bal

Associated Environments. Two deposits present at the deepest portion of the damage penetrations were analyzed *in situ* using EDS (Fig. 10). Each deposit contained major amounts of iron, chromium, and silicon. One deposit had a major amount of calcium, while the other had a trace quantity. Minor to trace amounts of sulfur, chlorine, potassium, nickel, and copper were present in both deposits. The iron, chromium, and a small percentage of the silicon are base metal constituents. The bulk of the silicon could have been a sand residue from the impeller casting process. The other deposit constituents could have come from the feedwater. The chlorine (chloride) increases the pitting potential within the preexisting shrinkage cavities. The copper and nickel are possible indicators of corrosion in water circuit heat exchangers.

Mechanical properties

Hardness. The bulk metal hardness of the impeller was measured using the Rockwell test. All of the bulk metal hardness readings were between 25 and 26 HRC, with an average of 25.7 HRC. This hardness converts to a tensile strength of approximately 855 MPa (124 ksi).

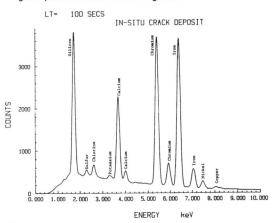
(a)
(b)

Fig. 8 Photomicrograph showing a typical example of some of the larger shrinkage cavities/casting voids observed in the damaged impeller. Etched in Vilella's reagent. 31×

(a)
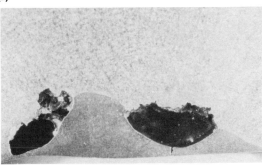
(b)

Fig. 9 Closeup views of a repair weld covering a casting void or shrinkage cavity. Etched in Vilella's reagent. (a) ~1.09×. (b) ~6.2×

(a)
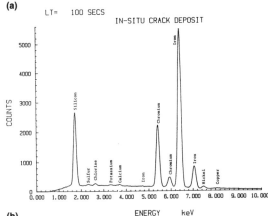
(b)

Fig. 10 Typical results of *in situ* EDS analyses at the base of shrinkage voids in the impeller

Conclusion and Recommendations

Most probable cause

Metallurgical examination of the damaged impeller indicated that the numerous visible cracks were shrinkage voids from the original casting process. The surface repair welds observed were used to fill in or cover over larger shrinkage cavities. The shrinkage voids examined were all relatively shallow and exhibited no evidence of enlargement or crack propagation.

Remedial action

More stringent criteria relative to shrinkage defects should be included in the purchase specifications for replacement impellers. These criteria could be implemented by requiring visual inspection of the castings and by adding supplementary nondestructive examination requirements to the purchase specifications, such as liquid penetrant inspection and/or magnetic particle inspection.

Fatigue Failure of an Aluminum Turbine Impeller

Anthony A. Tipton, Dresser-Rand Corporation, Wellsville, New York

An AMS 4126 (7075-T6) aluminum alloy impeller from a radial inflow turbine fractured during commissioning. Initial examination showed that two adjacent vanes had fractured through airfoils in the vicinity of the vane leading edges, and one vane fractured through an airfoil near the hub in the vicinity of the vane trailing edge. Some remaining vanes exhibited radial and transverse cracks in similar locations. Binocular and scanning electron microscope examinations showed that the cracks had been caused by high-cycle fatigue and had progressed from multiple origins on the vane surface. Structural analysis indicated that the fatigue loading probably had been caused by forced excitation, resulting in the impeller vibrating at its resonant frequency. It was recommended that the impeller design, control systems, and material of construction be changed.

Key Words			
	Airfoils	Aluminum-base alloys	Impellers
	Vanes	Fatigue failure	Turbines
Alloys	Aluminum-base alloy—7075		

Background

An AMS 4126 (7075-T6) aluminum alloy impeller from a radial inflow turbine fractured during commissioning.

Applications

The turbine is used to derive power at a natural gas letdown station. A finite-element model of the impeller is shown in Fig. 1. The turbine runs at 42,000 rev/min and drives a generator at 1500 rev/min through a reduction gear. The total output is 1600 kW.

Circumstances leading to failure

The turbine was tested at the manufacturing plant, at which time the rotor experienced an instability; the stability problem was corrected. The impeller, however, was not inspected. The turbine was then shipped to the site for commissioning tests at the rated speed. The machine ran two times for a total of approximately 35 min before it tripped. It could not be restarted because of a high level of vibration. Damage to the impeller, two fractured vanes, and numerous cracked vanes were discovered upon disassembly.

Specimen selection

The impeller was sectioned; one half was submitted for investigation, and the other half was retained by the manufacturer for independent analysis.

Fig. 1 Finite-element model showing geometry of turbine impeller

Fig. 2 AMS 4126 aluminum alloy impeller that fractured and cracked during commissioning. The impeller was sectioned before it was received for analysis.

Testing Procedure and Results

Surface examination

Macrofractography. Two adjacent vanes fractured radially/transversely through airfoils in the vicinity of vane leading edges, and one vane fractured transversely/radially through an airfoil approximately 19 mm (0.75 in.) from the hub in the vicinity of the vane trailing edge (Fig. 2 and 3); liberated portions of vanes were not recovered for investigation. Some of the remaining vanes exhibited radial and transverse cracks in similar locations (Fig. 4 and 5). The leading-edge portion of one vane was removed during machining to part the impeller before sending it to the laboratory. Visual examination of the impeller revealed no evidence of impact damage associated with any fracture or crack.

Scanning Electron Microscopy/Fractography. Binocular and scanning electron microscope (SEM) examination of fractures and fractures through cracks revealed a dark, discolored, predominantly crystallographic morphology with intermittent areas of intergranular propagation (Fig. 6 and 7); fractures exhibited progressive features. Initiation areas were along vane-to-hub blend radii approximately 1.6 mm (1/16 in.) from vane leading edges on the concave sides of vanes and on the convex airfoil sides of vanes in the vicin-

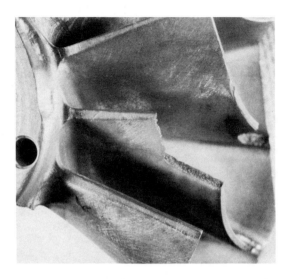

Fig. 3 Closeup view of impeller shown in Fig. 2, showing vane fracture in vicinity of the leading edge. 1.37×

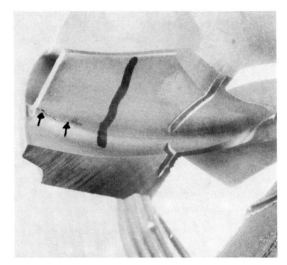

Fig. 4 Closeup view of a vane, showing location of radial crack (arrows). 1.5×

Fig. 6 Fracture through crack shown in Fig. 4, showing dark, discolored fracture path. Initiation appears to be from multiple sites along the vane-to-hub blend radius. 4×

Fig. 5 Closeup view of vane, showing transverse crack in vicinity of trailing edge. 1.14×

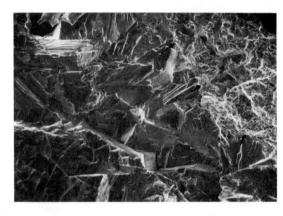

Fig. 7 Scanning electron micrograph of vane shown in Fig. 6, showing highly crystallographic morphology. Coarse-textured area is a fresh fracture created in the laboratory.

ity of trailing edges, depending on crack locations. "Mud cracking" was observed in selected areas of fracture surfaces. Energy-dispersive X-ray (EDX) spectrographic analysis of dark, discolored areas on fracture surfaces revealed primarily base metal oxides (Fig. 8).

Microstructural Analysis. Metallographic examination of sections through cracks revealed angular transgranular fracture paths (Fig. 9). The impeller microstructure exhibited copious amounts of insoluble phases along the grain boundaries (Fig. 10). EDX analyses identified the phases as being composed predominantly of magnesium and silicon (presumably Mg_2Si) and iron and aluminum (Fe_3Al). The structure otherwise appeared typical of properly solutioned and precipitation-hardened AMS 4126 aluminum-base alloy.

Chemical analysis/identification

Quantitative spectrographic analysis identified the impeller material as similar to AMS 4126 (Table 1).

Table 1 Results of chemical analysis

	Composition, %	
Element	**Impeller**	**AMS 4126 (7075) specification**
Silicon	0.09	0.40 (max)
Manganese	0.06	0.30 (max)
Chromium	0.22	0.18-0.35
Copper	1.68	1.2-2.0
Zinc	5.53	5.1-6.1
Magnesium	2.51	2.1-2.9
Titanium	0.01	0.20 (max)
Iron	0.16	0.50 (max)
Aluminum	bal	bal

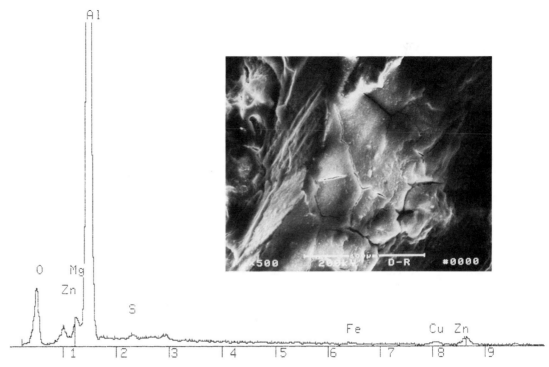

Fig. 8 Scanning electron micrograph of fracture surface, with EDX results. The mud-cracked area consists primarily of base metal oxides.

Fig. 9 Micrograph of section through a fracture surface, showing coarse, angular fracture path. 76×

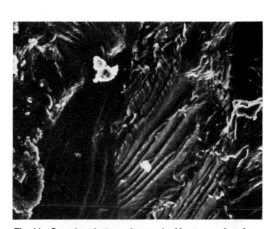

Fig. 10 Micrograph of section through impeller vane, showing extensive insoluble phases at grain boundaries. 76×

Mechanical properties

Ultimate tensile strength determined on a sub-size specimen excised from the impeller was 475 MPa (69 ksi); the specification requirement is 483 MPa (70 ksi) minimum. The small size of the test specimen precluded the use of extensometers; therefore, accurate and valid measurements of elongation and yield strength were not possible.

Room-temperature high-cycle bending fatigue testing was conducted on a specimen machined from the impeller at a maximum stress of 173 MPa (25 ksi) and R ratio of -1. The specimen failed after 9.5×10^6 cycles, which is typical of 7075-T6 aluminum alloy forgings. Binocular and SEM examination of the fatigue specimen fracture surface revealed relatively smooth-textured transgranular crack growth (Fig. 11); there was no evidence of crystallographic propagation.

Fig. 11 Scanning electron micrograph of fracture surface from fatigue specimen excised from impeller, showing typical high-cycle fatigue morphology for aluminum alloys tested in air. 266×

Discussion

The highly crystallographic fracture morphology of the impeller cracks indicates that the stress-intensity range was near threshold and that the value did not change significantly during fatigue crack propagation. Such crystallographic features have been observed on the fatigue surfaces of many metals and result from nonreversible slip on preferred planes. Slip bands of highly localized deformation are thus generated, resulting in extrusions and intrusions at the surface. These then penetrate along the length of the persistent slip bands to become sharp fissures. At some point the fissure breaks through the surface into the matrix, and a crack is developed.

The term "persistent slip band" refers to the fact that when the surface offsets are removed by polishing and the specimen cycled again, new cracks occur at the same sites, because the material within these persistent bands and below the surface is damaged. The depth below the surface to which the damage exists depends on the material characteristics and applied stress conditions.

Crystallographic crack propagation is referred to as stage I, indicating that the crack path is along slip planes and thus in the plane of maximum shear stress. Stage I growth occurs when the stress-intensity range is near threshold conditions. At higher stress-intensity ranges, stage II growth is observed, where the crack growth is normal to the applied stress.

As a crack grows as the result of a component vibrating at its natural frequency, the resonant frequency decreases, generally resulting in reduction of the vibratory amplitude. It is apparent that as the cracks propagated in the impeller, the reduction in the vibratory amplitude offsets the effect of the increase in crack lengths such that the stress-intensity range remained near threshold for most of the fatigue crack progression.

It is not possible to analytically calculate the vibratory amplitude of structures under resonant conditions without detailed knowledge of material damping, structural damping, and, most importantly, the appropriate magnifier for the vibratory response of the mode of interest. Industry has generally relied on strain gaging of components tested under specific operating conditions; however, such tests were outside the scope of this investigation. The fracture morphology observed on the fatigue specimen excised from the impeller indicates that it was subjected to a stress-intensity range above that of the cracks in the impeller. The presence of oxides on the fracture surface indicate that oxygen was present during the failure event. The end user indicated that oxygen-bearing compounds are added to the natural gas for various reasons, and initial test runs at the manufacturing facility were performed in air.

Conclusion and Recommendations

Most probable cause

Fracture of one vane and cracks in three additional vanes in the vicinity of vane leading edges were caused by high-cycle fatigue, which progressed from multiple origins along runout of the vane-to-hub blend radii on the concave airfoil side of the vanes. The cause of other vane fracture in the vicinity of the vane leading edge could not be determined because of extensive rub damage on the surface. Fracture of one vane and cracks in three additional vanes in the vicinity of vane trailing edges were due to high-cycle fatigue, which progressed from multiple origins on convex airfoil surfaces approximately 19 mm (0.75 in.) from the hub. The impeller material conformed to chemical composition requirements of AMS 4126. The ultimate tensile strength of the impeller material was slightly below specification requirements, but was not considered to have contributed to failure. No material or processing anomalies contributed to the failure. It could not be determined whether the cracks initiated at the manufacturing plant or at the operating site, because the impeller was subjected to sources of oxygen at both locations.

Remedial action

Recommended corrective actions include review and change of the impeller design, modification of the radial inflow turbine control system, and consideration of a different material of construction for the impeller.

Metallurgical Failure Analysis of Cracks in a Compressor Turbine Impeller

Edward V. Bravanec, Anderson & Associates, Inc., Houston, Texas

Cracking was discovered in an in-service, second-stage turbine impeller during a downtime inspection. The fabricated 4300 series low-alloy steel impeller was used in a compressor in an industrial petrochemical plant. It was also reported that a process upset had allowed a 10% NaOH solution to be ingested by the unit. Routine magnetic particle inspection revealed numerous cracks in the hub area and vane tips of the second-stage impeller. Additionally, the outside surface of the backing plate showed a cyclic pattern of cracks. An overview of a conventional, systematic metallurgical approach to failure analysis to confirm that the cracking was caused by a caustic stress-corrosion cracking mechanism is presented.

Key Words	Nickel-chromium-molybdenum steels	Compressors	Sodium hydroxide, environment
	Impellers	Chemical processing equipment	Compressor blades
	Stress-corrosion cracking	Caustic cracking	

Background

Severe cracking was discovered in a second-stage turbine impeller fabricated of low-alloy steel during a rebuild of a three-stage C-4501B compressor from a local petrochemical plant. The cracks were found by wet fluorescent magnetic particle inspection. The cracks were numerous and severe in the eye section. Other cracks were found on the surface of the backing plate near the bore, as well as at the toes of welds at some of the vanes. Cracks formed a stress pattern on the out-side surface of the backing plate from the bore to the vane tips. Although the cracks were numerous, they were very tight and difficult to detect. The second-stage impeller reaches a maximum temperature of 70 °C (160 °F) and turns at over 5000 rev/min. An operational upset was reported to have allowed a 10% sodium hydroxide (caustic) solution to enter the compressor impellers. The third-stage impeller had been scrapped. Effects were undertaken to determine the mode of cracking.

Visual Examination of General Physical Features

The second-stage impeller was examined and magnetic particle tested at a local machine shop. The eye, vane tip, and the stress pattern cracks on the outside surface of the backing plate were marked, and the impeller was shipped to a laboratory for analysis. A photograph of the crack pattern in the backing plate of the second-stage impeller, as well as an outline of three of the vanes, is shown in Fig. 1.

Testing Procedure and Results

Surface examination

Energy-Dispersive X-Ray Spectrometry. An area containing cracks was removed from the eye of the outside corner and notched from the inside, cooled in liquid nitrogen, and fractured open to enable analysis of a fracture surface. Standardless semi-quantitative scanning electron microscopy/energy-dispersive X-ray spectrometry (SEM/EDS) analysis revealed a high sodium content in the corrosion product (Fig. 2). The base metal elements were that of an AISI 4300 alloy steel (Fig. 3).

Element	wt%	at.%
Sodium	2.91	6.73
Aluminum	0.78	1.53
Sulfur	0.14	0.23
Chromium	1.05	1.07
Manganese	1.61	1.56
Iron	92.60	88.06
Nickel	0.91	0.82

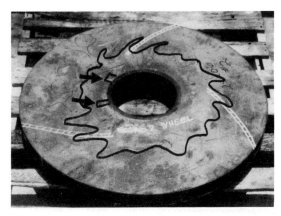

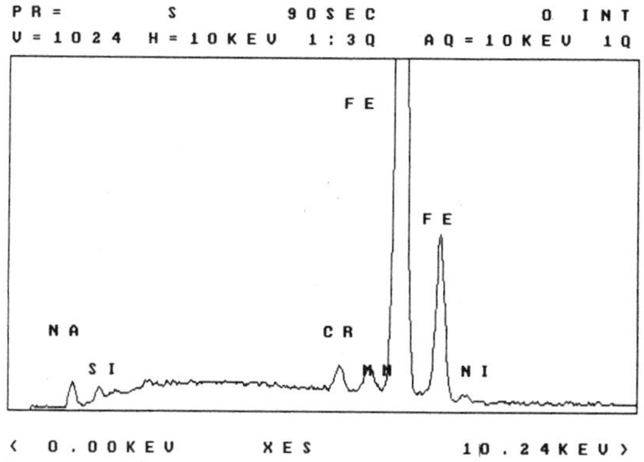

PR = S 90SEC O INT
V = 1024 H = 10KEU 1:30 AQ = 10KEU 10

FE

FE

NA CR

SI MN NI

< 0.00KEU XES 10.24KEU >

Fig. 1 Backing plate side of the second-stage impeller. Black paint outlines the stress pattern cracks. Hatched outline shows three of the vanes. Arrows indicate where metallographic samples were removed.

Fig. 2 Standardless semi-quantitative SEM/EDS element analysis data and graph of typical corrosion residue present in some of the cracks in the eye of the second-stage impeller. The high sodium content was believed to be the corrosion component for the stress-corrosion cracks.

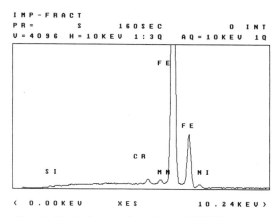

IMP-FRACT
PR =   S   160SEC   O INT
V=4096 H=10KEV 1:30   AQ=10KEV   10

F E

F E

C R

S I   M N   N I

< 0.00KEV   XES   10.24KEV >

Fig. 3 Standardless semi-quantitative SEM/EDS element analysis data and graph of a crack fracture surface. Base metal elements revealed that the impeller material was an AISI 4300 alloy steel.

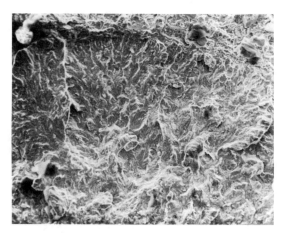

Fig. 4 SEM micrograph of a typical crack fracture surface form the second-stage impeller. Fracture mode is transgranular. 186×

Fig. 5 SEM fractograph of a typical area from the crack surface in Fig. 4, showing transgranular fracture mode and direction of propagation (arrow). 1860×

Scanning Electron Microscopy/Fractography. SEM examination of the laboratory-opened crack surface revealed a primary mode transgranular cracking, with some intergranular cracking identified as stress-corrosion cracking (SCC) (Fig. 4 and 5). The chemical component of the SCC was believed to be the sodium hydroxide from the operational upset.

Metallography

Samples were removed from the eye, the bore hub, and the backing plate area and were mounted and polished. Examination of unetched samples revealed numerous branched cracks perpendicular to the eye and extending to the eye surface (Fig. 6). Examination of etched samples revealed that the branched cracks were primarily transgranular (Fig. 7); some areas of the crack were intergranular. The cracks in the backing plate, hub, and the stress pattern areas of the backing plate were of the same mechanism as the hub cracks—that is, caustic stress-corrosion cracking (Fig. 8 and 9). The base material microstructure was composed of tempered martensite, with some retained austenite.

Fig. 6 Micrograph of an area containing longitudinal cracks from the eye section from the second-stage impeller. Cracks are numerous and very tight. Unetched. 81×

Element	wt%	at.%
Silicon	0.27	0.53
Chromium	1.10	1.18
Manganese	0.88	0.89
Iron	96.40	96.12
Nickel	1.35	1.28

Chemical analysis/identification

A sample was removed from the hub area of the impeller and subjected to chemical analysis by optical emission spectrometry (Table 1). Analysis revealed a 4300 series low-alloy steel.

Mechanical properties

Hardness. Rockwell B hardness indenta-

Fig. 7 Micrograph of cracks from the eye section shown in Fig. 6. The transgranular branched cracks were caused by a caustic SCC mechanism. 3% nital etch. 324×

Fig. 8 Micrograph of a typical crack from the stress pattern area in the backing plate of the second-stage impeller (see Fig. 1). Cracks are numerous and very tight. Unetched. 81×

Table 1 Results of chemical analysis

| | Composition, % | |
Element	Second-stage impeller sample	AISI E4325 specification requirement(a)
Carbon	0.22-0.27	0.22-0.28
Manganese	0.74	0.60-0.80
Phosphorus	0.009	0.025 (max)
Sulfur	0.007	0.025 (max)
Silicon	0.25	0.15-0.35
Nickel	1.82	1.65-2.00
Chromium	0.82	0.70-0.90
Molybdenum	0.20	0.20-0.30
Copper	0.10	NR
Aluminum	0.050	0.020 min for fine grain
Cobalt	0.030	NR
Iron	bal	bal

(a) NR, not required

tions were made in several areas of the second-stage impeller material; the average hardness was found to be 95 HRB.

Fig. 9 Micrograph of a crack from the stress pattern area shown in Fig. 8. The transgranular branched cracks were caused by a caustic SCC mechanism. 3% nital etch. 324×

Discussion   The results of the analyses, examinations, and tests revealed that the second-stage impeller contained numerous severe caustic stress-corrosion cracks. The caustic cracking was primarily transgranular, with some intergranular cracks. This is a non-standard caustic SCC mechanism, but the high stresses imposed on the impeller appear to have resulted in the transgranular crack propagation mode rather than the typical intergranular mode. The cracks on the outside of the backing plate followed a cyclic stress pattern. The backing plate cracks were very tight, but deep. Initiation and propagation of all cracks occurred in high-stress areas.

Conclusion and Recommendations

The decision to replace the second-stage impeller was sound in view of the severe caustic SCC, which was characterized by very tight cracks—some of them deep and all of them difficult to detect. It now becomes critical to evaluate the severity of the cracks in the first-stage impeller. All remaining cracks must be machined out. If they are not, they will propagate during operation and may result in a catastrophic failure. It is difficult to predict how rapidly the cracks will grow; however, this impeller material is tough and should resist sudden, brittle failure.

Failure Analysis of a Hydroturbine Shaft

Donald R. McCann and Michael V. Harry, A-C Siemens Power Corporation, West Allis, Wisconsin

A forged 4140 steel shaft that connected two runners in a hydroturbine failed catastrophically after approximately 5900 h of service. The runner and the mating section of the broken shaft were examined and tested by various methods. The results of the analyses indicated that the shaft failed by torsional fatigue starting at subsurface crack initiation sites. The forging contained regions of cracklike flaws associated with particles rich in chromium, manganese, and iron. Fracture features indicated that the fatigue cracks propagated under a relatively low stress.

Key Words	Chromium-molybdenum steels	Fatigue failure	Shafts (power)
	Hydroelectric generators	Forgings	Turbines
Alloys	Chromium-molybdenum steel—4140		

Background

A steel shaft that connected two runners in a hydroturbine failed catastrophically after approximately 5900 h of service over 27 months.

Applications

The runner shaft that failed was one of two shafts joined by a stub shaft and connecting four runners that provided the mechanical energy to drive a generator at 180 rev/min. Figure 1 shows a schematic of the hydroturbine unit.

Circumstances leading to failure

Review of photographs and on-site examination of the failed components disclosed that the first shaft failed initially adjacent to the No. 2 runner. The shaft then failed adjacent to the No. 1 runner by rapid brittle fracture caused by bending, probably as the shaft went through the casing. Subsequently, the second shaft failed adjacent to the No. 3 runner by torsional overload, probably

when the first shaft separated and jammed the No. 2 runner. The turbine was almost completely destroyed as the three runners, together with part of the first shaft, burst through the casings.

Pertinent specifications

The shaft was ordered as an SAE 4140 forging specified to be normalized, quenched and tempered, and stress relieved. No hardness or tensile requirements were specified.

Specimen selection

The No. 2 runner and the mating section of the broken shaft were chosen for failure analysis. The failed shaft was removed from the No. 2 runner by heating the runner hub and cooling the shaft while pressure was applied to the stub shaft end. The two as-received shaft pieces are shown in Fig. 2, with the shaft portion removed from the runner hub on the right.

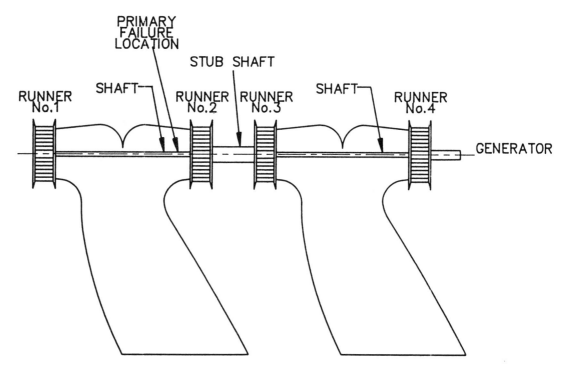

Fig. 1 Schematic of the hydroturbine plant

Visual Examination of General Physical Features

The fracture occurred at the transition between the 230 mm (9 in.) diam hub section and the 170 mm (6.5 in.) diam necked-down section of the shaft. Although the fracture surface on the necked-down section was completely destroyed by rubbing against the mating fracture surface, sufficient fracture features were present on the hub section that a failure analysis could be conducted.

Testing Procedure and Results

Nondestructive evaluation

Ultrasonic Examination. Both shaft pieces were ultrasonically inspected with a longitudinal-wave 2.25 MHz search unit from the diameters. The hub section was also inspected from the stub shaft end. Although the ultrasonic flaw detector was set to detect a 3 mm (0.1 in.) equivalent diameter flat-bottom hole, no flaws were detected in either piece.

Surface examination

Macrofractography. Figure 3 shows what remained of the fracture surface on the shaft hub section. Careful examination of the fracture surface revealed several crack initiation sites, all located below the shaft surface. Figure 3(a) shows four of the subsurface crack initiation sites (arrows), and Fig. 3(b) indicates that cracks propagated by torsional fatigue from these sites since the crack plane is about 45° to the shaft axis. The fracture surface was removed from the hub shaft section for microscopic examination by a saw cut through the cross section approximately 13 mm (½ in.) from the fracture surface. Another saw cut was made 25 mm (1 in.) from the first to obtain material for tensile testing.

Scanning Electron Microscopy/Fractography. The subsurface flaw shown in the middle of Fig. 3(a) (shaped like a half circle) was cut out for optical and electron microscope examinations. The flaw is shown in Fig. 4(a) after chemical cleaning to remove oil and rust. Ridges mark the transition between the subsurface flaw (bottom portion) and the fatigue crack (top portion). Enlargements on each side of the ridges are shown in Fig. 4(b) and (c). Higher-magnification views of the subsurface flaw near the ridges are shown in Fig. 4(d) and (e). Note the many pits and craters covering the surface.

Figure 5(a) shows the subsurface flaw features away from the ridges and toward the center of the half circle. The density of the pits is still high, and particles are present in the pits and craters (Fig. 5b). X-ray analysis revealed that the majority of the particles contained chromium, manganese, and iron and that some of the particles contained silicon. The chromium content of the particles was much higher than that of the surrounding metal. Higher-magnification views of the surface next to a crater and inside a crater are shown in Fig. 5(c) and (d). The texture of the surface was not typical of hydrogen embrittlement, as distinct cleavage river markings were not present. The texture of the crater was rough and contained smaller unknown particles, atypical of gas pockets or porosity. It appeared that this was an unusual forging defect.

Fig. 2 As-received pieces from the failed shaft. The shaft portion that was removed from the runner hub is at the right. 0.17×

(a)

(b)

Fig. 3 Fracture features. (a) Fracture surface of the hub piece, showing evidence of subsurface crack initiation sites (arrows). Most of the surface was damaged by pieces rubbing together. 0.37×. (b) Closeup view of subsurface flaws. Concentric markings around flaws indicate that crack propagation occurred by torsional fatigue. 0.63×

Metallography

Crack Origins/Paths. A cross section through the subsurface flaw adjacent to the right side of

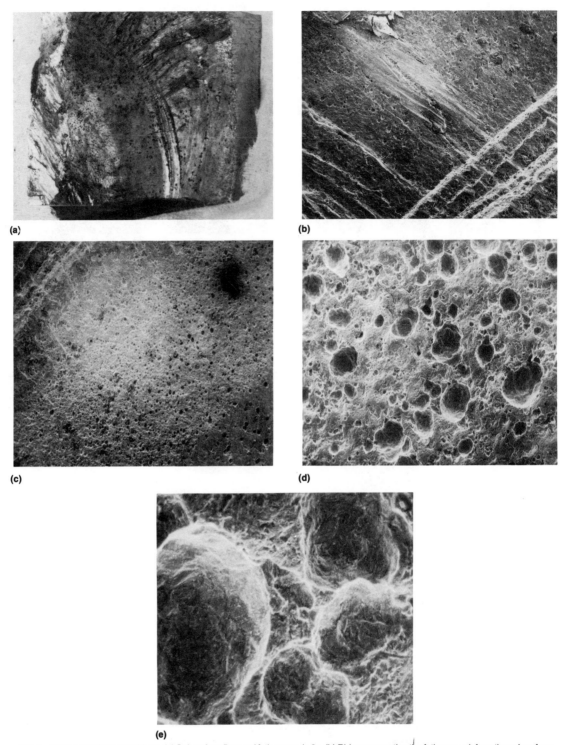

Fig. 4 Subsurface flaw near ridges. (a) Subsurface flaw and fatigue crack. 2×. (b) Ridges separating the fatigue crack from the subsurface flaw. 11.56×. (c) Fracture surface of the subsurface flaw near the ridges. Note the high density of pits and craters. 10.08×. (d) Higher-magnification view of fracture surface near ridge. 100.8×. (e) Higher-magnification view of fracture surface and craters. 504×

Fig. 4(a) is shown in Fig. 6(a). A portion of deformed metal was partially rolled into the flaw when the two surfaces rubbed together. Figure 6(b) shows that the subsurface flaw was intersected by many small, cracklike defects (arrows). Higher magnification of these defects (Fig. 6c and d) showed that they were filled with particles; X-ray analysis indi-

cated that the particles contained chromium, manganese, and iron.

Figure 6(e) shows additional subsurface flaws filled with similar chromium, manganese, and iron particles slightly below the point at which the fatigue crack initiated. Close examination of these flaws showed very tight cracks extending on both sides of the flaw.

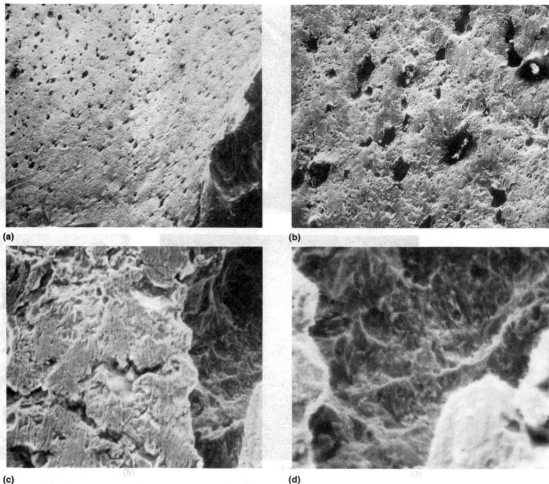

Fig. 5 Subsurface flaw away from ridges. (a) Fracture surface away from ridges and toward the center of the subsurface flaw in Fig. 4(a), showing many pits and craters. 12.58×. (b) Higher-magnification view of (a), showing pits and craters filled with particles containing chromium, manganese, and iron. 56.95×. (c) Higher-magnification view of fracture surface next to crater. Surface texture is not typical of hydrogen embrittlement. 578×. (d) Higher-magnification view of crater. Texture is very rough and not typical of gas porosity. Analysis of small particles within the crater was not possible. 1156×

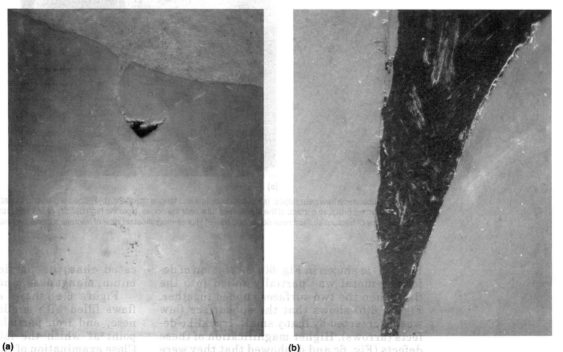

Fig. 6 Cross section through subsurface flaw. (a) Deformed metal has rolled against flaw. 2.88×. (b) Higher-magnification view of subsurface flaw, showing intersecting cracklike defects (arrows). 45×. **(continued)**

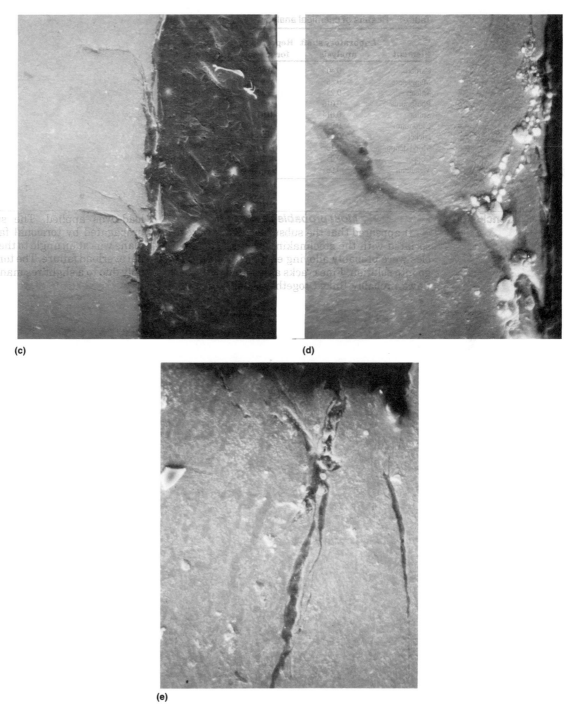

(c)

(d)

(e)

Fig. 6 Cross section through subsurface flaw. (c) Higher-magnification view of several intersecting flaws. 240×. (d) Intersecting flaws filled with particles rich in chromium, manganese, and iron. 950×. (e) Particle-filled subsurface flaws with fine cracks (arrows) in vicinity of fatigue crack initiation. 600×

Chemical analysis/identification

Results of the analysis of the chemical composition of No. 2 runner shaft are presented in Table 1.

Mechanical properties

The hardness of the shaft was determined to be 255 HB.

Tensile Properties. A tangential tensile specimen removed from a section 25 mm (1 in.) below the fracture surface was tested, with the following results: tensile strength, 928.8 MPa (134.7 ksi); 0.2% yield strength, 762.8 MPa (110.6 ksi); elongation, 12.5%; and reduction of area, 40%. These values are consistent with quenched and tempered material of the grade used in the section size utilized.

Table 1 Results of chemical analysis

Element	Laboratory shaft analysis	Reported by forge shop	SAE 4140 specification
Carbon	0.40	0.40	0.38-0.43
Manganese	0.86	0.84	0.75-1.00
Silicon	0.25	0.24	0.20-0.35
Phosphorus	0.016	0.012	0.035 (max)
Sulfur	0.014	0.011	0.040 (max)
Chromium	0.99	0.96	0.80-1.10
Nickel	0.04	0.05	...
Molybdenum	0.16	0.18	0.15-0.25
Copper	0.05	...	...
Vanadium	0.06	...	...

Conclusion

Most probable cause

It appeared that the subsurface flaws were associated with the steelmaking process; the particles were probably alloying elements that did not go into solution. Fine cracks associated with these flaws probably linked together when a torsional load was applied. The subsurface crack then propagated by torsional fatigue, since the crack plane was at an angle to the shaft axis, until the final overload failure. The torsional stress was probably due to a slight resonance condition.

Failure of a Crane Long-Travel Worm Drive Shaft

K.G. Wellington, New Zealand Aluminium Smelters, Invercargill, New Zealand

A crane long-travel worm drive shaft was found to be chipped during unpacking after delivery. Chemical analysis showed that the steel (EN36A with a case depth of 1 mm, or 0.04 inch did not meet specifications. Magnetic particle inspection revealed a crack on the side of the shaft opposite the chip. Metallographic examination indicated that the case depth was approximately 2 mm (0.08 in.) and that a repair weld of an earlier chip had been made in the cracked area. The chipping was attributed to excessive case depth and rough handling. It was recommended that the shaft be returned to the manufacturer and a replacement requested.

Key Words

Nickel-chromium steels
Cranes

Case depth
Chipping

Shafts (power)

Alloys

Nickel-chromium steel—EN36A

Background

A crane long-travel worm drive shaft was found to be chipped during unpacking after delivery.

Visual Examination of General Physical Features

On arrival, the keyway was found to be chipped in two locations (Fig. 1). Closer inspection revealed lighter colored areas on the side opposite the chipped area. Examination using a magnetic yoke and magnetic ink revealed a crack and outlined an area of magnetic permeability different from the base metal (Fig. 2).

Testing Procedure and Results

Metallography

Sections were cut from the keyway at the chipped area and the light-colored area. The microstructure showed a case-hardened structure at the surface and a tempered martensite matrix (Fig. 3 and 4). The case depth, specified as 1 mm (0.04 in.), was closer to 2 mm (0.08 in.).

The light-colored area did not etch with nital. Etching with Kalling's reagent showed the structure to be dendritic, like a cast or weld (Fig. 5). There also appeared to be a small crack under the light-colored metal (Fig. 6).

Chemical analysis/identification

Results of the chemical analysis of the shaft are shown in Table 1, along with chemical requirements for EN36A.

Pertinent specifications

The steel for the drive shaft was specified as EN36A, with a surface hardness at each end of 62 to 64 HRC and a 1 mm (0.04 in.) case-hardened depth.

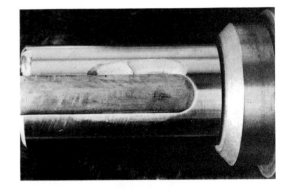

Fig. 1 Keyway in drive shaft, as received

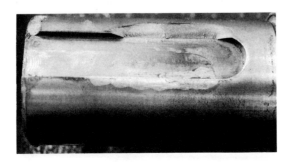

Fig. 2 Keyway after magnetic examination. A crack is visible as a line in the bottom of the keyway. Light-colored metal buildup is outlined by magnetic ink.

Fig. 3 Carburized zone. 2% nital etch. 47.5×

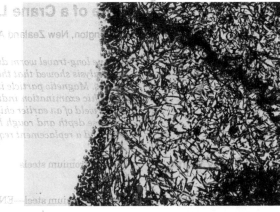

Fig. 4 Martensitic matrix structure of shaft. 2% nital etch. 47.5×

Fig. 5 Dendritic structure of weld buildup. Etched with Kalling's reagent. 95×

Table 1 Chemical analysis results

Element	Composition(a), % Shaft	EN36A
Carbon	0.16	0.10-0.16
Silicon	0.23	NR
Manganese	0.55	0.35-0.60
Nickel	0.75	3.0-3.75
Chromium	0.91	0.70-1.00
Molybdenum	0.01	NR

(a) NR, no requirement

Mechanical properties

Hardness was specified as 745 to 800 HV30. Testing produced the following values: surface near chipped area, 700 HV30; surface of light-colored metal, 246 HV30; 3 to 4 mm (0.12 to 0.16 in.) below surface, 317 HV30.

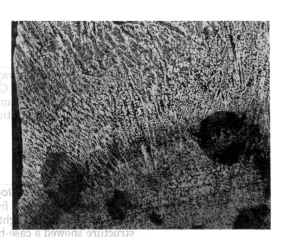

Fig. 6 Interface of weld buildup and base metal. Weld is on left (white, unetched), martensite on right (black and white, acicular). A crack is visible under the weld. 2% nital etch. 190×

Discussion The chipped area broke after carburizing and inspection by the manufacturer, probably because of rough handling. The opposite side of the keyway also chipped and was built up, probably with an austenitic steel, and then the keyway was machined. The crack in the keyway probably resulted from the welding stresses.

Conclusion and Recommendations The steel used, which was not as specified, was not particularly well suited for carburizing and was not strong enough for this application. Excessive case depth resulted in a brittle surface, causing the keyway to chip. The chipped keyway was built up with weld, and then the excess weld metal was machined away. During transport to the customer, the keyway broke on the opposite side. It was recommended that the shaft be returned to the supplier and a replacement requested.

Failure of a Service Water Pump Shaft

Hasan Shaikh, H.S. Khatak, and J.B. Gnanamoorthy, Metallurgy Division, Indira Gandhi Center for Atomic Research, Kalpakkam, India

A service water pump in a nuclear reactor failed when its shaft gave way. The fracture originated in the threaded portion of the sleeve nut on the drive-end side of the shaft. Results of the failure analysis showed that the cracking initiated at the thread root as a result of corrosion fatigue. Crack propagation occurred either by corrosion or mechanical fatigue. Evidence was found indicating high rotary bending stresses on the shaft during operation. The nonstandard composition of the En 8 steel used in the shaft and irregular maintenance reduced the life of the shaft. Recommendations included use of a case-hardened En 8 steel with the correct composition and regular maintenance of the pump.

Key Words		
Carbon steels	Shafts (power)	Corrosion fatigue
Fatigue failure	Pumps	Nuclear reactors
Cooling systems	Water cooling	
Alloys	Carbon steel—En 8	

Background

After 7 years of operation, a service water pump in a nuclear reactor failed when its shaft gave way. The fracture originated in the threaded portion of the sleeve nut on the drive-end side of the shaft.

Applications

Service water systems in nuclear reactors serve as a plant heat sink for various important components, such as preheating and emergency cooling, heat exchangers, nitrogen coolers, biological shield heat exchangers, diesel generators, air compressors, chiller condensers, and thermofluid heat exchangers. The heat is finally released to the atmosphere in an induced-draft cooling tower. The three service water pumps in this particular reactor take suction from the cooling water pit in the turbine building, and, after circulating water through respective heat exchangers, the heat is rejected to the atmosphere.

Pertinent specifications

The pumps operate at 30 hp, 1500 rev/min, with a flow rate of 175 m³/h (6180 ft³/h). They are negative-suction pumps (–1.5 M) with a discharge pressure of 2450 MPa (355 ksi). In a 12-day cycle, each pump would operate for 4 days with 8 days off. The pump shaft was made of En 8 steel.

Specimen selection

Specimens for fracture mode study were taken from the fracture surface. Specimens for microstructural and microhardness studies were taken some distance from the failed region.

Visual Examination of General Physical Features

Examination of the centrifugal pump revealed extensive rusting of the inside portion of the casing. Stagnant water was also found in the casing.

Only the drive-end side of the shaft was made available for analysis (Fig. 1). This component had also corroded extensively. The shaft failed at a distance of 240 mm (9.5 in.) from the drive-end side in the threaded portion of the sleeve nut, as shown in Fig. 2. The regions around the fracture location had diameters ranging from 31.4 to 34.7 mm (1.2 to 1.4 in.), indicating about 3.5 mm (0.14 in.) of mechanical wear of the shaft in the threaded portion of the sleeve nut. A 3 mm (0.12 in.) deep circumferential groove was visible 45 mm (1.7 in.) from the fracture surface.

Fig. 1 The drive-end side of the shaft

3.5 mm groove caused by corrosion

Keyway   Fracture   Keyway

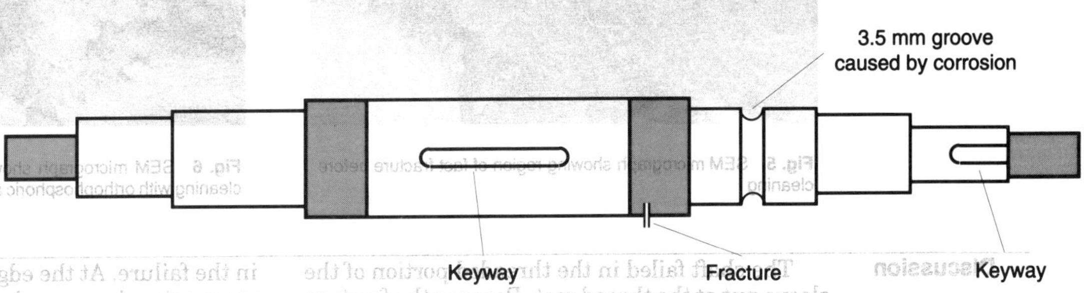

Fig. 2 Schematic of the shaft, showing the point of failure

Testing Procedure and Results

Surface examination

Visual. The fracture surface and the sides of the shaft were heavily corroded. Black corrosion deposits with yellowish brown patches at the periphery were found on the fracture surface. The fracture surface was helical in shape, smooth, and flat. Some portion of the fracture surface was found to have been rubbed. A cavity near the center was observed (Fig. 3). The presence of corrosion deposits masked the macroscopic fracture details. After cleaning with 50% orthophosphoric acid, more macroscopic details were observed, such as a corrugated periphery with steplike features called ratchet marks.

Scanning Electron Microscopy/Fractography. Examination of the cleaned fracture surface by scanning electron microscopy (SEM) showed multiple crack initiation points caused by the corrosion of the thread roots. One of the initiation points is shown in Fig. 4. Adjacent to the deeply corroded initiation point is the rubbed-off portion of the helical fracture surface. Figures 5 and 6 show the central cavity before and after cleaning with orthophosphoric acid.

Metallography

Microstructural Analysis. Optical microscopy studies were carried out after etching with 3% nital solution. Figure 7 shows the presence of pearlite (dark areas) in the ferrite matrix.

Crack Origins/Paths. Figure 8 shows the presence of multiple initiation points (darker areas at the edge), ratchet marks (brighter areas at the edge), and a wavy steplike appearance at the edge.

Chemical analysis/identification

Material. Chemical analysis of the En 8 steel indicated the presence of 0.19% C, 0.03% S, and 0.07% P.

Mechanical properties

Hardness. Microhardness measurements of the steel indicated a uniform hardness of 220 HV from the surface to the center.

Fig. 3 Fracture surface, showing the region of final fast fracture

Fig. 4 SEM micrograph showing crack initiation point adjacent to the rubbed area of the fracture surface

Fig. 5 SEM micrograph showing region of fast fracture before cleaning

Fig. 6 SEM micrograph showing region of fast fracture after cleaning with orthophosphoric acid

Discussion

The shaft failed in the threaded portion of the sleeve nut at the thread root. Because the fracture surface and the sides of the shaft were corroded, it was presumed that the environment played a role in the failure. At the edges of the fracture surface, more extensive corrosion products were observed at the initiation points than on the rest of the surface.

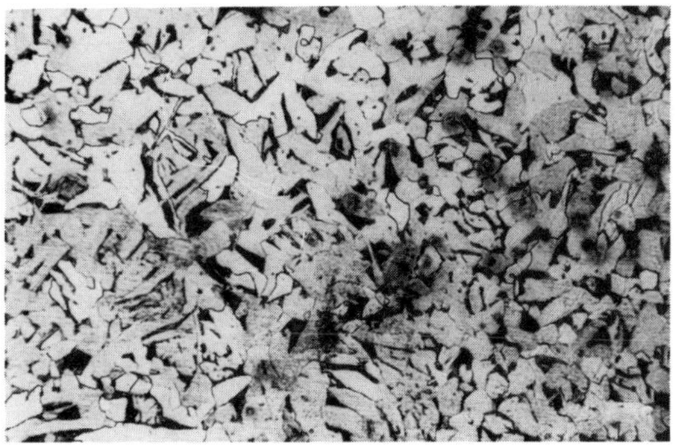

Fig. 7 Microstructure of pearlite in a ferrite matrix

Fig. 8 Stereomicrograph showing initiation points and ratchet marks

These yellowish brown corrosion deposits could have been hydrous ferric hydroxide. These products would form only if the area had experienced a stagnant environment for an extended period, thus implying crack initiation in those parts of the periphery of the shaft. This extended exposure was possible as a result of the stagnant water between the sleeve nut and the shaft. The ratchet marks around the corrugated perimeter of the shaft, coupled with the smoothness and flatness of the fracture surface, suggested a fatigue type of fracture, although no striations or tire tracks could be seen because of the corrosion of the fracture surface.

The helical shape of the fracture surface is explained by the ratchet marks. Ratchet marks are the result of multiple crack origins, which are typical of a corrosion fatigue failure, although they could occur even in mechanical fatigue failures under certain loading conditions. These multiple cracks do not initiate on one plane. Each crack produces a separate crack zone and propagates on planes inclined to the radial plane of the shaft. When two approaching crack fronts meet, a step called a ratchet mark is produced.

Multiple crack initiation points and ratchet marks indicated failure of the shaft due to stresses resulting from rotational bending. During each revolution of the shaft, both maximum and minimum loadings were exerted at each point around the perimeter of the shaft in the region of the maximum bending moment.

The presence of multiple cracks normally implies a relatively high applied load and rotational bending. When a shaft is rotated in only one direction, the crack advances asymmetrically, as evidenced by the fact that the final failure took place some distance from the center. The final failure in this case was characterized by a cavity a short distance from the center of the shaft diameter. The exact location of the final failure would depend on the magnitude of the stresses on the shaft, with the final failure moving toward the center with increas-

ing normal stress. The oval shape of the base of the final fast fracture region indicated that two mutually perpendicular and unequal bending stresses were present.

The multiple initiation points and extensive yellowish brown products at these initiation sites were evidence of the simultaneous action of the environment and the rotary bending stresses. However, the striations, tire tracks, or plateaus separated by tear ridges, which would indicate corrosion fatigue, could not be seen because of the corrosion of the fracture surface. Evidence at the periphery of the fracture surface, such as ratchet marks, corrugated periphery, and, more importantly, the multiple initiation points, suggested that the failure of this shaft was initiated by corrosion fatigue. The crack propagation might have occurred either by corrosion fatigue or by mechanical fatigue. Based on the appearance of the corrosion product on the fracture surface in the region some distance from the periphery, crack propagation by corrosion fatigue was unlikely. Crack propagation might have occurred by mechanical fatigue, and corrosion of the fracture surface might have occurred later.

The microstructure of the shaft material indicated the presence of 20 to 25% pearlite in a ferrite matrix. The chemical composition of the material indicated only 0.19% C, compared with the 0.4% C specified as the standard composition of En 8 steel. A lower carbon content results in a lower ultimate tensile strength. The fatigue resistance of a material is directly proportional to its tensile strength. This explains the higher susceptibility of the shaft to fracture under cyclic loading. Normally, the surface of a shaft is hardened to delay crack initiation and thus improve fatigue life. The effectiveness of the surface hardening is greater in applications where a high stress gradient exists, as in bending. The hardness profile of the failed shaft showed a uniform hardness from the edge to the center, indicating the absence of surface hardening.

Conclusion and Recommendations	Most probable cause	Remedial action
	The combined action of the environment and the rotary bending stresses on the shaft caused crack initiation by corrosion fatigue. Further crack propagation might have resulted by either corrosion fatigue or mechanical fatigue.	Use of En 8 steel with the correct composition and in the case-hardened condition was recommended. Regular maintenance of the shaft was also recommended.

Fatigue Failure of a Circulating Water Pump Shaft

Joyce M. Hare, Gelles Laboratories, Inc., Columbus, Ohio

A type 410 stainless steel circulating water pump shaft used in a fossil power steam generation plant failed after more than 7 years of service. Visual examination showed the fracture surface to be coated with a thick, spalling, rust-colored scale, along with evidence of pitting. Samples for SEM fractography, EDS analysis, and metallography were taken at the crack initiation site. Hardness testing produced a value of approximately 27 HRC. The examinations clearly established that the shaft failed by fatigue. The fatigue crack originated at a localized region on the outside surface where pitting and intergranular cracking had occurred. The localized nature of the initial damage indicated that a corrosive medium had concentrated on the surface, probably due to a leaky seal. Reduction of hardness to 22 HRC or lower and inspection of seals were recommended to prevent future failures.

Key Words			
Pitting (corrosion)		Martensitic stainless steels, corrosion	Stress-corrosion cracking
Fatigue failure		Shafts (power), corrosion	Electric power generation
Pumps			

Alloys Martensitic stainless steels—410

Background

A type 410 stainless steel circulating water pump shaft used in a fossil power steam generation plant failed after more than 7 years (66,000 h) of service.

Circumstances leading to failure

The 190 mm (7½ in.) outside diameter vertical shaft operated at 2550 hp in a temperature range of 15 to 35 °C (60 to 95 °F), not exceeding a maximum temperature of 40 °C (100 °F). Initial corrosive attack occurred at a localized region on the outside surface of the shaft, where a corrosive medium had been allowed to concentrate on one side of the shaft during storage. During service, the leakage of chlorine- and sulfur-bearing circulation water onto the damaged area of the shaft resulted in intergranular stress-corrosion cracking (SCC), which contributed to the initiation of the fatigue failure.

Pertinent specifications

The shaft material is specified to be ASTM A276 type 410 stainless steel (UNS S41000). The mechanical property design specifications varied from the ASTM standard by requiring a minimum yield strength of 520 MPa (75 ksi), a minimum tensile strength of 690 MPa (100 ksi), and a hardness ranging from 210 to 269 HB (approximately 99 HRB to 27 HRC).

Performance of other parts in same or similar service

Three similar shafts were operating in the same unit as the fractured shaft. It is not known whether these shafts were affected.

Specimen selection

Two sections of the failed shaft containing the mating fracture faces were submitted for investigation. Specimens were chosen from the least mechanically damaged face for scanning electron microscope (SEM) fractography, energy-dispersive spectroscopic (EDS) analysis, optical metallography, and spectrochemical analysis.

Visual Examination of General Physical Features

The fracture surface on the shaft is shown in the macroview in Fig. 1. The fracture surface was coated with a rust-colored oxide and displayed a flat region normal to the axis on the shaft. The prominent array of beach marks was visible, followed by a region of final fracture at a 45° angle to the plane of fatigue fracture. The fatigue zone covered a large proportion of the cross section, which is typical of a high-cycle/low-nominal-stress fatigue mechanism. Variations in color and width of the beach marks were apparent, indicating changes in frequency or loading pattern as well as oxidation of the crack during times of arrest (e.g., periods of shutdown).

The crack originated on the outside surface of the shaft; ratchet marks indicated that there was more than one initiation site. These sites, shown at higher magnification in Fig. 2, had a granular morphology and were covered with a dark scale.

Fig. 1 Macroview of the fracture surface. Beach marks typical of fatigue fracture originate at lower left. Samples A, B, C, and D (below C, not shown) were used in the study. Top left portion of fracture has sustained mechanical damage. 0.37×

(a) **(b)**

Fig. 2 Closeup views of the mating fracture surfaces in the region of the fracture origin. Arrows indicate suspected origin areas. 1.12×

Examination of the outside cylindrical surface of the shaft at the origin area revealed a thick, spalling, rust-colored scale and evidence of pitting. In comparison, the remainder of the cylindrical surface was relatively free of pitting and corrosion.

The locations of the samples used for SEM fractography, EDS analysis, and metallography are shown in Fig. 1. Samples A and B include crack initiations sites, sample C was taken opposite the origin area in the fast-fracture zone, and sample D (not seen in Fig. 1) was located immediately below sample C.

Testing Procedure and Results

Surface examination

Scanning Electron Microscopy/Fractography. The fracture initiation sites in samples A and B were examined in the SEM and subjected to EDS analysis in the as-received condition. Figures 3(a) and (b) show that the fracture surface in the initiation area had an intergranular morphology and was covered by a corrosion product. EDS analysis of the corrosion product on sample A revealed the presence of sulfur, calcium, aluminum, potassium, and chlorine (Fig. 3c). A similar analysis was found on the fracture surface of sample B. The most probable source of these contaminants was the water circulated by the pump.

The samples were cleaned ultrasonically and reexamined in the SEM. The intergranular nature of the fracture in the origin was more evident (Fig. 4a and b). The transition from intergranular cracking to transgranular fatigue cracking is shown in Fig. 4(c).

Examination of the outside machined cylindrical surface at the origin area on samples A and B revealed evidence of pitting, cracking, and intergranular attack (Fig. 5a and b, respectively). A higher-magnification view of the intergranular attack (Fig. 5c) shows corrosion product on the exposed grain facets. The initial attack, isolated to one side of the shaft, was probably the result of localized exposure to a corrosive media prior to service.

Metallography

Microstructural Analysis. Longitudinal metallographic sections were prepared from samples A, B, C, and D for optical microscopy. The specimens, initially examined in the as-polished condition, were etched with Vilella's reagent to reveal a microstructure of tempered lath martensite (Fig. 6). This microstructure was typical of all four specimens, which indicated that a microstructural defect was not a likely contributor to the failure.

Crack Origins/Paths. The as-polished microstructure of sample B (Fig. 7a), shows intergranular attack originating at the outside cylindrical surface, where large groupings of grains have dropped out. The intergranular cracking proceeded along prior-austenite grain boundaries and was accompanied by interlath attack, as can be seen by the delineation of the tempered martensite lath boundaries with corrosion product in Fig. 7(a). Figure 7(b), an etched view of the fracture surface of sample A, shows intergranular cracking along prior-austenite grain boundaries in a region of interlath corrosion, which appears darker than the unattacked region of tempered martensite at the bottom of the micrograph as a result of etching.

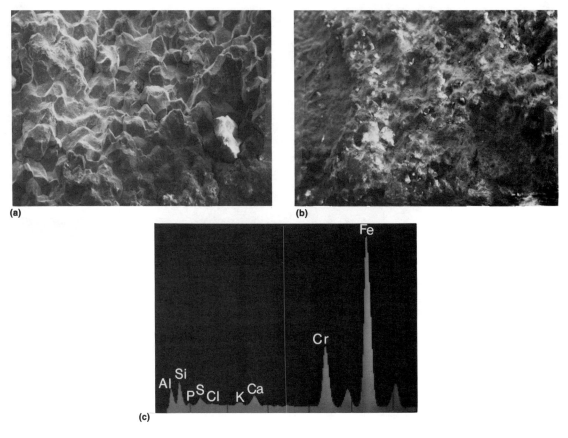

Fig. 3 SEM micrographs and EDS spectrum of the fracture initiation sites of samples A and B. (a) Sample A: intergranular fracture near the cylindrical surface of the shaft. (b) Sample B: lower-magnification view of the fracture surface near the cylindrical surface exhibiting intergranular fracture. (c) EDS spectrum corresponding to (a)

Chemical analysis/identification

Samples from areas B, C, and D (see Fig. 1) were subjected to spectrochemical analysis. The results of the analysis and the composition limits for type 410 stainless steel are summarized in Table 1. Comparison of the results with the standard composition for type 410 shows that the three areas are within specification and, more importantly, that there are essentially no compositional differences among the three areas of the shaft analyzed. Thus, a compositional inhomogeneity in the shaft did not contribute to the failure.

Mechanical properties

Hardness. Rockwell C hardness testing was carried out on metallographically prepared samples B, C, and D. The results, presented in Table 2, show that there is no significant variation in hardness among the three areas studied. The hardness values of samples B and D are within the specified range of 99 HRB to 27 HRC; the hardness of sample C, however, is slightly higher. Overall, the fracture area possessed no unique hardness attributes.

Discussion

Visual and macroexamination and SEM fractography clearly established that the circulating water pump shaft failed by fatigue. The origin of the fatigue crack, established by the same techniques as well as by optical metallography, was at a localized region on the outside cylindrical surface of the shaft that contained pitting and intergranular cracking.

The localized nature of the initial damage indicated that a corrosive medium had been allowed to concentrate on the surface. This probably occurred either as a result of a corrosive fluid that dripped on the shaft during storage or because of a leaky seal that allowed the treated circulating water to come into contact with the shaft during a shutdown period. EDS analysis of the corrosion product on the fracture surface revealed impurity lev-

els of calcium, sulfur, and chlorine, all of which can be attributed to the treated circulating water. Hence, the resulting stress concentration and initial crack propagation by SCC, in combination with the applied and cyclic stresses associated with a rotating shaft, produced the final fatigue fracture of the shaft.

The shaft material was found to be within specification with respect to composition. One sample examined was slightly high in hardness, although still typical of a quenched and tempered type 410 stainless steel and not related to the origin area. In addition, the microstructure appeared to be normal for a quenched and tempered martensitic stainless steel. In general, these observations were made at the fracture origin as well as at locations remote from the primary fracture, indicating

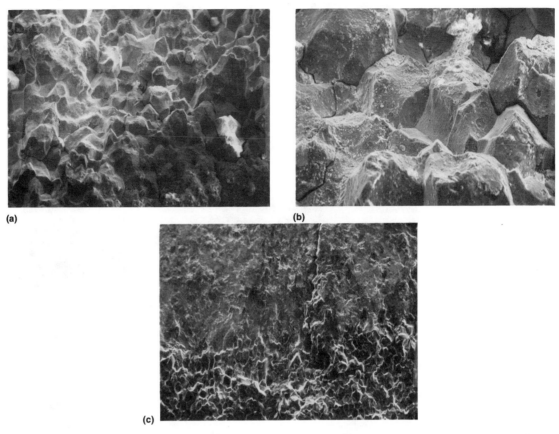

Fig. 4 SEM micrographs of the fracture surface after cleaning. (a) and (b) Views at increasing magnification of the crack origin area of sample A. (c) Transition in fracture morphology between the intergranular fracture of the origin area (bottom) and the fatigue zone (top)

Fig. 5 SEM micrographs of the machined cylindrical surface after cleaning. (a) Sample B: overview of the fracture surface (FS) intersecting the machined surface, showing localized pitting. (b) Sample A: overview of the machined surface, showing pitting, intergranular attack, and secondary cracking. (c) Higher-magnification view of the intergranular attack in (b)

Table 1 Results of spectrochemical analysis

Element	Composition, wt%			Chemical requirements for type 410 stainless steel
	Sample B	Sample C	Sample D	
Carbon	0.12	0.12	0.12	0.15 max
Manganese	0.42	0.42	0.41	1.00 max
Silicon	0.32	0.33	0.33	1.00 max
Chromium	12.47	12.30	12.33	11.5-13.5
Phosphorus	0.014	0.016	0.016	0.040 max
Sulfur	0.010	0.010	0.010	0.030 max

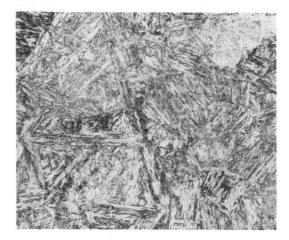

Fig. 6 Optical micrograph of the typical shaft microstructure, which consisted of tempered lath martensite. Longitudinal view. Etched in Vilella's reagent. 142×

Table 2 Results of hardness testing

Sample	Hardness, HRC	
	Range	Average
B	26-27	26.7
C	28-29	28.3
D	27	27

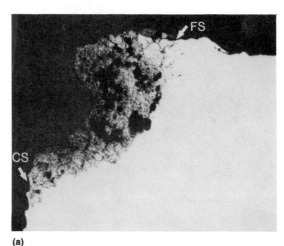

(a)

(b)

Fig. 7 Optical micrographs of samples A and B. Longitudinal views. (a) Overview of the cylindrical surface (CS) at the fracture surface (FS) of sample B, showing intergranular attack originating at the cylindrical surface. Groupings of grains have fallen out. As-polished. 17.75×. (b) Higher-magnification view of the fracture surface, showing cracking along prior-austenite grain boundaries in an area of interlath attack (top, darker etching) adjacent to unattacked tempered martensite. Etched in Vilella's reagent. 142×

that the material was homogeneous, appeared to have been properly heat treated to obtain the specified mechanical properties, and had no unique metallurgical problems that could have caused the fracture.

Martensitic stainless steels are relatively resistant to SCC in acid sulfide solutions when the hardness is equal to or below 22 to 25 HRC (Ref 1). The fractured shaft had an average hardness of approximately 27 HRC, a value at the high end of the specified range. Heat treating the shaft to produce a lower hardness and to reduce residual stresses will increase the resistance to cracking; however, the strength will be reduced. Use of type 410 stainless steel in this application requires either the elimination of the corrosive environment or a reduction in the strength to as low a value as practicable (Ref 2), such that the hardness is less than or equal to 22 HRC.

It is important to recognize that the failure of the shaft may have resulted from unusual circumstances: negligence in storing the shaft in a corrosive, aqueous environment and/or inadequate seal performance that allowed treated water to contact the shaft. If these two problems can be alleviated, the shaft may be able to be used at its present hardness.

Conclusion and Recommendations

Most probable cause

The shaft failed by fatigue under service stress. The crack initiated at a stress concentration on the outside cylindrical surface of the shaft, where pitting and intergranular attack resulted from localized exposure to a corrosive medium. The initial cracking was identified as intergranular SCC in the presence of chlorine- and sulfur-bearing species.

Remedial action

To prevent the failure of pump shafts in similar applications, it is recommended that the hardness be reduced to 22 HRC or lower, if it is possible to use a lower-strength shaft in this application.

Measures should be taken to alter the corrosive environment by eliminating the use of chlorine- and sulfur-bearing species in the water treatment if possible. In addition, the seals must work properly to prevent exposure of the pump shaft to treated circulating water.

The shaft should be protected from moisture or other aqueous solutions during storage. The packing used to isolate the shaft from the treated water should not contain sulfur. The shaft should be inspected prior to installation and periodically thereafter by nondestructive methods, such as visual inspection, dye penetrant testing, and wet fluorescence magnetic particle testing.

References

1. R.A. Lula, *Stainless Steel*, American Society for Metals, 1987, p 148-150
2. *Stress Corrosion Cracking and Hydrogen Embrittlement of Iron Base Alloys*, NACE-5, Unieux-Firminy, France, 12-16 June 1973, National Association of Corrosion Engineers, p 121-122

Fracture of a Coupling in a Line-Shaft Vertical Turbine Pump

I.B. Eryürek, A. Aran, and M. Capa, Istanbul Technical University, Istanbul, Turkey

A coupling in a line-shaft vertical turbine pump installed in a dam foundation fractured after a very short time. The coupling material was ASTM A 582 416 martensitic stainless steel. Visual, macrofractographic, and scanning electron microscopic examination of the coupling showed that the fracture was brittle and was initiated by an intergranular cracking mechanism. The mode of fracture outside the crack initiation zone was transgranular cleavage. No indication of fatigue was found. The failure was attributed to improper heat treatment during manufacture, which resulted in a brittle microstructure susceptible to corrosion. The crack initiated either by stress-corrosion or hydrogen cracking. It was recommended that the couplings in the system be examined for surface cracking and, if present, corrective measures be taken.

Key Words			
Martensitic stainless steels	Hydrogen embrittlement	Brittle fracture
Cleavage	Couplings	Turbine pumps
Transgranular fracture	Stress-corrosion cracking	Pumping plants
Intergranular fracture		

Alloys Martensitic stainless steel — A582 416

Background

A stainless steel coupling in a line-shaft vertical turbine pump installed in a dam foundation fractured after a very short time. Figure 1 shows a schematic of the pump, along with the location of the failed coupling. Dimensions of the coupling are given in Fig. 2.

Circumstances leading to failure

The coupling failed after 4300 h of service, during which it had been exposed to water at a temperature of 3 to 25 °C (38 to 77 °F) and a pH of 7 to 8. The pump had been stopped and restarted 180 times for various reasons. The failure was noticed during an attempt to restart the pump.

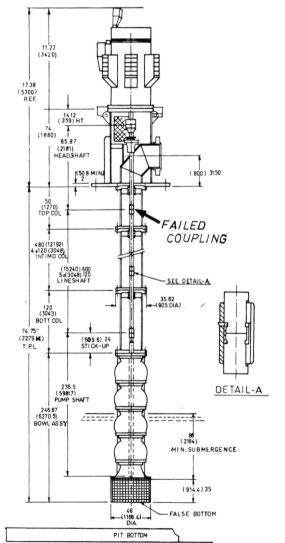

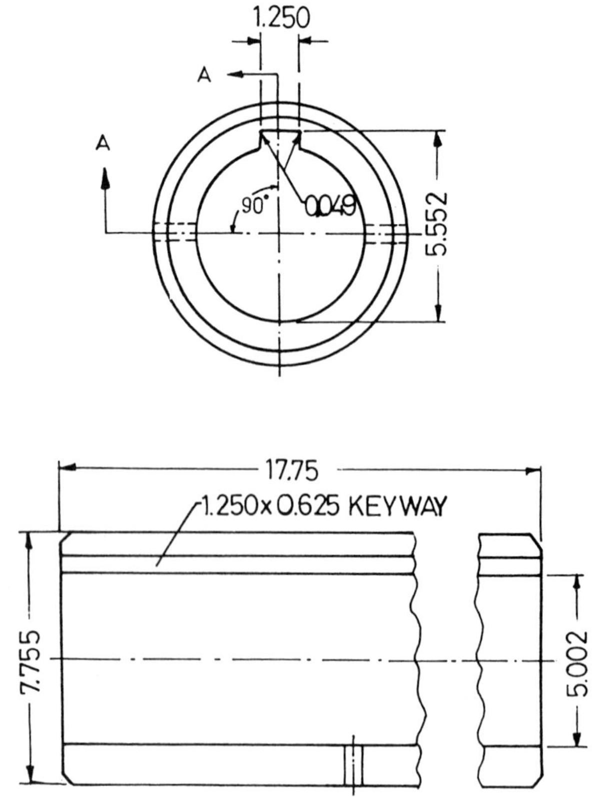

Fig. 1 Schematic of the pump assembly, showing location of the failed coupling

Fig. 2 Dimensions (inches) of the coupling

Pertinent Specifications

The material specified for the coupling was ASTM A 582 416 martensitic stainless steel.

| **Visual Examination of General Physical Features** | A general view of the fractured coupling is shown in Fig. 3. The spiral shape of the crack path | indicated that the fracture was brittle. This was also confirmed by surface examination. |

Testing Procedure and Results

Surface examination

Macrofractography. Studies of the chevron marks on the fracture surface indicated that the fracture originated at the location shown in Fig. 4.

Scanning Electron Microscopy/Fractography. Examination of the fracture surface by scanning electron microscopy (SEM) revealed that the fracture was initiated by an intergranular cracking mechanism (Fig. 5) and that the mode of fracture outside the crack initiation zone was transgranular cleavage. There was no indication of fatigue on the fracture surface. The microstructure of the fractured coupling material consisted of tempered martensite and ferrite.

Chemical analysis

Material. Specified and actual values of the major alloying elements of the material as determined by chemical analysis are given in Table 1.

Mechanical properties

Impact, hardness, and tensile property values for the coupling material are given in Table 2, along with the hardness values specified in ASTM A 582 for condition "H" (Ref 1).

Table 1   Results of chemical analysis

| Element | Composition, % | |
	Coupling material	Chemical specification (ASTM A 582)
Carbon	0.10	0.15 (max)
Chromium	11.8	12-14
Sulfur	0.32	0.15 (min)
Iron	bal	bal

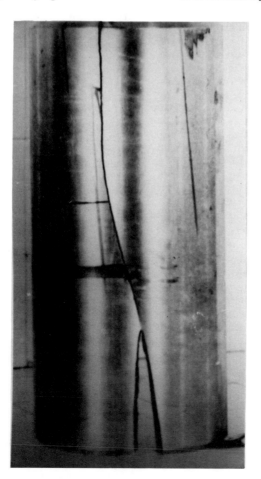

Fig. 3 Overall view of the fractured coupling. Arrow indicates the location of crack initiation.

Fig. 4 Fracture surface of the coupling

Table 2 Mechanical properties

Test No.	Yield strength MPa	Yield strength ksi	Tensile strength MPa	Tensile strength ksi	Elongation in 50 mm (2 in.), %	Reduction of area, %	Izod impact energy J	Izod impact energy ft · lbf	Hardness, HB
1	973	141	1063	154	10	33	12.8	9.4	287-328
2	1021	148	1055	153	13	36	13.3	9.8	287-328
3	987	143	1028	149	14	37	13.3	9.8	287-328
4	...	...	...	...	...	...	15.8	11.6	287-328
Condition "H" (ASTM A582)	...	...	...	...	...	...	...	...	293-352

Discussion

Measurement showed that the actual dimensions of the coupling were in accordance with the specified dimensions. The maximum value of the fluctuating shear stress at the inner corner of the keyway was computed as 71 MPa (10 ksi). The fatigue limit of the coupling material for fluctuating shear stresses is approximately 500 MPa (72 ksi). Considering factors such as surface finish, shape, and dimensions, the factor of safety for infinite fatigue life is estimated as 4.

Chemical composition of the material was in accordance with the specified composition, even through the amount of chromium was at the lower limit of the specified range. The hardness of the material corresponded to condition "H" described in Table 3 in Ref 1. The important consideration is that the material was brittle, with low ductility and low Izod impact values.

The most likely cause for brittleness in martensitic stainless steels is improper heat treatment. It is well known that tempering of martensitic stainless steels in the range of 370 to 565 °C (700 to 1050 °F) results in low impact properties and poor resistance to corrosion and stress corrosion (Ref 2). Low Izod impact values (9.5 to 15 J, or 7 to 11 ft · lbf) in this material are typical of those resulting from tempering treatments within this temperature range (Ref 3). Adequate Izod energy corresponding to the safe tempering range (150 to 370 °C, or 300 to 700 °F) is higher than 20 J (15 ft · lbf) (Ref 4).

The initial crack propagated by brittle intergranular cracking and acted as an initial surface flaw for fast cracking, which propagated by a transgranular cleavage mechanism. Fracture surface studies of Izod impact specimens confirmed

Fig. 5 Scanning electron micrograph showing appearance of fracture surface at the crack initiation zone

the transgranular cleavage mechanism. Thus, intergranular cracking at the crack origin appears to have been caused not only by material and stress conditions but also by environmental conditions conducive to either stress-corrosion or hydrogen cracking; high-strength martensitic stainless steels are subject to both types of cracking.

Actually, failure of this alloy under conditions that appear conducive to stress-corrosion cracking is similar to failure caused by hydrogen damage; it is often difficult to distinguish between the two mechanisms. The environments that cause failure are not specific. Even an environment as mild as fresh water at room temperature may cause failure in especially susceptible alloys (Ref 5).

Conclusion and Recommendations

Most probable cause

Improper heat treatment during manufacture of the coupling resulted in a structure with inadequate fracture toughness and increased susceptibility to corrosion. The surface crack initiated by either stress-corrosion or hydrogen cracking under service conditions. During a sudden loading this flaw propagated easily in the brittle material, resulting in complete failure.

Remedial action

Other couplings in the system may have undergone the same heat treatment and thus should be examined for surface cracking. If cracking is present, corrective measures should be taken.

References

1. "Specification for Free-Machining Stainless and Heat-Resisting Steel Bars, Hot Rolled or Cold Finished," A582-88, *Annual Book of ASTM Standards*, ASTM
2. *Metals Handbook*, 8th ed., Vol 2, American Society for Metals, 1964, p 245
3. J.G. Parr and A. Hanson, *Stainless Steel*, American Society for Metals, 1965.
4. *Metals Handbook*, 8th ed., Vol 1, American Society for Metals, 1961, p 421
5. *Metals Handbook*, 8th ed., Vol 10, American Society for Metals, 1975, p 220-221

VALVE FAILURES

Dezincification of a Chrome-Plated Cylinder Gas Valve

Richard L. Colwell, Air Products and Chemicals, Allentown, Pennsylvania

Two new chrome-plated CDA 377 brass valves intended for inert gas service failed on initial installation. After a pickling operation to clean the metal, the outer surfaces of the valves had been flashed with copper and then plated with nickel and chromium for aesthetic purposes. One of the valves failed by dezincification. The porous copper matrix could not sustain the clamping loads imposed by tightening the pressure relief fitting. The second valve failed by shear overload of the pressure relief fitting. Overload was facilitated by a reduction of cross-sectional area caused by intergranular attack and slight dezincification of the inner bore surface of the fitting. Dezincification and intergranular attack were attributed to excessive exposure to nonoxidizing acids in the pickling bath.

Key Words   Brasses, corrosion   Intergranular corrosion   Valves, corrosion
              Dezincification

Alloys   Brass—CDA 377

Background   Two new chrome-plated CDA 377 brass valves intended for inert gas service failed on initial installation. After a pickling operation to clean the metal, the outer surfaces of the valves had been flashed with copper and then plated with nickel and chromium for aesthetic purposes.

Applications

Brass valves are used extensively in the compressed gas industry. The CDA 377 alloy (60% Cu, 38% Zn, 2% Pb) is used because it is easy to forge into complicated shapes. Second-phase lead particles also allow it to be easily machined.

These valves are typically installed on compressed gas cylinders. There are two outlet ports: one feeds the regulator and the other is connected to a pressure relief fitting. Both are normally machined with external threads.

Circumstances leading to failure

Upon initial installation, both valves failed. Failure of the first valve occurred when the pressure relief fitting was installed. The threads on the relief valve port stripped when tightened by hand. The entire relief port sheared off the second valve.

Specimen selection

In addition to the two failed valves, four other valves were removed from the same lot for examination. A valve from another lot, shipped at the same time, was also examined.

Testing Procedure and Results

Surface examination

Visual examination of the first failed valve (Fig. 1 and 2) revealed that the threads on the pressure relief port were stripped and discolored. The color was that of dull red copper, and the metallic luster normally associated with brass was absent. The second valve did not exhibit discoloration; however, a fracture surface normally associated with shear failure was observed (Fig. 3).

Metallography

Microstructural Analysis. Metallographic examination of the bulk matrix revealed a two-phase $\alpha + \beta$ brass in all of the valves. No inhomogeneities were observed within the matrix. The outer surfaces exhibited various degrees of dezincification.

Corrosion Morphology. Metallographic analysis was performed on samples from the pressure relief fitting ports of both valves. Examina-

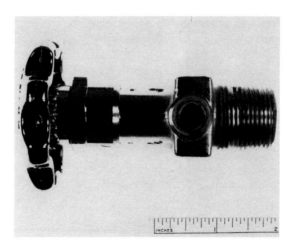

Fig. 1   First failed chrome-plated cylinder valve

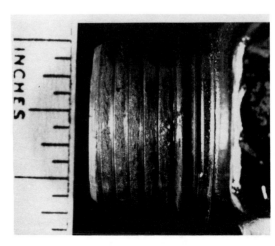

Fig. 2   Closeup view of first valve, showing damage and discoloration in the pressure relief fitting

Fig. 3 Shear failure on the pressure relief fitting of the second failed valve

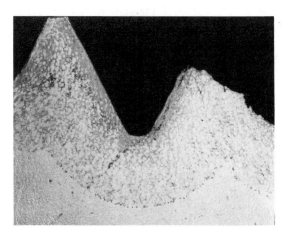

Fig. 4 Micrograph showing the depth of dezincification in the pressure relief valve threads of the first valve. 31.5x

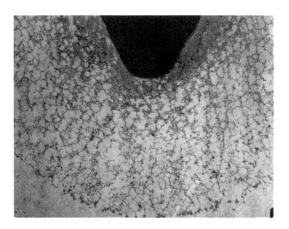

Fig. 5 Higher-magnification of Fig. 4. 63x

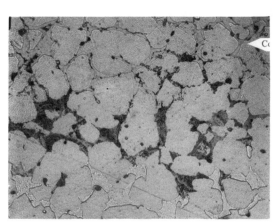

Fig. 6 Micrograph of first valve, showing preferential dealloying of the β phase along an interface. Etched with NH$_4$OH/H$_2$O$_2$. 315x

Fig. 7 Minute dezincification under the plating of the second valve. Etched with NH$_4$OH/H$_2$O$_2$. 630x

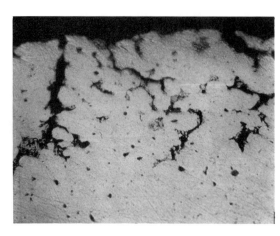

Fig. 8 Intergranular cracking and dezincification in the inner bore surface of the pressure relief part of the second valve. Unetched. 315x

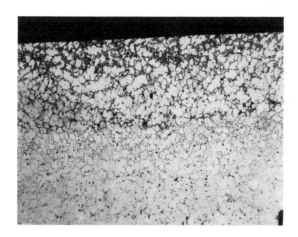

Fig. 9 Dezincification on the inner surface in the neck of the first valve. Etched with NH_4OH/H_2O_2. 63×

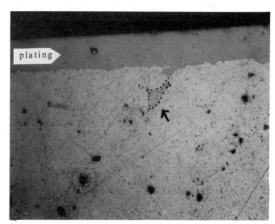

Fig. 10 Micrograph showing only traces of dezincification on the surface of the second valve. 630×

tion of the first valve showed that the red discolorations were the result of dezincification (Fig. 4 and 5). Attack appeared to be restricted to the discontinuous β phase (Fig. 6). The depth of attack in the threaded region averaged 0.53 mm (0.021 in.). The second valve exhibited less damage, but some dezincification was observed (Fig. 7). Copper enrichment/zinc depletion in the β phase was observed only to a depth of three grains on most of the outer surfaces. However, intergranular attack and dezincification were observed along the internal bore of the fitting port (Fig. 8). The depth of attack and dezincification was limited to approximately six grain diameters.

Sample sections were then taken from the threaded cylinder port and from various surfaces of the valve body. Chrome-plated and non-chrome-plated surfaces, both internal and external, were examined. Dezincification was observed in both valves. In the first valve, the maximum depth of attack was 1.4 mm (0.057 in.) (Fig. 9). Samples from the second valve showed only minute traces of dezincification (Fig. 10).

Pressure relief ports from five additional valves were sectioned. Examination showed that four valves, all taken from the same lot as the failed valves, exhibited some degree of dezincification. The fifth valve, taken from another lot, showed no dezincification.

Chemical analysis

Material. Semiquantitative chemical analysis of the valve material was performed using scanning electron microscopy/energy-dispersive spec-

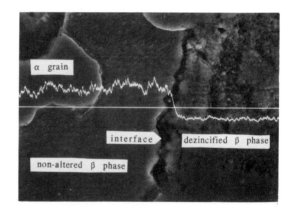

Fig. 11 EDS line scan for zinc. As the trace crosses the dealloying interface of the unaffected β grain, the line intensity increases, illustrating the selective removal of zinc from the β phase.

troscopy (SEM/EDS). Results indicated that the material was 62% Cu, 35% Zn, 1% Pb, and 2% other, a composition similar to CDA 377.

Corrosion or Wear Deposits. A polished cross section taken through a dezincified region of the first valve was examined using an EDS line scan. The scan shows a definite reduction in zinc concentration across the dealloying interface within the β phase (Fig. 11).

| **Conclusion and Recommendations** | One of the chrome-plated brass valves failed by dezincification; this resulted in inherently weak threads on the pressure relief fitting port. The other valve failed by shear overloading of the pressure relief port. The stresses required to cause shear overloading were reduced by the presence of intergranular cracks on the inner bore surface of the pressure relief port. The magnitude of dezincification was vast on one valve and limited on four other valves from the same | lot. It was completely absent on a valve from another lot. This condition can be attributed to exposure to a corrosive environment during manufacture, probably present in the pickling bath. Dezincification of brass is usually caused by exposure to water. However, exposure to nonoxidizing acids can also accelerate dealloying. It was suspected, and subsequently confirmed, that the severely dezincified valve had been in a pickling bath too long. |

Failure of Brass Hot Water Reheat Valves

Carmine D'Antonio, Department of Metallurgy and Materials Science, Polytechnic University, Brooklyn, New York

Two hot water reheat coil valves from a heating / ventilating / air-conditioning system failed in service. The valves, a 353 copper alloy 19 mm (¾ in.) valve and a 360 copper alloy 13 mm (½ in.) valve, had been failing at an increasing rate. The failures were confined to the stems and seats. Visual examination revealed severe localized metal loss in the form of deep grooves with smooth and wavy surfaces. Metallographic analysis of the grooved areas revealed uniform metal loss. No evidence of intergranular or selective attack indicating erosion-corrosion was observed. Recommendations included use of a higher-copper brass, cupronickel, or Monel for the valve seats and stems and operation of the valves in either the fully opened or closed position.

Key Words			
	Brasses, corrosion Air-conditioning equipment	Valves, corrosion Heating equipment	Erosion-corrosion
Alloys	Brasses—353, 360		

Background

Two hot water reheat coil valves from a heating/ventilating/air-conditioning (HVAC) system failed in service.

Circumstances leading to failure

Chlorinated, carbon-filtered, potable water was used in the reheat system at a supply temperature of 55 °C (130 °F). The valves had been failing at an increasing rate. The failures were confined to the stems and seats.

Specimen selection

Metallographic samples were cut from the two seats and stems and mounted to show the damaged surfaces in cross section. Drillings were machined for chemical analysis.

Visual Examination of General Physical Features

Figure 1 shows the failed stems and seats. Severe localized metal loss on the stem surfaces and facing seat surfaces was evident. The metal loss was in the form of deep grooves, the surfaces of which were smooth and wavy in texture. The grooved surfaces were free of corrosion products; the remaining surfaces were covered with a brown, adherent corrosion product.

Metallography

All of the metallographic samples exhibited essentially the same microstructure: a cold-worked alpha matrix with interspersed particles of lead. This is a normal structure for a free-machining brass. The profiles of the grooves showed uniform metal loss, with no evidence of intergranular or selective attack. Figure 2 shows a typical area of groove profile.

Chemical analysis/identification

Chemical analyses revealed that the 19 mm (¾ in.) valve was machined from copper alloy 353, and

Fig. 1 Stems and seats removed from two valves

Fig. 2 Micrograph showing the surface profile in one of the grooved areas. 62×

the 13 mm (½ in.) valve was copper alloy 360. Both are lead-bearing, free-machining, low-copper brasses.

Discussion	The features exhibited by the valve stems and seats are characteristic of erosion-corrosion: selected area gouging, the wavy texture of the gouged areas, the absence of corrosion product in the gouged area, the adherent, protective corrosion product on adjacent surfaces, and the uniform metal loss observed microscopically. Erosion-corrosion is a synergistic form of attack wherein liquid or gaseous erosion is assisted by corrosion and vice versa. The result is rapid, localized loss of metal. The greater the liquid velocity, the faster the deterioration. It is significant that the metal loss was localized in areas where liquid velocity was greatest. This velocity increased substantially when the valves were only partially opened (cracked).
Conclusion and Recommendations	The resistance of alloys to erosion-corrosion is usually directly proportional to their resistance to simple corrosion in a particular medium. Although alloys 360 and 353 have adequate corrosion resistance for this application, they are not the most corrosion-resistant alloys and thus are not the most erosion-corrosion resistant. Higher-copper brasses, such as alloys 260, 268, 270, or 330, would be a better choice. Ideally, however, an alloy such as cupronickel or Monel should be used. Monel is the most highly recommended. In addition, the valves should be used in the completely open or completely closed position when in service.

Failure of Three Valve Body Castings

George M. Goodrich, Taussig Associates, Inc., Skokie, Illinois

Three sprinkler system dry pipe valve castings (class 30 gray iron), two that had failed in service and one that had been rejected during machining because of porosity, were submitted for examination. The two failures consisted of cracks in a seating face. All three were from the same heat. Visual examination showed that the casting had cracked through a thin area in the casting sidewall. Evidence of a sharply machined corner at the fracture site was also discovered. Tensile testing and metallographic analysis revealed no metallurgical cause for the failure. It was recommended that the manufacturer work with the foundry to evaluate the criticality of core placement and to eliminate the undesired thin section.

Key Words	Gray iron	Castings	Valves
	Sprinklers	Casting defects	Foundry practice

Background

Three gray iron dry pipe valve castings were submitted for metallurgical analysis. Two of the valves had failed in service; the third had been re- jected by the manufacturer because of porosity found during machining. The valves were report- edly from the same heat.

Testing Procedure and Results

Surface examination

Visual. The submitted valves were visually ex- amined to determine the crack location. The crack in casting 1 was found in the drain area. The crack ex- tended radially through the seating surface and the internal casting wall adjoining the drain. Figure 1 il- lustrates the crack observed on the top of the seating surface in casting 1. Figure 2 documents the crack propagation through the seating surface to the cav- ity below; Fig. 3 illustrates the depth of the crack through the wall of the casting and location of the crack with reference to the casting drain.

The mating ring was removed from casting 2 to identify crack locations. Radial fractures were lo- cated in the seating surface in a location similar to that found in casting 1.

Macrofractography. A section containing the seating surface above the cavity and a portion of the adjacent casting wall were removed to ex- pose the fracture surface found in casting 1. The area removed displayed a reduced wall section at the fracture site. The minimum section thickness was 9.5 mm (⅜ in.) through the fractured area. Ad- jacent walls, by comparison, were 16 mm (⅝ in.) thick. Figure 4 illustrates the fractured surface in the thin section. The arrows indicate minimum

Fig. 1 Overall view of casting 1, showing the general location of the crack

Fig. 2 Detailed view, showing crack extending into the cavity

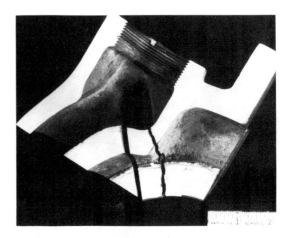

Fig. 3 Crack extending from the seating surface inner diameter through the wall of the casting into the drain area

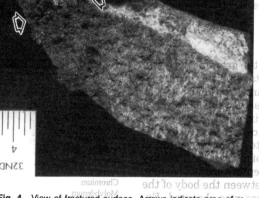

Fig. 4 View of fractured surface. Arrows indicate area of reduced wall thickness.

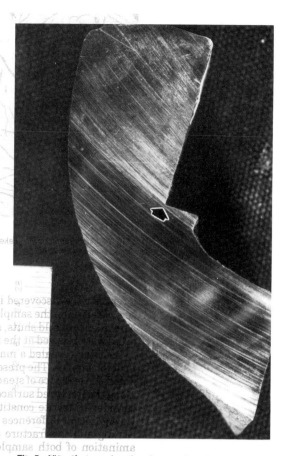

Fig. 5 View of removed section above cavity adjacent to drain. Arrow indicates sharp machined corner and outer diameter of seating surface.

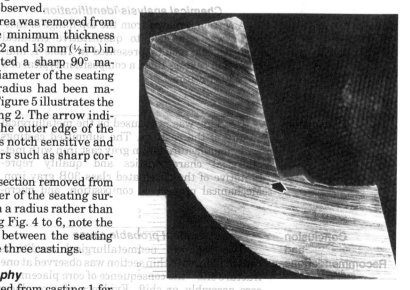

Fig. 6 Section removed from area over cavity adjacent to drain in valve 3. Arrow indicates corner with desired radius.

section thickness. The fractured surface revealed no identifiable fracture origin. No voids from shrinkage or misruns were observed.

The same internal wall area was removed from the other two castings. The minimum thickness was 16 mm (⅝ in.) in casting 2 and 13 mm (½ in.) in casting 3. Casting 2 exhibited a sharp 90° machined corner on the outer diameter of the seating surface. A more generous radius had been machined in the other casting. Figure 5 illustrates the section removed from casting 2. The arrow indicates the sharp corner at the outer edge of the seating surface. Gray iron is notch sensitive and can easily fail at stress risers such as sharp corners.

Figure 6 represents the section removed from casting 3. The outer diameter of the seating surface had been machined with a radius rather than a sharp corner. In comparing Fig. 4 to 6, note the variation in wall thickness between the seating surface and the cavity for the three castings.

Metallography

Two sections were removed from casting 1 for metallographic analysis. One section was taken from the top flange of the casting to represent the general microstructure of the casting. A second section was taken from the fracture site to be examined for detrimental conditions capable of inducing or contributing to cracking. The samples were mounted and metallographically prepared and were examined in the unetched and etched conditions.

The unetched structure in the sample taken from the fractured area contained flake graphite (ASTM type A, sizes 3 to 4). Type A flake graphite is preferred for optimum mechanical properties. Figure 7 illustrates the graphite type and size observed in the fractured sample. No significant dif-

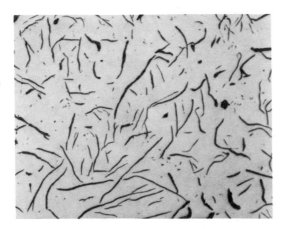

Fig. 7 Micrograph showing flake graphite observed in the fractured sample. Unetched. 62×

Fig. 8 Micrograph showing matrix microstructure of fractured sample. Dark constituent is pearlite; lighter constituent is steadite. Nital etch. 248×

ferences were discovered in the graphite type and size observed in the sample taken from the flange. No evidence of cold shuts, shrinkage voids, slag, or dross was observed at the fracture site.

Etching revealed a matrix of lamellar pearlite in both samples. The presence of steadite was also noted. No evidence of steadite networks was found along the fractured surface of the sample. Primary carbides or ferrite constituents were not visible. No significant differences between the body of the casting and the fracture area was observed. Examination of both samples revealed microstructures typical of class 30 gray iron with no metallurgical defects. Figure 8 illustrates the etched microstructure.

Chemical analysis/identification

A portion removed from the top flange of casting 1 was subjected to quantitative chemical analysis. The results, presented in Table 1, indicated that casting 1 had a composition typical of a class 30 gray iron.

Table 1 Results of chemical analysis

Element	Percent
Total carbon	3.49
Combined carbon	0.73
Manganese	0.58
Phosphorus	0.096
Sulfur	0.118
Silicon	1.59
Nickel	0.11
Chromium	0.14
Molybdenum	0.04
Copper	0.26
Tin	<0.02
Vanadium	0.01

Mechanical properties

Tensile Properties. A tensile test specimen was machined from the top flange of casting 1. Tensile strength was determined to be 205 MPa (29.8 ksi), which is characteristic of a class 30B gray iron casting with a wall thickness of 13 mm (½ in.).

Discussion

Cracking was not caused by the metallurgical quality of the castings. The submitted castings were manufactured from gray cast iron with metallurgical characteristics and quality representative of the designated class 30B gray iron. Mechanical properties, composition, and microstructure were judged to be typical of class 30 gray iron. Metallographic analysis also revealed no evidence of metallurgical defects, such as cold shuts, voids, or slags, in the body of the casting or at the fracture surface.

Conclusion and Recommendations

Most probable cause

Two separate nonmetallurgical conditions caused failures. A thin section was observed at one fracture site—the consequence of core placement, core assembly, or shift. Evidence of core placement differences was apparent in the three castings (Fig. 4 to 6). The thin section resulting from the core placement caused the failure of casting 1.

A sharp machined corner on the seating surface was responsible for the failure of casting 2.

Remedial action

It was recommended that the manufacturer work with the foundry to evaluate the criticality of core placement and to eliminate the undesired thin section. Avoiding sharp corners is also critical in a notch-sensitive material such as gray iron.

Fracture of a Spike Maul

Carmine D'Antonio, Department of Metallurgy and Materials Science, Polytechnic University, Brooklyn, New York

An AISI 9260 steel railroad spike maul failed after a relatively short period of service. The maul head fractured in two pieces when struck against a rail. Visual, fractographic, metallographic, and chemical analyses were conducted on sections taken from the maul head, which was found to have fractured across both sides of the eye. Failure occurred in at least three separate events: formation of two cracks immediately adjacent to the eye, extension of one of the original cracks over a portion of one of the eye sides, and abrupt extension of the original crack across the eye sides, resulting in separation into two halves.

Key Words   Railroads   Silicon-manganese steels   Hammers
Brittle fracture

Alloys   Silicon-manganese steel—9260

Background

A railroad spike maul failed after a relatively short period of service.

Circumstances leading to failure

The maul was used sparingly by a track crew of a metropolitan railway over a period of 6 months. The maul head fractured in two pieces when struck against a rail, injuring a worker. The temperature on the day of the accident was between –8 and –3 °C (17 and 26 °F).

Pertinent specifications

The maul was manufactured in accordance with American Railway Engineering Association (AREA) specifications for track tools.

Specimen selection

A section was cut from the maul head parallel to and 19 mm (3/4 in.) from the fracture surface on one of the halves. The removed section was then cut parallel to the fracture and approximately 3.2 mm (1/8 in.) below the fracture surface to produce a slab approximately 19 mm (3/4 in.) wide, 9.5 mm (3/8 in.) thick, and 6.4 mm (1/4 in.) long. Parallel flats were ground on this piece, which was then subjected to hardness testing and chemical analysis.

The remaining piece, containing the fracture surface, was further sectioned for analysis by scanning electron microscopy (SEM) and metallographic examination.

Visual Examination of General Physical Features

Figure 1 shows the failed maul head, which fractured across both sides of the eye.

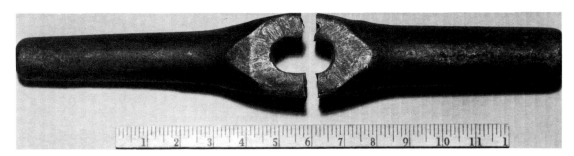

Fig. 1 Failed maul head

Testing Procedure and Results

Surface examination

Macrofractography. Figure 2 shows the mating fracture surfaces. Failure of the maul head occurred in at least three separate events. First, two cracks formed immediately adjacent to the eye. These initial cracks are shown as dark bands approximately 6.4 mm (1/4 in.) wide (arrows A). The texture of these bands was smooth, indicating brittle fracture, and covered with a black, adherent scale. These two cracks probably occurred at the same time.

Second, one of the original cracks extended over a portion of one of the eye sides. This zone is shown in Fig. 2 as a semicircular dark zone on the top fracture surface (arrows B). The texture of this portion of the fracture was typical of overload failure of a high-hardness steel. It was covered with a dark red corrosion product. This portion of the crack probably occurred during service and rusted over time.

The third event was the abrupt extension of the original crack across the eye sides, separating the

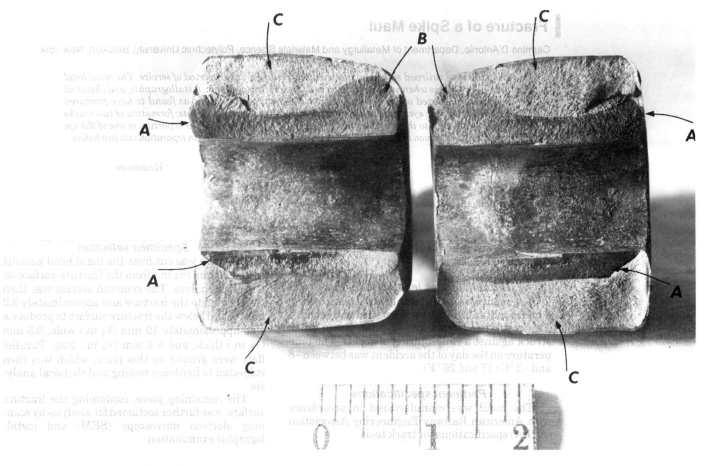

Fig. 2 Mating fracture surfaces

maul into two halves. These last fracture zones are indicated by arrows C. These zones were rust free and possessed a texture typical of overload failure of a hardened steel. It is possible that one side of the eye fractured during one blow of the maul and the other during a subsequent blow. The important point, however, is that each side completely fractured at once and in an abrupt manner.

Scanning Electron Microscopy/Fractography. SEM examination of the fracture surface corroborated that the failure mode in the final fracture area was typical of overload failure. The texture was primarily dimpled rupture, with some zones of cleavage.

Because of the thick scale on the initial crack zone (area A), no meaningful analysis could be made. The scale obliterated all of the features of this area, resulting in a bland topography.

Metallography

After SEM examination, two of the three fracture sections were mounted to show the fracture surface in cross section and metallographically prepared. The microstructure of the steel was tempered martensite. Of greater significance, however, was the presence of cracks stemming from the main fracture surface (Fig. 3). The cracks were branched, appeared to follow the prior-austenitic grain boundaries, and were oxide filled.

Chemical analysis/identification

Chemical analysis indicated that the material

Fig. 3 Micrograph showing a cross section of the fracture surface. Note the presence of branched, oxide-filled secondary cracks. 310×

was AISI 9260 steel. This grade was acceptable under the applicable standard.

Mechanical properties

Hardness tests showed that the maul possessed a uniform hardness of 51 to 52 HRC from the eye surface to the outside surface of the eye wall. This level of hardness was well within the limits specified for this application and was consistent with the microstructure and chemistry.

Discussion and Conclusions

The proximate cause of final fracture of the maul head was the presence of 6.4 mm (¼ in.) wide bands of initial fracture on both sides of the eye. These cracks caused severe stress concentration, a condition that induces abrupt overload failure under impact conditions, particularly in hardened steels. The low temperatures on the day of failure were also a contributing factor.

The texture of the initial fracture bands, the black oxide coating, and the presence of branched, scale-filled, secondary cracks indicated a quench cracking mechanism. It was concluded that the maul failed as the direct result of defects (quench cracks) in the head introduced during manufacture.

FASTENER FAILURES

Anomalous Fractures of Diesel Engine Bearing Cap Bolts*

G.H. Walter, R.M. Hendrickson, and R.D. Zipp, J.I. Case, Hinsdale, Illinois

Sudden and unexplained bearing cap bolt fractures were experienced with reduced-shank design bolts fabricated from 42 CrMo 4 steel, quenched and tempered to a nominal hardness of 38 to 40 HRC. Fractographic analysis provided evidence favoring stress-corrosion cracking as the operating transgranular fracture failure mechanism. Water containing H_2S was subsequently identified as the aggressive environment that precipitated the fractures in the presence of high tensile stress. This environment was generated by the chemical breakdown of the engine oil additive and moisture ingress into the normally-sealed bearing cap chamber surrounding the bolt shank. A complete absence of fractures in bolts from one of the two vendors was attributed primarily to surface residual compressive stresses produced on the bolt shank by a finish machining operation after heat treatment. Shot cleaning, with fine cast shot, produced a surface residual compressive stress, which eliminated stress-corrosion fractures under severe laboratory conditions.

Key Words			
	Chromium-molybdenum steels, mechanical properties		Bolts
	Intergranular fracture	Stress-corrosion cracking	Hydrogen sulfide
	Corrosion environments	Residual stress	Compressive stress
	Shot cleaning	Diesel engines	

Alloys	
	Chromium-molybdenum steel—42 CrMo 4

Background

When the International Harvester Company was in existence, sudden unexplained bearing cap bolt fractures were experienced in a line of diesel engines manufactured at one of its overseas facilities. Examination of engines at various plants, here and abroad, revealed that a small but significant percentage of the bearing cap bolts were fractured.

Pertinent specifications

The subject bolt (9/16 in.-12 UNC) was a reduced shank design with the shank diameter being smaller than the minor thread diameter. This type of bolt design had several proven advantages, especially in fatigue applications. The bolt was fabricated from an alloy steel, 42 CrMo 4, similar to AISI 4140, quenched and tempered to a nominal hardness of 38 to 40 HRC. The chemical analysis and mechanical properties required for this bolt are listed in Tables 1 and 2. The required minimum tensile strength (1,170 MPa, or 170 ksi) and yield strength (1,050 MPa, or 153 ksi) were somewhat higher than the corresponding values (1,030 MPa, or 150 ksi tensile; and 900 MPa, or 130 ksi, yield) specified for high-strength Type 8 bolts used in the United States. The high-strength bolt was commonly used in Germany by the automotive industry.

Performance of other parts in same or similar service

Bolts had been supplied by two vendors, A and B, and only bolts from vendor B were involved in the fractures. Also, only bolts from main bearing caps 2 and 4 of the four cylinder engine were involved in the fractures. These main bearing caps were the ones likely to be removed for bearing inspection after engine run-in subsequent to final assembly.

Visual Examination of General Physical Features

At first glance, broken bearing cap bolts (Fig. 1) gave the appearance of brittle fractures, primarily because of a distinct absence of deformation. However, this lack of ductility was not inherent to these bolts; under conditions of an increasing applied tensile or combined tension plus torsion load, the fractures obtained were accompanied by considerable prefracture deformation, as shown in Fig. 2.

Reprinted with revisions from SAE Paper #700528 presented at Mid-Year Meeting, 1970.

Fig. 1 Broken bolt removed from diesel engine. Absence of plastic deformation close to the fracture indicates a low ductility failure. 2×

Fig. 2 Comparison of overload failures with a typical engine fracture (right). Uniaxial tensile failure on left and an overtorqued failure (combined stress state) at center. Both overload failures exhibit ductility, with plastic deformation near the fracture regions reflected by a notable reduction of area. The engine failure (right) does not reflect a reduction in area, indicating a fracture of low ductility, unlike either overload condition. 0.95×

Fig. 3 Extensive cracking spread over the central shank portion of a nonfractured bolt detected by wet fluorescent magnetic particle inspection. Cracks are oriented circumferentially around the shank in planes of principal stress, resulting from combined torsional and tensile stresses. Regions of circumferential cracks were usually found on failed bolts in the vicinity of the fracture. The extent of cracking, however, was variable. In some instances, failed bolts were free of additional cracks other than that which initiated the fracture. 0.75×

Testing Procedures and Results

Nondestructive evaluation

Wet fluorescent magnetic particle inspection of failed bolts typically revealed regions of circumferential cracks present on the shank in proximity to the main fracture. These cracks and the fracture itself were oriented approximately 60° to the axis of the shank. This orientation appeared to correspond with the principal stress planes associated with the combined torsion and tensile stresses present in the shank when the bolt was tightened. The extent to which cracking occurred in some cases prior to failure is indicated in Fig. 3. In several cases, however, broken bolts were found to be completely free of these additional cracks.

Surface examination

Visual. The location of fractures and cracking on the bolt shank appeared to be quite random, occurring from near the head to close to the threads (Fig. 4). Crack origins in the shank corresponded to regions of maximum stress occurring in each particular bolt. The location of these regions varied among individual bolts as a result of nonuniform loads stemming from slight misalignments or geometric inconsistencies associated with component and assembly tolerances.

Macrofractography. Closer observation of the fracture surface in failed bolts (Fig. 5) revealed that usually three regions were present. At the bolt periphery, one or more rough areas (resolved to be crack origins) appeared brightly faceted and extended inward from the shank surface. In the multiple-origin fractures, these regions were usually present in a stepwise orientation around the outer surface. A central fibrous-appearing area, through which the multiple origins were joined, appeared adjacent to these faceted regions. In general, the faceted regions occurred on one side of the periphery. On the opposite side, the fibrous central area transformed to a shear lip region.

Electron fractography provided a closer examination of the fracture surface. The results of a fractographic analysis are depicted in Fig. 6 to 9. In this picture sequence, the crack propagation is traced from the intergranular origin area at the shank surface to the unstable crack growth region near the shank center. The crack originated intergranularly at the shank surface. It progressed slowly inward by this intergranular mode until the remaining cross section could no longer support the tensile load on the bolt, and sudden transgranular fracture occurred. The fracture mode of the transgranular crack was dimpled rupture.

Metallography

Crack Origins/Paths. The intergranular cracking originated at the surface is typical of several types of brittle fracture mechanisms. However, in Fig. 7 and 8, the presence of corrosion products on grain facets and networks of intergranular secondary cracking (appearing as black folds at grain boundaries) provided evidence to define the mechanism more explicitly. These two features are associated with stress-corrosion cracking in steels (Ref 2). Under conditions of high local stresses and an aggressive environment, a crack can initiate and propagate by stable intergranular growth.

The intergranular nature of these cracks is further represented by conventional microscopy in Fig. 10 and 11. In Fig. 11, the branching of secondary cracks can be readily seen.

Chemical analysis/identification

Associated Environments. Since fractographic analysis provided evidence favoring stress-corrosion cracking as the operative failure mechanism, the aggressive environment that initiated the cracks, needed to be identified. Initial testing consisted of a series of "steady stress-delayed fracture" tests, using production engine blocks. The test setup used standard bearing caps with small holes drilled through to the bolt chamber for the admission of test atmospheres.

Tests were conducted under static conditions, with bearing caps attached to the engine blocks by tightening the test bolts to specific torque values. The bolts were either dipped in a test fluid before tightening or subjected to a test atmosphere, introduced through the small hole in the bearing cap. Solutions that could come into contact with the bolts prior to or during the normal engine assembly procedure were tested. These fluids consisted of various cleaning solutions and oils used in engine production and assembly operations. The

Fig. 4 Circumferential cracks found in shank adjacent to threaded section. 1×

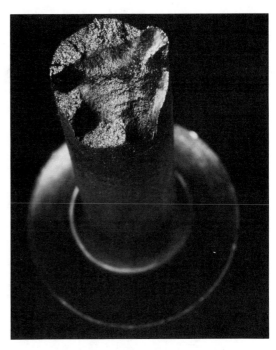

Fig. 5 Perspective view of fracture surface of a failed bolt. Several crack origin regions can be observed at the outer edge of the left side. These multiple origins occur in a stepwise fashion and appear as bright, faceted surfaces. The fibrous center region results from ductile overloading. At the right edge of the fracture surface are shear lips where final separation occurred. 3×

Fig. 6 Outer edge of a failure origin. This area reflects an intergranular mode of crack propagation. The intergranular region extends to the extreme outer surface of the fracture (some separation of the carbon replica is seen at this outer edge). 1900×

Fig. 7 Approximately 0.25 mm (0.010 in.) from origin edge showing intergranular mode of fracture and secondary intergranular cracking indicated by black folds. 1444×

Fig. 8 Approximately 0.51 mm (0.020 in.) from origin edge. Corrosion products as well as secondary cracking can be seen on the intergranular facets at this point. 1444×

Fig. 9 Nonembrittled region of fracture surface. In this area (about 3.8 mm, or 0.150 in., from origin) the mode of failure is equiaxed dimpled rupture. 1976×

Fig. 10 Longitudinal section through a failed bolt illustrating the propagation of a crack from the surface. Tool marks generated when machining the bolt shank acted as additional stress concentrators. Vilella etch, 87×

Fig. 11 A magnified view of Fig. 10, reflecting the intergranular path along prior austenite boundaries. Secondary cracks emanating from the primary crack are also evident. Vilella etch, 435×

Table 1 Chemical analysis

Element	Recommended analysis per DIN 267 (Ref 1)	Specified analysis for AISI 4140	Specified analysis for 42 CrMo 4
Carbon	0.19 to 0.52	0.38 to 0.43	0.38 to 0.45
Manganese	...	0.75 to 1.00	0.50 to 0.80
Surface	0.05 max	0.035 max	0.035 max
Phosphorus	0.04 max	0.04 max	0.035 max
Silicon	...	0.20 to 0.35	0.15 to 0.35
Chromium		0.80 to 1.10	0.90 to 1.20
Nickel	Σ ≥ 0.9	...	...
Molybdenum		0.15 to 0.25	0.15 to 0.25
Aluminum		...	...

original specified production torque of 163 to 167 J (120 to 123 ft · lb) was applied. Several bolts representing each condition survived for at least seven days without fracturing or cracking.

Negative results of the initial studies prompted consideration of other possible environments. In this group, one reactant, water with dissolved H_2S, precipitated cracks leading to delayed fracture. Warm moist gas was generated and vented into the bearing cap bolt chamber. To ensure the presence of the test environment, the test fixtures were periodically recharged. Positive results were obtained with H_2S within 20 h of testing when several bolt failures occurred, as shown in Table 3. Bolts from both suppliers (A and B) failed under these high H_2S concentration conditions, but with further testing at lower bolt tension, the frequency of failures was lower with vendor A bolts. Charging with dry bottled H_2S did not produce failures; it was necessary to have water with dissolved H_2S for failures to occur. The visual appearance of the H_2S fractures was quite similar to service failures (Fig. 12).

Engine oil is the most likely source of H_2S that could come into contact with bearing cap bolts. This likelihood was reinforced by reports that certain oil additives, such as zinc dithiophosphate (a corrosion and oxidation inhibitor), were capable of

releasing H_2S. In certain oils, a release temperature of 160 °C (320 °F) is possible in an operating engine. Some European oils were reported to have an even lower release temperature of 120 °C (250 °F). Laboratory tests confirmed that certain engine oils, mixed with a small quantity of water, could effect the release of H_2S. Further, in studies conducted by the petroleum industry, it was found that concentrations of H_2S as low as 0.1 ppm could develop stress-corrosion failures in steel (Ref 3). This low concentration of H_2S is well below the normal detectable odor level.

The probability of generating H_2S and water within the confined region of the bearing cap bolt chamber was increased by the bearing inspection procedure following engine run-in. After engine run-in, the engine was removed from the line and partially disassembled to facilitate inspection of one or more main bearings. Because the bearing caps of bearing 4 and 2 were easier to remove, these bearings were usually examined.

The time interval between run-in and inspection was minimal, resulting in the opening of hot engines to the surrounding atmosphere and thereby permitting the entrance and accumulation of moisture on internal parts (that is, on the bearing cap bolt well and chamber). During the run-in and bearing inspection, engine oil accumulated in the bolt wells. These two ingredients (moisture and engine oil), confined within the warm bearing cap bolt chamber, was all that was necessary to release H_2S.

Other possible environments resulting from breakdown of engine oil were investigated. Primary among these environments were N_2O, CO_2,

Fig. 12 Comparison of a laboratory H_2S stress-corrosion failure (left) with a broken bolt removed from an engine (right). Test conditions were high clamping load and moist H_2S-containing atmosphere. Similarity of fracture surfaces is evident. Arrows indicate origin regions. 1.0×

Table 2 Tensile properties: comparison of subject bolt and Grade 8 bolt

Property	Type 12.9 DIN 267 (Ref 1)	Grade 8 SAE J429d
Tensile strength		
max	1,380 MPa (200 ksi)	Not indicated
min	1,170 MPa (170 ksi)	1,030 MPa (150 ksi)
Yield strength, 0.2% offset		
min	1,050 MPa (153 ksi)	900 MPa (130 ksi)
Proof stress	930 MPa (135 ksi)	830 MPa (120 ksi)
Elongation, %		
min	8	12
Hardness, Brinell		
max	425	352
min	330	302
Hardness, HRC		
max	44	38
min	34	32

Table 3 Stress-corrosion tests: effect of bolt tension and vendor comparison

Vendor	No. of treatments with H_2S(a)	bolts tested	Torque J	Torque ft · lb	Results/ time, h
A	1	8	172	127	2 fail/8 to 20
A	4 to 6	4	172	127	2 fail/24 to 35
B	4 to 6	4	172	127	4 fail/25 to 30
A	4 to 6	4	138	102	1 fails/31
B	4 to 6	4	138	102	3 fail/24 to 27
A	6	4	102	75	0 fail/…
B	6	4	102	75	1 fails/33

(a) Each treatment consisted of a 60-s purge of generated warm moist H_2S gas into the bearing cap chamber once per hour.

Table 4   Properties of bolts from different vendors

Property	Vendor	
	B	**A**
Tensile strength		
MPa	1,230 to 1,300	1,190 to 1,230
ksi	178 to 188	173 to 179
Proportional limit		
MPa	1,090 to 1,120	1,100 to 1,160
ksi	158 to 163	160 to 179
Hardness, HRC	36 to 41	38 to 42

Table 5   Surface residual stresses on bolt shanks

Vendor	Condition	Compressive residual stress(a)	
		MPa	**ksi**
B	As received	76	11
A	As received	240	35
A	Stress-relieved at 538 °C (1000 °F) for 1 h	19	2.7
A	Shot-cleaned	510	74

(a) X-ray diffraction. Precision, ±55 MPa (8 ksi)

Table 6   Stress-corrosion tests: effect of shot cleaning

Vendor	Condition	No. of treatments with H$_2$S(a)	No. of bolts tested	Torque		Results/ time, h
				J	**fts · lb**	
B	As received	4 to 6	4	172	127	4 fail/25 to 30
B	As received	5	4	138	102	3 fail/7 to 27
B	Shot-cleaned	5	8	172	127	0 fail/…
B	Shot-cleaned	5	8	137	101	0 fail/…
B	Shot-cleaned	1	1	138	102	0 fail/—
A	Shot-cleaned	31	1	Yield point load(b)		1 fails/360

(a) Charged with H$_2$S, as in Table 3. (b) Tested in strain gated load cell

Fig. 13   Microstructure of a vendor B bolt. Structure consists almost entirely of tempered martensite. Prior austenite grain size ASTM 4 and 5. Vilella etch, 67×

Fig. 14   Fine tempered martensitic structure in a vendor A bolt. ASTM grain size 9 to 10. Vilella etch, 67×

SO$_2$, and formaldehyde. The various combinations of these fluids, mixed with water vapor, were introduced into a series of stress-corrosion load cells, using bolts at yield point loads. Each cell was charged three times per day, with no failures incurred until the gaseous atmosphere was heated to 60 °C (140 °F) prior to entering the load cells. Under these conditions, several failures were produced in the environment containing SO$_2$, but occurred at least 300 hours after test initiation. The failures obtained in this series, however, were heavily surface-corroded, which was not the case of service failures and H$_2$S test failures.

Mechanical properties

Hardness and Strength. Other important factors relevant to stress-corrosion cracking include the relative level of tensile stress imposed and the strength of the material. Studies of these factors in steel have shown that a relationship ex-

ists between them (Ref 4). With increasing material strength, the applied stress necessary to initiate stress-corrosion cracking decreases. At lower hardnesses, about 30 HRC, stresses in the range of the yield point are necessary to initiate cracking by H$_2$S stress corrosion. Plastic prestraining has a pronounced effect, by markedly increasing the sensitivity to H$_2$S stress-corrosion failures at hardness levels as low as 22 HRC.

Evaluation of the subject bolt steel strength characteristics, Table 4, showed that all failed and unfailed bolts were normal in complying with the specifications. There was no significant difference in hardness or strength level of the bolts supplied by the two vendors A and B.

Surface Finish and Residual Stress. Other factors found to have a significant effect on susceptibility to stress-corrosion cracking are associated with the fabrication of the bolt itself. The final machining process results in a rough shank surface

Fig. 15 Failure origin area with intergranular mode of fracture continuous to the surface

Fig. 16 Approximately 0.38 mm (0.015 in.) from origin. Mode of crack propagation is intergranular, with corrosion products and secondary cracking evident. 1444×

(Fig. 11). Surface finish requirements were quite liberal, allowing up to 4.5 µm (181 µin.). The resulting tool marks provided significant stress concentrators, which increased the maximum stress at the shank surface. In addition, the fabrication sequence, including heat treatment, provided an additional factor affecting stress-corrosion. All bolts manufactured by vendor B had been heat treated after final machining, leaving the bolts essentially free of surface residual stresses.

Bolts manufactured by vendor A were finish-machined after heat treatment, leaving a significant level of compressive residual stress at the surface. Stress-corrosion failures were nonexistent with vendor A bolts. The compressive residual stress effectively reduced applied tensile stress on the shank surface, and thereby reduced the sensitivity to stress-corrosion cracking. The values of surface residual stress measured by x-ray diffraction are listed in Table 5.

In further investigations of the effect of compressive residual stresses, stress-corrosion tests were conducted on shot-cleaned bolts. Vendor A bolts, received with a built-in residual compressive stress of 241 MPa (35 ksi), were shot-cleaned, which resulted in a stress of approximately 510 MPa (74 ksi). Bolts treated in this manner were highly resistant to stress-corrosion cracking in H_2S, as shown by the results of Table 6.

Metallurgical Characteristics. The role of metallurgical factors, such as microstructure and grain size, on H_2S stress-corrosion cracking were not well defined in this investigation. Microstructures of failed bolts reflected the normal, tempered martensitic structure characteristic of this material for the required strength. Significant grain size variations were observed between bolts supplied by the two vendors (Fig. 13 and 14). However, owing to the dominating factor of residual stress, definite conclusions with regard to the effect of

Fig. 17 Nonembrittled area of fracture in overload region, showing primarily a dimpled mode of fracture. 2964×

grain size could not be made.

Fractographic Analysis of Laboratory Test Samples. The fractographic analysis of stress-corrosion test failures induced in the laboratory was limited to the H_2S failed specimens because of the extensive corrosion of the fracture surfaces of SO_2 failures. A summary of this analysis is represented in Fig. 15 to 17, which illustrate the mode of H_2S stress-corrosion cracking propagation. Cracking originated at the bolt shank surface and progressed inward slowly along an intergranular path, accompanied by secondary cracks. When the effective cross section could no longer support the preload, rapid fracture by dimpled rupture completed the failure. The characterized failure by laboratory-induced H_2S stress-corrosion was very similar to the fracture mode of engine failures (Fig. 6 to 8).

Conclusion and Recommendations	***Most probable cause***	

Most probable cause

The unusual spontaneous fracture of high-strength diesel engine bearing cap bolts was the result of several concurring factors which established a mechanism of stress-corrosion cracking. Water containing H_2S was identified as the aggressive environment that precipitated the fractures in the presence of high tensile stress. This environment was generated by:

- The chemical breakdown of the engine oil additive, zinc dithiophosphate, which provided the H2S portion of the environment.
- Removal of certain bearing caps from the hot engine immediately after run-in, which introduced moisture into the normally sealed bearing cap chamber surrounding the bolt shank. The complete absence of fractures in bolts from one specific vendor is attributed primarily to surface residual compressive stresses produced on the bolt shank by a finish machining operation after heat treatment.

Remedial action

Shot cleaning, with fine cast shot, produced a high level of surface residual compressive stress, which eliminated stress-corrosion fractures under severe laboratory conditions.

Acknowledgment

The authors express their appreciation to their associates in the metallurgical, lubrication, and mechanical test laboratory at Engineering Research, to H. Michalek and his colleagues at Neuss Works for their technical assistance and guidance, and to E.J. Rusnak for help with the illustrations.

References

1. DIN 267, Part 3, German Standards Association, October 1967.
2. A. Phillips, V. Kerlins, and B.V. Whiteson, *Electron Fractography Handbook*, Technical Report ML-TDR-64-416, January 1965.
3. C.M. Hudgins *et al.*, "Hydrogen Sulfide Cracking of Carbon and Alloy Steels," *Corrosion*, Vol 22, August 1966.
4. D. Warren and G.W. Beckman, "Sulfide Corrosion Cracking of High Strength Bolting Materials," *Corrosion*, Vol 13, October 1957.

Cracking at the Threads of Stud Bolts Used for Lifting Plastic Mold Dies

W.B.F. Mackay, Materials and Metallurgical Engineering Department, Queen's University, Kingston, Canada

Two 38 mm (1.5 in.) diam threaded stud bolts that were part of a steel mold die assembly from a plastics molding operation were examined to determine their serviceability. Chemical analysis showed the material to be a plain carbon steel that approximated 1045. Visual examination revealed evidence of severe hammer blows to the clevis and boss areas and a gap between the die and the underside of the boss. Magnetic particle inspection showed cracks at the thread roots that, when examined metallographically, were found to contain MnS stringers. The cracking of the threads was attributed to a poor stud bolt design, which allowed a high stress concentration to occur at the base of the threads upon application of a lateral load. It was recommended that bolts of a new design that incorporated a stress-relieving groove be used. Threading of the bolt to eliminate the gap between the lower face of the boss and the die and an improved method of inserting or removing the bolt to avoid hammering (use of a wrench on a square or hexagonal boss) were also recommended.

Key Words			
Carbon steels	Cracking (fracturing)	Dies	
Bolts	Stress concentration	Studs	
Screw threads			

Alloys Carbon steel—1045

Background

Two 38 mm (1.5 in.) diam plain carbon steel threaded stud bolts were examined to determine their serviceability. They were part of a steel mold die assembly from a plastics molding operation.

Applications

The plastic molding company involved moves its steel dies by means of an overhead traveling crane. The stud bolts are threaded for their entire length, up to a round boss, to which a clevis attaches to a fulcrum pin in the boss. Each mold has tapped holes to receive two stud bolts. The molds are moved by engaging each clevis with a hook slung from the crane. Failure of the stud bolts could result in serious damage and/or injury.

It was reported that the two stud bolts were designed to lift a load of 10.4 Mg (23,000 lb) and that the maximum load encountered in service was 7.7 Mg (17,000 lb). The parts had been used an average of once per week over a period of 3 years. No specifications were available.

Specimen selection

Of the two stud bolts submitted for examination, one was relatively easily removed from the die, but the other proved too difficult to remove with available equipment.

Visual Examination of General Physical Features

Visual examination of both stud bolts showed evidence of severe hammer blows to the clevis and boss areas. The stud bolts were not fully screwed into the die, leaving a gap between the die and the underside of the boss.

Testing Procedure and Results

Nondestructive evaluation

Magnetic particle inspection revealed cracks at the root of several threads. This testing method did not provide much more information than did the visual (macroscopic) examination.

Surface examination

Visual. After degreasing and cleaning were conducted, a macroscopic examination revealed cracks in several adjacent threads. All were located at the root of the thread and progressed a short distance around the circumference. Similar cracks were observed in the same region on the diametrically opposite face of the stud bolt. The cracked threads were concentrated about 20 to 30 mm (0.8 to 1.2 in.) or more below the stud bolt boss. The exact location of a cracked thread would depend on how far the stud bolt had been inserted into the die block (see Fig. 1). The threads had been cut by turning on a lathe, rather than by roll forming. Rolled threads would have possessed a better surface finish and would have greatly lessened the notch effect at the thread root.

Microstructural Analysis. A specimen was taken from the base of a cracked thread root and viewed unetched (Fig. 2). A maximum crack depth of 4.8 mm (0.19 in.) was measured. Other cracks were observed in several adjacent threads. An etched specimen revealed manganese sulfide stringers in a relatively fine-grain, tempered martensitic matrix (Fig. 3).

Crack Origins/Paths. All cracks originated at the bottom of the threads.

Chemical analysis/identification

Material. A grinding wheel test was conducted on the stud bolt and the results were compared with known samples. The test indicated a plain carbon steel of medium-carbon content, approximating a 1045 steel.

Mechanical properties

Hardness tests on the stud bolt revealed an average hardness of 37 HRC.

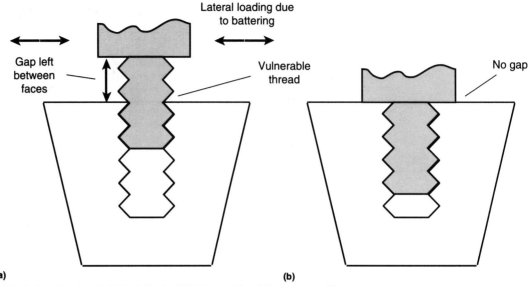

Fig. 1 Insertion of stud bolt into die block. (a) Existing condition. (b) Improved condition

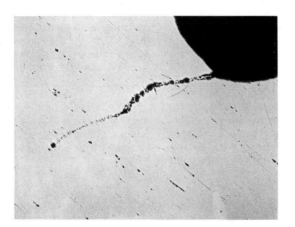

Fig. 2 Micrograph of stud bolt, showing crack at root of thread and nonmetallic stringers. Unetched. 38×

Fig. 3 General microstructure of stud bolt, showing gray MnS stringer inclusions running parallel to longitudinal bolt axis in tempered martensite matrix. Nital etch. 380×

Discussion

The battering of the clevis and boss when inserting or, more probably, when removing the stud bolt from the die would have caused a high stress concentration at the base of the threads. This situation was aggravated by the gap between the die and base of the boss (Fig. 1a). If the bottom of the boss had been in contact with the die (Fig. 1b), then the stress on the threads would have been diminished.

A further improvement in design is shown in Fig. 4. A stress-relieving groove would take the load off the upper, most vulnerable threads. The microstructure was satisfactory, although the MnS content may have been borderline, and the hardness was acceptable.

Another method of tightening and loosening the stud bolt must be devised in order to avoid hammering. A hexagonal or square-shaped boss might be rotated by a special wrench that would avoid contact with the clevis. A centrally located eyebolt might also be considered. Rolled threads would be preferred to the existing machine-cut

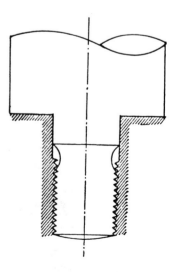

Fig. 4 Improved design with stress-relieving groove

threads. An increased lateral load could be produced when a wide stud bolt spacing is used with short slings.

| **Conclusion and Recommendations** | | |

Most probable cause

The cracks in the threads were probably caused by a high stress concentration at the thread root, which was due to a bending moment caused by hammering of the boss and clevis, especially with a gap between the boss and die. Machine-cut threads may have been a minor factor. The presence of MnS stringers could have aggravated the situation.

Remedial action

The following recommendations were submitted:

- Replace all stud bolts with those of a design similar to that shown in Fig. 4.
- Ensure that the inserted stud bolt has the lower face of the boss in contact with the die.
- Use roll forming rather than machine cutting for the threads.
- Devise a design that allows a wrench to engage the boss to allow insertion or removal of the stud bolt (for example, a hexagonal or square-headed boss).
- Conduct nondestructive testing of the stud bolts approximately every 6 months.
- Develop a cleanliness specification to avoid the presence of excessive MnS stringers and consider the use of an alloy steel.
- Ensure that the slings are not too short, especially when moving dies with a relatively wide spacing of the stud bolts.
- Do not, under any circumstance, batter a stud bolt or clevis when removing it from a die.

Failed Bolts From an Army Tank Recoil Mechanism

Victor K. Champagne, United States Army Research Laboratory, Materials Directorate, Watertown, Massachusetts

The heads of two AISI 8740 steel bolts severed while being installed into an Army tank recoil mechanism. Both broke into two pieces at the head-to-shank radius and the required torque value had not been attained nor exceeded prior to the failure. A total of 69 bolts from inventory and the field were tested by magnetic particle inspection. One inventory bolt failed because of a transverse crack near the head-to-shank radius. It was deduced that either a 100% magnetic particle inspection had not been conducted during bolt manufacturing, or the crack went undetected during the original inspection. Optical and electron microscopy of the broken bolts revealed topographies and the presence of black oxide consistent with quench cracking. The two bolts failed during installation due to the presence of pre-existing quench cracks. Recommendations to prevent future failures include: ensuring that 100% magnetic particle inspections are conducted after bolts are tempered; using dull cadmium plate or an alternative to the electrodeposition process, such as vacuum cadmium plate or ion-plated aluminum, to mitigate the potential for delayed failures due to hydrogen embrittlement or stress-corrosion cracking; ensuring that the radius at the shoulder / shank interface conforms to specifications; and replacing all existing bolts with new or reinspected inventory bolts.

Key Words		
Nickel chromium molybdenum steels	Cracking (fracturing)	Quench cracking
Magnetic particle testing	Stress corrosion cracking	Hydrogen embrittlement
Recoil mechanisms		

Alloys   Nickel chromium molybdenum steels — 8740

Background   The heads of two bolts severed while being installed into an an Army tank recoil mechanism.

Applications

Twelve bolts (4.13-in.) secure the recoil mechanism which absorbs the high impact forces induced by the 105-mm (4.13-in.) cannon. If these two bolts had failed during firing, the recoil mechanism may have been torn away from the tank, causing serious injury or death to the soldier operating the cannon.

Circumstances leading to failure

The bolts were previously stored in inventory, and were required to have been subjected to a 100% magnetic particle inspection during manufacturing. While being installed, both bolts broke into two pieces at the head-to-shank radius. It was reported that the required torque value of 163 to 190 N · m (120 to 140 ft · lbf) had not been attained nor exceeded prior to the failure.

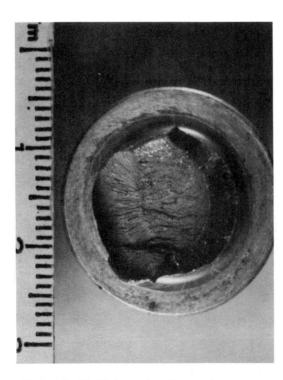

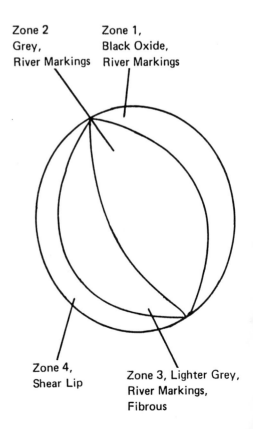

Zone 2, Grey, River Markings
Zone 1, Black Oxide, River Markings
Zone 4, Shear Lip
Zone 3, Lighter Grey, River Markings, Fibrous

Fig. 1   Bolt which failed magnetic particle inspection

Pertinent specifications

The nominal bolt size is 9/16 in.-18 (.5625-18 UNJ). The bolts were specified to be fabricated from AISI 8740 steel in accordance with MIL-S-6049A; quenched and tempered to a hardness of 39 to 43 HRC; and conform to the requirements of MIL-B-8831B, with respect to dimensions, thread rolling, grain flow, and mechanical properties. The parts were required to be electrolytically cadmium-plated in accordance with QQ-P-416E Type II, Class 2, and a post-bake at 190 °C (375 °F) was specified to prevent hydrogen embrittlement. A 100% magnetic particle inspection in accordance with MIL-I-6868 was required.

Performance of other parts in same or similar service

Bolts fabricated by the same manufacturer and from the same heat treat/shipping lots as the failed components had not failed during installation or in service.

Selection of specimen

The two broken bolts as well as others taken from the field and inventory were examined. These bolts represented the only two known suppliers, various heat treat lots, and a varied firing history.

Testing Procedure and Results

Nondestructive evaluation

Radius Measurement at the Shoulder/Shank Interface. The engineering drawing required the shoulder/shank radius to be 0.057 + 0.0000 to 0.0010 in. Approximately half the bolts inspected did not meet the specification requirement. The radius of these bolts were slightly sharper than specified, ranging from 0.048 to 0.055 in. The increased sharpness could provide crack initiation sites due to higher stress concentration.

Magnetic Particle Inspection. A total of 69 bolts from inventory and the field were subjected to magnetic particle inspection to determine the presence of discontinuities in accordance with MIL-I-6868. Only one of the inventory bolts failed due to the presence of a crack. Figure 1 shows evidence of this crack revealed by the test using black light photography. This bolt contained a transverse crack near the heat-to-shank radius, similar to the two failed bolts, which extended over two-thirds of the circumference of the shank. During manufacturing, a 100% magnetic particle inspection was required. Since this bolt failed the re-inspection, it was deduced that either a 100% inspection had not been conducted or the crack went undetected during the original inspection.

Surface examination

Visual. Figure 2 shows a fractograph representing a fracture half of one of the failed bolts.

Four distinct fracture zones are depicted schematically. A dark black area identified as Zone 1 located at the periphery of the circular fracture face was observed. This region resembled a crescent moon in geometry and contained coarse river markings indicative of crack growth. The origin of the black color encompassing Zone 1 did not appear to be the result of staining but rather a corrosion product or heat treat scale. Adjacent to Zone 1 was a grey area, Zone 2, which also contained coarse river markings. Zone 2 transitioned into another grey region, Zone 3, which contained more fibrous river markings. Final fracture occurred in Zone 4, a shear lip which encompassed much of the outer perimeter of the bolt.

Scanning Electron Microscopy/Fractography. Scanning electron microscopic examination of Zone 1 revealed an intergranular fracture surface within the black layer, as shown in Fig. 3. Energy dispersive spectroscopy on this surface showed those elements associated with the steel, as well as a high oxygen concentration (Fig. 4). The black matter was similar in appearance to a heat treat scale as opposed to simple atmospheric corrosion. EDS did not reveal any other surface contaminants or aggressive species. Zone 2 was characterized by a mixed intergranular and ductile dimpled topology (Fig. 5). Zone 3 also contained this mixed mode of fracture in a very fibrous manner but there was more ductile dimpling than

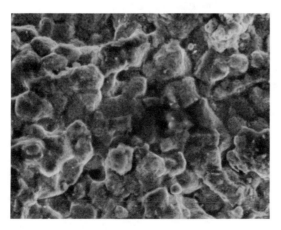

Fig. 2 Fractograph representing a fracture half of one of the failed bolts. 5×

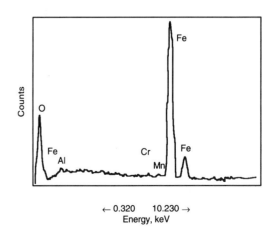

← 0.320   10.230 →
Energy, keV

Fig. 3 SEM of Zone 1 showing intergranular fracture surface with a covering of black oxide. 1000×

Fig. 4 EDS spectrum of Zone 1

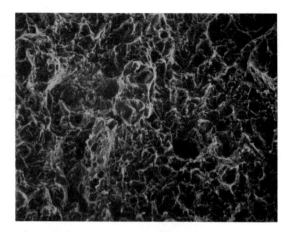

Fig. 5 Mixed intergranular and ductile dimpled morphology of Zone 2. 1000×

Fig. 6 SEM of the interface of Zone 3 (top) and Zone 4 (shear). 500×

Fig. 7 Representative micrograph of microstructure showing tempered martensite. 1% nital. 1000×

in Zone 2. Zone 4 displayed a typical shear/fast fracture morphology as evidenced by shear dimples (Fig. 6).

Metallography

Microstructural Analysis. Figure 7 shows the microstructure of the failed bolts, which consisted of tempered martensite, typical of a quenched and tempered low-alloy steel. This microstructure compared favorably to the microstructure of bolts from inventory and in the field. The material was clean with no major inclusions present.

Plating Thickness and Uniformity. Metallographic samples of the two failed bolts and bolts from inventory were prepared from the shoulder region to examine the cadmium plating. The cadmium plating on both specimens displayed good adherence. The first failed bolt conformed in average thickness (0.012 mm, or 0.00047 in.) to a class 2 plating as specified in QQ-P-416E. The bolts from inventory also conformed in average thickness (0.01 mm, or 0.00039 in.) to class 2. The second failed bolt contained a much thinner average coating (0.004 mm, or 0.00016 in.), which was less uniform. This coating did not conform to the class 2 requirement, but more closely to a class 3 coating.

Table 1 Chemical analysis (wt%) of failed bolts

Element	Bolt A	Bolt B	AISI 8740
Al	0.023	0.021	...
Cr	0.52	0.53	0.4 to 0.6
Cu	0.12	0.03	0.35
Mn	0.82	0.88	0.75 to 1.0
Mo	0.19	0.19	0.15 to 0.25
Ni	0.59	0.68	0.4 to 0.7
P	0.015	0.007	0.025
Si	0.21	0.27	0.2 to 0.35
S	0.01	0.006	0.04
C	0.39	0.43	0.38 to 0.43

Chemical analysis/alloy identification

Material. Samples from both failed bolts were analyzed for chemical composition. The chemistry, as listed in Table 1, satisfied the requirements of AISI 8740 steel.

Mechanical properties

Hardness. Knoop microhardness profiles were conducted on samples prepared from both failed bolts, as well as bolts from the field and inventory. There was evidence of a slight hardness gradient. The bolts from inventory exhibited hardness values of 39.8 to 42.5 HRC in conformance with the governing specification (39 to 43 HRC). However, the two failed bolts and the bolts with a prior in-service firing history taken from the field exhibited slightly higher hardness values (44.0 to 45.1 HRC).

Ultimate Tensile Load. A total of eleven bolts were chosen randomly to determine the ultimate tensile load, as required by MIL-B-8831B. Each bolt was inserted into a threaded fixture at one end while the head of the bolt was constrained. The bolt was then pulled in tension to failure and the ultimate tensile load recorded. The values were between 190 and 203 kN (42,750 and 45,750 lbf) exceeding the minimum requirement of 174 kN (39,200 lbf).

Torque Testing. Torque testing was conducted on eight bolts from the field and eight bolts from inventory, employing a calibrated wrench fitted with a heavy-duty socket. The torque-to-failure of each bolt fell within 373 to 610 N·m (275 to 450 ft·lbf). These values exceeded the minimum torque requirement of 163 to 190 N·m (120 to 140 ft·lbf). Seven of the bolts failed at the beginning of the threaded round section while the remaining bolts failed in the center of the threaded region. Since none of the bolts failed at the heat-to-shank radius, the sharper than specified radius did not inde-

pendently contribute to premature failure of the bolt. Generally, the bolts from the inventory exhibited higher torque failure values (averaging 546 N·m, or 403 ft·lbf) when compared to the bolts from the field (averaging 469 N·m, or 346 ft·lbf).

Stress Durability Testing. To investigate the possibility that hydrogen may have been introduced into the bolt during the electrolytic cadmium-plating operation, and may not have been adequately removed by the low-temperature embrittlement relief treatment (causing hydrogen embrittlement), a stress durability test was carried out in accordance with MIL-STD-1312, test 5. Eleven new bolts from inventory and six field bolts were loaded to 80% of the ultimate tensile strength and in this loaded condition subjected to 200-h stress durability test (MIL-STD-1312 specifies a duration of 23 h). None of the bolts fractured, and transverse cracks were not observed during NDT inspection of the bolts after testing. Twelve additional bolts from inventory were also tested with the duration of the test extended to 400 hours. There were no failures after stressing at 80% of the ultimate tensile strength for 400 h.

Discussion

Quench cracks in steel result from stresses produced during the austenite-to-martensite transformation, which is accompanied by an increase in volume. The observed cracks in the failed bolts met the following characteristics of quench cracks (Ref 1): the crack extends from the surface toward the center of mass; the crack grows and exhibits a shear lip at the outer surface; the crack does not exhibit evidence of decarburization in a microscopic examination; and as a result of tempering after quenching, the fracture surface is blackened by oxidation. Further support of a quench crack fracture mechanism may be found in the examination of both light optical and electron microscopic fractographs of typical quench cracks in a 4340 steel (Ref 2). These fractographs show the quench crack crescent where the crack is intergranular in nature. Comparable fractographs of the two failed bolts (Fig. 2 and 3) show the same features; the quench crack crescent designated in Zone 1, and

the intergranular fracture mode in Zone 1. Fractographic examination of the bolt which failed magnetic particle inspection showed features very similar to those observed on the two bolts which failed during installation. This also supports the contention that the cracks were pre-existing quench cracks and not due to the service environment.

Although fractographic examination of the failed bolts showed intergranular fracture origins (Zone 1) which can occur as a result of hydrogen embrittlement (HE) and stress-corrosion cracking (SCC), it is unlikely that HE or SCC caused these failures. The stress durability test showed that new bolts were not susceptible to HE. Neither HE nor SCC produces the black oxide observed. The only known mechanism which would form this oxide is thermal growth during heat treatment, since this part was not exposed to high temperatures except during fabrication.

Conclusion and Recommendations

The two bolts failed during installation due to the presence of pre-existing quench cracks. Future failures may be prevented if the 100% magnetic particle inspection of the bolts after tempering is ensured. Also, dull cadmium plate or an alternative to the electrodeposition process, such as vacuum cadmium plate or ion-plated aluminum, should be utilized to mitigate the potential for delayed failures due to HE or SCC. After plating, a

24-hour embrittlement relief baking at 190 °C (375 °C) to remove and redistribute hydrogen within the bolt should prevent HE failures. In addition, the radius at the shoulder/shank interface must conform to specification requirements. Finally, all existing bolts should be replaced with new or reinspected inventory bolts to prevent the possibility of undetected small quench cracks growing under firing loads.

References

1. *ASM Metals Handbook, Vol 10: Failure Analysis and Prevention*, 8th ed., American Society for Metals, Metals Park, OH, 1975, p 74.

2. *ASM Metals Handbook, Vol 9: Fractography and Atlas of Fractographs*, 8th ed., American Society for Metals, Metals Park, OH, 1974, p 308.

Failed Main Rotor Pitch Horn Bolt from an Army Attack Helicopter

Victor K. Champagne, Marc Pepi, and Gary Wechsler, United States Army Research Laboratory, Materials Directorate, Watertown, Massachusetts

One of the two AISI 4340 steel pitch horn bolts from the main rotor hub assembly failed while in service. Optical microscopy revealed evidence of corrosion pitting in regions adjacent to the fracture. Fractographic examination utilizing a scanning electron microscope revealed multiple crack origins which assumed a "thumbnail" shape and displayed surface morphologies which resulted from intergranular decohesion. Many of the crack sites initiated from corrosion pits. Energy dispersive spectroscopy performed on areas within the crack initiation site showed the presence of chlorides. The failure was attributed to stress-corrosion cracking. Short- and long-term recommendations to prevent future failures are given.

Key Words

Fasteners	Bolts	Brittle fracture
Pitting (corrosion)	Stress-corrosion cracking	Crack initiation
Nickel chromiummolybdenum steels corrosion		

Alloys Nickel-chromium molybdenum steels—4340

Background

A main rotor pitch horn bolt was found to be broken during a maintenance procedure on an Army attack helicopter.

Applications

The bolt is illustrated in Fig. 1. Two pitch horn bolts are inserted into the pitch horn assembly which controls the pitch of the helicopter blades during flight. If one of these bolts were to fail in service, the load is transferred to the remaining component, which drastically reduces its fatigue life. Failure of both pitch horn bolts would prevent the pilot from properly controlling the helicopter during flight and when landing.

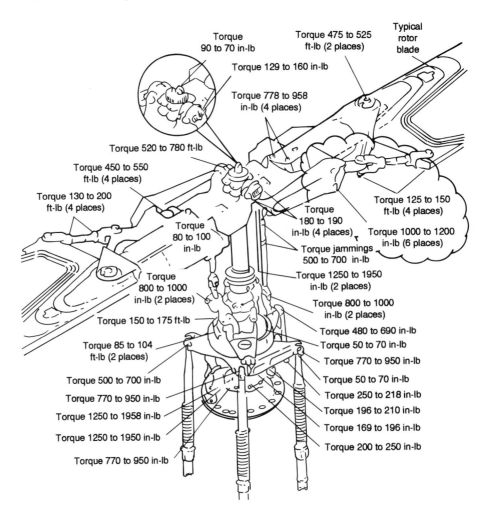

Fig. 1 Schematic of pitch horn bolt **(continued)**

Table 1 Chemical analysis of a main rotor pitch horn bolt

Material	Composition, wt%								
	C	Mn	Si	P	S	Cr	Ni	Mo	Cu
Bolt	0.387	0.71	0.28	0.007	0.002	0.82	1.76	0.28	0.067
AMS 6414	0.38 to 0.43	0.60 to 0.90	0.15 to 0.35	0.015 max	0.015 max	0.70 to 0.90	1.65 to 2.00	0.20 to 0.30	0.35 max

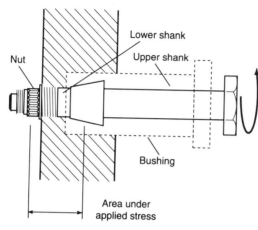

Fig. 1 Schematic of pitch horn bolt

Fig. 2 Component in the as-received condition. .81×

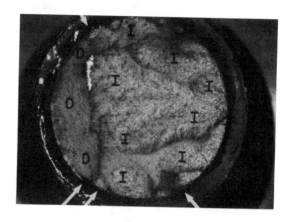

Fig. 3 Fracture surface showing the various crack propagation zones (represented by arrows). "I" represents intergranular, while "D" represents ductile. 5.33×

Pertinent specifications

The pitch horn bolt was fabricated from AISI 4340 steel in accordance with AMS 6414, and subsequently heat treated to attain an ultimate tensile strength range of 1241 to 1379 MPa (180 to 200 ksi). The bolts were electrolytically cadmium plated per QQ-P-416. The bolts were then subjected to 100 percent magnetic particle inspection per MIL-I-6868.

Performance of other parts in same or similar service

Bolts fabricated by the same manufacturer and from the same heat treat/shipping lots as the failed component did not have a history of breaking in service.

Specimen selection

The two broken sections of the failed pitch horn bolt were subjected to a metallurgical examination.

Testing Procedure and Results

Surface examination

Surface Finish. The surface finish of the failed bolt was measured. The engineering drawing specified a surface finish of 125 RMS, except at the lower shank to conical section radius where it was required to be 63 RMS. Measurements taken lengthwise on the bolt, 120° apart, revealed average readings of 64, 68 and 70 RMS, well within the drawing requirements. However, accurate measurements could not be taken at the lower shank to conical section radius, because the failure occurred in this region.

Visual. Figure 2 shows the component in the as-received condition. The failure occurred at the radius located between the conical section and the lower shank. Further examination of surfaces near the failure revealed evidence of corrosion pits and machining marks. A representative

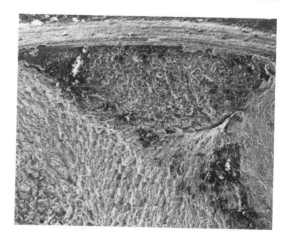

Fig. 4 Scanning electron micrograph showing a "thumbnail" crack. 47.25×

Fig. 5 Scanning electron micrograph showing corrosion pits adjacent to the fracture (designated by arrows). 31.5×

Fig. 6 Scanning electron micrograph of a failure mode transition from intergranular to ductile dimples. 126×

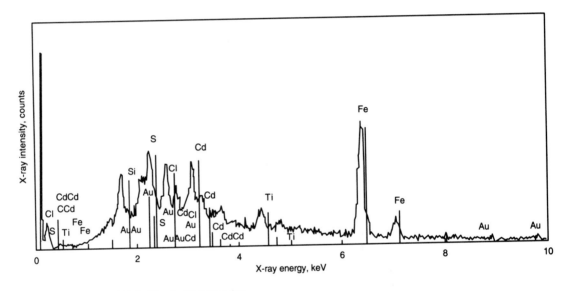

Fig. 7 EDS spectrum within the "thumbnail" crack region

macrograph of the fracture surface of the failed bolt is shown in Fig. 3. The cadmium plating circumventing the fracture was blistered and worn away entirely in numerous locations, exposing a bare metal surface to the environment. This condition is ideal for stress-corrosion cracking (SCC). The arrows on the macrograph represent the numerous fracture origins. These areas assumed a geometry indicative of a "thumbnail" crack. The fracture path was complex and not easily discernable. The material did not exhibit much plasticity during fracture.

Fractography. Figure 4 shows one of the "thumbnail" cracks covered with a dark layer of corrosion. The failure occurred at the machined radius, which contained corrosion pits, as shown in Fig. 5. Many of the thumbnail cracks initiated from corrosion pits. The morphology within the thumbnail crack was different than the surrounding fracture surface. The failure mode within this region was intergranular, with some evidence of secondary cracking. The area beyond the thumbnail cracks exhibited a mixed topography of ductile dim-

Fig. 8 Representative tempered martensite microstructure of the pitch horn bolt. 245×

ples and intergranular cracking, before the onset of final fast fracture, as shown in Fig. 6. Energy dispersive spectroscopy (EDS) was used to characterize the composition of the dark corrosion layer which covered the thumbnail crack regions. EDS analysis revealed chlorides (Fig. 7) which were detected in large concentrations, and have been known to accelerate the corrosion rate of carbon steel.

Metallography

Microstructural Analysis. A 1% nital etch applied to metallographically prepared cross-sections of the bolt revealed a typical tempered martensitic structure, as shown in Fig. 8. No regions of unusual precipitation or coagulation of carbides existed. Cross-sectional samples of material containing machine marks and corrosion pits were also taken. The machine marks were approximately 0.1 to 0.2 mm (0.004 to 0.008 in.) deep while some pits were as much as 0.6 mm (0.025 in.) deep. In addition, the material was free of large inclusions or inherent material defects.

Plating Thickness and Uniformity. Metallographic samples were utilized to determine the cadmium plating thickness. Where measurable, the plating exhibited an average thickness (0.046 mm, or 0.0018 in.) in conformance with QQ-P-416.

Chemical analysis/alloy identification

Material. A sample of the failed bolt was analyzed for chemical composition, and the results (Table 1) were within specifications for AISI 4340 steel.

Mechanical properties

Hardness. Macrohardness measurements were taken on a transverse section of the bolt. The results (43.7 HRC) were slightly higher than the specified limits (39 to 43 HRC).

Discussion

The material did not exhibit much plasticity during fracture, which was anticipated from an embrittlement-type failure. Multiple crack origin sites are a common feature associated with SCC, whereas a failure attributed solely to hydrogen embrittlement (HE) would normally consist of a single large crack. In addition, HE fracture surfaces tend to be free of heavy oxides and corrosion, while SCC surfaces are not. SCC failures are environmentally induced, the result of combined interaction of mechanical stress and corrosion, where neither factor acting independently or alternately would initiate and propagate a crack until final catastrophic failure has occurred. The stresses required to cause SCC are often times quite minimal, especially in high strength steels which have a considerably low value of K_{ISCC}.

Conclusion and Recommendations

The failure of the pitch horn belt was attributed to SCC based upon the results of the investigation.

Short-term recommendations include the assurance that surface coatings remain intact to combat the formation of corrosion pits, which act as crack initiation sites. However, if plating discontinuities are revealed during depot overhaul and repair, the parts should not be replated electrolytically and placed back into service. Instead, the bolts should be discarded because it is not cost effective to strip and replate the bolts. Finally, the surface finish of the bolts should be visually inspected and compared to a standard. Bolts should be rejected if they contain deep machining marks because they form stress concentration regions and act as SCC initiation sites.

Long-term solutions to the problem of HE or hydrogen-assisted SCC would be to redesign the bolt to minimize stress concentration areas. In addition, the component could be fabricated from an alternative material that has a higher value of K_{ISCC}. It was strongly recommended that vacuum-deposited cadmium be used in place of electrolytic cadmium plating. The vacuum deposition process will prevent HE of the material during plating that can be induced by the electrolytic plating process.

Failure Analysis of a Helicopter Main Rotor Bolt

Mohan D. Chaudhari, Columbus Metallurgical Services, Inc., Columbus, Ohio

A helicopter main rotor bolt failed in the black-coated region between the threads and the taper section of the shank during assembly. The torque applied was approximately 100 N·m (900 lbf · in.) when the bolt sheared. No other bolts were reported to have failed. The failed bolt material conformed to AISI E4340 steel, as specified. The microstructure was tempered martensite, with hardness ranging from 41 to 45 HRC. Failure was in the shear ductile mode. The crack initiated in the area of slag inclusions. Inspection of other bolts from the same shipment was recommended.

Key Words		
Nickel-chromium-molybdenum steels	Nonmetallic inclusions	Bolts
Helicopters	Ductile fracture	

Alloys   Nickel-chromium-molybdenum steel—4340

Background

During assembly, a helicopter main rotor bolt made of AISI E4340 steel failed while being torqued. The torque being applied reached approximately 100 N· m (900 lbf · in.) when the pitch horn bolt sheared and shot across the hangar. No other bolts were reported to have failed during assembly.

Visual Examination of General Physical Features

The rotor bolt fractured in the black-coated region between the threads and the taper section of the shank. Figures 1 to 3 show the as-received bolt pieces. The head, the straight portion of the shank, and the nut were gold colored. The fracture surfaces were first cleaned with soap and water and were then ultrasonically cleaned to remove surface debris. Longitudinal and transverse sections were prepared for optical microscopy. The bolt surface was also inspected for identification of coatings and surface treatments.

Fig. 1   As-received failed rotor bolt

Fig. 2   Closeup view of failure area

Fig. 3   Macrograph of mating fracture surfaces

Testing Procedure and Results

Surface examination

Optical Microstructure. The bolt had been heated to obtain a tempered martensitic microstructure. The metallographic specimen showed a significant amount of nonmetallic inclusions. Figure 4 shows a pore, probably a pipe defect in the original bar stock.

Scanning Electron Microscopy/Fractography. Figure 3 shows the two mating halves of the fracture. Figure 5 shows closeup views of the fracture initiation region. The white inclusions were nonmetallic particles that were high in calcium. Figure 6 illustrates a ductile mode of failure elsewhere on the fracture surface. The ductile dimples exhibit a combination of shear and tensile modes of fracture propagation.

X-Ray Analysis. Various spectra and X-ray maps were collected. The X-ray maps illustrated

Fig. 4 Closeup view of a pore in a matrix of tempered martensite. 378×

Table 1 Results of chemical analysis

Element	Bolt material	AISI E4340 chemical requirement(a)
Carbon	0.41	0.38-0.43
Manganese	0.64	0.65-0.85
Phosphorus	0.006	0.025 (max)
Sulfur	0.002	0.025 (max)
Silicon	0.27	0.15-0.30
Copper	0.093	NR
Nickel	1.87	1.65-2
Chromium	0.75	0.70-0.90
Molybdenum	0.25	0.20-0.30
Aluminum	0.034	NR
Calcium	0.0012	NR

(a) NR, no requirement

(a)

(b)

Fig. 5 SEM micrographs showing segregation of inclusions along edge where failure probably initiated. (a) Backscattered electron image. (b) Secondary electron image

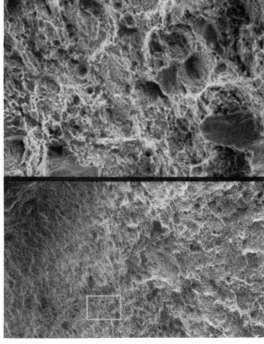

Fig. 6 SEM fractographs showing ductile mode of failure. (a) Higher-magnification view of boxed area in (b)

the presence of calcium in the white spots. These spots were thus interpreted to be slag inclusions in the steel. Areas containing the inclusions were cleaned with a soft brush several times to ascertain that the X-ray indications were not debris that collected on the fracture surface during storage and handling after the failure.

The gold-colored areas of the bolt were found to have been cadmium plated, followed by a chromate treatment. The black-coated surfaces contained molybdenum and sulfur, indicating a MoS$_2$ treatment.

Chemical analysis/identification

Table 1 compares the actual specimen composition, as determined by optical emission spectrometry, with nominal specifications for AISI E4340 steel.

Mechanical properties

Hardness. Knoop microhardness measurements were conducted using a 500 g load. Longitudinal specimens had an average hardness of 421 HK (~41 HRC). Transverse specimens had an average hardness of 441 HK (~45 HRC).

Conclusion and Recommendations

Most probable cause

It was initially suspected that hydrogen embrittlement caused crack initiation. It was concluded, however, that the failure was caused by a high concentration of nonmetallic inclusions.

Remedial action

Inspection of other bolts from the same shipment was recommended.

Fracture of a Trunnion Bolt Used To Couple Railway Cars

Carmine D'Antonio, Department of Metallurgy and Materials Science, Polytechnic University, Brooklyn, New York

A trunnion bolt that was part of a coupling system in a metropolitan railway system failed in service, causing cars to separate. The bolt had been in service for more than 10 years prior to failure. Visual examination showed that the failure resulted from complete fracture at the grease port and surface groove located at midspan. Drillings machined from the bolt underwent chemical analysis, which confirmed that the material was AISI 1045 carbon steel, in accordance with specifications. Two sections cut from the bolt were subjected to metallographic examination and hardness testing. The fracture origin was typical of fatigue. The ultimate tensile strength of the bolt was in excess of requirements. Wear patterns indicated that the bolt had been frozen in position for a protracted period and subjected to repeated bending stresses, which resulted in fatigue cracking and final complete fracture. It was recommended that proper lubrication procedures be maintained to allow free rotation of the bolts while in service.

Key Words			
Carbon steels	Axles		Lubrication
Bolts	Fatigue failure		Couplings
Railroads			

Alloys Carbon steel—1045

Background

A trunnion bolt that was part of a coupling system in a train failed in service, causing cars to separate.

Applications

The trunnion bolt was used in a metropolitan railway system and thus was subjected to repeated bending impact loads. Under such service conditions, the presence of a stress concentration (e.g., a fatigue crack) would be conducive to overload failure.

Circumstances leading to failure

The bolt had been in service for more than 10 years prior to failure. Failure occurred in January when the temperature was near –7 °C (20 °F).

Pertinent specifications

It was specified that the bolt be machined from AISI 1045 steel bar in the normalized and tempered condition. A minimum ultimate tensile strength of 610 MPa (88 ksi) was required.

Specimen selection

Drillings were machined from the bolt for chemical analysis. Two sections were cut from the bolt. One was mounted and prepared for metallographic examination. The other was ground under coolant to provide parallel surfaces for determination of hardness near the surface and at the core.

Visual Examination of General Physical Features

Figure 1 shows the as-received bolt. The fracture origin and a specific wear pattern on the bolt surface are indicated. Figure 2 shows one of the two mating fracture surfaces.

Failure resulted from complete fracture at the grease port and surface groove located at midspan. As shown in Fig. 2, fracture initiated at two small zones of fatigue located at the entry point of the grease port into the bolt body (arrows). Fracture

Fig. 2 One of the mating fracture surfaces. Fracture propagated from the two small zones of fatigue around the grease port hole. The dark zone at the bottom of the fracture surface is merely a shadow resulting from curvature of the surface. Note the old grease in the grease port.

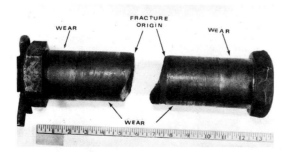

Fig. 1 Trunnion bolt showing the fracture origin and surface wear pattern

propagated from the fatigue zones across the cross section of the bolt, as indicated by the fracture texture. Also visible in Fig. 2 is hardened grease in the grease port, indicating that the system had not been lubricated for some time prior to failure.

Testing Procedure and Results	*Surface examination* **Macrofractography.** Visual and low-power stereomicroscopic examination showed that the zones at the fracture origin were typical of fatigue; there was no evidence of plastic deformation, and the fracture texture was flat and featureless. These zones were covered with a thin, adherent film of corrosion product that could not be removed. Therefore, scanning electron microscopy was not conducted. *Metallography* The microstructure consisted of nodular pearlite plus equiaxed ferrite, a typical structure for	AISI 1045 steel in the normalized and tempered condition. *Chemical analysis/identification* Wet chemical analysis of the sample drillings showed that the bolt material conformed to the specified requirements for AISI 1045 steel. *Mechanical properties* **Hardness** was uniformly 20 to 21 HRC from the core to near the surface. This level was consistent with the chemistry and microstructure of the steel and also indicated an ultimate tensile strength well above that required.
Discussion	The fatigue initiation mechanism that led to complete fracture could be inferred from the wear patterns on the bolt. As Fig. 1 shows, wear occurred on one side at either end of the bolt and on the opposite side over its center span. The bolt had apparently been frozen in this orientation for a protracted period and thus had been subjected to repeated bending stresses. These stresses placed the oil port hole in cyclic tension, leading to fatigue cracking around the hole. If the system had been	properly lubricated, the bolt would have rotated during service, spreading the cyclic tensile stresses around the circumference of the bolt, and fatigue would not have occurred. The fatigue cracks acted as stress concentrators, rendering the bolt susceptible to overload failure—which occurred from normal service impacts. The low temperatures at the time of failure were also a contributing factor.
Conclusion and Recommendations	*Most probable cause* The bolt material was normal with respect to chemistry, microstructure, and hardness. No defects were present to contribute to failure. Improper lubrication procedures caused the bolt to be frozen in an orientation such that the grease port was subjected to repeated tensile loading. This led to fatigue cracking and final complete fracture.	*Remedial action* The failure did not appear to be systemic. However, proper lubrication procedures should be maintained to allow free rotation of the bolts while in service.

Hydrogen-Assisted Stress Cracking of Carburized and Zinc Plated SAE Grade 8 Wheel Studs

Roy G. Baggerly, Kenworth Truck Co., Kirkland, Washington

Several case-hardened and zinc-plated carbon-manganese steel wheel studs fractured in a brittle manner after very limited service life. The fracture surfaces of both front and rear studs showed no sign of fatigue beach marks or deformation in the form of shear lips that would indicate either a fatigue mechanism or ductile overload failure. SEM analysis revealed that the mode of fracture was intergranular decohesion, which indicates an environmental influence in the fracture mechanism. The primary fracture initiated at a thread root and propagated by environmentally-assisted slow crack growth until final fracture. The natural stress concentration at the thread root, when tightened to the required clamp load concomitant with the presence of cracks in the carburized case, was sufficient to exceed the critical stress intensity for hydrogen-assisted stress cracking (HASC). The zinc plating exacerbated the situation by providing a strong local corrosion cell in the form of a sacrificial anode region adjacent to the cracked thread. The enhanced generation of hydrogen in a corrosive environment subsequently lead to HASC of the wheel studs.

Key Words

Fasteners
Stress-corrosion cracking
Zinc plating

Wheel studs
Hydrogen embrittlement

Carbon manganese steels
Case hardened fasteners

Alloys

Carbon manganese steels—1340

Background

Several wheel studs fractured in a brittle manner after very limited service life

Applications

Most heavy truck wheels are clamped to steer axle and drive axle hubs using ten spherical ball mount wheel nuts. The wheel stud thread sizes are typically 1-1/8-in. - 16 for front hubs and 3/4-in. - 16 for rear hubs. Torque levels of 542 to 630 N · m (400 to 465 ft · lb) are specified for both front and rear studs. These high torque levels are typically achieved by using air impact wrenches. Actual clamp load levels attained in the field may vary due to inexperience in the use of impact wrenches, thread contamination which changes the coefficient of friction, as well as the possibility of inaccurate torque values.

Components for use in this application must be able to withstand the severe environment to which they are subjected. This environment includes corrosive conditions from road salt and acid-forming air pollutants. In the last several years, the truck industry has upgraded the strength level of wheel studs from SAE Grade 5 to SAE Grade 8, which is satisfied by the chemistry and heat treatment of medium-carbon alloy steels. The basic configurations for the studs involved are shown in Fig. 1.

Circumstances leading to failure

A report of four wheel stud failures was re-ceived 6 weeks after sourcing the components from a new supplier. This supplier had developed a proprietary material processing specification for the wheel studs, and was having them manufactured to their specification. Early indications were that the failures could be described as brittle, and that fatigue was not a factor. Several other failures followed these initial reports, and fractured samples from rear studs as well as front studs were soon available for analysis. Failures were reported for studs that had been exposed to service ranging from only a few miles to as many as 25,000 miles.

Pertinent industry specifications and manufacturer's requirements

Threaded fasteners used in the automotive industry are specified to SAE Standard J429, "Mechanical and Material Requirements for Externally Threaded Fasteners." This standard lists different levels of fastener strength which have requirements for chemistry, core and surface hardness, and level of allowable surface discontinuities. SAE Grade 8 requirements are given in Table 1.

SAE Standard J1102 for wheel hardware, "Mechanical and Material Requirements for Wheel Bolts," employs a serrated wheel stud configuration that is carburized. The core hardness is consistent with an SAE Grade strength level. Serrated truck wheel studs, prior to around 1984,

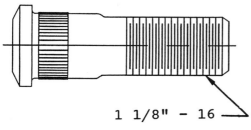

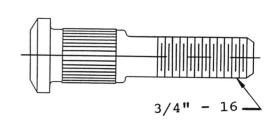

Fig. 1 Wheel stud configurations. (a) Front 1-1/8-in. stud. (b) Rear 3/4-in. stud

Table 1 Specifications and requirements for wheel studs

Specification	Ultimate tensile strength(a)		Yield strength(a)		Elongation (a), %	Hardness, HRC		Chemistry
	MPa	ksi	MPa	ksi		Core	Surface	
SAE Standard J429, Spec. Grade 8	1034	150	896	130	12	33 to 39	39 max (58.6 HR30N)	Medium carbon alloy steel, quenched and tempered
SAE Standard J1102, Spec. Grade 5	827	120	634	92	14	25 to 34	34 min (77 HR15N)	Medium carbon alloy steel, quenched and tempered after carburization to a case depth of 0.10 to 0.30 mm (0.004 to 0.012 in.)
Supplier (proprietary)	1034	150	896	130	12	32 to 38	42 to 47	SAE 1340 manganese alloy steel, carburized to a case depth of 0.10 to 0.30 mm (0.004 to 0.012 in.). Zinc plated per Federal Specification QQ-Z-325, Type II, Class 2

(a) Minimum properties

conformed to this grade. See Table 1 for requirements.

The specification required by the supplier for the serrated wheel studs described in this analysis are also given in Table 1.

Specimen Selection

The samples that initially failed were available for detailed examination. Additional stud samples that had not been placed in service were also obtained and evaluated using various mechanical tests, including normal and slow-strain-rate tensile testing as well as sustained-load clamp tests, with and without a 5% salt solution.

Visual Examination of General Physical Features

The appearance of both front and rear stud failures were similar in that the fracture surfaces showed no sign of fatigue beach marks or deformation in the form of shear lips that would indicate either a fatigue mechanism or ductile overload failure. Areas of corrosion on the fracture surfaces suggested a corrosive environment may have been influential in the failure mechanism. Figure 2 shows the characteristic appearance of the fracture surfaces of the studs.

Testing Procedure and Results

Surface examination

Scanning Electron Microscopy Fractography. The fracture surface from several failed studs were examined in greater detail using scanning electron microscopy (SEM). SEM analysis revealed that the mode of fracture was by intergranular decohesion which indicates an environmental influence in the fracture mechanism. This fracture mode occurred across the entire cross-section of the stud in the slow crack growth region. Figures 3 and 4 show characteristic fractographs typical of the failed studs examined in the SEM. Regions of corrosion were chemically analyzed using energy dispersive spectroscopy (EDS) and the elements potassium and sulfur were found to be present. An EDS spectrograph from the corrosion products of a typical stud is shown in Fig. 5.

Metallography

Microstructural Analysis. Metallographic analysis indicated the microstructure was tempered martensite, which is consistent with a heat-treated medium-carbon steel. The near surface region revealed a carburized and hardened layer.

Crack Origins/Paths. The dominant fracture initiated at a thread root and propagated by envi-

(a)

(b)

Fig. 2 Fracture surface appearance studs. Approximately 1×. (a) 1-1/8-in. stud. (b) 3/4-in. stud

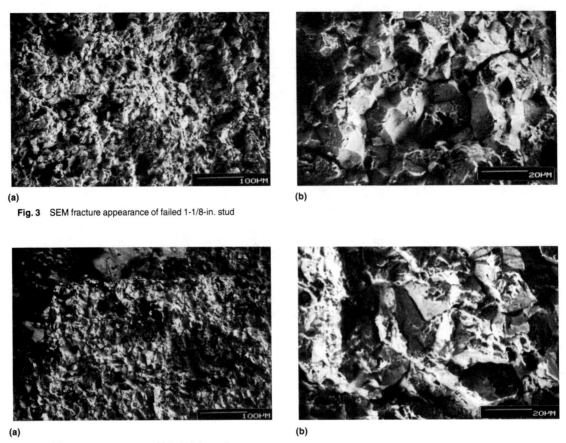

Fig. 3 SEM fracture appearance of failed 1-1/8-in. stud

Fig. 4 SEM fracture appearance of failed 3/4-in. stud

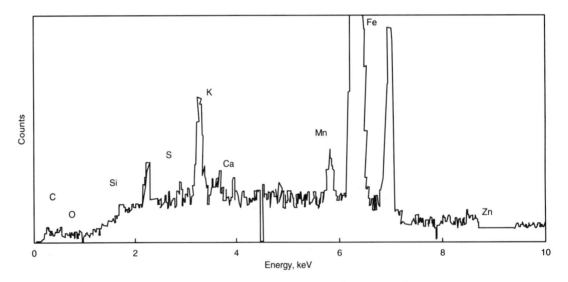

Fig. 5 EDS Kα spectrograph from fracture surface corrosion products

ronmentally-assisted slow crack growth until final fracture. Thread roots adjacent to the dominant crack also contained cracks that were initiated in the carburized case but arrested at the case/core interface. Figure 6 shows a crack that initiated at the carburized thread root but subsequently arrested in the softer and tougher core material.

Chemical analysis/identification

A chemical analysis using an optical emission spectrograph indicated that all of the steel samples tested were within the requirements for SAE 1340 steel. Table 2 lists the required steel chemistry limits and the chemical analysis from three failed studs.

Coatings or Surface Layers. A zinc plating of

Table 2   Results of chemical analysis

	Composition, wt%			
Element	Stud 1	Stud 2	Stud 3	SAE 1340 steel
Carbon	0.39	0.40	0.38	0.38 to 0.43
Manganese	1.72	1.73	1.71	1.60 to 1.90
Phosphorus	0.015	0.015	0.014	0.035 max
Sulfur	0.015	0.014	0.014	0.040 max
Silicon	0.22	0.21	0.21	0.15 to 0.35

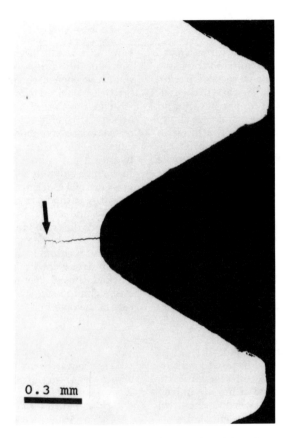

Fig. 6   Arrested crack at the case/core interface of the carburized thread root region

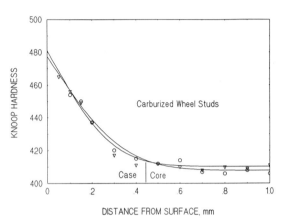

Fig. 7   Variation of microhardness with distance from the thread root

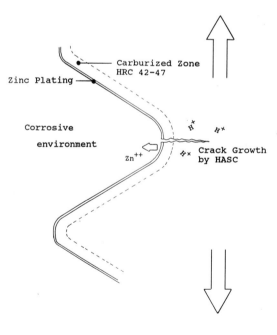

Fig. 8   Interaction of a corrosive environment with a carburized and zinc plated high strength fastener under stress
Anode reaction $Zn \leftrightarrow Zn^{++} + 2e^-$;
Cathode reaction $H2O \leftrightarrow H^+ + OH^-$

approximately 12.7 µm (0.0005 in.) was applied to the wheel stud surface and the thickness was confirmed by metallographic measurements. Records from the plater/heat treater confirmed that all parts had undergone a 4-h bake at 204 °C (400 °F) within 1 h of plating; this was required to help reduce the propensity for hydrogen embrittlement from the cleaning and zinc plating processes.

Mechanical properties

Hardness. The average core hardness of the failed studs tended to be at the higher end of the SAE Grade 8 range. These studs tested at 36 to 38 HRC; the required range is 32 to 38 HRC. The surface hardness was measured using a Knoop microhardness diamond indenter. A typical microhardness survey as a function of distance from the thread root surface is shown in Fig. 7. The stud, including the threaded region, had been carburized, and the equivalent hardness at the root of the thread is 45 HRC. These values for surface and core hardness were within the manufacturer's

specification.

Tensile Properties. The average tensile and yield strength from testing the 3/4-in. studs showed they were within the required range for SAE Grade 8. These studs were tested at a constant deflection rate of 10^{-2} min. Machined samples with smooth gage lengths were also tested at two different deflection rates, 10^{-2} min. and 10^{-5}

min. The slower strain rate tests showed a small drop in strength, but the statistical accuracy was not sufficient to state unequivocally that residual hydrogen content had an effect.

Sustained Clamp Load Test. Twelve unused studs were tightened to a torque level of 881 N · m (650 ft · lbf) and retightened every 24 h. Six of the studs were tested in the dry condition and six were exposed to a 5% salt solution. At the end of 72 h, the thread roots were examined for cracks using a ste-reo microscope and subsequently pulled to failure in tension to expose the extent of any cracks that appeared to be present. Two of the six studs exposed to the salt solution contained significant cracks that had formed during this 72-h exposure and had extended over approximately 30% of the stressed area. The fracture mode was intergranular decohesion, and the fracture surface appeared very similar to the failures that occurred in the field.

Discussion

An initial response to these failures was to determine why both fasteners failed in the same manner, since the clamp loads were so disparate with equivalent torque levels. The ratio of the tensile stress area was 2.38 to 1 for the 3/4-in. and 1-1/8-in. studs, respectively. The worst case clamp load for the 3/4-in. stud will be in the range of 13,636 to 18,182 kg (30,000 to 40,000 lb) at the specified torque values. This results in tensile stresses of approximately 738 MPa (107 ksi) for the 3/4-in. wheel stud and 310 MPa (45 ksi) for the 1-1/8-in. wheel stud. The elastic stress concentration factor at the root of the thread has K_t values of approximately 4.1 and 4.4 for the two studs. This implies a localized thread root stress of 3027 MPa (439 ksi) and 1365 MPa (198 ksi), respectively, when the rear and front wheel studs are tightened to the maximum torque value. These high values of localized stress may be realized, since the thread root area has a high hardness and negligible duc-tility as a result of the carburization treatment. Since torque values are usually achieved using air impact wrenches, the probability of fracturing the brittle carburized thread root region is very high.

The zinc plating on the carburized fastener creates a source of hydrogen when the fastener is exposed to corrosive conditions. Zinc will act as a sacrificial anode during corrosion, with the result that hydrogen atoms will be liberated at the cathodic region in a galvanic cell between zinc and iron. The presence of a crack at the root of a thread produces a nearby cathodic region between the zinc-plated threads and steel core at the tip of the thread root crack. A diagram depicting the condition of a cracked, zinc-plated thread root under stress in a corrosive medium is shown in Fig. 8. Hydrogen-assisted stress cracking can then occur from the cracked thread root under the sustained high-stress clamp load, with subsequent ultimate fracture of the fastener.

Conclusion and Recommendations

Most probable cause

The potential for cracking a hardened carburized case in a high-strength stud is significant, considering the common use of either air impact wrenches or the possibility of exceeding the proof stress with a wrench. A cracked case enables moisture to penetrate to the tip of the crack, which will become cathodic with respect to the zinc plating. Hydrogen charging from corrosion fosters embrittlement and sustained slow crack growth at the tip of the crack, due to the residual stress from the high clamp load and HASC. A high clamp load is required to prevent fatigue and restrain the wheels from slippage movement on the hub.

Remedial action

The requirement for a carburized case was eliminated from the manufacturing specification.

How failure could have been prevented

These failures could have been prevented by not requiring carburization of the stud. The design of wheel studs requires consideration of the effect of nut tightening techniques, the required clamp loads, and how these assembly procedures affect the thread root region of the fastener.

Acknowledgments

The author gratefully acknowledges Kenworth Truck Co. for encouraging publication of this failure analysis investigation.

Hydrogen Embrittlement Delayed Failure of a 4340 Steel Draw-In Bolt

Eli Levy, Chief Metals Technology Group, de Havilland, Inc., Canada

The draw-in bolt and collet from a vertical-spindle milling machine broke during routine cutting of blind recesses after a relatively long service life. The collet ejected at a high rotational speed due to loss of its vertical support and shattered one of its arms upon impact with the work table. SEM fractography and metallographic examinations conducted on the bolt revealed hairline indications along grain facets on the fracture surface and stepwise cracking in the material, both indicating failure by hydrogen embrittlement. Similar draw-in bolts were discarded and replaced with bolts manufactured using controlled processes.

Key Words	Milling cutters	Nickel-chromium-molybdenum steels	Bolts
	Machine tools	Hydrogen embrittlement	
Alloys	Nickel-chromium-molybdenum steel—4340		

Background

The draw-in bolt and the collet from a vertical-spindle milling machine broke during routine cutting of blind recesses after a relatively long service life. Having lost its vertical support, the collet ejected at a high rotational speed (23,600 rev/min), shattering one of its arms upon impact with the work table.

Applications

To use a collet, a collet sleeve, having both an external and an internal taper, is first placed in the spindle hole. The collet is then placed inside the sleeve. A draw-in bolt is inserted into the opposite end of the spindle and screwed into the threaded end of the collet. As the draw-in bolt is tightened by means of a handwheel, it pulls the collet inward against the tapered sleeve and causes the jaws to tighten against the work (Fig. 1).

Visual Examination of General Physical Features

The fracture surface of the draw-in bolt displayed oyster shell-like marks resembling crack growth bands (Fig. 2). Such marks have been known to represent slight alternate changes in the plane of crack propagation due to abrupt changes in load levels.

The fracture surface of the collet (Fig. 3) displayed an appreciable shear lip, which is characteristic of ductile overload failure (the cone part of a cup-cone fracture). The broken piece contained a deep gouge on its side, probably inflicted by the drill-bit tool end.

Based on these preliminary findings, it appeared that the draw-in bolt was the first to fracture. Following failure of the bolt, the rotating collet struck the work table, shattering one of its arms.

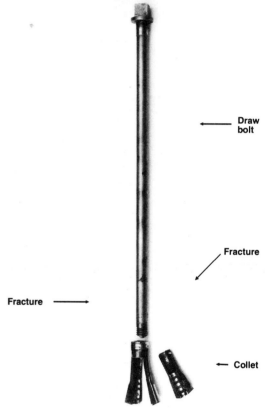

Fig. 1 Overall view of the broken assembly. ~1×

Testing Procedure and Results

Surface examination

Scanning Electron Microscopy/Fractography. The fracture surfaces of both the bolt and the collet arm were examined with a scanning electron microscope (SEM). Clear evidence of hairline cracks were apparent on grain facets throughout most of the fracture surface of the draw-in bolt (see Fig. 2). Intergranular indications such as hairline cracks are often associated with the hydrogen-embrittled material. The fracture surface of the collet contained equiaxed dimples, characteristic of failure by ductile overload (see Fig. 3).

Metallography

The microstructure of the bolt consisted of fine, uniform tempered martensite, which is normal for a 4340 steel heat treated to an ultimate tensile strength of 1520 MPa (220 ksi). Evidence of hydrogen-induced stepwise cracking was observed along the fracture surface and at the thread roots (Fig. 4).

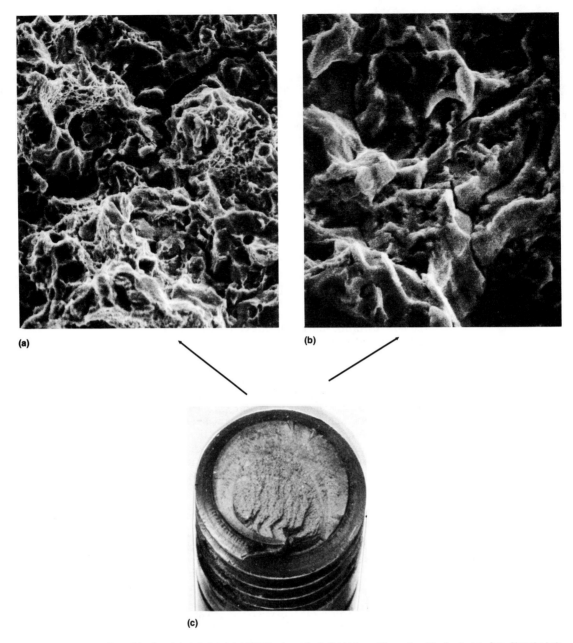

Fig. 2 Fracture surface of the draw-in bolt (bottom). (a) SEM fractograph of a field taken at the center of the fracture surface. Note the hairline cracks along the grain facets (hydrogen embrittlement) and the ductile dimples (tension overload). 1215×. (b) SEM fractograph of a field taken near the threads on the fracture surface. Note the hairline cracks along the grain facets. Such cracks are often associated with hydrogen embrittlement. 2430×

Conclusion and Recommendations

Most probable cause

The presence of hairline indications along grain facets on the fracture surface, coupled with stepwise cracking in the material, points to hydrogen embrittlement. It appears that fracture in service progressed transgranularly to produce delayed failure under dynamic loading. The source of hydrogen generation on the surface of the bolt was unknown; the manufacturer was requested to scrutinize the manufacturing procedure with the aim of identifying the hydrogen source.

Remedial action

All similar unused and used draw-in bolts were discarded. The new replacement bolts were manufactured using controlled processes. No further hydrogen embrittlement failures were encountered.

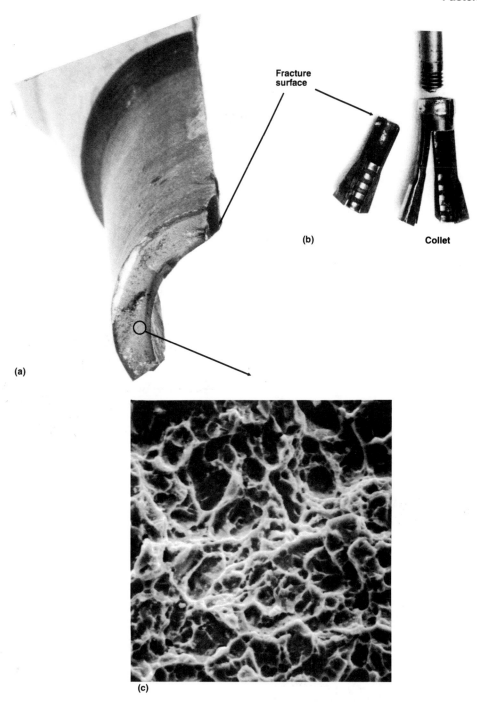

Fracture
surface

(b)

Collet

(a)

(c)

Fig. 3 SEM fractograph (a) of a field on the fracture surface shown in (b). The topography throughout the fracture surface was of ductile dimples, indicative of failure by tension overload.

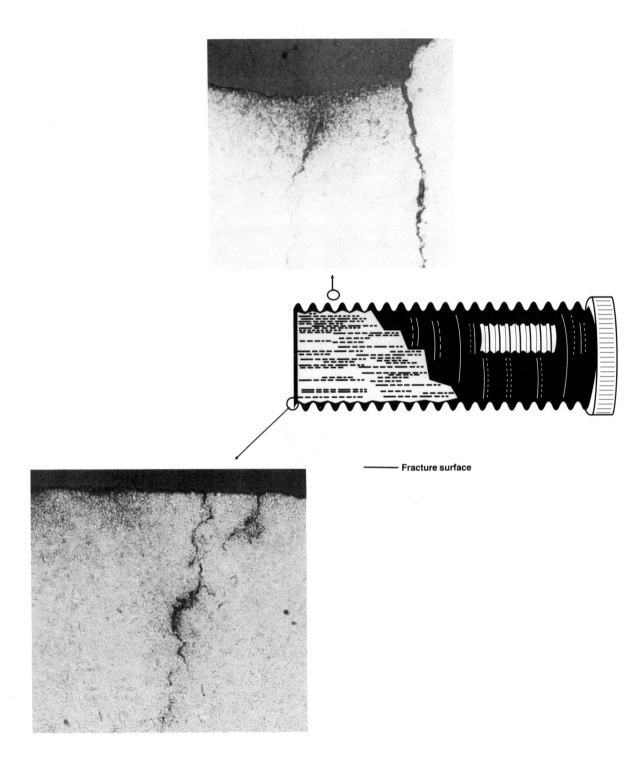

Fracture surface

Fig. 4 Fractured draw-in bolt (right). (a) Optical micrograph through a thread root, showing stepwise cracks indicative of hydrogen embrittlement. Similar cracks were noted in most of the thread roots. Etched with 2% nital. 400×. (b) Optical micrograph through the fracture surface, also showing cracks caused by hydrogen embrittlement. Similar cracks were observed throughout the fracture surface. Etched with 2% nital. 200×

ELECTRICAL EQUIPMENT FAILURES

Arcing Fault Burndown in Low Voltage Residential Service Entrance with Aluminum Conductors

Howard F. Prosser, Roch J. Shipley, and Peter C. Bouldin, Engineering Systems, Inc., Aurora, Illinois

Three instances involving the failure of aluminum wiring at the service entrance to single-family homes are discussed. Arcing led to a fire which severely damaged a home in one case. In a second, the failure sequence was initiated by water intrusion into the service entrance electrical box during construction of the home. In the third, failure was caused by a marginal installation. Strict adherence to all applicable electrical codes and standards is critical in the case of aluminum wiring. Electrical components not specifically designed for aluminum must never be used with this type of wiring. All doors, panels and similar portions of electrical boxes should be secured to prevent damage to surroundings in the event of an electrical fault. If symptoms of arcing are observed, professional service should be sought. The latest designs of connectors for use with aluminum wiring are less susceptible to deviations in installation practice.

Key Words	Electrical wire	Electrical connectors	Electrical arcs
	Aluminum base alloys	Wiring	Fire hazards
Alloys	Aluminum-base alloys—1350		

Background

Application

This case study concerns the failure of aluminum wiring at the service entrance to a single-family home. As illustrated in Fig. 1, the service entrance is the point at which the electrical lines are connected to the main circuit breakers of the building. These electrical lines come from the power company transformer, through the electrical meter.

It must be emphasized that the building circuit breakers are on the customer side of the service entrance. Thus, the building breakers cannot protect against an electrical fault in the service entrance. An example of a fault would be a short circuit between the power lines or a short circuit to ground.

Circumstances leading to failure

Three instances of service entrance arcing faults associated with aluminum wiring are known to the authors.

Incident 1. A fire occurred at a residence. Associated with the fire was an electrical outage in the homes on the same side of the street. As evidenced by clocks in the home and power company records, the outage occurred approximately fifteen minutes before the fire was reported to the fire department.

The homeowner had recently replaced a circuit breaker and had left the breaker panel cover off after replacing the breaker. The fire origin was determined to be underneath the electrical service panel.

Incident 2. The fire investigator for Incident 1 examined the 23-year-old service entrance panel in his own residence and discovered aluminum service entrance conductors. The insulation on one was severely melted back from the main circuit breaker connection, and the plastic case of the circuit breaker was badly eroded from heat generated at the connection terminal. The case erosion was sufficient to inhibit circuit breaker operation. The service panel, located in the basement, had reportedly been completely submerged by flooding when the house was under construction.

Incident 3. The homeowners noted flickering and dimming of lights occurring at random for ap-

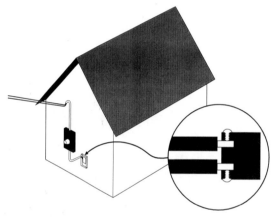

Fig. 1 Schematic showing location of service entrance

proximately one week. Then, electrical power was lost to approximately half of the circuits in the home. The electric utility company investigated and found that one of the two "hot" aluminum conductors in the meter box had become detached from its terminal. While there was evidence of arcing, it had not been severe enough to damage anything outside of the box.

As described above and shown in Fig. 1, the meter box is on the power company side of the service entrance. To understand the symptoms noted by the homeowner, note that nominal 110-volt circuits in residential service are typically provided through one of the two "hot" lines providing nominal 220-volt service. The 220-volt service requires both "hot" lines or phases. Equipment such as electric appliances requiring 220-volt service will usually not operate if supplied with only one side of the 220-volt service.

Pertinent specifications

In 1968, Underwriters Laboratories published standards for wiring terminals. These have evolved to the present standards:
- ANSI/UL 486A *Wire Connectors and Soldering Lugs for Use with Copper Conductors*

- ANSI/UL 486B *Wire Connectors for Use with Aluminum Conductors*
- ANSI/UL 486E *Equipment Wiring Terminals for Use with Aluminum and/or Copper Conductors*

Standards 486B and 486E require that terminals intended for use with aluminum shall be coated to inhibit oxidation and contact with copper. The coating must be electrically conductive. Tin has been found to be acceptable. These standards also prescribe heat cycling tests, tightening torques for various types of screws, test currents, and pullout tests. The above ANSI/UL standards are in accordance with the National Electrical Code (NFPA 70). The wire used for aluminum conductors is typically alloy 1350.

Performance of other parts in same or similar service

Aluminum wiring is used for service in literally millions of installations. The incidents described here represent exceptions to a record of satisfactory performance. However, aluminum wiring is more sensitive than copper wiring to deviation in installation, either original installation or repairs/remodeling.

Visual Examination and General Physical Features

The badly burned electrical panel from Incident 1 is shown in Fig. 2 and Fig. 3. Three aluminum wires entered the panel through the hole indicated by the arrow. These aluminum wires connected the power meter to the breaker panel. The badly melted and arced stubs of these wires protruded approximately 25 mm (1 in.) into the breaker panel. A close-up of one of the wires is shown in Fig. 4.

The service entrance panel from Incident 2, with the damaged insulation on the aluminum conductor, is shown in Fig. 5. The insulation was melted to within a few millimeters (fractions of an inch) from the conduit shown by the arrow in Fig. 6. While there was evidence of arcing between the aluminum conductor and the circuit breaker terminal, it was not severe enough to damage anything outside of the service entrance panel. Further melting of the insulation would have resulted in an arcing fault as in Incident 1.

Fig. 2   Electrical panel and surrounding area, Incident 1

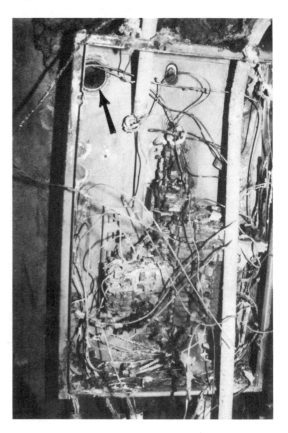

Fig. 3   Close-up of electrical panel in Incident 1. Arrow indicates where aluminum wires entered the box.

Fig. 4 Close-up of one of the melted and arced wires in Incident 1

Testing Procedure and Results

Surface Examination

Macrophotography. Figure 7 shows a close-up front view of the breaker assembly and the aluminum wires at the input terminal (service entrance) from Incident 2. Approximately 90 mm (3.5 in.) of insulation was missing from the right-hand wire. Figure 8 shows the underside of the breaker assembly from another angle. Damage to the right hand breaker (a hole in the outside wall) is evident. The double wall separating the two individual breakers had also been compromised. Figure 9 shows a closer view of this damage. Figure 10 compares the damaged breaker with an undamaged assembly. The yoke assemblies on the damaged breakers showed evidence of rusting. There was also what appeared to be dried mud in several locations on the breaker housing.

Fig. 5 Electrical panel in Incident 2. Arrow indicates aluminum wire with deteriorated insulation.

Fig. 6 Close-up of aluminum wire with deteriorated insulation. Arrow indicates where deteriorated portion of wire was close to conduit.

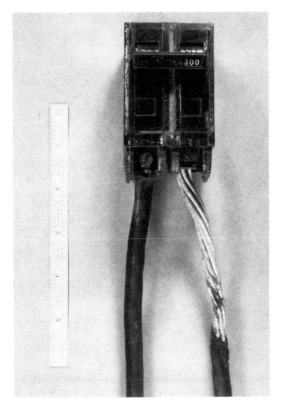

Fig. 7 Aluminum wire in Incident 2 after removal from electrical box

Fig. 8 Close-up of aluminum wire and breaker assembly in Incident 2

Fig. 9 Close-up of damage to breaker in incident 2

Fig. 10 Damaged breaker from Incident 2 (left) and new breaker

Discussion

Synthesis of evidence

Problems with aluminum wiring can be understood through consideration of aluminum oxide. Aluminum oxide is nonconductive. In fact, it is a very good electrical insulator. When present in significant thickness at an electrical connection, this oxide increases the resistance of that connection. The increased resistance causes increased heat.

The heat is harmful in several ways. First, thermal expansion and contraction cycles loosen the mechanical clamping of the wire in the terminal. Second, increased temperature promotes oxidation. Both effects lead to further deterioration of the connection and more heating. As the temperature of the connection continues to increase, the wire may deform plastically, leading to even further loosening of the connection. Ultimately, the connection becomes so loose that there is no longer true electrical contact and series arcs are drawn, leading to rapid deterioration of the wiring and terminal. This sequence of events occurred in Incident 3.

Heat buildup in the electrical connection also degrades the insulation on the aluminum wires and eventually permits wire-to-wire or wire-to-ground contact, resulting in electric arcs between

phases or from phase to ground. This type of arcing is more severe than the series arcing, and can easily generate quantities of molten aluminum (at 593 °C, or 1100 °F) which can start fires, especially if not confined. Significant deterioration of insulation was observed in Incidents 1 and 2. In Incident 2, the problem was discovered in time to prevent more serious consequences.

Arcing at the service entrance will not trip the main circuit breaker in a home or business since the arcing fault is on the upstream (power company) side of the circuit breaker. In Incident 1, the arcing was so severe that it eventually overloaded the fuses in the primary side of the supply transformer which provided service to several homes in the neighborhood. Unfortunately, the cover to the electrical box had been left off, so the arcing also started a fire.

In the other incidents, the problems were discovered and rectified without any damage beyond the point of series arcing itself. Fortunately, the electrical boxes were secure and confined the arcing.

The wiring and terminal involved in Incident 1 were badly damaged by the arcing and subsequent fire. A deviation in the original installation resulted in a connection of increased electrical resistance. This initiated the cycle of heating, oxidation, and loosening described above. Over the years, improvements have been made in the electrical components used with aluminum wiring to provide more secure mechanical clamping, and therefore, a connection of minimal electrical resistance.

Galvanic effects between dissimilar metals are also a concern with aluminum wiring. UL 486E indicates that tin is acceptable as a coating for a terminal in contact with aluminum wire. Cadmium and zinc are other candidates that might be considered from a corrosion standpoint, as these are usually satisfactory in direct contact with aluminum. The most critical consideration in electrical applications is the damage to aluminum that can occur when it is in contact with copper or copper alloys. Electrical joint compounds are available to inhibit corrosion in applications involving aluminum wire. Commercially available products consist of a grease-like base in which zinc particles are suspended.

Conclusion and Recommendations

Most probable cause

The cause and origin of the fire in Incident 1 was due to arcing in the aluminum service entrance wires, most probably caused by improper tightening of the terminal screws. A contributing cause to the spread of the fire was the fact that the cover to the breaker panel was intentionally left off by the homeowner. Molten aluminum at approximately 593 °C (1100 °F) sputtered out of the service panel and ignited combustibles located on the floor below.

Improper tightening was also the most probable cause in Incident 3 and may have contributed to Incident 2 as well. However, water ingress into the electrical box was the most probable cause of Incident 2.

Remedial action

All doors, panels, and similar portions of electrical boxes should always be secured to prevent damage to surroundings in the event of an electrical fault. If any of the symptoms of arcing are observed, professional service should be sought. The latest designs of connectors for use with aluminum wiring are less susceptible to deviations in installation practice.

How failure could have been prevented

In the case of Incident 2, the failure most likely would not have occurred if the contractor had not allowed water to flood the basement and electrical box during construction. The water and dissolved impurities which it contained promoted oxidation. In the other incidents, greater care should have been taken in installation of the aluminum wiring.

Acknowledgments

The assistance of Messrs. Steve Smith and Robert Franzese with the preparation of the graphic illustrations and photographs is gratefully acknowledged.

Corrosion Fatigue Failure of Stainless Steel Load Cells in a Milk Storage Tank

Daryl C. Collins, ETRS Pty Ltd., Australia

Two type 420 martensitic stainless steel load cell bodies, which had been installed under two of the four legs of a milk storage tank failed in service. The failure occurred near a change in section and involved fracture of the entire cross section. Examination showed a brittle fracture that was preceded by a small fatigue region. Pitting corrosion was evident at the fracture origin. The areas around the load cells had been subjected to regular washdowns using high-pressure hot water, and the pitting was attributed to crevice corrosion between the load cell and the holddown bolts. Prevention of such corrosion by the use of a flexible sealant to eliminate the crevice was recommended.

Key Words			
Martensitic stainless steels	Brittle fracture	Agricultural equipment	
Bolted joints	Pitting (corrosion)	Corrosion fatigue	
Pressure cells	Stress-corrosion cracking		

Alloys   Martensitic stainless steel—420

Background

Two type 420 martensitic stainless steel load cell bodies, which had been installed under two of the four legs of a milk storage tank, failed in service. The failure occurred near a change in section and involved fracture of the entire cross section.

Applications

The two legs without load cells were hinged, enabling accurate measurement of the volume of milk in the tank to be made by the weight registered on the legs containing the load cells (Fig. 1). The tank was filled to capacity each day, and the load cells were exposed to a 5 tm (5.5 t) load with each filling. The areas around the load cells were subjected to regular (twice daily) washdowns using high-pressure hot water.

Circumstances leading to failure

Two load cells from the same tank failed within several days of each other. Both cells had become difficult to zero and had shown calibration drift.

Performance of other parts in same or similar service

Load cells on a similar tank that was used less often had not failed and were relatively easy to calibrate.

Fig. 1 View of load cell installation

Visual Examination of General Physical Features

Each load cell operated as a cantilever beam and the deflection of the beam was measured by strain gages within a cavity in the load cell body. Failure had occurred at the forward holddown bolt, within the wider section of the body. Considerable deflection of the body had occurred in service (Fig. 2).

Testing Procedure and Results

Nondestructive evaluation

Liquid Penetrant Testing. Dye penetrant testing of the failed load cells revealed no additional cracks (other than the failure). The load cells from the second tank showed no cracks.

Surface examination

Visual. The load cell body had rust stains on the top and bottom surfaces. The holddown bolts

Fig. 2 Closeup view of a load cell, showing holddown bolts

Fig. 3 Fractured load cell

Fig. 4 Bottom view of load cell, showing fracture at holddown bolt hole

Fig. 5 Fracture surfaces of load cell, indicating brittle fracture

Fig. 6 Fracture origin (top left corner) and fatigue region after cleaning in inhibited acid

and washers were stainless steel, while the base plate was painted steel. There was no evidence of impact to the body (Fig. 3 and 4).

Macrofractography. Examination of the fracture surface showed areas typical of fatigue on each side of the forward holddown bolt. These areas were present at the top edge of the bolt holes and extended for 10 mm (0.4 in.) along the top of the load cell and in the bore of the hole. The balance of the fracture surface was typical of a brittle fracture (Fig. 5 and 6).

Corrosion Patterns. Minor pitting corrosion was present at the fracture origin. Rust stains were present within the bore of the holddown bolt hole (Fig. 7).

Metallography

Microstructural Analysis. The load cell body had a microstructure of tempered martensite, with a second phase (either carbide or ferrite) present as a grain-boundary network and within grains (Fig. 8).

Corrosion Morphology. The corrosion at the crack origin was present as deep pitting. Just behind the origin, a crack extended from the base of a deep pit (Fig. 9 and 10).

Chemical analysis

Material. The bodies of the load cells were of the following composition:

This composition is comparable to AISI type

Element	Content, %
Carbon	0.38
Manganese	0.54
Silicon	0.86
Sulfur	<0.01
Phosphorus	0.02
Chromium	14.00

(continued)

Fig. 7 Area opposite fracture, showing pitting at top right side

Element	Content, %
(continued)	
Nickel	0.24
Molybdenum	0.09
Copper	0.03
Titanium	<0.05
Aluminum	0.01
Niobium	<0.05
Vanadium	0.24

420 stainless steel.

Mechanical properties

Hardness. Vickers hardness testing of metallographic specimens yielded results of 520, 523, and 527 HV30, with an average of 523.

Discussion

The load cells were manufactured from type 420 stainless steel and were fully hardened. At the observed hardness levels, an ultimate strength of 1720 MPa (250 ksi) and a yield strength of 1480 MPa (210 ksi) would be expected. A martensitic stainless steel was considered to provide the best combination of strength and corrosion resistance.

The service conditions were not considered highly corrosive; this was confirmed by the limited corrosion on the painted steel base plate. In addition, the martensitic condition provides the greatest corrosion resistance for type 420 stainless steel.

Failure of the load cell was caused by fatigue crack growth and the fatigue was initiated by corrosion pits. The fatigue had propagated only slightly when catastrophic brittle fracture occurred.

Conclusion and Recommendations

Most probable cause

Corrosion had occurred under the head and washer of the holddown bolt. The crevice at this location, together with regular wetting due to washing of the area, had resulted in loss of passivation of the 420 material. The pitting observed was typical of that seen on stainless steels.

Corrosion alone would not have affected the operation of the load cell. The strength of the 420 material and the load cell design were also suitable for the service conditions. However, when combined, the applied cyclic loading and the pitting resulted in fatigue crack initiation and growth.

Remedial action

Further failure of these load cells was prevented by sealing the crevice with a flexible silicone sealant. The sealant was applied during load cell installation.

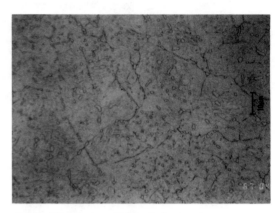

Fig. 8 Microstructure of load cell, showing tempered martensite containing carbides and/or ferrite. Etched with Kelling's reagent

Fig. 9 Pitting at origin of fatigue crack (fracture face at left). Unetched

Fig. 10 Crack extending from pitting (fracture face at left). Unetched

Failure of Nickel Anodes in a Heavy Water Upgrading Plant

T.P.S. Gill and J.B. Gnanamoorthy, Metallurgy Division, Indira Gandhi Centre for Atomic Research, Kalpakkam, India
V. Rangarajan, Madras Atomic Power Station, Kalpakkam, India

Nickel anodes failed in several electrolysis cells in a heavy-water upgrading plant. Dismantling of a cell revealed gouging and the presence of loosely attached black porous masses on the anode. The carbon steel top plate was severely corroded. An appreciable quantity of black powder was also present on the bottom of the cell. SEM/EDX studies of the outer and inner surfaces of the gouged anode showed the presence of iron globules at the interface between the gouged and the unattacked anode. The chemical composition of the black powder was determined to be primarily iron. Cell malfunction was attributed to the accelerated dissolution of the carbon steel anode top, dislodgment of grains from the material, and subsequent closing of the small annular space between the anode and the cathode by debris from the anode top. Cladding of the carbon steel top with a corrosion-resistant material, such as nickel, nickel-base alloy, or stainless steel, was recommended.

Key Words

Carbon steels, corrosion	Nickel	Anodes
Electrolytic cells	Nuclear reactor components	Heavy water

Background

Several electrolysis cells in a heavy-water upgrading plant began malfunctioning because of arcing between the anode and the cathode.

Applications

Upgrading of heavy water is achieved by the electrolysis of alkaline heavy water containing low to high concentrations of heavy water, which has been collected from various points in a nuclear reactor and processed in a cleanup section to remove impurities and corrosion products. The process of upgrading is based on the principle that the discharge kinetics of the reduction of deuterium ions on the cathode is sluggish compared with that of hydrogen ions.

The upgrading of depleted heavy water is carried out in batch operations. In a typical heavy-water plant, there are 14 banks of upgraders, each bank containing 10 cells. The upgrading of heavy water takes place at successively higher levels as the electrolyte is concentrated in each bank until the desired concentration is reached.

A schematic diagram of an electrolytic cell is shown in Fig. 1. It consists of a cylindrical nickel anode with a carbon steel top. The anode is surrounded by two concentric carbon steel cylindrical cathodes. The annular space between the anode and cathode is 3.5 mm (0.14 in.). Both the inner and outer cathodes are cooled by flowing water. The electrolyte to the cell is fed from the bottom, whereas the outlet is at the top, but at a level well below the nickel-to-carbon steel joint.

Circumstances leading to failure

Each cell is provided with a voltage indicator. Any malfunctioning in the electrolysis is reflected in the voltage indicator lamps. In one such cell, the indicator showed a low potential difference. Dismantling of the cell revealed that the nickel anode was gouged, and in some places a black porous mass was loosely attached to the unattacked nickel. The carbon steel top plate was severely corroded. An appreciable quantity of black powder was present at the bottom of the cell. Subsequent to these findings, many other cells were also dismantled because of disruption of the electrolysis process. The conditions leading to failure appeared to be the same in all the cells, although the extent varied.

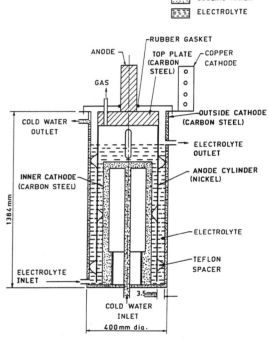

COOLING WATER

ELECTROLYTE

ANODE

GAS

COLD WATER OUTLET

INNER CATHODE (CARBON STEEL)

1384 mm

ELECTROLYTE INLET

RUBBER GASKET

TOP PLATE (CARBON STEEL)

COPPER CATHODE

OUTSIDE CATHODE (CARBON STEEL)

ELECTROLYTE OUTLET

ANODE CYLINDER (NICKEL)

ELECTROLYTE

TEFLON SPACER

3.5 mm

COLD WATER INLET

400 mm dia.

Fig. 1 Schematic of the electrolysis cell

Pertinent specifications

The materials of construction were:

Anode cylinder	3.2 mm (0.125 in.) thick nickel plate conforming to ASTM B-162
Anode top plate	50 mm (2 in.) thick carbon steel conforming to IS-2062
Outer and inner cathode shells	Carbon steel (A519—MT 1010)

The operating parameters were:

Electrolyte composition	5 to 10 wt% potassium hydroxide in the feedwater
Cell voltage	2.5 to 3.5 V dc
Current	5000 A
Temperature	Electrolyte side: 50 °C (120 °F)
	Cooling water side: 45 °C (110 °F)

Specimen selection

Several specimens from the damaged portion of the nickel anode were collected for examination.

A sufficient quantity of the black powder was also obtained for characterization.

Visual Examination of General Physical Features	Severe dissolution of the carbon steel top plate of the nickel anode (Fig. 2) was visible, as was the presence of an adherent deposit (1.5 to 2.5 mm, or 0.06 to 0.098 in., thick) on the adjacent cathode. The nickel anode surface showed the presence of interference colors, and the rubber gasket separating the nickel anode from the cover plate (Fig. 1) was charred.

A sizable amount of black powder was found at the bottom of the cell. The nickel anode was gouged, mostly at the bottom, and the nickel mass that had melted and resolidified was sticking to the nickel anode (Fig. 3). Fine metallic globules were sticking to the unattacked portion of the nickel anode.

Testing Procedure and Results	**Surface examination** **Scanning Electron Microscopy/Fractography.** Scanning electron microscopy (SEM) and energy-dispersive X-ray (EDX) studies were carried out on the outer and inner surfaces of the gouged anode. EDX analysis of the unattacked nickel matrix showed the presence of nickel only (Fig. 4a). Analysis of the outer surface of the gouged resolidified nickel, however, revealed the presence of nickel, iron, aluminum, silicon, chlorine, potassium, and so on (Fig. 4b) and indicated that the inner side contained nickel and iron, but iron appeared to be present in higher concentrations (Fig. 4c). SEM micrographs showing the details of the outer surface of the resolidified nickel mass and the presence of iron globules at the interface between the gouged and the unattacked nickel anode are shown in Fig. 5 and 6, respectively. **Chemical analysis/identification** **Corrosion Products.** The chemical composi-

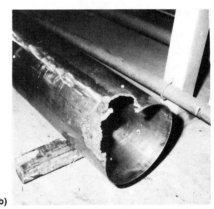

(a)

(b)

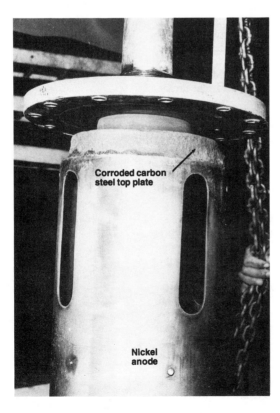

Fig. 2 Anode assembly and corroded steel plate

Corroded carbon steel top plate

Nickel anode

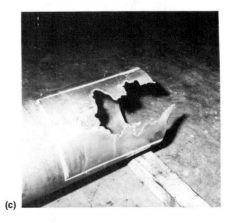

(c)

Fig. 3 Typical damage to the nickel anodes. (a) Initiation of the process. (b) and (c) Badly damaged anodes

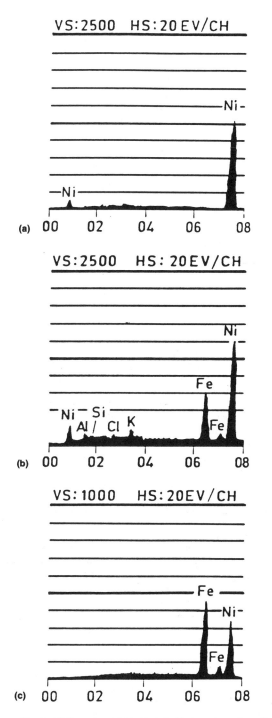

Fig. 4 EDX spectra. (a) Unaffected nickel anode surface. (b) Outer surface of gouged anode. (c) Inner surface of gouged anode

Fig. 5 SEM micrograph showing details of the outer surface of the gouged nickel anode. 32×

Fig. 6 SEM micrograph showing the presence of iron globules at the edge of unattacked nickel anode. 256×

tion of the black powder was determined using atomic absorption spectrometry. It consisted primarily of iron (77.6%), with copper, chromium, nickel, and manganese present in small amounts. X-ray diffraction (XRD) of the powder was carried out to determine its crystal structure. Results indicated the presence of alpha-iron (Fig. 7). The electrical conductivity of the powder was measured, and it was found to be a good conductor of electricity.

Discussion

During operation of the heavy-water upgrading plant, the electrolyte came in contact with the carbon steel top of the anode. Under the influence of applied potential, steel can dissolve in the alkaline medium. Dissolution can be uniform or nonuniform, depending on the potential applied and on the microstructure of the material. Because the potential difference between the anode and cathode was maintained between 2.5 and 3.5 V, the dissolution of the steel should have been nonuniform, with grain-boundary attack predominating over attack on the remaining grains. This kind of attack led to the release of ferrous ions in the electrolyte and to the dislodgment of grains from the matrix. Both of these factors appear to have contributed to the ultimate failure of the nickel anodes.

When the first of the affected cells was disman-

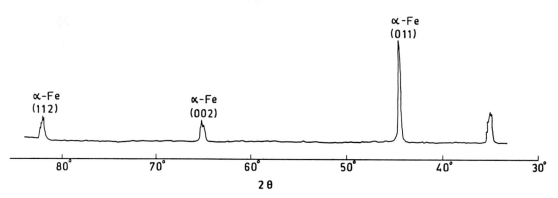

TARGET: CuK$_\alpha$
VOLTAGE: 40 kV
CURRENT: 25 mA
SCAN SPEED: 1° 2θ min^{-1}

Fig. 7 Diffractometer trace of black powder collected from the bottom of the electrolysis cell

tled, the appearance of a thick deposit on the upper portion of the outer cathode adjacent to the steel anode top suggested the possibility of iron deposition from the reduction of ferrous ions in the electrolyte. It is possible that the increasing thickness of the deposit was closing the gap between the electrodes, producing arcing that resulted in localized heating of the upper portion of the nickel anode. This phenomenon produced the interference colors and the charring of the rubber gasket.

As the dissolution of the steel anode top continued, iron grains were dislodged from the matrix, and a portion of the grains may have settled at the bottom of the same cell, whereas some were carried to the next cell by the electrolyte stream. The black powder found at the bottom of the cells consisted of debris or grains from the steel top plate material. The presence of highly electrically conducting powder in substantial quantities at the bottom of a good cell must have shorted the electrodes, thus disrupting the electrolytic process.

The passage of high current through a narrow cross section might have led to melting of the nickel anode. The presence of the electrolyte might have quenched the molten nickel, and the solidified globules might have been thrown onto fresh surfaces where they stuck, thus narrowing the gap between the electrodes. As such, the process of nickel anode decay became self-propagating.

Evidence supporting this mechanism was provided by SEM/EDX studies of the unattacked nickel—the outer and inner surfaces of the resolidified nickel mass. Pure iron globules were found sticking to the nickel anode surface on the edge of resolidified nickel (Fig. 6). These iron globules may have originated from dislodged grains of the steel top or from the steel cathode. Inspection of the cathode bottom could not be carried out because of its unapproachability. Even the resolidified nickel surface showed substantial amounts of iron contamination. SEM/EDX studies of the inner surface of the gouged nickel indicated the presence of iron and nickel.

Conclusion and Recommendations

Most probable cause

Electrolytic dissolution of the carbon steel top plate of the nickel anode led to the disruption of the upgrading process and to the decay of the nickel anode.

Remedial action

The easiest way to avoid the dissolution of the steel top plate of the nickel anode would be to prevent the electrolyte from approaching the steel plate by modifying the operating parameters, such as the flow rate of the electrolyte. Because control of the electrolyte level was difficult to attain in

practice, cladding of the steel plate with a corrosion-resistant alloy, such as nickel, nickel-base alloy, or stainless steel, was recommended. As an immediate measure to halt the anode decay, placing strainers in the inlets of the cells was suggested to filter the iron powder.

The corroded carbon steel top plates were clad using E NiCrFe-2 electrodes and returned to service. Some of the cladding was also performed using E 309 stainless steel electrodes. Both cladding materials have completed 9 years of service without recurrence of the problem.

Fatigue Failures of Connecting Bolts in a Truck Crane

Yoshio Kitsunai and Yutaka Maeda, National Research Institute of Industrial Safety, Kiyose, Japan

JIS SCM435 steel bolts that connected the slewing ring to the base carrier on a truck crane failed during the lifting of steel piles. The bolts were double-ended stud types and had been in operation for 5600 h. Failure occurred in the root of the external thread that was in contact with the first internal thread in the slewing ring. Examination of plastic carbon replicas indicated that failure was the result of fatigue action. Failure was attributed to overloading during service and increased stress concentration on a few bolts due to nonuniform separations around the slewing ring. A design change to achieve equal separation between bolt holes was recommended.

Key Words

Bolting	Studs	Piles
Chromium-molybdenum steels	Fatigue failure	Gears
Bolts	Cranes	

Alloys   Chromium-molybdenum steel—SCM435

Background

JIS SCM435 steel bolts that connected the slewing ring to the base carrier on a truck crane failed during the lifting of steel piles.

Applications

Truck cranes are widely used in construction and hoisting operations because of their mobility. The requirements for such mobility demand that the structural components and associated equipment in the crane often be reduced in weight. In contrast, operating rates of truck cranes have tended to increase over those of ten years ago. Accordingly, the margin for the strength of the components is not always sufficient, and damage or failure sometimes occurs. It has been shown that most truck cranes are damaged within ten years (Ref 1). Failure or damage is usually associated with fatigue, because truck cranes are generally subjected to severe cyclic loading in service.

Circumstances leading to failure

When steel piles being driven into the ground at a construction site were drawn out using a truck crane with a hoisting capacity of 300 kN (34 tonf), the crane suddenly collapsed, and the upper revolving structure, including a boom and an operation room, fell to the ground. The operator was not injured. The crane had been in use for approximately 10 years. Five years before, it was found that several connecting bolts had failed and fallen out from the base carrier. At that time, all connecting bolts were replaced with new ones. After replacement of the bolts, the crane was operated approximately 800 days (5600 h) before failure.

The crane collapsed because of the fracture of the bolts that connected the slewing ring to the base carrier. The upper revolving structure rotated on a steel slewing gear ring, which was bolted to the base carrier with 32 high-strength bolts. The structure of the slewing ring system is shown in Fig. 1. The bolts that failed, a double-ended type of stud, were 20 mm (0.8 in.) in diameter (M20), had an effective thread of 40 mm (1.6 in.), and had a total length of 175 mm (7 in.). The pitch of the screw of the bolt was 1.5 mm (0.06 in.) on one side and 2.0 mm (0.08 in.) on the other. The bolts failed at the root of the external thread that was in contact with the first internal thread in the

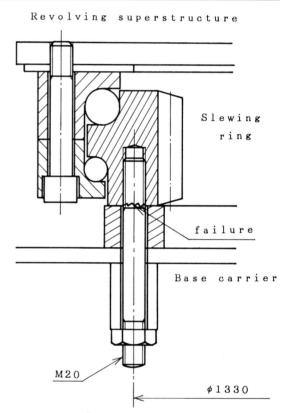

Fig. 1 Schematic of the slewing ring of the truck crane. Location of failure of the bolts is shown.

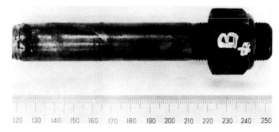

Fig. 2 Failed connecting bolt

slewing ring. The connecting bolt failure is shown in Fig. 2.

Testing Procedure and Results

Surface examination

Visual. The fractured surfaces of some of the failed bolts were covered completely with beach marks, which are characteristic of fatigue cracking (Fig. 3). The other fracture morphology of the bolts was terminal tensile fracture, as indicated by the characteristics of sudden shear overloading. Short fatigue cracks were found in most of the bolts that finally failed by shear (Fig. 4).

The relationship between the area percentage of fatigue cracking in each bolt and its location on the slewing ring is shown in Fig. 5. The bolt number on the abscissa, which shows the location of the bolt, was defined as the number when counting clockwise from the bolt located at approximately 12 o'clock on the slewing ring (Fig. 6). The bolts with area percentages of fatigue cracking more than 40% are in the first and second quadrants in Fig. 6, while the bolts that failed mainly by shear fracture are located primarily in the third and fourth quadrants. These results suggested that the fatigue failures of several bolts occurred in the

Fig. 3 Macroscopic view of the fracture surface of a bolt that failed completely by fatigue cracking

Fig. 4 Macroscopic view of fractured surface of a bolt that failed by fatigue cracking and terminal shear fracture. Areas F and H show fatigue crack and shear fracture zones, respectively.

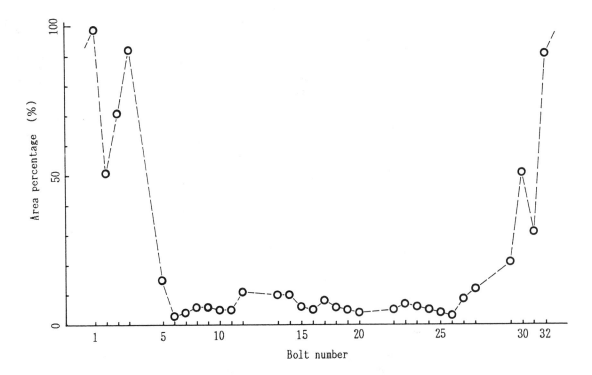

Fig. 5 Relationship between area percentage of fatigue cracking and bolt location on the slewing ring

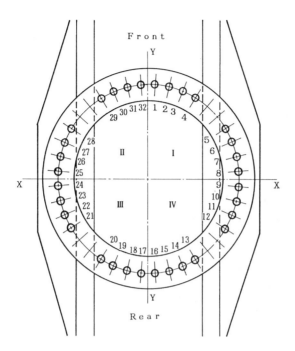

Fig. 6 Relationship between bolt number and location of the bolt on the slewing ring. Note the separations of the boltholes. For example, the separation between the No. 4 and 5 boltholes is larger than that between the No. 5 and 6 boltholes.

Fig. 7 Fractograph showing striations formed on the fracture surface of a bolt

Fig. 8 Fractograph showing striations and dimples formed on the fracture surface of a bolt. Areas S and L indicate small and large striations, respectively.

Fig. 9 Microstructure of the failed bolt

first and second quadrants and that the remaining bolts failed mainly by tensile shear fracture because of the resulting overloading.

Electron Microscopy. The fracture surfaces of the bolts were replicated by the plastic carbon method, and the replicas were examined with a transmission electron microscope (TEM). The fracture surfaces were also observed using a scanning electron microscope (SEM). Examination revealed abundant fatigue striations (Fig. 7), except for the bolts that failed mainly by shear. This verified that the primary cause of the failures was fatigue action. A large scatter of striation spacings was observed. Small striations with regular spacings and large striations with dimples were often alternately formed on the fracture surface (Fig. 8), which suggested that the connecting bolts were subjected to large variable-amplitude loading during operation of the crane.

Metallography

Microstructural Analysis. The microstructure of a typical bolt, tempered martensite, is shown in Fig. 9. No microstructural defects were observed.

Chemical analysis/identification

The chemical composition of a failed bolt was determined by spectral analysis. The results—0.45% C, 0.17% Si, 0.68% Mn, 0.011% S, 1.05% Cr, and 0.18% Mo—satisfied the requirements of JIS SCM435.

Mechanical Properties

Hardness. Microhardness testing using a 300 g load was conducted along the longitudinal section of a failed bolt. Hardness ranged from 357 to 381 HV.

Tensile Properties. Based on the Vickers hardness results, the tensile strength of the bolt material was approximately 1200 MPa (175 ksi).

Fatigue Properties. Fatigue testing of an unfailed bolt was performed under tension-tension cyclic loading with a stress ratio of 0.05 (Fig. 10). The fatigue limit was approximately 50 MPa (7 ksi).

Stress analysis

It has been shown that the striation spacing in various materials is given by the following relationship (Ref 2):

$$S = 9.4 (1 - v^2) (\Delta K/E)^2 \qquad \text{(Eq 1)}$$

where S is striation spacing, and v is Poisson's ratio, ΔK is stress-intensity factor range, and E is Young's modulus.

The stress amplitude applied to the bolts was estimated using Eq 1 and 2. To determine S, small striations with regular spacings were measured. These striations were chosen only because the spacing of larger striations with dimples varied widely.

The stress-intensity factor range for a surface crack in a bolt subjected to cyclic tension was estimated using the relation of Murakami et al. (Ref 3):

$$K = 0.650 \, \Delta \sigma m \sqrt{\pi (A)^{1/2}} \qquad \text{(Eq 2)}$$

where $\Delta \sigma m$ is the total stress range in a given stress cycle and A is the projected area of the crack.

The total stress range calculated by Eq 1 and 2 is shown in Table 1. The crack depth, a, in Table 1 was measured from the root of the thread. The total stress range from striation spacing was about 300 MPa (44 ksi). Therefore, the stress amplitude was about 150 MPa (22 ksi), which exceeded the fatigue limit of the bolts (about 50 MPa, or 7 ksi), as shown in Fig. 10.

The bolts in the slewing ring were subjected to tensile or compressive loading, depending on the revolving direction of the boom. To measure the stresses acting on the bolts in the slewing ring, two-element close-strain gages were glued onto

bolts at diametrically opposite locations (Fig. 11). The bolts with strain gages on them were attached to a slewing ring of the same type of truck crane as the failed one. Before measurement, the initial tightening force was adjusted to 470 ± 20 MPa (68 ± 3 ksi) in all the bolts.

Measurements were carried out under the following two conditions: (1) hoisting a dead weight of 103 kN (12 tonf) by the crane, which corresponds to the rated load at the working radius of the crane, and (2) a similar operating condition, that is, draw-

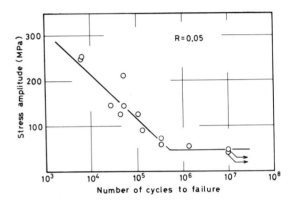

Fig. 10 S-N curve for the same type of bolts as those that failed in service

Table 1 Stress range estimated from striation spacing on a failed bolt

Crack depth (a)		Striation spacing (S)		Stress-intensity factor range (ΔK)		Total stress range ($\Delta \sigma m$)	
mm	in.	mm	in.	MPa√m	ksi√in.	MPa	ksi
1.5	0.06	9.92×10^{-5}	0.39×10^{-5}	22.2	20.2	306	44
6.0	0.2	1.74×10^{-4}	0.069×10^{-4}	29.4	26.8	292	42

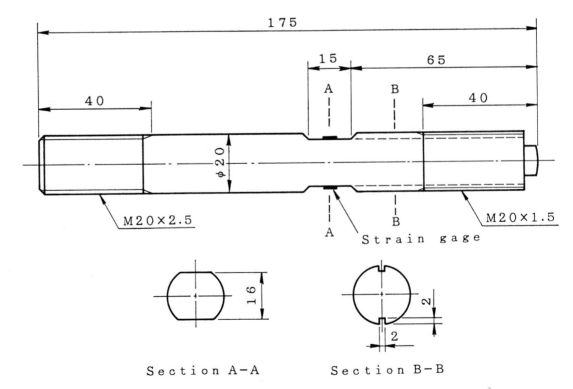

Fig. 11 Schematic of the bolt used for stress measurement

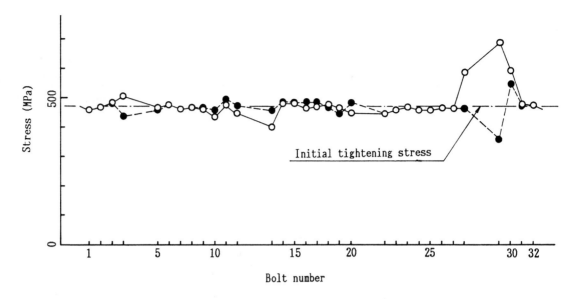

Fig. 12 Results of the stress measurements of bolts in the slewing ring of a truck crane while the rated load was hoisted. O, stress measured as the weight was hoisted at the rear of the crane. ●, stress measured as the weight was hoisted in front of the crane

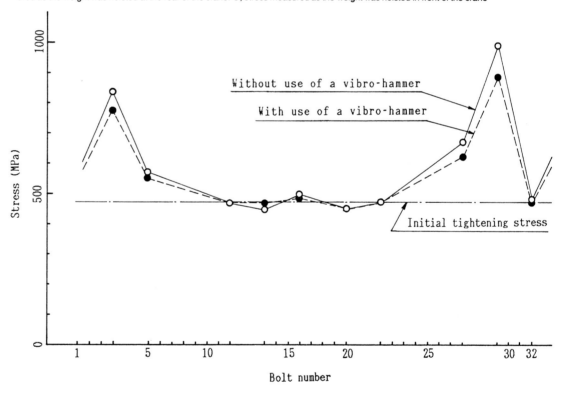

Fig. 13 Results of the stress measurements of bolts in the slewing ring of a truck crane under simulated service conditions

ing a steel pile from the ground by the crane. When the dead weight was hoisted by the crane, the stress acting on each bolt was found to be as shown in Fig. 12. The No. 29 bolt was subjected to a tensile stress of about 220 MPa (32 ksi), and a compressive stress of about 120 MPa (17 ksi) in addition to the initial tightening stress. The stress amplitude acting on the No. 29 bolt reached about 170 MPa (25 ksi), which exceeded the 50 MPa (7 ksi) fatigue limit of the bolt, even when just the

rated load was applied. Therefore, it was again confirmed that the bolt failed by fatigue.

Figure 13 shows the stress of the bolts measured when the pile located at the rear of the crane was drawn out from the ground. Stress measurements were conducted both with and without use of a vibrohammer, a machine used to vibrate the pile. It was difficult to measure the stresses on all the bolts, because the instruments needed for the dynamic strain measurements were limited in number. Remarkably high stresses, reaching approximately the tensile strength of the bolts, acted on the No. 4 and 29 bolts while the pile was drawn out directly without the use of the vibrohammer. The locations of these bolts corresponded to the locations of bolts with a high area percentage of fatigue cracking, as shown in Fig. 5. There was a larger separation between the No. 4 and 5 bolts and between the No. 28 and 29 bolts, as compared with the separation between the bolts subjected to relatively low stresses, as shown in Fig. 6. These larger separations were due to the structure of the truck crane, because the carrier frame prevented the use of boltholes. When a pile was drawn out using a vibrohammer, the stress acting on the bolt was found to decrease slightly compared with the stress obtained without the use of a vibrohammer, as shown in Fig. 13. When the crane collapsed, a vibrohammer was not being used.

Conclusion and Recommendations	**Most probable cause**

Most probable cause

The collapse of the truck crane was caused by fatigue failures of the bolts connecting the slewing ring to the base carrier. The fatigue failures resulted from overloading during service and from increased stress concentrations on a few bolts due to nonuniform separations around the slewing ring.

Remedial action

To prevent such failures, the pitch of the bolthole should be at the same distance on the slewing ring to avoid the stress concentration for some of the bolts. Moreover, the stiffness of the base carrier attached to the slewing ring should be increased. Although only a slight decrease in stress was obtained by using a vibrohammer, its use may be advantageous under other hoisting conditions.

References

1. "Survey Report on Standards of Welding Procedure for Cranes," 1987-3, Japan Crane Association, 1987 (in Japanese)
2. H. Kobayashi, H. Nakamura, and H. Nakazawa, *Proc. 3rd ICM*, Vol 3, 1979, p 529
3. Y. Murakami and S. Namat-Nasser, *Eng. Fract. Mech.*, Vol 17, 1983, p 193

Formation of Refractory Films on Metal Contacts in an Electrical Switchgear

John R. Hopkins, Consultant, West Hartford, Connecticut

During routine quality control testing, small circuit breakers exhibited high contact resistance and, in some cases, insulation of the contacts by a surface film. The contacts were made of silver-refractory (tungsten or molybdenum) alloys. Infrared analysis revealed the film to be a corrosion layer that resulted from exposure to ammonia in a humid atmosphere. Simulation tests confirmed that ammonia was the corrodent. The ammonia originated from the phenolic molding area of the plant. It was recommended that fumes from molding areas be vented outside the plant and that assembly, storage, and calibration areas be isolated from molding areas.

Key Words

Silver-base alloys, corrosion
Electric contacts, corrosion
Electrical resistance

Ammonia, environment
Circuit breakers

Switchgears
Atmospheric corrosion

Background

During routine quality control testing, electrical switchgear devices exhibited high contact resistance and, in some cases, insulation of contacts by a surface film.

Pertinent specifications

Many of the contacts in the molded-case circuit breakers are made of silver-tungsten or silver-molybdenum alloy. Figures 1 and 2 show typical microstructures. The type of device and its rating, application, and cost determine the type of plastic used in the molded cases and covers. The working parts are mounted or encased in molded plastic for structural reasons and for electrical insulation of current-carrying components.

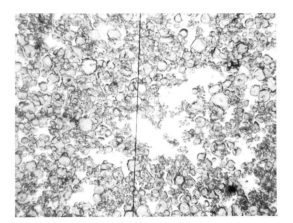

Fig. 1 Typical microstructure of a 50Ag-50Mo contact. The light areas are silver, the dark areas molybdenum. 380×

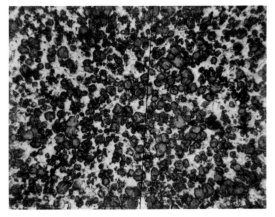

Fig. 2 Typical microstructure of a 27Ag-73W contact. The light areas are silver, the dark areas tungsten. 380×

Visual Examination of General Physical Features

The surfaces of the electrical contacts exhibited a visible film (Fig. 3), which appeared to be a corrosion film or possibly excess brazing flux that had not been properly removed during the washing operation.

Testing Procedure and Results

Chemical analysis/identification

Coatings on Surface Layers. Infrared analysis was performed on the visible surface films. The resulting spectra revealed complex metal hydrates of ammonia—specifically, ammonium heptamolybdate tetrahydrate on silver-molybdenum contacts and a complex ammonium tungstate hydrate on silver-tungsten contacts.

Associated Environments. Many manufac-

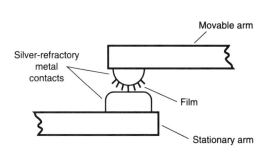

Fig. 3 Schematic showing the silver-refractory metal contacts

turers use wood-flour-filled phenol formaldehyde molding material when making molding cases for small circuit breakers. The phenolic resin contains an accelerator, hexamethylene tetramine, which aids in setting the resin. Heat supplied by the molding press causes the powder to solidify under pressure, but at the same time releases water vapors, phenol, ammonia, and several other gases into the air in the factory. Some of these vapors and gases are retained in the wood flour in the cases and covers.

Discussion

The formation of films can occur during assembly and testing if atmospheric ammonia vapors condense on surfaces when the temperature of the circuit breaker components are below the dew point. Films can also form during service in humid atmospheres when trapped ammonia gases diffuse out of the phenolic case and cover and attack the contacts.

The degree of heating of the contacts is determined by the location, composition of the film, film thickness, and current passing through the contacts. It is possible in some cases for the film to be so thick that no current will flow at all. Overheating occurs when thinner films are present. The arcing process (when the circuit breaker is interrupting current) develops a temperature so high that the films break down.

Conclusion and Recommendation

Most probable cause

Ammonia fumes were carried by air-circulating fans throughout the plant. The relative humidity at the time was high. When the devices were transported to the calibration room, whose dew point was lower than the rest of the factory, moisture condensed on the contacts and the ammonia dissolved in the moisture, forming the refractory films.

Remedial action

It was recommended that fumes from molding areas be vented outside the plant and that assembly, storage, and calibration areas be isolated from molding areas.

Stress-Corrosion Cracking in Stainless Steel Heater Sheathing

P. Muraleedharan, H.S. Khatak, and J.B. Gnanamoorthy, Metallurgy Division, Indira Gandhi Centre for Atomic Research, Kalpakkam, India

Cracking occurred in type 304L stainless steel sheaths on nichrome wire heaters that had been in storage for about 5 years in a coastal atmosphere. The cracks were discovered when the heater coils were removed from storage in their original polyethylene packing materials and straightened for use. Fractography established that fracture occurred by stress-corrosion cracking. The cracks originated at rusted areas on the cladding that occurred under iron particles left on the surface during manufacture. High hardness values indicated that solution annealing following cold working had not been carried out as specified. It was recommended that the sheathing material be fully annealed and that the outer surface be pickled and passivated.

Key Words

Austenitic stainless steels, corrosion
Sheaths, corrosion
Stress-corrosion cracking

Electric heating elements
Heat-distributing units
Nuclear reactor components

Marine atmospheres
Rusting

Alloys

Austenitic stainless steel — 304L

Background

Cracking was discovered in type 304L stainless steel sheaths on nichrome wire heaters that had been in storage for about 5 years in a coastal atmosphere.

Applications

Such heaters are wound over stainless steel pipes to heat sodium inside the pipes.

Circumstances leading to failure

Five hundred heaters were supplied by the manufacturer in the form of coils in polyethylene packing material and transported by sea to the site. When the coils were straightened for use after about 5 years of storage in their original packing, about 20% of them broke into pieces. Some of the packings were also damaged.

Pertinent specifications

The heaters use nichrome wire as the heating element, magnesia powder as insulating material, and type 304L stainless steel as sheathing material. The heaters were in coils of various lengths (5 to 30 m, or 15 to 100 ft); the minimum diameter of the heater wire was about 4 mm (0.15 in.). The thickness of the stainless steel sheath varied from 0.2 to 0.3 mm (0.0080 to 0.010 in.) where the diameter was minimum. The heaters were specified to be annealed, pickled, and passivated at the final stage of manufacturing.

Specimen selection

Several broken heaters were sent for failure analysis. Fractographic examination required cutting the defective portions of heaters into lengths that could be accommodated by a scanning electron microscope (SEM). Longitudinal sections of the heaters were mounted in molds of quick-setting resin and prepared for metallographic examination.

Visual Examination of General Physical Features

Reddish brown rust spots were observed in several places on the failed heaters. Cracks oriented in a longitudinal direction originated from the rusted areas. When the brown corrosion product was removed, it was discovered that the stainless steel had been attacked beneath the rust deposits.

Testing Procedure and Results

Surface examination

Scanning Electron Microscopy/Fractography. Figure 1 shows an SEM fractograph of a portion of a heater where severe cracking was evident. At high magnification, typical transgranular stress-corrosion cracking (SCC) fractures with fan-shaped patterns were apparent (Fig. 2 and 3). Figure 3 also shows a small amount of dimple fracture near the inner surface of the sheathing. This indicated that the area near the inner surface was the last to break by mechanical means.

Metallography

Microstructural Analysis. The mounted sections of the failed heaters were polished to diamond finish and etched. Optical microscopic examination revealed a "step structure," indicating that the material was not sensitized.

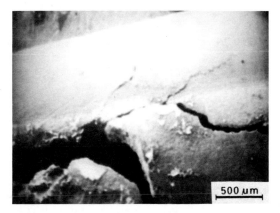

Fig. 1 SEM fractograph revealing severe cracking in a portion of a failed heater

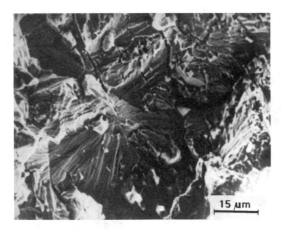

Fig.2 SEM micrograph showing typical transgranular SCC fracture in a failed heater

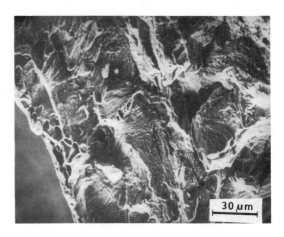

Fig. 3 SEM micrograph showing transgranular SCC and dimple fracture regions

Crack Origins/Paths. Optical microscopic examination of a cracked portion revealed underdeposit attack; that is, corrosion of the sheath material below the deposits on the surface. Figure 4 shows a typical microstructure of a wide underdeposit attack. Cracks initiated from these regions and penetrated across the wall thickness (Fig. 4). Some cracks propagated in the longitudinal direction.

Chemical analysis/identification

Corrosion or Wear Deposits. Energy-dispersive X-ray analysis of the brown spots on the heater surfaces indicated that the deposits were composed of iron and chlorine. This indicated that the brown spots formed after the iron particles embedded on the surface during manufacture rusted.

Mechanical properties

Hardness. Several microhardness measurements were made on polished and unetched metallographic samples. The hardness of the material was in the range of 190 to 232 HV, increasing to-

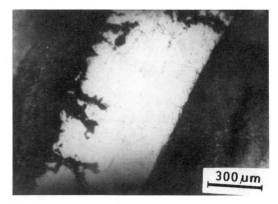

Fig. 4 Optical micrograph showing crack initiation in a region of underdeposit attack

ward the outer surface. This measured hardness was higher than the reported value for solution-annealed type 304L stainless steel (~150 HV).

Discussion	Fractography established that the fracture occurred by SCC originating at regions where the cladding had rusted. Underdeposit attack had also occurred. The rusting resulted from incomplete removal of iron particles embedded on the heater surface during manufacture. The material was not sensitized—an important observation, because sensitization is a common cause of failure of stainless steel components during storage in coastal areas. However, the hardness of the material was higher than that of solution-annealed type 304L stainless steel, indicating that the sheathing was cold worked. It appeared that full solution annealing had not been performed, even though it was specified.

Because pickling and passivation were not carried out properly, iron impurities on the heater surfaces had not been removed completely. The iron particles on the surface rusted during exposure to the humid atmosphere. Rust generally contains ferric ions and also collects moisture and chloride ions from the atmosphere. The presence of ferric ions raised the electrochemical potential of the stainless steel, creating conditions for the underdeposit attack and SCC. Stress present in the material because of improper annealing may have been augmented by stresses arising from the wedging action of the corrosion products formed during the underdeposit attack.

Conclusion and Recommendations

Most probable cause

The primary cause of the fracture was SCC originating at rusted regions on the surface of the heaters. The presence of ferric ions in the rust and absorption of chloride ions and moisture from a coastal atmosphere created the necessary environmental conditions for underdeposit attack and SCC. Residual stress in the material in conjunction with stresses resulting from the wedging action of corrosion products facilitated SCC.

Remedial action

It was recommended that proper pickling and passivation of the heaters be conducted as the final stage in the manufacturing process to provide a surface free from surface impurities. Annealing of the heaters would reduce the hardness of the material and relieve the residual stress.

Termination Delamination of Surface-Mount Chip Resistors

Jude M. Runge-Marchese, Taussig Associates, Inc., Skokie, Illinois

Several surface-mount chip resistor assemblies failed during monthly thermal shock testing and in the field. The resistor exhibited a failure mode characterized by a rise in resistance out of tolerance for the system. Representative samples from each step in the manufacturing process were selected for analysis, along with additional samples representing the various resistor failures. Visual examination revealed two different types of termination failures: total delamination and partial delamination. Electron probe microanalysis confirmed that the fracture occurred at the end of the termination. Transverse sections from each of the groups were examined metallographically. Consistent interfacial separation was noted. Fourier transform infrared and EDS analyses were also performed. It was concluded that low wraparound termination strength of the resistors had caused unacceptable increases in the resistance values, resulting in circuit nonperformance at inappropriate times. The low termination strength was attributed to deficient chip design for the intended materials and manufacturing process and exacerbated by the presence of polymeric contamination at the termination interface.

Key Words   Electronic devices   Delaminating

Background

Several surface-mount chip resistor assemblies failed during monthly thermal shock testing and in the field.

Circumstances leading to failure

The resistor, the only electrically active component in the final assembly, exhibited a failure mode characterized by a rise in resistance out of tolerance for the system. The failures occurred during monthly testing after 100 h of thermal shock (from −65 to 125 °C, or −85 to 255 °F, 30 min at each extreme for 100 cycles) and "suddenly" in the field (e.g., overnight).

As part of a quality control test program, a die shear test was performed on the resistors after solder attachment to the carrier strips. The most recent attachment shear forces varied from 0.4 to 2 kg (0.8 to 4.5 lb). The resistors with low force values reportedly exhibited entire termination failure. The attachment never broke within the solder joints. Rather, the termination consistently came apart from the substrate.

Pertinent specifications

A chip detail drawing (Fig. 1) indicated that the chip substrate was high-purity alumina. The resistive element was a sintered thick film that was coated with a protective glass film after laser trimming and finished with an epoxy coating. Continuity through the resistive element was established by solder attachments to the assembly lead frame through edge terminations.

The terminations were of a lamellar design. Initial resistive element/termination contact was at the top land termination. The wraparound portion of the terminations coated the top land termination and "wrapped around" the ends of the substrate. No material was specified for the termination layers; however, it was specified that the wrap-around or edge terminations were to be coated with a nickel barrier layer (electroplated) and finished with electroplated tin deposit for solderability (Fig. 1).

Termination/carrier strip attachment was reportedly performed in a belt furnace. The carrier strip was stamped from coin silver strip. Solder paste (Pb-8Sn-2Ag) and type RMA flux were used

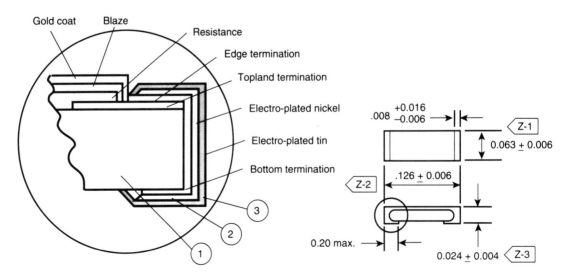

Fig. 1 Chip detail schematic, indicating chip construction as having both a top and bottom termination. Actual construction and sales literature indicated only a top land termination. The edge termination has a wraparound configuration. Dimensions given in inches

to make the attachment. The reflow profile was such that the temperature ramped up to a maximum of 330 °C (625 °F), which is 40 °C (70 °F) higher than the liquidus temperature of the solder. The dwell time above the liquidus temperature was 110 s.

The carrier strip/chip assembly was overmolded with Valox resin at a temperature of 250 °C (480 °F); a recommended operating range of 240 to 260 °C (465 to 500 °F) was specified. Valox is a General Electric polymer blend of Lexan, a GE polycarbonate, and polybutylene terephthalate.

Changes in resistance were often measured after solder attachment and encapsulation. Changes were also determined after thermal shock and were measured on returns from the field.

Specimen selection

Representative samples from each step in the manufacturing process were selected for analysis. Additional samples were selected representing the various resistor failures. Samples submitted for analysis were divided into the following groups:

Group No.	Description
1	As-received resistor chips from storage
2	Chips exposed to the solder reflow temperature of 330 °C (625 °F) for a dwell time of 60 to 90 s
3	Chips soldered on a silver carrier strip and washed in methylene chloride after attachment in order to remove flux residue
4	Carrier strip/chip assemblies as encapsulated with Valox resin
5	Encapsulated resistor assembly that exhibited a failure rise in resistance after thermal shock. A void in the solder fillet was noted.
6	Assembly that reportedly exhibited a cracked solder joint
7	Assembly that exhibited an increase in resistance after thermal shock
8	Several resistor chips that exhibited low shear force values during die shear testing. Some exhibited termination separation from the resistor die.

Testing Procedure and Results

Surface examination

Visual. The chips were soldered with the resistive element toward the carrier plate. The resistor/carrier assemblies were encapsulated with Valox and cut from the carrier strip.

The resistor chips were rectangular with metallic end terminations. The resistive element was coated with a gold-colored polymeric material. The material appeared soft, but could not be indented with a fingernail. The gold coating apparently turned brown when exposed to the heat of soldering (Fig. 2).

Examination of various group 8 resistors, which failed by die shear, revealed two different types of termination failures. The first, and apparently most prevalent, type was that of total delamination (Fig. 3). The end terminations peeled from the substrate, exposing the ceramic ends and fracturing the polymer coating. It was noticed that the polymer coating on the failed resistors appeared darker in color than those in the as-received condition. The gold polymer coating also appeared to have melted in the area of the delamination.

The second type of failure noted was that of par-

Fig. 2 Group 1 (left) and Group 2 resistors. The group 2 resistor had been exposed to the heat of soldering for 90 s. Note the dark appearance of the top polymer coating. This is due to the pyrolysis, or breakdown, of the epoxy coating. 19.25×

Fig. 3 Delaminated group 8 resistor that exhibited a low pushoff strength. Note the puckered appearance of the gold polymer coating at the left termination. This was caused by the heat of soldering exceeding the T_g of the epoxy. 21.6×

Fig. 4 SEM micrograph of a resistor with a partially delaminated termination. The light-appearing portion is the lead-tin alloy on the surface of the termination. The dark-appearing area beneath the lead-tin alloy is exposed alumina.

tial delamination. The gold-colored polymer coating did not appear as dark on these failures. The failed termination appeared to fracture at the end and around the bottom of the resistor.

Fig. 5 SEM micrograph of the delaminated termination. Note the gray scale difference between the surface of the delaminated termination and that of the alumina substrate in Fig. 6. This is because delamination occurred at the cermet/thick-film polymer interface.

Fig. 6 Higher-magnification examination of the surface of the delaminated portion of the resistor termination. Note the absence of evidence of any type of fracture or tearing on the surface.

Scanning Electron Microscopy/Fractography. In order to determine at which interface the resistors were delaminated, several samples from group 8 were examined using a scanning electron microscope (SEM). Examination of the fractured termination confirmed visual findings that the fracture occurred at the end of the termination, exposing the ceramic substrate. The top land termination adjacent to the resistive element was apparently intact (Fig. 4). Examination of the delaminated termination revealed no evidence of fracture. The edge termination appeared to have peeled from the end of the resistor (Fig. 5 and 6).

Energy-dispersive x-ray spectrographic (EDS) analysis was performed on the delaminated surface and determined the presence of zirconium, silver, silicon, lead, manganese, magnesium, and aluminum. The aluminum was most likely from the alumina resistor substrate. Silicon, zirconium, magnesium, and manganese oxide are various glass formers and metal oxides used in forming a ceramic-to-metal bond on chip components, most likely the top land termination. The silver component may have been the metallic portion of the metallization, which enabled subsequent electrodeposition of the nickel barrier and solderable tin coatings.

Metallography

Microstructural Analysis. In order to determine the construction integrity and material failure mode of the resistor chips, transverse sections of samples from each of the groups were metallographically prepared. Examination was performed with a calibrated metallurgical microscope with magnification capabilities up to 2000×.

Examination of the group 1 resistors, those which reportedly were in the as-received condition, revealed the following termination construction. The top land termination contacted the resistive element and ended at the edge of the substrate. It exhibited a typical cermet (ceramic-metal) metallization that had been fired to fuse the termination to the alumina substrate. The gold-colored polymer that coated the resistive element often overlapped the cermet top land termination.

The top land termination was the only portion of the termination that exhibited actual ceramic-to-met-

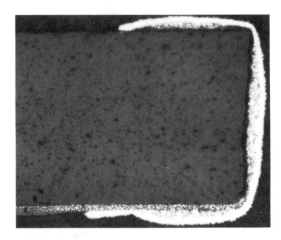

Fig. 7 Chip cross section from group 1, those samples taken in the as-received condition prior to processing. As polished. 87×

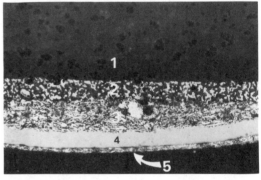

Fig. 8 An interesting detail on the group 1 resistors was the presence of an amorphous dark-appearing phase dotting the interface between the cermet metallization and the thick-film polymer conductive ink. Fourier transform infrared analysis determined that this dark-appearing phase was actually resin separation from the epoxy binder in the thick-film ink. The following features can be discerned: area 1 is the alumina substrate; area 2 is the cermet metallization; area 3 is the thick-film polymer ink; area 4 is the electroplated nickel barrier; and area 5 is the lead-tin surface alloy. 31.0×

al bonding. The wraparound portion of the termination was achieved with what appeared to be conductive (metal-loaded) polymer-base thick-film ink. Small, dark-appearing amorphous areas dotted the interface between the thick-film ink and the cermet metallization. The termination appeared to have been plated with two distinct metallic layers (Fig. 7).

Examination of the metallization interface determined that each of the layers composing the termination was separate and distinct. There was no evidence of intermetallic formation between the cermet and thick-film ink. The presence of the dark-appearing amorphous phase at this interface was confirmed (Fig. 8). Samples from group 2, those that had been processed through the soldering profile, exhibited similar characteristics.

In order to understand the metallurgical ramifications of a reported failure, a transverse section through a sample from group 6 was metallographically prepared. It was reported that the sample exhibited a cracked solder fillet. Examination of the cross section revealed gross delamination of the terminations (Fig. 9). So severe was the delamination that the encapsulant had flowed into the interface between the cermet metallization and the polymer thick-film ink. This expanded the plating and thick-film ink layers from the substrate to the point of rupture at the top of the termination.

Metallographic examination of the group 6 samples and samples from the remaining groups determined the presence of consistent interfacial separation between the cermet and polymer thick-film layers. The condition ranged in severity from its presence as an amorphous layer of contamination to that of total delamination. Gross delamination was consistently observed in the group 3 samples (Fig. 10), indicating that the solder attachment of the chip to the carrier strip was the point in the manufacturing process that produced the failure condition.

Representative sections from groups 2 and 3 were analyzed using electron probe microanalysis (EMA) to determine the constituents of the resistor materials and possibly to identify the contaminating layer present at the metallization interface. EMA of the group 2 sections (see Fig. 11) determined that the resistor substrate was high-purity alumina. The metal portion of the cermet metalli-

zation was silver rich; the ceramic portion was zirconium rich. Significant amounts of silicon were also detected, most likely present as an oxide or

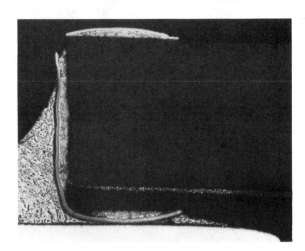

Fig. 9 Delaminated termination on a group 6 sample, showing Valox (the pellet encapsulant) intruding on the cermet/polymer thick-film metallization interface. Metallization forced the expansion of the termination, rupturing the nickel barrier and giving the solder joint a cracked appearance.

Fig. 10 Consistent gross delamination in the group 3 samples, those that had been soldered to the carrier plate. Measurable separation was noted at the metallization interface. 124×

Fig. 11 SEM micrograph of a group 2 resistor termination. Areas of interest are as follows: area 1 is the alumina substrate; area 2 is the cermet thick-film metallization; area 3 is the polymer thick-film ink; area 4 is the nickel barrier layer; and area 5 is the lead-tin surface coating. Arrows indicate the critical interface. Note at this interface, even in the group 2 components, the appearance of a dark thin layer at the cermet/polymer thick-film interface.

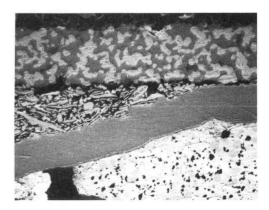

Fig. 12 Interface on a group 3 resistor. Note the increased thickness at the critical interface. This was caused by the forces exerted by the expansion and contraction of the silver carrier plate during soldering. The polymer contamination (most likely resin separation from the epoxy thick-film ink) at the surface of the polymer metallization layer created a nonadherent film at the interface.

glass former. Segregation due to diffusion and mass transport, typical for a ceramic-to-metal bond, was observed. The heaviest concentration of the oxides and glass formers was toward the substrate, whereas the silver remained at the surface for subsequent metallic joining operations.

EDS analysis of the wraparound metallization determined that the metallic portion was silver. The dark phase was carbon rich, supporting the visual observation that the binder in the thick-film ink was polymer based. EDS analysis of the electroplated layers confirmed that the barrier layer was nickel and that the surface layer was tin-lead alloy. The gold-colored polymer coating over the resistive element was a metal-oxide-loaded polymer with a significant amount of chlorine, probably an epoxy.

EMA of the sections of the group 3 samples that exhibited delamination (see Fig. 12) confirmed the metallographic analysis results by determining the presence of an amorphous contaminant at the cermet/polymer thick-film interface. Higher-magnification examination of this layer revealed the presence of voids within the amorphous material. EDS analysis of the material determined that it was organic in nature, with a significant amount of chlorine, and probably was an epoxy.

Chemical analysis/identification

Fourier Transform Infrared Analysis. In order to discern the compositions of the polymer constituents of the resistors, Fourier transform infrared (FTIR) analysis was performed. Specifically, the gold coating, the polymer-base thick-film ink, and the interfacial contamination layer were compared to try to determine the nature of the amorphous layer consistently observed at the metallization interface.

FTIR analysis of the gold coating material determined that it was an anhydride-cure epoxy. The significant chlorine content determined by way of EDS suggested that the curing agent for the epoxy was chlorendic anhydride. This material, thermosetting in nature, will not melt when exposed to elevated temperatures. Instead, the epoxy exhibits a glass transition temperature, or T_g. This is the point at which the amorphous phase of the polymer changes from a hard brittle condition to a viscous rubbery condition. This could explain the ductile, puckered appearance of the gold coating on the delaminated resistors.

The T_g for most anhydride epoxies is about 130 °C (265 °F); some formulations (specifically, dianhydride cure epoxies) exhibit T_gs as high as 280 °C (535 °F). It is never advisable to subject epoxies to temperatures in excess of the T_g for an extended period of time.

The polymer binder from the thick-film ink on several delaminated samples was removed by squeezing the silver flakes of the metallization together and extruding the binder between them. FTIR determined that this material consisted of polyamide and an epoxy similar to the polymer epoxy gold coating. These results were consistent for the binder in the thick-film ink on resistors from groups 1 and 2. These consistencies indicated that the epoxy contained a polyamide curing agent. This material, if fully cross-linked, would not melt, but would also exhibit a T_g.

The metallization interface was analyzed by preparing samples from group 3, the group that exhibited consistent gross delamination. Samples were prepared for transmission FTIR by replication in amyl acetate on potassium chloride blanks. During preparation, it was noted that a significant amount of a dark yellow, oily substance was present at the interface. FTIR analysis determined that it was a triglyceride. Triglycerides are often added to epoxy formulations as drying oils. They are also typically found in animal or vegetable fats and can be attributed to handling. However, glycidyl groups are the most common constituents of epoxy resin functions; exposure to acids from soldering flux or residual cleaning solutions can convert un-cross-linked resin to triglycerides by way of nucleophilic substitution. This suggested that the amorphous contamination observed at the metallization interface was the result of epoxy degradation by way of resin separation from the thick-film ink.

To determine how early in the manufacturing process resin separation was apparent in the chip resistors, samples from groups 1 and 2 were prepared for analysis by forcing delamination and replicating the surface as described above. Smaller amounts of an oily substance, not yellow in color, were extracted from the samples. FTIR analysis again determined the oily substance to be a triglyceride, suggesting that degradation was present in chips in the as-received condition.

Analysis of the delaminated surface on the group 8 samples revealed the presence of human skin cells. These skin cells could certainly contribute to delamination failures; however, they could also be the result of improper handling of the resistors after failure and prior to receiving them for analysis.

Discussion

Metallographic analysis determined that the failure mode exhibited by the resistors, that of an unacceptable increase in resistance during test and during service, was due to delamination of the wraparound or edge portion of the terminations. The focus of the failure analysis thus turned to determination of the nature and cause of the delamination.

The consistent presence of an amorphous, apparently polymeric layer at the metallization interface required FTIR analysis. By determining the nature of the layer via FTIR, the source for failure was indicated as both material and process related. In other words, the resistor termination design, because it incorporated polymeric materials and thus limited metallurgical bonding, was not appropriate for the recommended soldering and encapsulation profiles. This inadequacy was amplified by the presence of the layer of polymer-based contamination at the metallization interface.

In order to support this conclusion, it is necessary to understand the failures. Two types of failures were observed: total and partial delamination.

Partial delamination occurred at the metallization interface, confined to the portion of the wraparound termination away from the top land termination. This was primarily due to the absence of a ceramic-to-metal bond in this portion of the termination. Termination fracture occurred when die shear forces exceeded the mechanical bond force between the ink and the substrate during die shear testing. During manufacturing, fracture occurred when the resistor assembly was manufactured with the top land termination up and shear forces exerted by the thermal expansion and contraction of the silver carrier plate during soldering and encapsulation exceeded the strength of the mechanical bond between the termination and the resistor substrate. These failures were the easiest to cull out, because intermittent performance or abrupt failure was often revealed during thermal shock testing.

The more subtle and most prevalent delamination type of failure also occurred at the metallization interface. The primary source for total delamination was also the absence of a true ceramic-to-metal bond at this interface. Exacerbating this lack of bond was the presence of interfacial polymer-based contamination.

The consistency of a concentrated film or layer of amorphous polymer film at the metallization interface suggested that the epoxy binder in the thick-film ink was neither homogeneous nor stable. Resin separation from the epoxy binder created an oily, nonadherent film at this critical interface. This meant that not only was there no ceramic-to-metal bond fixing the termination, but there also was no epoxy adhesive bond. The amount of resin separation varied from chip to chip, with those with worst-case adhesion exhibiting the most resin separation and those off the

shelf with apparently good termination strength displaying only a small amount. It was observed that this layer became more pronounced as the resistors were exposed to the heat of processing. The dark yellow color of the oily residue observed at the metallization interface of the group 3 samples was most likely due to thermal breakdown of the resin during attachment to the carrier strip.

Gold coat epoxy overlapped the top land termination of many of the samples. While this was probably a process control problem during component manufacture, the dimpled, reflowed appearance of the material on failed components indicated that the heat of processing the resistor assemblies exceeded the T_g for the gold coat. Its softening would certainly undermine termination strength in the area, aiding thermal expansion forces in pulling the terminations from the resistor substrate.

In the resistor assembly manufacturing process, as the components are soldered to the silver carrier plate, they are soaked above 290 °C (555 °F), peaking at 330 °C (625 °F) for 110 s. Even assuming the highest possible T_g for the epoxy systems in the chip, 280 °C (535 °F), visual evidence supports the conclusion that the material certainly became viscous during this operation. Any polyamide curing agent or uncured resin that had not reacted during the cure cycle for the polymer-base thick-film ink degraded. Thermal expansion forces exerted by the carrier strip on the metallization interface pulled the terminations apart. Because nothing other than the wraparound termination was making contact with the top land termination and the resistive element, an open condition or measurable increase in the resistance value for the assembly was created.

Conclusion and Recommendations

Most probable cause

Analysis of several groups of surface-mount chip resistors determined that sudden unacceptable increases in the chip resistance values both during testing and in the field were caused by inadequate wraparound termination strength. The source of the low termination strength was twofold. The first and most prevalent cause appeared to be inadequate component design for the intended manufacturing process. The second cause was polymeric contamination at the termination metallization interface.

Remedial action

In order to manufacture a resistor assembly with chips of this design, the attachment solder and reflow profile must remain below the T_g for the epoxy systems manufactured into the component. A new encapsulant and encapsulation procedure are needed that will not exceed the critical T_g.

Variations in strength due to the presence of contamination at the metallization interface or degradation of the polymer constituents should be determined by nondestructive die shear testing. Chips soldered to carriers should be subjected to a nominal force for a period of time, released, and retested for shifts in the resistance values. Failed chips would be those that exhibit a shift in resistance out of the accepted range.

Rather than alter the materials and add tests to the current process, it is also recommended that a resistor that exhibits a complete ceramic-to-metal wraparound termination be used. Manufactured properly, the resistor should be able to withstand the heat of soldering and encapsulation without delaminating.

It is imperative when considering a new resistor design that the processing recommendations by the chip manufacturer be addressed. It still may be necessary to alter the assembly process and materials to reflect the temperature requirements of the chip.

MISCELLANEOUS FAILURES

Failure of a Stainless Steel Hip Fracture Fixation Device

Carmine D'Antonio, Department of Metallurgy and Materials Science, Polytechnic University, Brooklyn, New York

A type 316L stainless steel "Jewett nail" hip implant failed after 2 months of service. Fracture occurred through the first of five screw holes in the plate section. Microscopic examination of mating fracture surfaces showed that failure had initiated at the outside (convex) surface of the plate and proceeded through its thickness. The fracture morphology was characteristic of fatigue. A beveled area on the inside surface of the plate indicated that the implant had been fractured for some time prior to removal. Metallographic examination of samples cut from the plate section revealed a series of hidden repair welds on the inside surface of the plate in the vicinity of the fracture. Comparison of the microstructure in the area of the fracture with that in an area away from the weld indicated that the repair welding had resulted in the creation of an annealed, softened zone. Manufacturers should never attempt to salvage this type of critical device by welding or any other procedure that might compromise its integrity.

Key Words	Austenitic stainless steels	Repair welding	Surgical implants
	Fatigue failure		

Alloys Austenitic stainless steel—316L

Background

A "Jewett nail" hip implant failed after 2 months of service. Fracture occurred through the first of five screw holes in the plate section.

Applications

The "Jewett nail" implant is widely used to assist in immobilizing the femoral head in cases where femoral neck fracture has occurred. The nail portion of the appliance is inserted through the neck, across the fracture, and into the head. The plate portion rests on the shank of the femur and is affixed to it with screws.

Pertinent specifications

The applicable specification at the time of manufacture was ASTM F55-66T, which pertains to stainless steel bars and wires for surgical implants.

Specimen selection

Three samples were cut from the plate portion of the implant, two in the vicinity of the fracture and one away from the fracture. These samples were mounted in a cold-setting resin and used for metallographic examination and microhardness testing. A small section was cut from the plate section for chemical analysis.

Visual Examination of General Physical Features

Figure 1 shows the as-received device. Failure resulted from fracture of the plate section across the first screw hole. This is the area of maximum cyclic bending stress during service.

Testing Procedure and Results

Figure 2 shows the mating fracture surfaces. It is evident that failure initiated at the outside (convex) surface of the plate and progressed through its thickness. The fracture morphology is characteristic of fatigue, in that it is flat and brittle in texture.

The inside surface of the plate section on the nail side is shown in Fig. 3. Note the beveled, burnished area in the final fracture zone caused by rubbing of the screw after fracture occurred. This burnishing indicates that the implant had been fractured for some time prior to its failure, which occurred within a relatively short time after implantation. The cyclic stresses in the fracture area were very high relative to the strength of the material, and failure required relatively few stress cycles.

Metallography

Microstructural Analysis. Metallographic examination of samples cut from the plate section of the implant revealed a series of repair welds on the inside surface of the plate in the vicinity of the fracture. These welds had been ground and polished and were not apparent until the implant was sectioned. Figure 4 shows two metallurgical mounts

Fig. 1 As-received "Jewett nail" implant

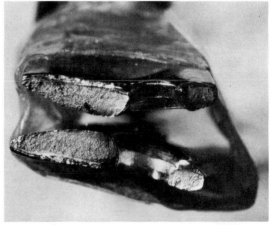

Fig. 2 Mating fracture surfaces. 14.24× Failure initiated on the outside (convex) surface and progressed across the plate thickness by low-cycle fatigue. ~15×

of samples taken near the fracture. The repair welds are clearly delineated. The fracture surface is on the right in Fig. 4(a) and on the left in Fig. 4(b) and is approximately 2.5 mm (1/10 in.) from the welds. The microstructure in one of these semilenticular areas is shown in Fig. 5.

Figures 6 and 7 depict the microstructures observed away from the weld (and fracture) and near the weld (and fracture), respectively. Figure 6 shows the structure observed approximately 13 mm (1/2 in.) from the fracture shown in Fig. 4(b). This structure was unaffected by the heat of welding and was generally present in the plate material (strain-hardened austenite). The hardness of this material was determined to be 30 to 34 HRC, which is consistent with the observed microstructure.

The structure shown in Fig. 7 is recrystallized austenite. This structure was adjacent to the welds and at the fracture surface and was the direct result of repair welding, which created an annealed, softened zone. The hardness of this material was 80 to 86 HRB.

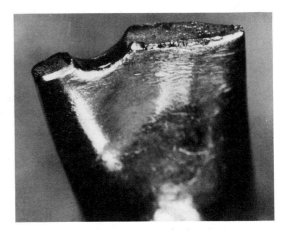

Fig. 3 Inside surface of the plate section on the nail side. The beveled zone was caused by rubbing of the screw head after

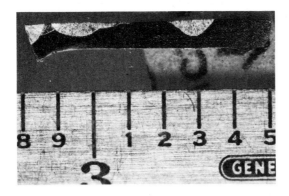

(a)

(b)

Fig. 4 Metallurgical mounts of samples cut from the plate section at the fracture. The fracture surface is on the right in (a) and on the left in (b). The divisions on the scale are 2.5 mm (1/10 in.).

Fig. 5 Microstructure of one of the semilenticular areas in Fig. 4(a), clearly showing it to be a weldment. 61×

Fig. 6 Microstructure of material away from the repair welds. 122×

Chemical analysis/identification

The chemical composition of the implant material was analyzed as type 316L stainless steel, which was within specification for ASTM F55-66T.

Discussion

The process of repair welding changed the condition of the material adjacent to the welds and substantially reduced yield strength, tensile strength, and attendant fatigue strength. The welds were placed at the point on the plate that experienced the most severe bending stresses in service. Thus, the plate was made weakest in the area where it should have been strongest.

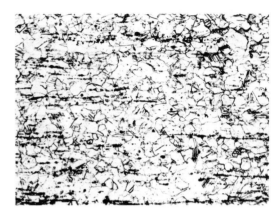

Fig. 7 Microstructure of material adjacent to the weld and at the fracture surface shown in Fig. 4(b). 122×

Conclusion and Recommendations

Most probable cause

The hip implant contained a series of hidden repair welds located on the inside of the plate section between the first and second screw holes. The plate material was recrystallized by the heat of welding, causing a severe loss in strength that led to low-cycle fatigue failure.

How failure could have been prevented

Manufacturers should never attempt to salvage this type of critical device by welding or any other procedure that might compromise its integrity.

Fatigue Fracture of Titanium Alloy Knee Prostheses

Jeremy L. Gilbert, Division of Biological Materials, and S. David Stulberg, Department of Orthopaedic Surgery, Northwestern University, Chicago, Illinois

Total knee prostheses were retrieved from patients after radiographs revealed fracture of the Ti-6Al-4V ELI metal backing of the polyethylene tibial component. The components were analyzed using scanning electron microscopy. Porous coated and uncoated tibial trays were found to have failed by fatigue. Implants with porous coatings showed significant loss of the bead coating and subsequent migration of the beads to the articulating surface between the polyethylene tibial component and the femoral component, resulting in significant third-body wear and degradation of the polyethylene. The sintered porous coating exhibited multiple regions where fatigue fracture of the neck region occurred, as well as indications that the sintering process did not fully incorporate the beads onto the substrate. Better process control during sintering and use of subsequent heat treatments to ensure a bimodal microstructure were recommended.

Key Words

Titanium-base alloys   Ceramic coatings   Surgical implants
Sintering (powder metallurgy)   Fatigue failure

Alloys

Titanium-base alloy—Ti-6Al-4V

Background

Total knee prostheses were retrieved from patients after radiographs revealed fracture of the Ti-6Al-4V extra low interstitial (ELI) metal backing of the polyethylene tibial component. Porous-coated and uncoated tibial trays had failed.

Applications

Titanium alloy knee prostheses are used to replace the articulating surface of diseased or damaged knee joints to restore as much natural function as possible. A typical knee prosthesis consists of a metal femoral component, generally made from a cobalt-chromium or titanium alloy; a polymeric wear surface made of ultrahigh molecular weight polyethylene (UHMWPE, ASTM F648); and a metal tibial component to which the polyethylene is attached.

The femoral component articulates against the UHMWPE tibial cup, which in turn is anchored to the tibia by means of a Ti-6Al-4V tibial tray. The tibial tray is anchored to the tibia either by a polymeric grouting agent known as acrylic bone cement (an admixture of polymethyl methacrylate/styrene copolymer beads, methyl methacrylate monomer polymerized in place, and barium sulfate, which serves as a radiopacifier) or by the growth of bony tissue into the interstices of metal beads sintered onto the metal substrate. This porous structure allows bone tissue to grow into and mechanically lock the device in place.

Circumstances leading to failure

The prostheses were implanted into patients whose average age was 63.4 years and whose aver-

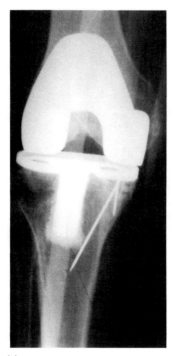

(a)

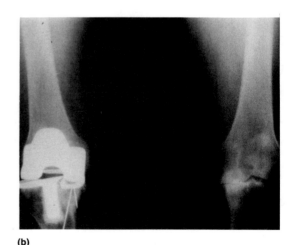

(b)

Fig. 1 X-ray radiographs. (a) Implanted total knee prosthesis. Femoral component is at top, tibial component is at bottom, and polyethylene, which cannot be seen with X-rays, is between the two components. (b) Failed tibial component after 3.5 years of implantation. Fracture is evident on the medial side.

age weight was 90 kg (200 lb). These patients were active and fairly heavy. Typically, failure is observed in patients with poor underlying support for the tibial tray. Often, upon implantation of the component, insufficient bone stock is present to adequately fix the device in place. In such cases, a bone graft is placed in the site to fill in the defect and provide support. If the graft resorbs or if one side of the tibia (usually the medial side, toward the centerline of the body) loses underlying support for the tibial component, then an asymmetric cantilever loading configuration results. This cantilever geometry results in tensile fatigue stresses on the superior (top) aspect of the metal tibial tray. Repeated loading caused by walking ultimately results in the formation of a fatigue crack that eventually propagates to failure.

A radiograph of a total knee prosthesis immediately after implantation is shown in Fig. 1(a). A radiograph taken after approximately 3.5 years, at which point failure of the tibial tray was observed, is shown in Fig. 1(b). Under the failed component are two bone anchoring pins, which were used to support a bone graft. An average individual will cyclically load his or her leg about 1.5 million times per year; thus, 3.5 years represents a high-cycle fatigue regime.

Pertinent specifications

ASTM F 136 and ASTM F 620 govern the compositional, microstructural, and mechanical properties of Ti-6Al-4V used in implant devices. Microstructurally, these specifications suggest that the alloy be used in the bimodal form obtained by thermomechanical treatment in the $\alpha + \beta$ phase field. According to these specifications, no coarse α platelets should be present.

Performance of other parts in same or similar service

Although this study focused on the failure of titanium alloy devices, two cast Co-Cr-Mo (ASTM F 75) tibial components were also retrieved after failure. Titanium alloys are used more frequently for the tibial component.

Specimen selection

Specimens were collected from patients after revision surgery was performed to replace the failed components. Those devices that best represented the nature and origin of the failure are discussed in this study. Some of the retrieved implants were severely damaged from postfracture abrasion and from removal.

Visual Examination of General Physical Features

Failure of a tibial tray component (Fig. 2) occurs at or near the junction of the stem of the component, which extends into the tibia (coming out of the photograph), and the tray that is used to hold the polyethylene articulating surface (polyethylene is transparent to X-rays, see Fig. 1a). Also present in the tibial tray is a U-shaped notch. This recess is needed to preserve the posterior cruciate ligament of the knee in order to maintain stability of the joint and minimize the changes in anatomy required for implantation. These geometric features, besides being located at the maximum cantilever position, act as stress concentrations and were the sites of initiation for several of the failures observed. Figure 2 shows that the crack propagated to an elliptical hole in the tray. This site of fatigue crack initiation was common in prostheses with cruciate-preserving notches.

The porous-coated components did not have a single stem; rather, they had three pegs which, along with the underside of the component, were coated with beads (Fig. 3). The device in Fig. 3 had several other features worthy of mention. Adjacent to the fatigue crack (right side of photograph) was another crack that did not quite propagate to failure. Again, the U-shaped cruciate-preserving notch influenced where the failure occurred (top of Fig. 3). The secondary fatigue crack can be seen emanating from this site.

The inferior side of the device in Fig. 3 was cov-

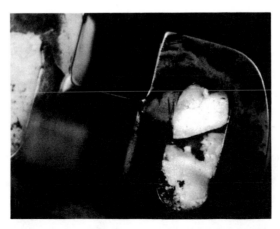

Fig. 2 Bottom of failed tibial tray. Fracture emanates from cruciate-preserving notch and propagates to elliptical hole used to lock implant to bone with bone cement.

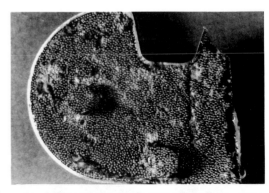

Fig. 3 Underside of porous-coated implant that failed by fatigue. Two fatigue cracks are present: the failure and a secondary crack parallel to the fracture surface. Porous beads fractured in the region adjacent to each fracture and migrated into the joint cavity.

ered with beads of titanium sintered onto the metal substrate. In the vicinity of each fatigue crack, it was apparent that the beads had broken off the component. Once free, these beads migrated throughout the joint capsule and became embedded in the polyethylene articular cup. Figure 4 shows a severely worn polyethylene cup with titanium beads

embedded in the surface. The particulate debris resulting from this wear process can have a detrimental effect on the surrounding tissues.

Testing Procedure and Results

Surface examination

Scanning electron microscopy (SEM) was performed on several of the failed implants. Two general areas were investigated: the fracture surfaces and the bead-tray junction.

Fracture Surfaces. Figure 5 is a low-power SEM micrograph of the initiation site for the fatigue fracture shown in Fig. 3 and 4. This was the region where the polyethylene cup was held in place. The right side of Fig. 5 shows the capture lip protruding from the raised rim of the tray.

A higher-magnification view (Fig. 6) shows that the initiation region consisted of several flat-faceted regions, where the initial fatigue damage occurred. These individual fracture sites grew and linked to form the final fracture surface. Figure 7 shows fatigue striations perpendicular to the direction of crack propagation, which occurred from the upper left to lower right. These striations were located in the base of the tray and clearly indicated the fatigue nature of the failure process.

Figure 8 shows another fractured tibial tray at the fatigue initiation region. Again, this site was the polyethylene capture lip for the device. However, the initiation site was not at the top of the lip, but rather at the base—as indicated by the convergence of the fracture lines to a point at the bottom of Fig. 8. At higher magnification (Fig. 9), the convergence of the fracture lines is more evident, indicating that the fracture initiated at the surface near the base of this region.

Figure 10 shows a higher-magnification view of the initiation region. A large prior-β grain was present, resulting in a large region of similar orientation that was favorable for the accumulation of fatigue damage. A secondary crack through the initiation site was also present.

Bead-Tray Junctions. Porous coatings have been developed for orthopedic use primarily to pro-

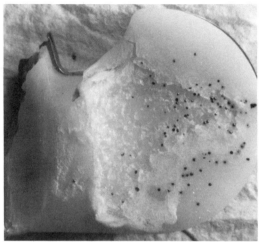

Fig. 4 Top surface of implant shown in Fig. 3. Note the severe wear pattern and the presence of beads in the polyethylene surface.

Fig. 5 SEM micrograph of polyethylene retaining lip of titanium tibial tray. Fatigue crack initiated at upper right region.

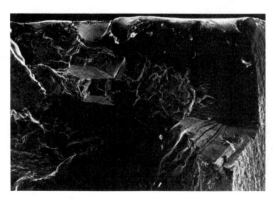

Fig. 6 Higher-magnification view of initiation region in Fig. 5. Localized fracture occurred at facets and eventually linked to form final fracture surface.

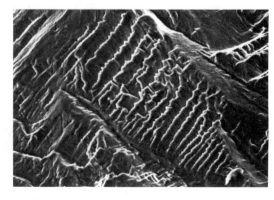

Fig. 7 Fatigue crack propagation region of device in Fig. 6. Crack propagated from upper left to lower right.

Fig. 8 Low-magnification micrograph of crack initiation region at base of capture lip in a titanium alloy tibial component

Fig. 9 Higher-magnification view of initiation region in Fig. 8

Fig. 10 Higher-magnification view of Fig. 9. Note appearance of large prior-β grain in initiation region.

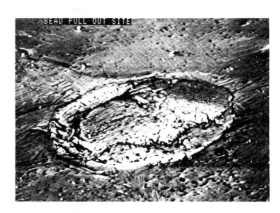

Fig. 11 Micrograph of surface of titanium alloy tibial tray where a sintered bead has fractured off. Note incomplete fusing of the bead to the tray.

vide a space into which bone can grow and mineralize, locking the device into place. This is termed biological fixation. During loading, however, this porous network is subjected to variable and nonintuitive stresses (that is, tensile stresses on a nominally compressive stress region), which may cause fracture or loss of the porous coating.

The beads used to create the interstitial space for bone ingrowth are sintered onto the surface of the tibial tray. This is accomplished by heating into the beta-phase field and holding for a sufficient time to allow for the formation of sinter necks between beads and between beads and tray. These sinter necks were observed by sectioning the bead-coated trays with a diamond saw, polishing, and imaging with back-scattered electrons. The sinter necks resulting from this process are not always uniform or fully developed.

Figure 11, a micrograph of the tibial tray surface, shows where a bead was pulled out from the surface. The coarse beta microstructure resulting from the sintering heat treatment is just visible adjacent to this site. It is evident that the bead was not fully attached or incorporated onto the tray substrate, but rather was attached only at local regions about the bead circumference.

Figure 12 shows a region of a bead where an adjacent bead has fractured off. Fractures are present on the circumference of this bead junction, which penetrate into the bead itself. The sintered neck junctions are regions of stress concentration and can give rise to notch-sensitive behavior of the beads.

Small fatigue fractures were noticed in several regions between the beads and the tray, as well as at interbead junctions. These fractures occurred away from the actual fatigue fracture and were sometimes present even in the case of "well-fixed" implants. Figure 13, a micrograph of a bead-tray junction, shows a fatigue crack propagating into

the tray surface. Fracture of a bead-bead junction is shown in Fig. 14. Several cracks are present, traversing obliquely across the interbead junction. Fracture of the beads will result in their migration to regions where third-body wear and degradation processes can accelerate the failure of other components. Again, Fig. 13 and 14 show the coarse lamellar microstructure associated with a beta sintering heat treatment.

Fig. 12 Micrograph of sinter neck of titanium bead. Note appearance of fracture lines about the circumference of the bead neck.

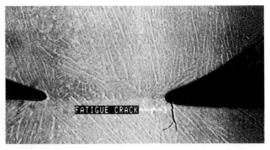

(a)

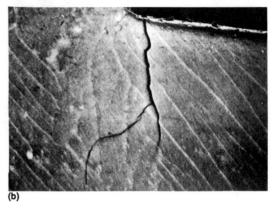

(b)

Fig. 13 (a) Micrograph of polished section of titanium tray-bead junction. Note fatigue crack penetrating into tray. (b) Higher-magnification view of (a)

Discussion

The evidence was generally fractographic in nature. However, these knee prostheses revealed several aspects of materials failure. First, the metal backing used to support the polyethylene cup was subject to fatigue failure if there was incomplete support under one portion of the device. Fatigue cracks initiated on the superior aspect and propagated through the tray. If beads for porous ingrowth were present, the progressing tray fracture sometimes loosened and broke off the beads, which then migrated to other parts of the joint and created severe wear problems.

Multiple fractures at bead-bead junctions and bead-tray junctions were observed in the porous-coated implants. The neck regions between beads and at bead-neck junctions resulted in severe stress-concentration effects, which can be detrimental because of the notch sensitivity of titanium in these implants. These fractures were present in regions remote from, as well as adjacent to, the macroscopic fatigue fractures and were sometimes present even in well-fixed implants. Also, the sintering process did not result in the complete formation of sinter necks, thus making these regions highly susceptible to fracture processes. The process of sintering transforms the α and β bimodal microstructure to a β microstructure that is known to be less resistant to fatigue initiation.

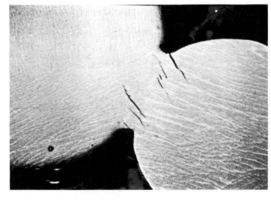

Fig. 14 Micrograph of interbead junction. Note fatigue cracks at oblique angle to neck.

Conclusion and Recommendations

Most probable cause

Porous-coated and uncoated tibial trays failed by a fatigue process in which cracks initiated on the superior aspects of the trays. The porous coating sintering method resulted in a microstructure that was not as fatigue-initiation resistant as the bimodal microstructure recommended by ASTM F 136. Geometric constraints on the design, due to anatomic considerations, and asymmetric loading contributed to the ultimate failure of the components.

Remedial action

Because these components are subjected to high-cycle fatigue loading, a bimodal (α + β) micro-

structure with equiaxed primary α should be used to resist fatigue crack initiation. Porous coating processes or subsequent heat treatments that yield a coarse β microstructure should not be used. Also, better process control should be exerted to minimize the formation of incomplete sinter necks, which serve as stress-concentration sites and fracture-initiation regions.

Design modifications to minimize fatigue failure might include thickening the tray cross section in order to lower stresses or redesigning the cruciate-retaining "U" to minimize stress concentrations. Anatomical constraints limit the extent of modification. For instance, thickening of the tray can be accomplished either at the expense of polyethylene cup thickness or by the removal of greater amounts of bone. Alternative surgical techniques to inhibit asymmetric loading patterns might be employed, such as the use of metallic spacers or modular components instead of bone grafts to fill bone defects.

The sintering process appears to be highly variable in these devices, resulting in irregular sinter necks, which are highly susceptible to failure. Also, the sintering temperature is high enough to result in a β transformation of the microstructure. This microstructure decreases the fatigue-initiation resistance of the material and can degrade the high-cycle fatigue performance of the component. Alternate bonding techniques that do not result in a β transformation of the microstructure, such as diffusion bonding, should be explored.

Acknowledgment   Thanks are directed to Dr. A. Tsao, Department of Orthopaedics, Northwestern University, for providing the retrieved implants for this analysis.

Fracture of a Cast Stainless Steel Femoral Prosthesis

Carmine D'Antonio, Department of Metallurgy and Materials Science, Polytechnic University, Brooklyn, New York

A cast stainless steel femoral head replacement prosthesis fractured midway down the stem within 13 months of implantation. Visual examination showed severe "orange peel" around the fracture on the concave side. This effect was not observed on the convex side, which suggested fatigue fracture. Metallographic examination of samples revealed an extremely large grain size and corroborated fatigue fracture. Chemical analysis indicated that the material conformed to the requirements for type 316L stainless steel. Substandard-size tensile bars machined from another prosthesis from the same manufacturer showing identical grain sizes were used for mechanical testing. Tensile tests indicated that the material did not meet the manufacturer's stated strength criteria in the portion of the stem that fractured. The failure was attributed to low strength, which resulted in fatigue. The extremely coarse grain size was considered a major factor in strength reduction.

Key Words	Austenitic stainless steels	Castings	Surgical implants
	Fatigue failure	Prosthetics	Grain size
Alloys	Stainless steel—316L		

Background

A cast stainless steel femoral head replacement prosthesis failed midway down the stem within 13 months of implantation.

Applications

This type of prosthesis is used to replace the femoral head and neck in cases of arthritic or other degenerative diseases of the hip.

Pertinent specifications

The manufacturer specified a minimum yield strength of 240 MPa (35 ksi) and a minimum tensile strength of 480 MPa (70 ksi).

Performance of other parts in same or similar service

The substitution of cast stainless steel for the traditional wrought stainless steel or cast cobalt-chromium alloy in this application caused a dramatic reduction in strength and an attendant increase in the likelihood of mechanical failure. Cast stainless steel possesses roughly half the yield and tensile strengths of the other two materials.

Specimen selection

Two samples were cut from the lower section of the stem, one at the fracture surface and the other from the bottom end. The samples were mounted to show longitudinal cross sections. Drillings from the lower section of the stem were used for chemical analysis.

Another prosthesis with identical grain sizes was obtained from the same manufacturer. Two substandard-size tensile bars were machined, one from the bottom of the stem and the other from the body of the stem where the subject prosthesis fractured.

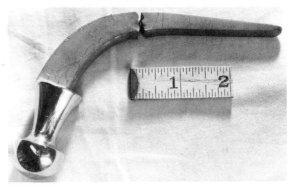

Fig. 1 Failed femoral head prosthesis

Visual Examination of General Physical Features

Failure of the implant occurred approximately midway down the stem (Fig. 1). Severe "orange peel" was present around the fracture on the concave (compression) side of the stem. This effect (Fig. 2, see arrows) is indicative of an underlying coarse grain size. The surface markings on the portion of the stem on the right in Fig. 2 were made by the tool used to remove the stem from the femur after fracture.

Figure 3 shows the mating fracture surfaces. The fracture texture was coarse and faceted, again reflecting the underlying internal structure. Note that the "orange peel" was confined to the concave side of the stem. There was a complete absence of this effect on the convex (tensile) side of the fracture (indicated by arrows). Fracture in this area initiated and propagated via fatigue, culminating

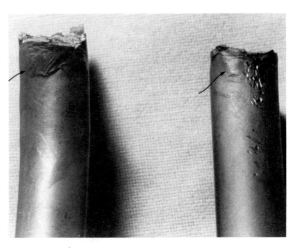

Fig. 2 Concave side of stem, showing gross "orange peel" surface (arrows)

in overload failure of the remainder of the cross section accompanied by gross plastic deformation (orange peel).

Testing Procedure and Results	**Metallography**

Metallography

Microstructural Analysis. Metallographic examination of the two samples cut from the stem verified that the implant was a casting and that the grain size was extremely large. Figure 4 shows the mounted sections in the etched condition. There were only four grains through the entire section at the fracture.

Figure 5 shows a secondary crack propagating from the convex side of the stem 13 mm (½ in.) from the main fracture. There was no plastic deformation in the material adjacent to this crack, characteristic of fatigue. The general microstructure consisted of a matrix of austenite with particles of

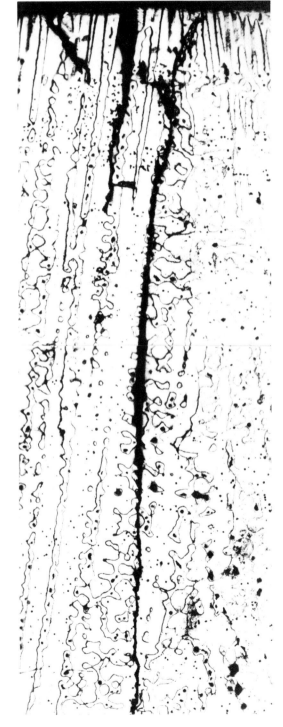

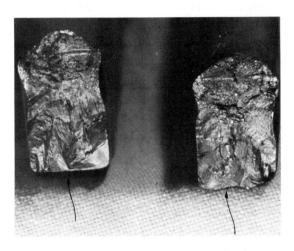

Fig. 3 Fracture surfaces. Fatigue fracture initiated on the convex (tensile) side of the stem (arrows).

(a)

(b)

Fig. 4 Macrographs of the metallographic samples taken at the fracture (a) and stem end (b). The fracture surface is on the right in (a). 3×

Fig. 5 Micrograph showing a secondary crack adjacent to the main fracture at the tensile surface. The crack is typical of fatigue. The general microstructure consists of a matrix of austenite with particles of ferrite. 81×

ferrite, a structure commonly found in cast stainless steel.

Chemical analysis/identification

Chemical analysis of the prosthesis showed that the material conformed to the requirements for type 316L stainless steel. It did not conform to any of the cast stainless steel alloys.

Mechanical properties

Hardness. Standard Rockwell tests conducted on the stem of the fractured prosthesis showed the hardness to be 70 to 71 HRB both at the end of the stem and near the fracture.

Tensile Properties. Tensile tests conducted on samples machined from the sample prosthesis supplied by the manufacturer yielded the following results:

Property	Sample taken from stem end (finer grain)	Sample taken from midstem (coarser grain)
0.2% offset yield strength, MPa (ksi)	211 (30.6)	173 (25.1)
Ultimate tensile strength, MPa (ksi)	460 (66)	418 (60.6)
Hardness, HRB	70-71	70-71

The prosthesis did not meet the manufacturer's stated strength criteria in the portion of the stem that fractured. It is interesting to note that the hardness did not reflect the difference in strength between the coarser- and finer-grained materials.

Discussion

It was evident that the prosthesis was incapable of sustaining the high cyclic stress imposed during service and failed via fatigue. Loads on the hip joint increase up to five times body weight during the heel-strike phase of walking. In addition, abnormal activity of the patient and/or deterioration of the fixation cement can result in increased stresses. Therefore, it is imperative that biocompatible materials of the highest possible strength be used for prostheses so that the highest possible stresses can be tolerated. In this instance, a class of materials was chosen that was probably marginally biocompatible but that possessed less than half the strength of the wrought stainless or cast cobalt-chromium alloys traditionally used.

The casting procedure produced an inordinately large grain size in the area that experienced the highest stress, further exacerbating the danger of failure. The well-known relationship between grain size and strength was effectively illustrated by the tensile tests performed on the sample prosthesis. Moreover, the tests showed that neither the fine- nor the coarse-grained material conformed to the manufacturer's own strength requirements.

Conclusion and Recommendations

Most probable cause

The conditions of service are unknown, and the question as to whether or not the prosthesis would have failed even if a higher strength material had been used cannot be answered. Therefore, the most probable cause of failure was the low strength of the prosthesis material, which resulted in fatigue failure in a relatively short period of service.

The large grain size in the stem also must be considered a major factor in the reduced strength.

How failure could have been prevented

Whether failure would have occurred if the grain size had been less coarse is arguable. However, the class of material used for this application was mechanically weaker than those traditionally used.

Analysis of Music Wire Springs Used in a Printer Mechanism

Michael Neff, SEAL Laboratories, El Segundo, California

Music wire springs used in a printer return mechanism failed near the bend in the hook portion of the spring during qualification testing. Samples were examined in a scanning electron microscope equipped with an energy-dispersive X-ray microprobe. Fatigue fractures originated at rub marks on the inside edge of the spring. An investigation of loads encountered in service indicated that the springs had been loaded to a large fraction of the yield strength. Redesign of the spring mechanism was recommended.

Key Words   Springs (elastic)   Fatigue failure   Woody fracture

Background

Music wire springs used in a printer return mechanism failed near the bend in the hook portion of the spring during qualification testing.

Applications

Music wire springs were specified for a new printer application. The springs were 21-coil music wire springs wound from 0.76 mm (0.030 in.) diam wire, which had been zinc plated (Fig. 1).

Pertinent specifications

The springs were manufactured from spring-quality music wire per ASTM A 228, which specifies an ultimate tensile strength of 2275 to 2515 MPa (330 to 365 ksi) for the 0.76 mm (0.030 in.) diam wire.

Visual examination of general physical features

All the springs had fractured in the hook portion (Fig. 1). Although the wire diameter was quite small, optical examination at 30× indicated flat fracture regions initiating on the inside of the hooks, with several different overload fracture

Fig. 1 Optical photograph of several of the spring failures, showing the typical failure locations

morphologies. Some details of the fracture surface could be examined optically, but much better photographs were obtained with scanning electron microscopy (SEM).

Testing Procedure and Results

All the samples were examined in an SEM equipped with an energy-dispersive X-ray (EDX) microprobe. Metallographic cross sections were also prepared in the longitudinal direction through the origin on the ends of four broken spring samples.

Surface examination

Scanning Electron Microscopy/Fractography. Figure 2 shows the typical failure geometry for all the springs examined. The fractures included a fatigue fracture, which was a flat portion on the inside of the bend extending approximately halfway through the wire thickness, and an overload fracture, which appeared intergranular toward the outside of the bend. In many of the springs examined, the inside of the hook had a rub mark on the surface at the origin of the fatigue fracture. The mark resulted from rubbing of the spring against the mating surface (Fig. 2). In one of

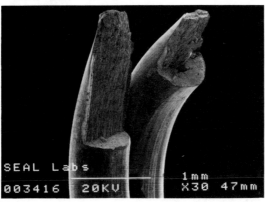

(a)

(b)

Fig. 2 SEM micrographs showing a rub mark at the fatigue origin. (a) 21.6×. (b) Detail of (a). 92.3×

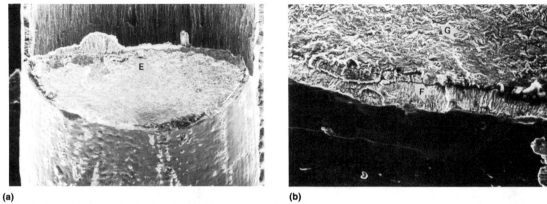

(a)   **(b)**

Fig. 3   SEM micrographs showing an inclusion at the origin. (a) 81.9×. (b) 491.4×

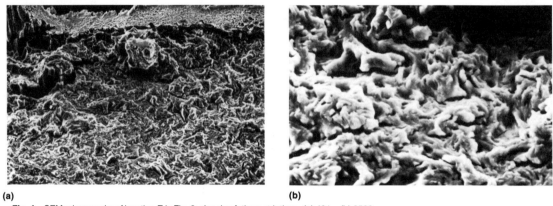

(a)   **(b)**

Fig. 4   SEM micrographs of location E in Fig. 3, showing fatigue striations. (a) 491×. (b) 2520×

the springs examined, there was no rub mark, but rather a small, nonmetallic inclusion at the origin of the fatigue fracture (Fig. 3). The average fatigue striation spacing was approximately 0.6 µm near the overload fracture, indicating that at least 500 load cycles were associated with the fatigue crack progression (Fig. 4). EDX analysis of the small inclusion found in one of the samples showed that it contained a small amount of oxygen, carbon, and chlorine—elements not present in the base metal, except for a very small amount of carbon and oxygen (Fig. 5 and 6).

The overload portion of the fracture followed the elongated grains (a "woody" fracture) (Fig. 7) and should not be confused with intergranular fracture. Because the springs had been zinc plated, hydrogen embrittlement, which normally produces an intergranular fracture, was a specific concern. However, this did not occur; all the spring fractures initiated in a fatigue mode.

Metallography

The metallographic cross sections showed a similar microstructure and crack profile for all the samples (Fig. 8). Although one of the fatigue fractures initiated at an inclusion (Fig. 3), the as-polished cross sections showed a low concentration of microstructural anomalies, such as inclusions, which may have caused stress concentrations. The etched microstructure, typical for music wire, was an elongated structure characteristic of the heavily worked high-carbon steel used in many types of commercial springs. In most of the springs, the fa-

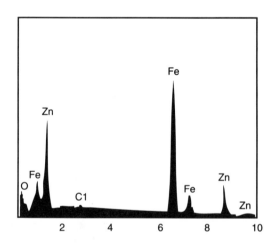

Fig. 5   EDX microprobe spectrum obtained at location F in Fig. 3. A 20 kV beam was used.

tigue portion of the fracture was less than one-half of the wire thickness (Fig. 8).

Chemical analysis/identification

The chemical analysis met the requirements of music wire per ASTM B 228, that is, 0.70 to 1.00% C, 0.20 to 0.60% Mn, 0.10 to 0.30% Si, 0.025% max P, and 0.030% max S.

Mechanical properties

Hardness. Microhardness measurements were obtained on the longitudinally cross-sectioned wires (Table 1). The longitudinal indentations were obtained in the center of the wire, indicating an average Knoop hardness of 617 HK, which corresponds to 54 HRC—well below the properties specified for 0.76 mm (0.030 in.) wire (Fig. 9a). However, microhardness measurements obtained with the diamond pyramid indentation oriented transverse to the wire axis showed an average of 815 HK, or 63.5 HRC, which, when converted to a tensile strength of 2358 MPa (342 ksi), meets the tensile requirements for spring-quality music wire in the 0.76 mm (0.030 in.) diameter (2275 MPa, or 330 ksi, minimum) per ASTM A 228 (Fig. 9b). These results showed considerable plastic anisotropy in this material, which would be expected because of the heavily cold-worked (fibrous) microstructure. A detailed discussion of the effect of anisotropy on hardness testing can be found in Ref 1.

Because of the small size of the springs, tensile testing was not performed. It was concluded, based on the hardness measurements, that the spring wire material met the strength requirements per ASTM A 228.

Stress analysis

An investigation of the loads encountered in service indicated that these springs had been loaded to approximately 90% of the yield strength and thus were not properly specified for this application.

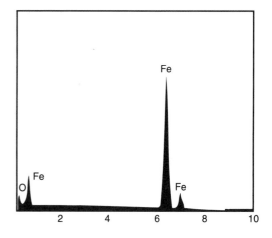

Fig. 6 EDX microprobe spectrum obtained at location G in Fig. 3. A 5 kV beam was used.

Discussion

Fractography clearly established failure by fatigue as a result of the high loads caused by improper design of the springs. Improper design for fatigue is often encountered in qualification testing. In this analysis, it was important to rule out hydrogen embrittlement and material defects as the cause of failure, in order to proceed with a redesign. In these types of springs, the highest loads

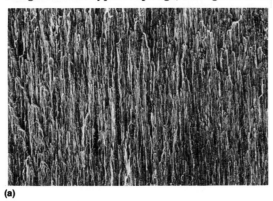

(a)

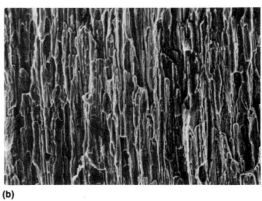

(b)

Fig. 7 SEM micrographs of the overload fracture, showing a woody fracture appearance. (a) 126×. (b) 630×

(a)

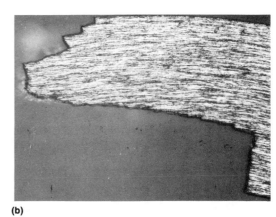

(b)

Fig. 8 Optical micrographs of the cross section of one of the spring fractures. (a) 31.5×. (b) 2% nital etch. 63×

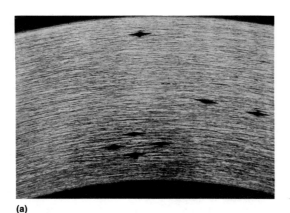

(a)

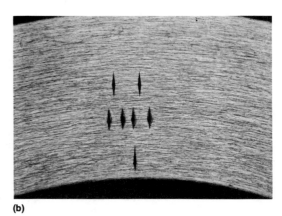

(b)

Fig. 9 Optical photographs showing the microhardness indentations on the longitudinal spring sections. (a) Longitudinal orientation (low hardness). 63×. (b) Transverse orientation (acceptable hardness). 63×

the cause of failure, in order to proceed with a redesign. In these types of springs, the highest loads normally occur near the inside of the hook, and most spring fractures initiate at this area. Therefore, the fracture surface morphology near the inside of the hook is often the area of primary interest in spring failure analysis.

Conclusion and Recommendations

The results showed that in all the springs fatigue fractures had initiated at rub marks on the inside surface of the hook, except for one spring, in which fracture had initiated at an inclusion on the inside surface of the hook. The rub marks were caused by rubbing of the spring hooks against the edge of a hole to which they were attached. This is a normal location for fatigue crack initiation in this type of coil spring, because the highest tensile stresses are present on the inside surface of the hook. It was determined that the springs had been loaded beyond the fatigue strength (in bending) of the spring wire diameter used. A redesign using a spring with a larger wire diameter solved the problem.

Reference

1. J.H. Westbrook and H. Conrad, Ed., *The Science of Hardness Testing and Its Research Applications*, American Society for Metals, 1973, chap 11

Table 1 Microhardness measurements obtained on the longitudinal sections of the springs
Different readings were obtained with different Knoop pyramid indenter orientations.

Location	Length in microns	Hardness HK	Hardness HRC
Longitudinal in center			
	107	622	54.5
	108	610	54.0
	107.5	616	54.5
	107.0	622	54.5
	108.0	610	54.0
Average			54.5
Longitudinal on the edges			
	108.5	605	53.5
	107.0	622	54.5
	107.0	622	54.5
	108.5	605	53.5
	107.5	616	54.5
Average			54.9
Transverse in center			
	92	841	65
	93	841	64
	95	789	62.5
	95	789	62.5
	104		
Average			63.5
Transverse on the edge			
	92	841	65.0
	93	823	64.0
	95.0	789	62.5
	95.0	789	62.5
Average			63.5

Failed Missile Launcher Detent Spring

Gary Wechsler, Victor K. Champagne, Marc Pepi, United States Army Research Laboratory, Watertown, Massachusetts

A missile detached from a Navy fighter jet during a routine landing on an aircraft carrier deck because of a faulty missile launcher detent spring. Visual inspection of Inconel 718 detent spring assembly revealed that four of the nine spring leafs comprising the assembly were plastically deformed while two of the deformed leafs did not meet minimal hardness or tensile requirements. Liquid penetrant testing revealed no cracks or other surface discontinuities on the leaf springs. Material sectioned from the soft spring leafs was heat-treated according to specifications in the laboratory. The resultant increase in mechanical properties of the re-heat-treated material indicated that the original heat treatment was not performed correctly. The failure was attributed to improper heat treatment. Recommendations focused on more stringent quality control of the heat-treat operations.

Key Words	Leaf springs, Mechanical properties Plastic deformation Yield strength	Superalloys, Mechanical properties Hardness Elongation	Nickel base alloys, Mechanical properties Tensile strength Heat treatment
Alloys	Nickel base alloys—Inconel 718		

Background

The detent spring missile launcher failed in service during routine landing of a Navy fighter jet.

Applications

The detent spring assembly is comprised of nine spring leafs, as shown in Fig. 1. The detent spring assembly is utilized on a missile launcher. The leaf springs deflect under a load exerted by the missile, fixing it in place on the aircraft missile launcher rails. The applied load is distributed throughout all nine spring leafs. The leafs return to equilibrium when the load is removed. Deficiencies in the mechanical properties of any of the spring leafs causes an undue stress in the remaining components, especially during takeoff and landing when the greatest loads are encountered.

Circumstances leading to failure

A missile became detached from a Navy fighter jet during a routine landing on an aircraft carrier deck. The missile bounced on the deck several times without detonating before plunging into the ocean. Visual examination of the detent spring assembly revealed that several of the leaf springs had been deformed, allowing the missile to become free.

Pertinent specifications

The spring leafs were fabricated from Inconel 718 steel in accordance with specification AMS 5596. This specification requires components to be heat-treated to 1035 MPa (150 ksi) minimum yield strength, 1240 MPa (180 ksi) minimum ultimate tensile strength, 12% minimum elongation, and a hardness of 38 to 44 HRC. A 100% liquid penetrant inspection is performed in accordance with MIL-STD-6866, Type 1, Method B, Level 3 after fabrication.

Performance of other parts in same or similar service

These components do not have a history of failures.

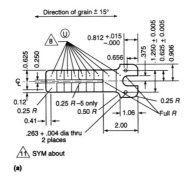
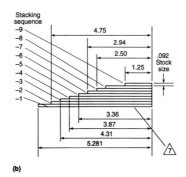
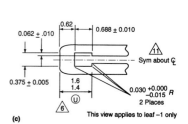

Fig. 1 Engineering drawing of the detent spring

Specimen selection

The detent spring assembly was removed from the missile launcher. All nine spring leafs comprising the assembly were examined metallurgically, but emphasis was placed on the deformed leafs.

Testing Procedure and Results

Nondestructive testing

Liquid Penetrant Testing. There was no evidence of cracks or other surface discontinuities detected on the leaf spring surfaces, as determined by liquid penetrant inspection in accordance with MIL-STD-6866, Type 1, Method B, Level 3.

Surface examination

Visual. The detent spring contained four bent leafs as received. These were identified as leafs 1, 2, 6, and 7 (Fig. 2). Leafs 3, 4, 5, 8, and 9 were relatively flat and parallel to each other, as required.

Metallography

Microstructural Analysis. Specimens taken from each of the nine leafs were sectioned and metallographically prepared to determine the direction of grain flow. The engineering drawing stated that the direction of grain flow shall not vary more than 15° from the lengthwise direction of each leaf. Metallographic examination revealed MC carbides arranged preferentially (Fig. 3) in the direction of grain flow. In all nine leafs the grain flow was within 15° of the lengthwise direction. In addition, there was no significant difference in grain size between the leafs. It was revealed that specimens from leafs 6 and 7 etched lighter than those taken from the remaining leafs. The dark etching characteristic indicates that niobium is tied in the γ'' phase, the strengthening mechanism of Inconel 718.

Chemical analysis/alloy identification

Material. Samples from five leafs were analyzed for chemical composition, and the results as indicated in Table 1 satisfied the requirements of Inconel 718 per AMS 5996.

Mechanical properties

Hardness. Eight HRC measurements were taken on each of the nine leafs. Each leaf conformed to the specified hardness of 38 to 44 HRC,

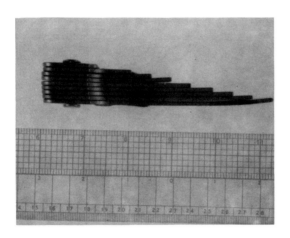

Fig. 2 Failed detent spring in the as-received condition. 0.2×

Fig. 3 Representative macrograph showing direction of grain flow. 7.88×

Table 1 Chemical analysis

Element	Specification AMS 5596 range, wt%	Composition of individual leafs, wt% Leaf Number				
		1	2	3	6	7
Nickel	50-55	53.1	52.3	52.6	52.4	52.7
Chromium	17-21	18.4	18.6	18.5	18.6	18.5
Columbium + Tantalum	4.75-5.50	5.13	5.25	5.14	5.18	5.15
Molybdenum	2.80-3.30	3.01	3.08	3.05	3.08	3.05
Titanium	0.65-1.15	1.05	1.07	1.08	1.07	1.07
Aluminum	0.20-0.80	0.55	0.56	0.56	0.57	0.59
Cobalt	1.00 max	0.18	0.17	0.17	0.17	0.17
Carbon	0.08 max	0.045	0.044	0.043	0.044	0.046
Manganese	0.35 max	0.11	0.10	0.11	0.11	0.11
Silicon	0.35 max	0.18	0.17	0.18	0.18	0.18
Phosphorus	0.015 max	0.011	0.011	0.011	0.011	0.013
Sulfur	0.015 max	0.003	0.003	0.003	0.003	0.003
Boron	0.006 max	0.005	0.005	0.005	0.005	0.005
Copper	0.30 max	0.028	0.029	0.028	0.029	0.028
Iron	remainder	18.2	18.6	18.5	18.6	18.4

Table 2 Tensile properties

Leaf spring coupon number	Yield strength, 0.2% offset		Ultimate tensile strength,		Elongation,	Modulus of elasticity,	
	MPa	ksi	MPa	ksi	%	GPa	10⁶ psi
1(a)	1140	165	1450	210	27.2	202	29.3
1(b)	1165	169	1455	211	27.0	220	31.9
2(a)	1200	174	1455	211	23.0	177	25.6
2(b)	1220	177	1470	213	22.1	190	27.6
3(a)	1200	174	1450	210	25.7	181	26.2
3(b)	1220	177	1475	214	25.2	197	28.5
4(a)	1220	177	1490	216	22.5	190	27.6
4(b)	1225	178	1505	218	23.6	192	27.8
5(a)	1220	177	1490	216	27.9	199	28.9
5(b)	1225	178	1495	217	24.6	189	27.4
6(a)	470	68	950	138	49.8	186	27.0
7(a)	455	66	950	138	49.8	189	27.4
Drwg Min.	1035	150	1240	180	12	...	...

Table 3 Tensile properties after heat treatment

Leaf spring coupon number	Yield strength, 0.2% offset		Ultimate tensile strength,		Elongation,	Modulus of elasticity,	
	MPa	ksi	MPa	ksi	%	GPa	10⁶ psi
6(b)	1055	153	1400	203	28.0	176	25.5
7(b)	1089	158	1385	207	28.0	177	25.6
Drwg Min.	1035	150	1240	180	12	...	...

except for leafs 6 and 7 (these were two of the four parts that were bent). The two remaining bent leafs (1 and 2) met the specified hardness requirement. The average hardness of leaves 6 and 7 was 93.5 HRB, indicating that they were in the annealed condition.

Tensile Properties. Two tensile specimens were fabricated from each leaf. Tensile coupons from leafs 1 through 5 met the mechanical requirements of the engineering drawing (Table 2). One coupon from each of the softer leafs (6a and 7a) failed to meet the drawing requirements.

Heat Treatment. To determine if the mechanical properties of the softer material could be improved to meet the drawing requirements, cou-

pons 6(b) and 7(b) were given an age-hardening heat treatment in accordance with the governing specification (AMS 5996). This treatment consisted of initially heating the coupons to 720 °C (1325 °F) for 8 h and then cooling the furnace to 620 °C (1150 °F) until a total heat treatment time of 18 h was obtained.

Coupons 6(b) and 7(b) conformed to the required hardness values as a result of heat treatment. Coupon 6(b) averaged 39.8 HRC, while coupon 7(b) averaged 39.1 HRC, meeting the hardness requirements of AMS 5596.

The heat-treated coupons satisfied the required tensile properties listed in Table 3, meeting the hardness requirements of AMS 5596.

Discussion

A detent spring operates by deflecting under a load and returning to equilibrium with the removal of the load. The assembly is composed of nine leafs of various lengths. The missile contacts leaf 9, and the load is transferred from leaf 9 and distributed throughout the remaining leafs in descending order. The total force exerted by the missile will dictate the amount of deflection the entire leaf spring assembly will undergo. During takeoff and landing these forces tend to be the greatest (Ref 1). The detent spring assembly failed to restrain the missile during a landing exercise on an aircraft carrier. Since leafs 6 and 7 were deficient in strength, the remaining leafs were subjected to higher stresses to such an extent that four leafs plastically deformed. This caused the missile to detach from the aircraft.

Two major variables that can affect the mechanical properties of the leaf spring material are the service environment and prior manufacturing processes. There were no signs of environmental effects, such as corrosion or wear, revealed during this investigation. However, hardness and mechanical test data, laboratory heat treating, and metallographic examination indicated that leafs 6 and 7 were not heat-treated properly.

Conclusion and Recommendation

The detent spring failed by plastic deformation because the mechanical property deficiency in two leaf springs reduced the maximum load capacity of the detent spring assembly below the normal service load. Leafs 6 and 7 were not heat-treated properly before the detent spring was placed into service. This was substantiated by the increase in strength, the remaining leafs were subjected to mechanical properties obtained by properly heat treating samples of material from the deformed leaf springs in the laboratory.

It was recommended that a more stringent quality control be placed on the heat treatment process and that defective leafs be removed from service. This could be accomplished by screening leafs representing all of the heat treat lots, using a portable hardness testing unit and/or conductivity testing, if deemed feasible, to identify defective material.

Reference

1. R. Biederman, Unpublished research, Worcester Polytechnic Institute, Worcester, MA, 1993

▌Brittle Failure of a Titanium Nitride-Coated High Speed Steel Hob

Alan Stone, ASTON Metallurgical Services Company, Inc., Chicago, Illinois

Recurring, premature failures occurred in TiN-coated M2 gear hobs used to produce carbon steel ring gears. Fractographic and metallographic examination, microhardness testing, and chemical analysis by means of EDS revealed that the primary cause of failure was a coarse cellular carbide network, which created a brittle path for fracture to occur longitudinally. As the cellular carbide network must be dispersed and refined during hot working of the original bar of material, the hobs were not salvageable. Minor factors contributing to the hob failures were premature wear resulting from lower matrix hardness and high sulfur content of the material, which contributed to lower ductility through increased nucleation sites. It was recommended that the hob manufacturer specify a minimum amount of required reduction for the original bar of tool steel material, to provide for sufficient homogenization of the carbides in the resultant hob, and lower sulfur content.

Key Words	High speed steel tools	Hobbing cutters	Brittle fracture,
	Carbides	Wear	Microstructural effects
	Titanium nitride, Coatings	Coating	Sulfur
Alloys	High speed tool steels—M2		

Background

A gear manufacturer reported recurring premature failures of titanium nitrided M2 tool steel gear hobs. The hobs were used to produce carbon steel ring gears.

Applications

Gear reduction motors drive the rotating hobs into the steel stock which is radially indexed to cut the circumferences of the rings.

Circumstances leading to failure

A new off-the-shelf hob was inserted into a gear cutter. The gear manufacturer stated that the hob teeth "peeled" after two passes. No unusual circumstances regarding stock, alignment or feeds and speeds were noted.

Sample selection

A failed hob 10 cm (4.0 in.) in diam by 10 cm (4.0 in.) long, was submitted for failure analysis. A representative section containing three teeth was subsequently removed for more intensive investigation.

Visual Examination of General Physical Features

The hob exhibited excessive wear and fracturing on the leading surfaces of the cutting teeth, as shown in Fig. 1 and 2.

Fig. 1 Overall view of titanium nitride-coated hob. Approximately .1×

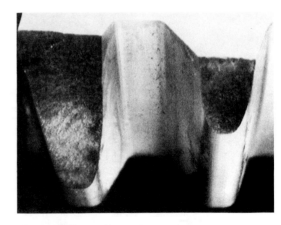

Fig. 2 Closer view of fractured cutting edges. Approximately 5×

Testing Procedure and Results

Surface examination

Macrofractography. The hob was cross-sectioned in a water-cooled abrasive saw to provide smaller samples for further examination. Surprisingly, the clamping force of the vise and stress from sectioning was sufficient to cause additional multiple fractures.

Scanning electron microscopy/EDS. A Leica S-200 scanning electron microscope (SEM) equipped with a light element detecting energy dispersive X-ray spectrometer (EDS) was used to examine the freshly fractured surfaces, fractured teeth and TiN surface. The EDS detector is capable of detecting elements with atomic numbers of sodium and above. Elements boron and above are detected in light element mode. The typical minimum detection limits are 0.1/0.5%. The minimum volume analyzed is typically 1 μm (40 μin.) in diam/thickness.

The fracture was intergranular and highly directional with exposed carbide and sulfide stringers, as shown in Fig. 3 and 4. No elements associated with corrosion of TiN were detected on the coated surface. The EDS spectra is shown in Fig. 5.

Metallography

Transverse and longitudinal cross-sections were prepared in accordance with ASTM E3-86 and etched with 10% nital in accordance with ASTM E407-89. The matrix consisted of tempered martensite with fine spheroidal carbides as well as a coarse cellular carbide network. Localized regions of higher and lower carbon contents resulting from segregation were observed (Fig. 6 to 8). No visual evidence existed of eutectic melting, retained austenite, untempered martensite, grain coarsening, surface decarburization or localized burning associated with the TiN coating at the surface. The cellular carbide networking is indicative

Fig. 3 Electron micrograph showing the fractured cutting edge. Note the directional features going across the face of the tooth. 12×

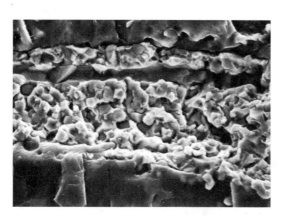

Fig. 4 Electron micrograph showing the brittle intergranular fracture surface. Carbides and inclusions are present on the fracture. 1260×

Fig. 5 Higher magnification view showing tempered martensite, sulfide stringer inclusions and carbide networking. 315×

Table 1 Microhardness Readings

| Location distance | | Hardness, | Hardness, |
mm	in.	KHN	HRC (conversion)
0.025	0.001	917	68
0.050	0.002	842	65
0.076	0.003	907	68
0.102	0.004	863	66
0.127	0.005	907	68
0.254	0.010	1017	>70
0.254	0.010	863	66
0.254	0.010	842	65
Carbide-		766	62
deficient		749	61
region		755	61
		759	61
Carbide-		795	63
rich		842	65
region		885	67
		945	69

Table 2 Chemical analysis of high speed steel hob(a)

Element	wt%
C	0.94
Mn	0.22
P	0.025
S	0.098
Si	0.28
Cu	0.12
Ni	0.26
Cr	4.17
Mo	4.77
W	6.53
V	2.09
Co	0.40

(a) Balance Fe

Fig. 6 Photomicrograph showing the cellular carbide networking. Unetched. 100×

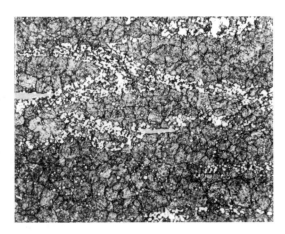

Fig. 7 Photomicrograph showing coarse cellular carbide networking. 10% nital. 50×

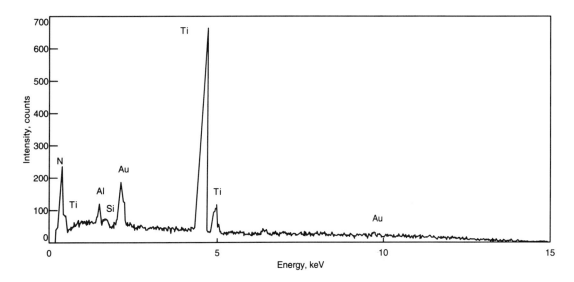

Fig. 8 Energy dispersive spectra showing titanium and nitrogen on the surface. Sample after gold sputter coating. 500×

of insufficient hot working to disperse, refine and homogenize the carbide structures formed when producing the original bar of material.

Mechanical properties

Hardness readings taken in accordance with ASTM E18-89 indicate the bulk/core hardness to be 64 HRC.

Knoop readings were taken in accordance with ASTM E384-92 with a 100-g (3.5-oz.) load at 400 × in carbide-rich and deficient locations as well as adjacent to a prematurely worn surface. Knoop hardness values were converted to HRC, and are reported in Table 1. Significant variances in relative hardness were observed.

Chemical analysis/identification

Material. Chemical analyses were performed in accordance with ASTM E572-88, E353-89, and E1019-88. The results are given in Table 2, and indicate the alloy to be an M2 tool steel with a high sulfur content.

Discussion   The final austenizing temperature prior to quenching for high speed tool steels such as M2 is typically 1177 to 1218 °C (2150 to 2225 °F). This temperature is near the eutectic melting point. Slight overheating will cause localized melting and grain coarsening, although no such overheating was observed. Overheating and poor quenching may result in retained austenite. Careful heat treatments include sub-zero treatments to transform retained austenite to martensite prior to final double tempering. Otherwise, martensite may form subsequent to tempering and result in untempered martensite in the final product. No retained austenite or untempered martensite was observed. Localized burning from the titanium nitriding was also not observed.

The coarse cellular carbide network and carbon segregation resulted from insufficient hot working of the original bar of material. High speed tool steels require hot working to break apart, refine and homogenize the carbides. The hob exhibited carbide networking, which results in low energy crack paths as well as locally depleting carbon from the matrix to reduce local hardness. The high sulfide content reduces ductility by creating additional fracture nucleation sites at inclusions.

Conclusions and Recommendations

Most probable cause

There are three contributing causes to these failures. The primary factor is the coarse cellular carbide network which creates a brittle path for fracture to occur. A minor factor is that premature wear may result from lower matrix hardnesses. Additionally, the high sulfide content contributes to lower ductility through increasing nucleation sites.

Remedial action

The hobs are not salvageable. The cellular carbide network must be dispersed and refined during hot working, which requires heavy mechanical working at high temperatures. Conventional heat treatment, performed near the melting point of the alloy, is insufficient to adequately refine and homogenize the high-temperature carbides.

How failure could have been prevented

The choice of a high-performance coating combined with a hard, abrasion- and temperature-resistant base material should provide for long tool life. The hob manufacturer should specify a minimum amount of required reduction to provide for sufficient homogenization of the carbides. A cleaner material, low in sulfur, would also be beneficial.

Cracking on the Parting Line of Closed-Die Forgings

Anthony J. Koprowski, Consultant, Elmhurst, Illinois

An investigation was conducted to determine the factors responsible for the occasional formation of cracks on the parting lines of medium plain carbon and low-alloy medium-carbon steel forgings. The cracks were present on as-forged parts and grew during heat treatment. Examination revealed that areas near the parting line exhibited a large grain structure not present in the forged stock. High-temperature scale was also found in the cracks. It was concluded that the cracks were caused by material being folded over the parting line. The folding occurred because of a mismatch in the forgings and from material flow during trimming and / or material flow during forging.

Key Words			
Carbon steels	Low-alloy steels	Die forgings	
Cracking (fracturing)	Closed-die forging		

Background

Cracks occasionally formed on the parting lines of medium plain carbon and low-alloy medium-carbon steel forgings.

Applications

Closed-die forging is one method for producing near-net-shape parts. A steel billet is heated to forging temperature (1100 to 1260 °C, or 2000 to 2300 °F) and placed into successive preforms in the bottom impression of the die. The upper die is then dropped or propelled downward onto the bottom impression. This hammering forces the hot metal to flow and conform to the die impression. The finishing impression is generally the center section of the die. It is used to produce the final shape before trimming. Closed-die forgings can be produced in as few as one blow or as many as two dozen blows. Those discussed in this study normally are produced by 8 to 15 blows. Trimming of the finished platter is kept as close to the part as possible. This reduces the amount of finishing needed in this area of the part. Trimming can be done hot or cold. This analysis deals with hot trimming.

Circumstances leading to failure

The cracks were generally found after heat treatment, but were occasionally found in the as-forged parts. This resulted in production losses and delays because of the testing required after the cracks were discovered, including magnetic particle inspection of the lots.

Testing Procedure and Results

Nondestructive evaluation

Magnetic particle inspection of as-forged parts showed that there were small (<6 mm, or ¼ in.) indications present. Inspection after heat treatment showed that the cracks on the parting line grew. There was no correlation between the initial length of the indication and the final length after heat treatment.

Surface examination

Visual inspection of as-forged parts revealed suspect lines. After heating to the austenitizing temperature, these parts had been quenched in hot oil and then tempered. The quenched parts cracked along the parting line. Examination of the fracture surface revealed an area of discoloration, which was determined to be an oxide coating from the thermal processing. The discoloration indicated that the fracture was open to the atmosphere during one of the heating cycles and was not produced by the quenching operation.

Macrofractography. The fracture surfaces exhibited discoloration from oxidation during thermal processing. This was true of both as-forged and heat-treated parts. There were no radial lines or beach marks to pinpoint one origin (Fig. 1), indicating that the origin may have been the entire discolored area.

Scanning Electron Microscopy/Fractography. The fracture surface of the heat-treated parts exhibited a transgranular fracture on the discolored surface. There were indications of oxidation. The oxidation was probably caused by the

Fig. 1 Section of discolored part of fracture surface showing no clear origin. Dark area is discoloration from thermal processing. 1.24×

thermal processing that the parts underwent during forging. Bar seams and folds were eliminated as a possible source.

Metallography

Microstructural Analysis. Cross sections were taken of an as-forged part and a heat-treated part. The samples were mounted and polished. The microstructure—lamellar pearlite and ferrite—was typical of the medium-carbon steel used for the product. The grain size near the parting line of the forging was larger than the grain size in the surrounding area. This indicated that either this area of the forging was not worked as much as the rest of the forging; there was uneven flow, or this area received more heat, all of which can cause the grains to grow larger.

Crack Origins/Paths. The cross section of an austenitized and quenched part showed that the fracture was intergranular and branched (Fig. 2). High-temperature scale was present in the cracks. This scale was probably produced on the surface of the bar during its heating before forging. The scale was then folded into the material during the forging operation. The part exhibited partial decarburization at the crack surface, indicating that it was oxidized during the heating or annealing operation. Austenitizing and quenching are done in a neutral atmosphere and would not produce this condition. Oxidation was present on both the as-forged and heat-treated parts. It was probably produced during the heating of the bar for forging. The partial decarburization was not present on any of the machined surfaces, which further supported the assumption that the decarburization was produced during forging.

Chemical analysis/identification

The material used varied depending on the parts being forged, but generally was medium plain carbon steel or low-alloy medium-carbon steel. The steels of different forgings were analyzed to determine whether there was a correlation between elements in the steel and the cracks found on the parting lines. Correlation of chemistry and cracking indicated a greater chance for formation of cracks if the steel was vanadium grain refined and if the vanadium content was less than

0.06%.

Stress analysis

Experimental. Parts were forged with material that was aluminum grain refined with an aluminum content of 0.022%, vanadium grain refined with a vanadium content of 0.03%, and vanadium grain refined with a vanadium content of greater than 0.06%. Samples of each were examined after forging. The aluminum-grain-refined parts and those parts forged with more than 0.06% V showed fewer crack indications than the parts forged with 0.03% V.

A second experiment was performed with material that was aluminum grain refined, but used an opened trimmer to allow more clearance. The samples from this group did not show any indications after forging, and none of the parts cracked upon quenching.

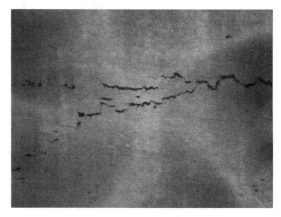

Fig. 2 Cross section of suspect area, showing a typical intergranular and branched fracture. 76×

Discussion Examination of the parts in some of the cases showed reverse flow lines in the forgings. This would cause the material to fold at the parting line area. The fold would also be fully or partially decarburized at this point.

The closeness of the trim dies caused a smearing or pulling at the part surface (Fig. 3). This was accentuated by a slightly larger grain size in the trim area than in the rest of the forging. The tear-

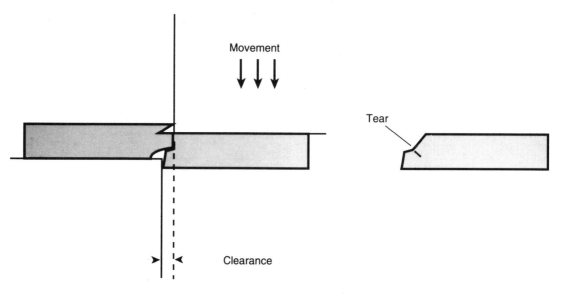

Movement

Tear

Clearance

Fig. 3 Possible tearing mechanism during trimming. The punch pulls part of the material with it, causing stresses at the grain boundaries. Grains may separate and tear if the elastic limit is reached.

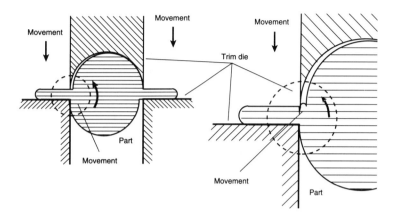

Fig. 4 Effects of die mismatch during forging, which can produce a fold or lap during trimming. The force of the punch moving down causes the material to flow to the empty portion of the die; at the same time, material flows into the area from the trim line.

ing occurred as the part was trimmed. Because the trimmer was fitted so closely to the part geometry, the tear went below the surface of the part. This caused the material to separate at the grain boundaries and produced the cracks found in both the as-forged and heat-treated parts.

Although the mismatch and the closeness of the trimming appeared to be the major cause of the cracks, alloys grain refined with 0.03% V seemed to suffer more cracking than aluminum-grain-refined material or the alloy grain refined using 0.06% V.

Mismatch during forging caused material to flow in the wrong directions during both the forging operation and the trimming operation (Fig. 4). At the trimming temperature (approximately 1100 °C, or 2000 °F), plastic deformation still occurred. As the trim die closed, it pushed the mismatched edge of the part in the direction of the empty trim die cavity. This occurred at the same time the trimmed material was being pushed up.

The combination of the trimming and material flow produced a fold at the top of the trim line. Microstructural examinations confirmed that partial decarburization was present on the surfaces of internal cracks. This also explained the scale found on the crack surface.

During the investigation, it was discovered that areas near the parting line showed a large grain structure. This structure was not present in the forged stock and was not typical of the forging process. The suspected cause was uneven flow of material during the forging process, which resulted in stagnant areas where grain growth occurred. Microscopic examination of the parts revealed large grain areas near the parting line. The rest of the forging showed a uniform grain structure. This stagnant or nonuniform material flow was caused by mismatch of the forging dies, which allowed some areas to be worked and others to remain stagnant. This lack of work at forging temperature produced rapid grain growth.

Conclusion and Recommendations

Most probable cause

Magnetic particle indications were found in as-forged products. These indications took the form of folds and, in some cases, cracks. They occurred on the top side of the parting line at the original thickness of the flash line. The grain structure in the forgings was uneven; that is, there were both large- and small-grain areas. Large grains were found near the parting line; small, uniform grains occurred elsewhere. This indicated that the material flow and/or the temperature in these areas were different, perhaps due to mismatch or improper working of the material during forging.

The failures were from three sources. The first was folds in the forging produced during the forging process. These were related to poor maintenance on the hammer, inexperienced operators, and so forth.

The second cause was trimming of the material very close to the geometry of the part, which produced tearing along the parting line during the trimming operation. This tearing extended into the part.

Finally, abnormal material flow produced folds or laps. This was caused either by design problems in the die or by the mismatch of the dies due to improper alignment, poor hammer maintenance, and so on.

All of these conditions can produce cracks that open during subsequent thermal processing. The occurrence of cracks was most prevalent in material that was vanadium grain refined with a vanadium content of less than 0.06%.

Remedial action

When parting line separation occurs, the forging process should be examined to ensure that the parts are properly worked and formed and that the material is produced with proper alloying. This means checking to ensure that the dies are properly aligned and matched in the hammer. This can be done simply by cutting the finished platter before trimming the parts and measuring for mismatch. Also, the hammer should be checked for proper clearance and alignment.

Trimmers should be opened as much as possible to reduce the problem of tearing. This can be done, in most cases, without additional expense or operations.

The material used, if vanadium grain refined, should have a minimum of 0.06% V to reduce the susceptibility of the parts to parting line separation. Finally, the material flow and grain size, especially in the parting line area, should be examined. If there is a problem with flow or grain growth, the die design should be examined and changed to help enhance proper flow.

Spalling of a Ball-Peen Hammer

Carmine D'Antonio, Department of Metallurgy and Materials Science, Polytechnic University, Brooklyn, New York

A carbon steel ball-peen hammer ejected a chip that struck the user's eye. Failure occurred when two hammers were struck together during an attempt to free a universal joint from an automotive drive shaft. Two samples were cut from the face of the hammer, one through the chipped area on the chamfer and the other from the undamaged area on the chamfer. The shape and texture of the fracture surfaces were typical of spalling. The fracture was conchoidal and exhibited a complete lack of plastic deformation. White etching bands that intersected the face and chamfer were revealed during metallographic examination. Fracture occurred through a white band. Failure was attributed to formation of envelopes of untempered martensite under the chamfer that ruptured explosively during service.

Key Words	Carbon steels	Hammers	Spalling

Background

A carbon steel ball-peen hammer ejected a chip that struck the user's eye.

Circumstances leading to failure

The user was attempting to free a universal joint from an automotive drive shaft. The shaft was secured in a vise and two ball-peen hammers were being used, one face up and the other face down. The hammers were struck together, causing sparking of the hammer faces and ejection of a chip from the striking hammer.

Pertinent specifications

The hammer was manufactured to conform to ANSI B173.2-1978, "Safety Requirements for Ball Peen Hammers."

Specimen selection

Two samples were cut from the face of the failed hammer, one through the chipped area on the chamfer and the other from an unchipped area on the chamfer. Both were mounted in cold-setting resin to show the chamfer (and chipped area), face, and bell in cross section. A third piece was cut from the cheek for chemical analysis.

Visual Examination of General Physical Features

Figure 1 shows the chipped area at the chamfer of the hammer and the ejected chip. The shape and texture of fracture surfaces were typical of those on spalled hammers. The fracture was conchoidal and exhibited a complete lack of plastic deformation. These features were caused by the underlying microstructure.

Testing Procedure and Results

Metallography

The general microstructure of the hammer material under the face was tempered martensite. Of greater significance, however, was the presence of white etching bands that intersected the face and chamfer. Figures 2(a) and (b) show a network of these bands found in the specimen taken from the undamaged area. Note that a portion of one of the bands is cracked.

Figure 3 shows a cross section of the chipped area. It is evident that the fracture occurred through a white band.

Chemical analysis/identification

The results of chemical analysis are presented in Table 1. The hammer material conformed to the chemical requirements for ball-peen hammers (ANSI B173.2-1978).

Mechanical properties

Hardness. Microhardness tests (Knoop, 1 kg load) conducted on the metallographic samples showed that the hardness of the material under

Fig. 1 The face and bell of the ball-peen hammer, showing the area where the chip was ejected. The chip in shown in the insert at the right. The fracture was conchoidal and brittle in appearance.

Table 1 Results of chemical analysis

	Composition, %	
Element	Hammer	Chemical requirement (ANSI 173.2-1978)
Carbon	0.85	0.45-0.88
Manganese	0.71	0.30-0.90
Phosphorus	0.020	0.04 (max)
Sulfur	0.026	0.05 (max)
Silicon	0.24	0.10-0.30

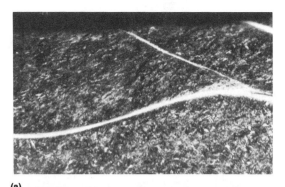

(a)

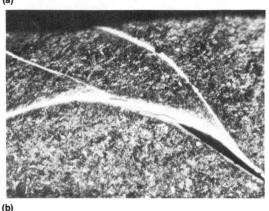

(b)

Fig. 2 White etching bands of untempered martensite that intersected the face and chamfer of the hammer in a zone away from the chipped area. A crack is present in one of the bands. 34.65×

Fig. 3 Micrograph taken at the chipped area. Fracture occurred through a white band. 315×

the face (converted to the Rockwell C scale) was 51 to 52 HRC, well within the allowable range of 45 to 60 HRC. Microhardness tests were also performed on the white bands and the material immediately adjacent to them. These tests were conducted using lighter loads of 200 and 500 g. The hardness of the bands was 64 to 65 HRC, and the adjacent matrix material had a hardness of 42 to 46 HRC.

Discussion

The mechanism of chipping is generic to striking (or struck) tools. When a hardened steel tool is struck against a material as hard or harder than itself, conchoidal envelopes (appearing as white bands when viewed in cross section) of impact-induced untempered martensite can form at the striking surface, particularly under the chamfer. This envelope of brittle material can fracture explosively, expelling a chip either as the result of a tensile wave reflected back from the free surface of the hammer head or by a shock wave caused by a subsequent impact.

The mechanism of formation of white bands has been described by investigators as zones of intense adiabatic shear. Zener and Holloman (Ref 1) proposed that the bands resulted from high strain rates that caused shear instability in an otherwise uniformly straining solid. These high strain rates can cause deformation to change from isothermal to adiabatic, resulting in an instability where deformation cannot be homogeneous. They concluded that the white bands were martensite formed by the high temperatures resulting from the concentration of shear strain in the band regions, which changed the structure to austenite. The austenite was then rapidly quenched by the surrounding material.

This phenomenon was probably responsible for the hammer failure. The high hardness of the white bands, the evidence of incipient cracking in the bands, and the fracture path in the chipped area through a white band all support this conclusion. Additional proof that formation of the white martensite was caused by localized heating is provided by the lower hardness values of the material adjacent to the bands, which indicate further tempering of these zones.

Conclusion and Recommendations

Most probable cause

When the two hammers were struck together, spalling occurred. Because the hammers possessed the same hardness, envelopes of untempered martensite were formed under the face and chamfer. One of these envelopes fractured, ejecting a chip.

Remedial action

The applicable standard sets forth comprehensive safety requirements and limitations of use, one of which states that a ball-peen hammer should never be struck against hard or hardened objects. All hammers must be marked with a warning to that effect as well as a caution to wear safety goggles. The warnings notwithstanding, in most cases it is difficult to determine the hardness of an object without hardness data. In this instance, however, the user should have known that both hammers were of similar hardness and should have worn safety glasses. It should also be stressed that in this instance the hammer had been repeatedly used to strike hard objects. This is indicated by the damage present on the hammer face. Such conditions of use greatly increase the likelihood of this type of failure.

Reference

1. Zener and Holloman, *J. Appl. Phys.*, Vol 15, 1944, p 22

Brittle Fracture in a Large Grain Storage Bin

William T. Becker, University of Tennessee, Knoxville, Tennessee

A 22 m (72 ft) diameter filled grain storage bin made from a 0.2% carbon steel collapsed at a temperature of –1 to 4 °C (30 to 40 °F). Failure analysis indicated that fracture occurred in a two-step process: first downward, by ductile failure of small ligament from a bolt hole near the bottom of the tank to create a crack 25 mm (1 in.) long, and then upward, by brittle fracture through successive 1.2 m (4 ft) wide sheets of ASTM A446 material. Site investigation showed that the concrete base pad was not level. Chemical analysis indicated that the material had a high nitrogen content (0.02%). The allowable stress based on yield was estimated using four different design criteria. Correlation among those results was poor. The different criteria indicated that the material was loaded from the maximum allowable to approximately 30% less than allowable. Nevertheless, at this stress level, fracture mechanics indicated that the 25 mm (1 in.) starter crack exceeded or was very near the critical crack length for the material. Additional factors not taken into account in the design equations included cold work from a hole punching operation, thread imprinting in bolt holes, and an additional hoop stress created by forcing an incorrectly formed panel to fit the pad base radius. These factors increased the nominal design stress to a sufficiently large value to cause the critical crack length to be exceeded.

Key Words

Low carbon steels,
 Mechanical properties
Crack propagation
Bolted joints
Fracture toughness

Galvanized steels,
 Mechanical properties
Stresses
Ductile fracture

Tanks
Brittle fracture
Design
Fracture mechanics

Alloys

Low carbon steels—A446 Grade B   Low carbon steels—A446 Grade F

Background

A large grain bin failed cataclysmically soon after erection (Fig. 1, 2). Failure occurred while the bin was in the static loaded condition. The temperature at the time of failure was –1 to 4 °C (30 to 40 °F), and there was no strong wind, snow loading, or seismic loading.

Investigation of the failure started approximately 6 weeks after collapse of the silo, and at that time cracking associated with bolt holes was found in the two other storage bins of similar construction located at the site. A survey of the concrete pad after the debris was removed indicated that the pad was not level and that not all stiffeners were at the same vertical elevation. Some question was raised as to the accuracy of the survey to determine pad elevation, but there was no disagreement that the pad was not level. Additionally, it was determined that the side panels had been formed to fit erection of a 16 m (52 ft) diameter tank. These smaller radius panels were forced to fit the base ring of the larger 22 m (72 ft) design. This would introduce tensile bending stresses on the inside of the tank wall, and these stresses would add to the hoop stress acting in the same direction created by the material in the tank.

Applications

The failed tank was 22 m (72 ft) in diameter with a capacity of 9690 m³ (275,000 bushels). The wall was fabricated from 24 circular sections of nominal 1.2 × 2.6 m (4 × 8 to 12 ft) galvanized corrugated mild steel sheets. The 1.2 m (4 ft) dimension was vertical. Twenty-four vertical stiffeners (2 per sheet) were used. The vertical section of the tank was 25 courses high. Sidewall thickness varied from 8 gage (4.11 mm, or 0.162 in.) at the base to 18 gage (1.25 mm, or 0.049 in.) at the top. Sheets were fastened to the stiffeners using No. 5, 9.5 mm (3/8 in.) diameter bolts. Bolts were threaded the complete length of the shank.

The tank sat on a concrete foundation consisting of a 2.1 m (7 ft) wide ring with a 37° inverted conical hopper having a depth of 7.3 m (24 ft). The vertical walls were bolted to a base plate on the concrete pad on a bolt circle located 19.2 mm (0.75 in.) above the bottom of the first course. The first bolt hole in the stiffener was located 125 mm (5 in.) above the bottom of the first course.

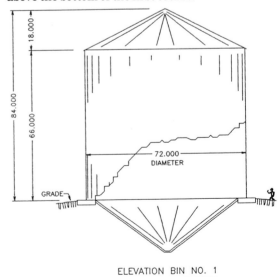

Fig. 1 Accident scene showing collapsed silo and two adjacent silos

Fig. 2 Schematic diagram of silo

The grain bin was first filled in the spring after erection the previous fall. Within the next month, the bin was emptied and refilled with wheat of 12.5% moisture. Approximately five months later a crack was observed in the bottom course of the side wall, with bulging of the sidewall observed above the crack. The decision was made to empty the tank and to perform a repair. Some 14 h later, and before the emptying had started, the crack grew into the second course. Courses were nominally 1.2 m (4 ft) high, so the crack was more than 1.2 m (4 ft) long at that time. Approximately 3 h later, and still before emptying had been started, the fracture propagated still further and the tank collapsed. Temperature of the steel at the time of collapse was estimated to be –1 to 4 °C (30 to 40 °F).

Pertinent specifications
Material: ASTM A446 grade B or F steel sheet, zinc-coated by the hot-dip process, structural quality

Design Code: German: Silo code DIN 1055 E and DIN 1055 F. U.S.: Janssen theory

Performance of other parts in same or similar service
The design and erection of smaller 16 m (52 ft) diameter tanks has a successful history. Six tanks of the new larger 22-m (72-ft) diameter were erected in sets of three tanks at two different locations at about the same time. One of these tanks had three vertical stiffeners per side sheet, as used on the 16 m (52 ft) diameter tanks, and the other five used a new design of two stiffeners per side sheet. Galvanized corrugated steel for the side sheets for the six tanks, having a nominal yield strength of 345 MPa (50 ksi), came from three different suppliers and was not separated by tank. During the accident investigation, cracks were found in bolt holes in the second and third tanks at the site of the collapsed tank. No cracks were found in the three tanks at the second site.

Visual Examination of General Physical Features

Failure started near the bottom of the first course of sheeting in a bolt hole located 19 mm (0.75 in.) from the bottom of the first course and then propagated upward perpendicular to the corrugations and parallel to the axis of the bin (Fig. 3-5). Visual examination of the crack showed the presence of shear lips on the ligament between the bolt hole and the bottom of the tank and flat fracture above the bolt hole. The crack above the bolt hole propagated completely through the first course (approximately 4 ft). Examination of the failed bolt hole, A, and an adjacent bolt hole, C, showed distortion in hole C parallel to the tearout in hole A (Fig. 4, 5c). This would occur if the right-hand sheet shifted downward and to the right. Examination of these bolt holes with a hand lens showed thread impressions in the holes along the line E-E in Fig. 4b.

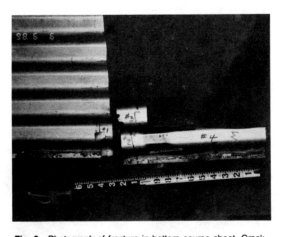

Fig. 3 Photograph of fracture in bottom course sheet. Crack propagated upward completely through the first course. A section of material containing the crack has been removed from the larger left panel.

Fig. 4 Closeup of section containing fracture surface in failed silo. Crack propagated normal to corrugations and normal to concrete base pad. Two ruptured bolt holes (a,b), one distorted bolt hole (c). Bolt in hole A connects to vertical stiffener. Bolts in B and C connect to base ring on concrete pad. Region D was removed before section received.

Testing Procedure and Results

Testing conducted by various parties involved in the investigation included chemical analysis, optical metallography, tensile testing, subsize Charpy impact testing, Rockwell hardness testing, microhardness testing, and scanning electron microscopy (SEM) fractography.

Surface examination
Macrofractography. As indicated above, visual examination showed that the flat fracture started at the base ring bolt hole B at the thread imprints and propagated upward in a brittle manner (Fig. 5b). When examined with the naked eye, the fracture behind the bolt hole appeared to be essentially flat with the exception of a darker area at one side of the fracture surface (which was triangular in nature, about 1 mm, or 0.04 in., wide and 4 mm, or 0.16 in., long). This area was inclined to the remainder of the fracture surface.

The width of the specimen was measured along the fracture surface and compared to the width of the specimen about 13 mm (1/2 in.) away from the

Fig. 5 (a) Closer view of fracture surface and bolt hole B. Flat fracture region is nominally perpendicular to tensile hoop stress. Slant fracture surface is inclined to tensile hoop stress. (b) Macro view of fracture, including bolt hole B, showing chevrons indicating brittle crack propagation upward from hole B starting near (1). Ductile slant fracture region below hole B is also visible . (c) Macro view of bolt hole C. Bolt hole is distorted. A line drawn from bolt hole A through bolt hole C (line E-E in Fig. 4b) shows common direction of distortion.

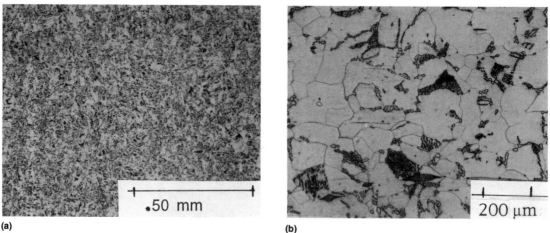

Fig. 6 (a) Microstructure of galvanized sheet. Longitudinal section. Microstructure consists of pearlite and primary ferrite. Note coarse pearlite and divorced cementite. (b) Higher magnification of Fig. 6(a)

fracture surface. Aside from a very small area near the thread imprints, the width was constant. (This constancy of width was also subsequently verified by examination of the specimen at low magnification in the SEM. When examined on the side of the specimen, the small area that showed a decrease in width was a circular imprint, most likely cre-

ated by a washer between the bolt head and the sheet.)

When viewed nominally perpendicular to the flat fracture surface with an 8× hand lens, chevrons could be seen behind the bolt hole that pointed toward the bolt hole (Fig. 5b). When viewed with the same hand lens, but almost paral-

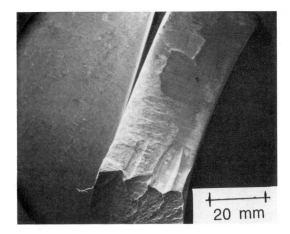

Fig. 7 Schematic view of fracture surface for SEM fractographs. Region 1, Fig. 8, 10-12. Regions 2 and 3, Fig. 9. Region 4, Fig. 13

Fig. 8 (a) Low-magnification fractograph showing bolt hole B. Note stamping marks and thread impressions inside. hole. (b) Higher magnification of Fig. 8(a)

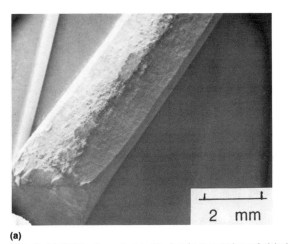

(a)

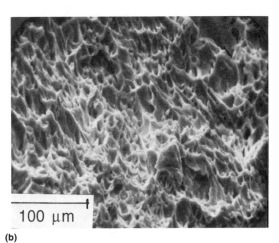

(b)

Fig. 9 (a) SEM fractograph of ductile slant fracture region at 6 o'clock in hole B (regions 2 and 3 in Fig. 6). (b) Higher magnification of Fig. 9(a) showing shear dimples and direction of crack propagation. Crack propagation direction indicated

lel to the fracture surface, two different sets of radial marks could be seen. One set radiated out from a region in the lighter fracture surface, and ridges in the pattern coincided with the geometry of the thread imprints. A different set of these lines ended abruptly along the boundary between the light and dark areas on the fracture surface. A second set of radial lines located in the darker material pointed back to a thread imprint at the juncture of the light and dark areas on the fracture surface.

Visual examination with a hand lens of the material between the bottom edge of the sheet and the bolt hole showed that there were two fracture surfaces inclined to the surface of the sheet (Fig. 9a). All of the above features were subsequently documented using the SEM, and several higher-magnification views were taken to determine the microscopic scale features of the fracture surface. A low-magnification fractograph is shown in Fig. 10 and is used to reference the higher-magnification fractographs.

Higher-magnification fractographs taken in material between the bottom of the sheet and the bolt hole showed ductile elongated shear dimples

pointing away from the bolt hole (region 2 in Fig. 6 and Fig. 9b). This indicates that the ductile crack started at the bolt hole and propagated downward to the bottom of the sheet.

The light and dark areas of the fracture surface behind the bolt hole were examined in some detail. Examination of the nonrubbed area from the thread imprints back to the origin of the radial marks showed that it consisted primarily of cleavage (some fracture facets contained river patterns), with some tear ridges visible together with small, isolated patches of microvoid coalescence (Fig. 10-12). Thus, the fracture is essentially brittle over this total area.

Higher magnification of the dark slant fracture region showed it to be a rubbed region, which unfortunately obscured details of the fracture surface. There definitely appeared to be evidence of brittle fracture (cleavage), but there was also some indication of microvoid coalescence (Fig. 11).

Metallography

Optical metallography showed the material to have a microstructure consisting of primary ferrite and pearlite. High magnification showed that it

Fig. 10 (a) Low-magnification fractograph showing fracture surface oriented to emphasize fracture surface rather than thread imprints. (b)Schematic of fracture surface in Fig. 10a with features labeled. A, Crack initiation site and associated radial marks at thread imprint on edge of bolt hole. B, Crack initiation site and associated radial marks. Higher-magnification view in Fig. 12(a). C, Rubbed zone. Appears as dark region in Fig. 11. Contains crack initiation site A. Rubbed surface inclined to nonrubbed area. D, Partial set of radial marks along boundary between rubbed and nonrubbed material. E, Very thin rubbed ridge along edge of specimen. F, Location for examination at higher magnification in Fig. 12(b). G, Location for examination at higher magnification in Fig. 12(c). T, Thread imprints at edge of bolt hole

Fig. 11 (a) Higher magnification of interface between rubbed and nonrubbed areas in Fig. 10. Rubbed region is darker and to the right. Thread imprint visible at lower left. (b) Higher magnification of Fig. 11(a)

also contained divorced cementite. The amount of pearlite appeared to be high for a 0.2% carbon steel in the hot-rolled condition (Fig. 6a, b). When examined unetched, the volume fraction inclusions present appeared to not be excessive (Fig. 6a, b).

Chemical analysis/identification

Chemical analysis on specimens removed from the three bins indicated that the material in question was a nominal 0.2% carbon steel having an abnormally high nitrogen content (0.018 to 0.02%), a low aluminum content (0.005%), and a manganese content of approximately 0.8%.

Mechanical properties

Hardness. Microhardness (0.1 kgf load, 136 diamond pyramid hardness or DPH) traverses taken radially from the punched bolt holes showed the hardness to decrease from approximately 295 DPH to approximately 260 to 280 DPH in about 200 μm (8 mil) and to 242 DPH in 255 μm (10 mil) away from the hole.

Tensile Properties. Tensile specimens tested by various parties and removed from all three bins indicated a yield strength of 347 to 421 MPa (50.3 to 61.1 ksi), a tensile strength of 485 to 542 MPa (70.3 to 78.7 ksi), and a tensile elongation of 20.9 to 26.3%.

Impact Toughness. Subsize Charpy specimens were removed from panels in the second and third bins. There was considerable variation from panel to panel in energy absorption. Data from one panel showed a rapid drop in absorbed energy between −18 to −29 °C (0 and −20 °F), while data from the other panel did not show a rapid drop in absorbed energy until the temperature was reduced to below 0 °C(−50 °F).

Stress analysis

Considerable effort was made to determine the loading intensity in both the side walls and in the bolts by structural engineers who had design experience with storage bins. Several different formulas were applied with varying results. Some design equations indicated that the side wall panels were loaded to the design maximum, while others indicated that loading was somewhat less (e.g., 30%), neglecting any temperature stress and bending of the sheets to fit the larger tank diameter. This would introduce tensile bending stresses on the inside of the tank wall, and these stresses would add to the hoop stress acting in the same direction created by the

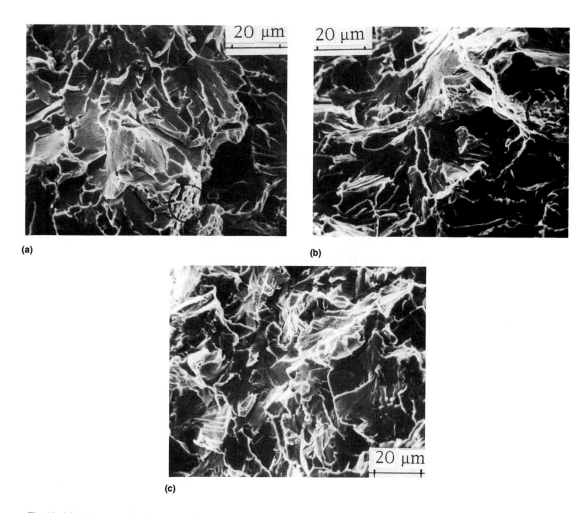

Fig. 12 (a) Higher magnification of region B in Fig. 10. Cleavage with some river patterns, tear ridges, and microvoid coalescence visible. Fracture is predominantly cleavage. One region of microvoid coalescence is circled. (b) Higher magnification of region F in Fig. 10. Fracture features are similar to those in Fig. 12(a). Fracture is essentially cleavage. One region of microvoid coalescence is circled. (c) Higher magnification of region G in Fig. 10. Fracture features are similar to those in Fig. 12(a) and (b) except no region of microvoid coalescence is visible.

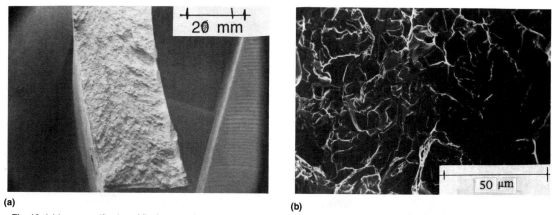

Fig. 13 (a) Low magnification of flat fracture above thread imprints (regions 1 to 4 in Fig. 6) . (b) High magnification of Fig. 10(a) showing brittle fracture. Fracture features are similar to those in Fig. 12. Corresponds to region 4 in Fig. 6. Area is approximately 50 mm (2 in.) away from the bolt hole.

material in the tank. No estimate was made of the bending stress required to force the curved panels to fit the larger diameter. The temperature stress was later estimated to be about 28 MPa (4 ksi).

Discussion   Several well-known factors (loading rate, section thickness, carbon content, microstructure, service temperature) may increase constraint to a level where brittle crack propagation occurs for a given level of the ratio of fracture toughness to applied nominal stress. However, brittle fracture in a hot-rolled steel of 0.2% C, in a thin section (4.06 mm, or 0.160 in., in this instance) at a service temperature of −1 to 4 °C (30 to 40 °F) is somewhat surprising. It should also be noted that no decrease in width of the specimen at the fracture surface behind the bolt hole could be measured with a point micrometer, nor could any be seen when viewed at the proper orientation in the SEM. This implies that for some reason, the material above the bolt hole fractured under conditions of plane strain.

Various design criteria used to size the cylindrical section, to determine the number of required stiffeners, and to size the bolts, produced inconsistent results (e.g., 30% in the hoop stress at the base ring of the tank). Discussion with structural engineers indicated that the state of stress at a bolt hole in a vertical stiffener is not well understood quantitatively, which is confirmed by the difference in results obtained from the four design equations used. Nevertheless, there is clearly a nominal hoop stress due to the grain loading, which was estimated to be from 145 to 210 MPa (21 to 30 ksi) at the bottom of the first course, using the design equations. The temperature stress of about 28 MPa (4 ksi) must be added to the nominal hoop stress. The issue, then, is whether this stress is sufficient to cause fracture that was observed to be both macro- and microscopically brittle in material above the bolt hole, or whether additional factors must be considered. These factors would include:

- Additional tensile hoop stresses on the inside of the sheet due to forcing the material rolled for a 16 m (52 ft) diameter tank to fit the actual 22 m (72 ft) diameter tank
- An increase in the applied stress due to the downward and lateral shifting of the sheet, as evidenced by the bolt hole distortion
- The presence of cold work adjacent to the bolt hole from the hole stamping operation
- The additional cold work and high local stress concentration created by imprinting the threads in the bolt hole
- The abnormally high nitrogen content

The first two factors increase the stress at the bottom of the tank. The increased cold work adjacent to the bolt hole is expected to increase the transition temperature and decrease the fracture toughness (Ref 1).

The stress concentration factor at the root of the imprinted threads is significantly higher than the stress concentration at the bolt hole without the thread imprint.

The ability of nitrogen to increase the ductile-brittle transition temperature is pronounced, and there is an indication that high nitrogen will also decrease the fracture toughness (Ref 2). Although some aluminum was detected in the chemical analysis, it is insufficient to negate the effect of the nitrogen.

Fracture mechanics predictions

It is not unreasonable to assume that the fracture started in a ductile manner by tearing downward from the bolt hole to the bottom of the first sheet. Both the macroscopic and microscopic evidence indicates ductile fracture in this region.

When the continuation of the failure is considered, some difficulties arise in semantics. Should the macroscopic and microscopic brittle fracture above the bolt hole be considered a continuation of the ductile fracture event, or a situation in which a second fracture is initiated by a brittle mechanism and propagated by a brittle mechanism? A second initiation event appears to be the better choice, partially because the bolt hole can, in some respects, be considered a crack arrester on a macroscopic scale, and also because, as discussed below, there is some uncertainty as to the chain of events after the ductile crack below the bolt hole was created.

If a second initiating event is considered, there is a pre-existing crack (first ductile crack plus the bolt hole diameter) that is about 25 mm (1 in.) long. Using an approximation published by Barsom and Rolfe (Ref 3), the plane strain fracture toughness (K_{Ic}) is estimated to be about 65 MPa√m (60 ksi√in.). If this value of K_{Ic} is used with the pre-existing crack length of 25 mm (1 in.), the allowable stress is about 205 MPa (30 ksi), assuming that the relationship between the fracture toughness, nominal stress, and crack length is given by the following (Ref 4):

$$K_{Ic} = \alpha \, \sigma \sqrt{a}$$

where $\alpha \approx 2$. One set of design equations predicted a nominal stress of 145 MPa (21 ksi) at 20 °C (70 °F), which is increased to about 165 MPa (24 ksi) at −1 to 4 °C (30 to 40 °F). To this must be added the additional stress of unknown magnitude created to force the material to fit the larger bin diameter geometry, and the unknown but degraded fracture toughness of the cold-worked, thread-imprinted hole.

This discussion raises the issue of whether material of this thickness can be loaded in plane strain. The oft-quoted criterion that a thickness of $2.5 \, (K_{Ic}/\sigma_{ys})^2$ is needed to obtain plane strain conditions indicates that the assumed fracture toughness must be considerably in error. However, it is important to remember that this expression is a conservative guideline to estimate the required thickness to achieve test conditions, such as those defined by ASTM E399, to determine the plane strain fracture toughness. It is incorrect to assume that plane strain conditions cannot be met for thinner material, and in fact there is both visual and quantitative data in this situation to show that the fracture above the bolt hole was plane strain.

The sequence of events leading to complete fracture in the sheet is somewhat clouded by the lack of clear fractographs in the rubbed area. Two fracture initiation sites are visible, one of which is in the rubbed area. The rubbed area is also inclined at an angle to the nonrubbed area and the hoop stress. The inclination of the fracture surface tempts one to assume that the rubbed area failed under conditions of plane stress and that the fracture in the nonrubbed area failed under conditions of plane strain. (Analytical expressions suggest that the constraint is higher in areas further removed from the bolt hole than it is at the bolt hole.) However, although the loading may macroscopically be plane stress, it would not be inconsistent if the fracture mechanism were brittle. If the ductility inherent to the material has been used by the local plastic deformation from the thread imprinting and the additional amount created by the washer imprint on the side of the specimen, there may be almost no ductility remaining in the material. The fracture mechanism must then be brittle. The plastic zone size can be estimated according to:

$$r_p = \frac{1}{2\pi}\left(\frac{K_{Ic}}{\sigma}\right)^2 \text{ (Plane Stress)}$$

but this has little significance if the material is incapable of straining plastically due to the local prior cold work.

It is not possible to determine whether the rubbed area failed first or the nonrubbed area failed first after the crack propagated across the bottom ligament. Nor is it possible to determine whether there was a significant time delay between the two initiation events. The most likely situation appears to be a second initiation in the rubbed area, followed by a third initiation in the nonrubbed area, but whether there was a significant time delay between the two events is unknown. The end result in either scenario is macroscopic and microscopic brittle fracture over almost the total 1.2 m (4 ft) width of the sheet once the crack started above the bolt hole. The trigger for the total failure, however, was ductile fracture in the ligament below the bolt hole. Using the direct observations of the macro and micro fracture appearance, it is concluded that the actual stress exceeded the stress for no brittle fracture in the width of the sheet above the lower bolt hole.

Thus, within these approximations, it was concluded that the allowable stress for no brittle fracture was at or above that necessary for brittle fracture.

As noted above, nitrogen not only increases the ductile-brittle transition temperature (DBTT), but also likely decreases the fracture toughness. The magnitude of these effects is unknown, but it is not unreasonable to conclude that temperature at failure may correspond to a temperature below that at which full shelf toughness is achieved. This would further decrease the allowable nominal stress.

Conclusions and Recommendations

Several factors contributing to a loss in fracture toughness and/or allowable nominal stress have been enumerated. It is the author's experience in doing failure analysis that seldom can a single factor be identified as the cause for failure, especially when the stress affects a suboptimal microstructure/chemistry. Clearly the thread imprinting and cold work were detrimental, as was the high nitrogen content.

There was clearly an assembly defect due to the nonlevel concrete pad and the forced fit of the incorrect radius of the steel panels. It is likely that shifting of the bin panels initiated the short ductile crack from the bottom bolt hole to the bottom of the first course. This crack was then sufficiently long, in conjunction with the actual stresses at this location and the fracture toughness of the material, to initiate brittle fracture in the remaining width of the panel. It is difficult, however, to assume that collapse of the structure was due only to nonmetallurgical issues. It should be remembered that brittle cracking at bolt holes was found in the two similar storage bins adjacent to the failed bin, but these bins did not collapse.

As noted above, there was a considerable variation in fracture toughness of the subsize Charpy specimens tested from the other two storage bins.

One possibility, impossible to eliminate during the investigation, was whether the sheeting in the failed tank came from more than one supplier and was therefore of presumably different nitrogen content. Each supplier presumably marked its material, but after cutting the sheets to length, it was impossible to determine the supplier of a sheet with no marking. A second issue is whether the other bins containing cracks sat on level pads. If the pads were level on the other two bins, there may not have been a sufficient increase over the nominal stress to create a critical crack length at a critical location (e.g., at the bottom of the bottom course where the nominal stress is the largest). It should also be remembered that the three tanks located at the second site did not show any evidence of cracking around the bolt holes. The downward and lateral shift of the failed tank visible in Fig. 4 is considered to be a primary cause for initiation of the ductile crack.

In summary, it is believed that the loading conditions, macroscopic and local, were the primary causes for failure in a material whose fracture toughness and transition temperature may have been significantly impaired by the high nitrogen content as well as by the detrimental effects of cold work locally created at the bolt hole.

Acknowledgment

The author would like to thank Don Otis, metallurgical consultant, for supply of Fig. 1 to 3.

References

1. P.K. Liaw and J.D. Landes, *Metall. Trans.*, Vol 17A, 1986, p 473
2. *Properties and Selection: Irons and Steels, Metals Handbook*, 9th ed., Vol 1, American Society for Metals, 1980, p 691, 697, 698
3. J. Barsom and S.T. Rolfe, ASTM STP 466, 1966, cited in R.W. Hertzberg, *Deformation and Fracture Mechanics of Engineering Materials*, 2nd. ed., John Wiley, 1983, p 344
4. R.W. Hertzberg, *Deformation and Fracture Mechanics of Engineering Materials*, John Wiley, 1983, p 281

Chloride-Induced Stress-Corrosion Cracking of Cooling Tower Hanger Rods

Wendy L. Weiss, Radian Corporation, Austin, Texas

One-quarter inch diameter 304 stainless steel cooling tower hanger rods failed by chloride-induced stress-corrosion cracking (SCC). The rods were located in an area of the cooling tower where the air contains droplets of water, below the mist eliminators and above the flow of water. The most extensive cracking was observed in the rod nuts and in the portions of the rod which were covered by the nuts. Cracking was transgranular with extensive branching, and some corrosion occurred along the crack paths. The clamping force from the nuts used on both sides of the supported member and residual stresses from thread rolling likely contributed to the stresses for the cracking mechanism, along with the stresses induced by the supported load. The external surfaces of the hanger rods were reportedly exposed to a chloride-containing atmosphere, likely due to the biocide. Type 304 stainless steel is not a suitable material for this application, and materials that resist SCC, such as Inconel, should be considered.

Key Words		
Austenitic stainless steels, Corrosion Cooling towers—hangers	Stress-corrosion cracking Water chemistry	Chlorides Biocide treatment

Alloys Austenitic stainless steels—304

Background

Four hanger rods removed from a cooling tower at a southeastern utility were examined for the cause of failure. The rods were reportedly 6 mm (0.25 in.) diam and constructed from 304 stainless steel. The maximum temperature in the cooling tower is 130 °F. The rods are located in an area of the cooling tower where the air contains droplets of water, below the mist eliminators and above the flow of water. An unfailed rod, a failed rod, a rod broken during repair, and a cracked but not completely failed rod were submitted for analysis.

Testing Procedures and Results

Surface examination

Visual and SEM examination of the fracture surfaces did not reveal any relevant information due to the degree of surface corrosion (Fig. 1).

Metallography

The failed rod and its nuts were cut longitudinally in the threaded area. The cut sections were mounted and prepared for metallurgical examination by grinding, polishing, and etching. The prepared sections were examined on a metallurgical microscope for evaluation of the failure mode, the microstructure, and the threaded region, especially under the nut.

Crack Origins/Paths. A macrograph showing the section containing the nut is shown in Fig. 2. The most extensive cracking was observed in nuts and in the portions of the rod which were covered by the nuts. Figure 3 shows photomicrographs of the cracking. The cracking is transgranular (through grain), with extensive branching and some corrosion along the crack paths.

Chemical analysis/identification

Material. A sample from one of the hanger rods was analyzed for chemical composition and was confirmed to be ASTM SA-580 Type 304 stainless steel.

Corrosion and Wear Deposits. One half of the fracture surface on the failed rod was examined in a scanning electron microscope (SEM) using energy dispersive X-ray spectroscopy (EDS) to determine the elemental constituents. The deposits on the fracture surface contained a significant amount of iron and chromium, with minor to trace amounts of silicon, sulfur, molybdenum, calcium, chlorine, nickel, potassium, and aluminum.

Fig. 1 An example of the fracture surface observed on the hanger rods.

Fig. 2 A macrograph of the prepared section. The arrow points to an area of cracking.

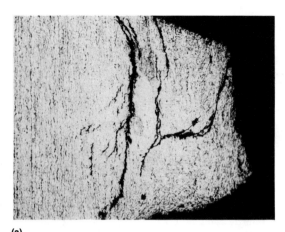
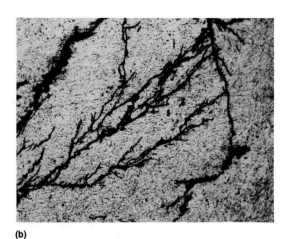

(a)   (b)

Fig. 3 Examples of the cracking observed in the hanger rods.

Discussion

The cracking in the hanger rods was due to chloride-induced stress-corrosion cracking. Aqueous chloride solutions were the most likely corrosive media promoting the stress-corrosion cracking, because the rods have crevices (threads) which trap and hold liquid. The highest concentration of liquid would likely be under the nuts on the hanger rods. The clamping force from the nuts used on both sides of the supported member and residual stresses from thread rolling likely contributed to the stresses for this cracking mechanism, along with the stresses induced by the supported load.

Conclusion and Recommendations

Most probable cause
The external surfaces of the hanger rods are reportedly exposed to a chloride-containing atmosphere, which is likely caused by the biocide. If the potential continues to exist for water droplets to come in contact with the external surfaces of the hanger rods, Type 304 stainless steel cannot be considered a suitable material for this application.

Remedial action
There are several possible alternate materials that could be used to replace the 304 stainless steel hanger rods. One possibility is a nickel-iron-chromium alloy such as Inconel 625 or Incoloy 825. High-nickel alloys such as these have a good resistance to general corrosion and stress-corrosion cracking. Switching to a nonmetallic material might also solve the problem. While the above-mentioned materials would resist stress-corrosion cracking much better than a 300-series stainless steel, issues such as mechanical properties, stresses to which the components will be subjected, and the load the components will have to support should also be considered in choosing an alternate material.

Failure of a Laminated-Paper Food Cooking Tray

David O. Leeser, Consulting Engineer, Scottsdale, Arizona

A laminated-paper microwave food tray collapsed with hot food in it. Microscopic examination of the failed tray revealed no structural or material defects. Five additional trays of like construction were also tested to determine the conditions necessary to simulate the permanent deflection of the tray handles that had occurred in the failed tray. Full distortion of the handles was obtained experimentally only by dropping a full hot tray on its end onto the floor. The test results indicated that the tray had slipped from the hand of the user.

Key Words	Food packaging	Paper products	Laminates
	Deflection	Bending	Microwaves

Background

A claim was made that a microwave food tray made of laminated paper was defective and collapsed, spilling hot food on the user.

Circumstances leading to failure

A laminated-paper tray containing a cooked microwaveable dinner was claimed to have "given way" and collapsed upon removal from a microwave oven. It was necessary to determine whether structural or material defects existed in the tray. Tests were conducted to identify the cause of the accident and apparent failure.

Specimen selection

The tray involved in the mishap was submitted as tray 1. Tray 2, with the contents still in it, was submitted for testing by the husband of the claimant. He stated that he had cooked another microwaveable dinner the preceding evening. He had removed the tray containing the food without problems. Specifically, the handles (flat sections at each end and around the tray) did not collapse (Fig. 1).

Information indicated that such trays are purchased from about five different vendors and that the same type and style of tray is used for other mi-crowave-cooked dinners, regardless of the source. Accordingly, four more packages (trays 3 to 6) were bought randomly to compare their performance properties. The handling and mishandling of the trays at temperatures are described in Table 1.

Fig. 1 Condition trays 2-5, showing normal position of all handles after cooking and removal from microwave oven

Table 1 Cooling and Handling History for Microwave Trays

Tray No.	Source	Microwave/handling procedures(a)
1	Evidence (claimant)	Cooked per instructions on package. See text for accident details and description.
2	Comparison (husband)	Cooked per manufacturer instructions. No problems encountered.
3	Purchased by investigator for test	Overcooked 10 min(b). Handles partially distorted by deliberate forceful bending at cooking temperature.
4	Same as tray 3	Same as tray 3.
5	Same as tray 3	Overcooked 10 min(b). No distortion by normal handling procedures.
6	Same as tray 3	Overcooked 10 min(b). Ends of tray (handles) were completely flattened, just like tray 1, when dropped on floor while still hot.

(a) Coupon(s) cut from each tray after emptying for deflection testing at elevated temperatures, with exception of tray 6, which had not yet been cooked. (b) Cooking time extended to ensure worst possible conditions

Testing Procedure and Results

Surface examination

Visual. Microscopic examination of the surfaces and cross sections of the trays showed a laminated construction. Adhesive bonding held the laminations together, and a plastic coating covered the outside surfaces. Bonded laminated sheets are known to be stronger than monolithic, or one-layered, sheets of the same thickness.

Tray 1, submitted as evidence of the accident, is shown in Fig. 2. The tray was examined, section by section, under a binocular microscope at magnifications up to 30×. No structural or material defects were found that could have caused the side of the tray to give way and spill hot meat and gravy down the arm and hand of the claimant.

Scanning Electron Microscope/Energy-Dispersive X-ray Analysis of the laminations showed that silica (SiO_2) was the only inorganic

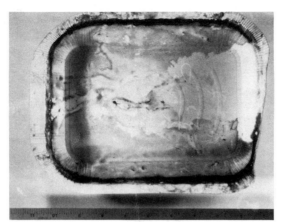

Fig. 2 Condition of the failed tray, showing one of the handles bent completely upright from its original flat position after cooking and removal from microwave oven

constituent. Silica is a common ingredient in paper products used as containers.

Simulation tests

Three different microwave ovens were used to compare the effects of heating on the six laminated-paper trays. The intent was to ensure the representation of a good cross section of microwave heating sources, because there were no problems associated with tray 2, which was cooked in the claimant's oven. One quarter cup of cold water was placed in separate glass custard cups at each of the four corners in the center section of each oven. Using a stopwatch, the time required for the water to begin boiling was recorded for each cup position. The experiment was repeated four times for each of the three ovens. All test results were within experimental accuracy.

To measure and verify the cooking temperatures of microwaveable dinners, sheathed-thermocouple meat thermometers were used at various stages in the cooking process. The thermocouples were put into the meat and gravy immediately after opening the microwave oven door. It took only a short period for the thermocouple to reach equilibrium. The maximum temperature the food could reach during microwave cooking was close to the boiling point of water. It took excessive force, by pushing with the hands, to distort the tray handles at this temperature, resulting in only limited permanent deformation.

A procedure was devised to determine the temperature at which permanent deformation would occur at low loads, similar to the load encountered when lifting the tray of food. The net weight of the food was 794 g (28 oz), and the tray weighed 42.5 g

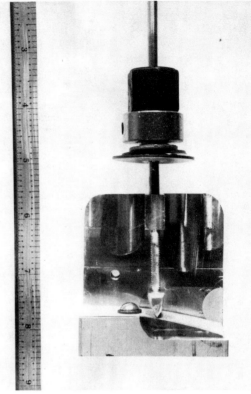

Fig. 3 Test fixture developed and used at elevated temperature to obtain permanent deformation on cantilevered coupons cut from individual trays

(1.5 oz). With a 99 g (3.5 oz) loss of water in cooking, the weight of the food-containing tray being removed from the microwave oven was 795 g (24.5 oz). The claimant's fingers under the hot pads would measure about 64 mm (2.5 in.). Hence, when both hands were supporting the 0.64 mm (0.025 in.) thick tray, as would be the case when lifting it out of the oven, the load on each hand/handle would be 90 kPa (13 psi).

A cantilever test fixture, in which specimens are supported at one end only, was built to apply this load at rising temperatures. The fixture is shown in Fig. 3 after its use in, and removal from, a convection oven where temperatures can be controlled to within ±1.1 °C (2 °F).

No measurable permanent deflection could be detected at 96 °C (205 °F), the maximum temperature obtained by the meat and gravy during microwave cooking. There also was no deflection at 167 °C (331 °F). It was not until a temperature of 205 °C (400 °F) was reached that a measurable permanent set could be observed.

Conclusion and Recommendations		

Most probable cause

The tray was not defective, nor did it collapse with the hot food in it. Rather, the tray began to slip from the right hand of the claimant, which caused it to upend and spill its hot contents on her arm and hand, whereupon she dropped the tray. Full distortion of the tray handles, which could

compare directly with the tray presented as evidence, was obtained experimentally only by dropping a full, hot tray on its end onto the floor. Test results indicated how closely the individual trays performed with respect to one another, despite the fact that they were obtained from random paper tray shipments and food-lot numbers.

Failure of a Squeeze-Clamped Polyethylene Natural Gas Pipeline

Robert E. Jones Jr, The University of Texas at San Antonio, San Antonio, Texas
Walter L. Bradley, Texas A&M University, College Station, Texas

A high-density polyethylene (HDPE) natural gas distribution pipe (Grade PE 3306) failed by slow, stable crack growth while in residential service. The leak occurred at a location where a squeeze clamp had been used to close the pipe during maintenance. Failure analysis showed that the origin of the failure was a small surface crack in the inner pipe wall produced by the clamping. Fracture mechanics calculations confirmed that the suspected failure process would result in a failure time close to the actual time to failure. It was recommended that: materials be screened for susceptibility to the formation of the inner wall cracks since it was not found to occur in pipe typical of that currently being placed in service; pipes be re-rounded after clamp removal to minimize residual stresses which caused failure; and a metal reinforcing collar be placed around the squeeze location after clamp removal.

Key Words

Polyethylenes, Mechanical properties
Stable crack propagation
Natural gas pipes

Plastic pipe, Mechanical properties
Clamping damage

Ductile fracture
Residual stress

Background

A high-density polyethylene (HDPE) natural gas delivery pipe leaked in service leading to an accumulation of gas under a residence and a serious explosion.

Applications

The 10-cm (4-in.) nominal diameter pipe had been in service for approximately ten years prior to failure. The pipe was operated as an intermediate pressure line, with a nominal pressure of 0.2 MPa (30 psi). It had been attached to the 10-cm (4-in.) steel main to extend the main. Smaller service lines carried gas from the mains to the residences. A common maintenance practice on PE pipelines is the use of a bar clamp to squeeze the pipe flat whenever the flow of gas is to be stopped for downstream maintenance. This greatly reduces the need for valve installations and eliminates the venting of large quantities of gas to clear a pipe prior to maintenance.

Circumstances leading to failure

The pipe had been in service four years before it was squeeze-clamped to stop the flow of gas and permit the replacement of a leaking T-fitting downstream. Six years and three months later the pipe began leaking at a high rate from a crack located at the site of the pinch clamp application.

Pertinent specification

The pipe was extruded from ASTM grade PE 3306 polyethylene in accordance with ASTM standard D 2513. This is a high-density polyethylene with an approximate degree of crystallinity of 0.5. It was elevated temperature strength class A and had a standard diameter ratio (SDR) of 11.5. For this SDR, the specified outer diameter is 11.43 ± 0.0229 cm (4.5 ± 0.009 in.) and the wall thickness is 1.04 ± 0.124 cm (0.409 ± 0.049 in.).

Performance of other parts in same or similar service

There have been no general problems with the performance of squeeze-clamped polyethylene pipe, though there has been some concern in the industry about the possibility of unforeseen long-term effects of the procedure.

Specimen selection

A 170-cm (66.9-in.) long section of pipe containing the longitudinal through-wall crack responsible for the leak was provided for analysis. A similar length of the predominant pipe material in new installations, ASTM PE 2306 grade, was also provided for comparison.

Visual Examination of General Physical Features

The pipe showed permanent deformation where the squeeze clamp had been applied (Fig. 1). The crack appeared to follow the permanent crease in the pipe caused by the clamping procedure.

Testing Procedures and Results

Nondestructive evaluation

Outer diameter measurements at nine points well away from the clamped section yielded an average diameter of 11.34 cm ± 0.0178 cm (4.46 in. ± 0.007 in.). This is slightly below the specified diameter of 11.43 cm ± 0.0229 cm (4.5 in. ± 0.009 in.) but should not have been a problem. Thickness measurements indicated a wall thickness between 1.02 and 1.1 cm (0.40 and 0.43 in.) with minimum thickness at the crack of 1.04 cm (0.41 in.). The minimum

Fig. 1 Photograph of as-received pipe failure. Arrows indicate ends of crack at outer surface.

allowable thickness of ASTM D 3035-83 is 1.04 cm (0.41 in.). Thus, most measurements were above the minimum allowable value and all measurements near the failure were within specified limits.

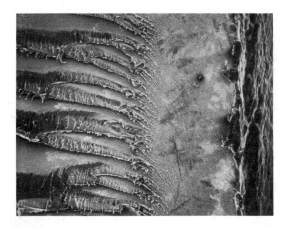

Fig. 2 Overall view of fracture surface. 15×

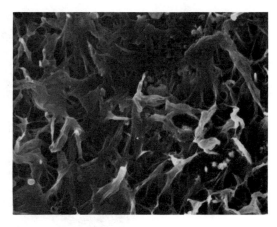

Fig. 3 First region of crack growth, zone A. 1500×

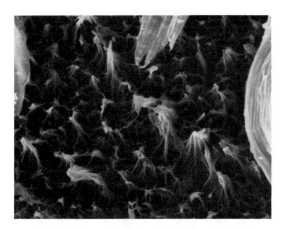

Fig. 4 Second region of crack growth, zone B. 1500×

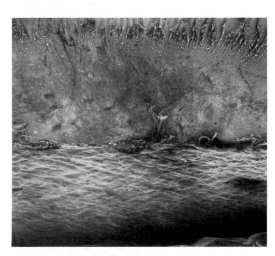

Fig. 5 Bands of crack growth indicating origin of failure—pipe inner wall at bottom. 15×

Surface examination

Macrofractography. The section containing the failure was cut from the pipe, cooled in liquid nitrogen and then broken open to expose the fracture surface. The liquid nitrogen fracture provided a clear delineation of the pre-existing service crack and the crack induced by the opening process. The service crack did not propagate completely from the inside to the outside of the pipe wall along its entire length. The through crack length on the outside of the pipe wall was 9.5 cm (3.74 in.) compared to a length of 15 cm (5.91 in.) on the inside pipe wall. The crack surface showed pronounced ridges beginning at the outer diameter of the pipe. These ridges could be traced back to a small region near the inside wall of the pipe.

The inner wall of the pipe in the clamped region contained longitudinal bands of wrinkled material and small white bubble structures in the two opposed areas where maximum strain was induced by the clamp.

Scanning Electron Microscopy. The fracture surface of the through crack could be divided into three regions (Fig. 2). The first (zone A) was a band of long fibrils and large voids (Fig. 3) which extended from the inside surface of the pipe across approximately one third of its thickness. There

was a gradual transition into a second region (zone B) of shorter fibrils and smaller voids with pronounced ridges of long tears interspersed among them (Fig. 4). The ridges pointed back to a single location, presumably the initiation site, along the inner wall of the pipe. The total growth of these two regions was 0.5 cm (0.20 in.) deep and 7.6 cm (2.99 in.) long. In the final fracture region (zone C and Fig. 5) the surface was duplex in appearance with pronounced ridges separating the two different types of surface features. One band continued with the fibrils of decreasing size while the other was relatively smooth, indicating a more brittle type fracture. Published literature indicated that fibril and void size decrease with decreasing crack growth rate. Thus the crack appears to have slowed as it propagated through the stable growth region.

The direction of the drawn PE (Fig. 4) indicated that the failure initiated at the inside surface. Examination of the indicated region of the inner wall disclosed an elliptical feature in the area of the surface damage caused by the pinch clamp (Fig. 6). This feature has characteristics of both ductile fracture and an impact or high rate failure. The material near the surface appears to have bubbled

Fig. 6 Elliptical flaw, origin of the failure. 100×

Fig. 7 Surface bubble (A) and thumbnail crack (B) in lab clamped pipe. 15×

outward, and is in fact associated with a surface bubble. There was no evidence of an inclusion, impurity concentration or manufacturing defect. No similar structures were found upon examination of the rest of the inner edge of the fracture surface. This feature appears to be the flaw which initiated the failure.

Chemical analysis/identification

Comparison of the dynamic mechanical spectra of the material near the failure and that at a distance indicated that there was no local variation in the structure of the material. Identical spectra were obtained and both samples melted at 130 °C (266 °F). Likewise, the specific gravity of material did not vary significantly. Near material had a measured density of 0.942 g/cm^3 (0.03403 lb/in.^{3}) while the distant material density was 0.940 g/cm^3 (0.03396 lb/in.^{3}). Both values are at the lower end of the range for this class of polyethylene.

Mechanical properties

Tensile Properties. Standard uniaxial tests were performed utilizing dogbone specimens specified by ASTM D 638-82a. This pipe material had a yield strength of 21.3 MPa (3.08 ksi) and an elongation at failure of >500%. The modulus was estimated to be 1034 MPa (150 ksi). These values all fall within the published ranges for these materials.

Fracture Toughness. Fracture toughness was measured using fatigue precracked compact tension specimens and the resistance curve J integral approach of ASTM E 813-81. The critical value of J, J_{IC}, at crack propagation was found to be 3.08 kJ/m^2 (3.52 ft · lbf/in.^{2}).

Stress analysis

There are several loads which could contribute to the stress state causing crack growth. These include internal residual stresses from processing, stress due to the internal pressurization of the pipe, stresses due to bending and ground loading, and residual stresses induced by the clamp application and removal.

Analytical. The hoop stress due to internal

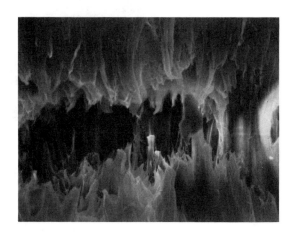

Fig. 8 Interior of ruptured bubble. 1000×

pressure is easily calculated from the relationship $\sigma_h = Pr/t$, where P is the internal pressure, r is the nominal pipe radius, and t is the pipe wall thickness. Substitution of the appropriate values yields a tensile stress in the pipe wall of 1.08 MPa (157 psi).

Experimental. Williams and Hodgkinson (Ref 1) suggest a simple means of estimating the residual stresses present in extruded plastic pipe. Rings, 2.54 cm (1.0 in.) wide were cut from the pipe and then cut diametrally into two halves. The change in diameter of the halves was measured and used to estimate the residual stresses present in the pipe due to the manufacturing process. This procedure gave an estimated maximum tensile stress at the inner wall of 1.9 MPa (276 psi).

The stresses from other sources cannot be estimated from the available information. The permanent deformation at the inner wall caused by the clamping procedure would suggest that the material at the inner wall will be in a tensile stress state upon clamp removal. This would be added to the residual stresses from manufacturing and hoop stresses due to internal pressure.

Fracture Mechanics Predictions. A J_{IC} fracture toughness has been determined for this material using fatigue precracked specimens. Since the field failure occurred with little bulk yielding around the crack tip, an equivalent K_{IC} can be found from the measured J_{IC} and linear elastic fracture mechanics used for failure prediction.

An equivalent plane strain K_{IC} is defined by the equation:

$$K_{IC} = \sqrt{J_{IC} \cdot E} \tag{Eq 1}$$

Using the measured J_{IC} of 3.08 kJ/m^2 (3.52 ft · lbf/in.^{2}) and the material modulus of 1034 MPa (150 ksi) gives a K_{IC} of 1.78 MPa$\sqrt{m}$ (1.62 ksi$\sqrt{in.}$).

If the crack was assumed to have grown stably until the duplex surface (zone C) where ductile and brittle regions were found side by side, the approximate crack dimensions at the onset of unstable growth were:

a = 0.308 cm (0.121 in.) (depth)
$2c$ = 7.62 cm (3.0 in.) (width)
t = 1.016 cm (0.4 in.) (pipe wall thickness)

The standard relationship for K_{IC} is:

$$K_{IC} = c\sigma\sqrt{\frac{a\pi}{Q}} \tag{Eq 2}$$

For an elliptical surface flaw with the above dimensions, Hertzberg (Ref 2) gives $Q = 0.8$ and Rooke (Ref 3) gives $c = 3.0$. The critical stress can then be found by rearranging and solving for σ. The resulting estimated critical stress is 4.2 MPa (609 psi). This is only slightly larger than the previously calculated total tensile stress of 2.98 MPa (432 psi) at the inner wall of the pipe. The additional stress was probably easily provided by stresses due to the clamp-imposed deformation.

The development of the crack prior to the onset of instability can also be modeled with fracture mechanics. Chan and Williams (Ref 4) proposed a relationship for stable crack growth rate as a function of stress intensity in PE piping materials:

$$K_C = 91.7(\dot{a})^{0.25} \tag{Eq 3}$$

where K is in MPa$\sqrt{m}$ and $\dot{a}$ is in m/s. Since this was a surface crack in the pipe we may substitute the previous relationship for K_C. Using the appropriate dimensions for the initial elliptical flaw at the pipe inner wall, $c = 1.12$, $a/2c = 0.4$, $Q = 2$, solving for $\dot{a}$, separating variables, and integrating gives:

$$\frac{1}{a_0} - \frac{1}{a_C} = 5.5 \times 10^{-8} t\sigma^4 \tag{Eq 4}$$

For a lifetime of 6 years (time between clamp application and failure of the pipe) $t = 1.9 \times 10^8$ s. If a service stress of 3.0 MPa (435 psi) is assumed and a critical flaw size a_C of 0.5 cm (0.20 in.) is used, then a_0 is 0.09 cm (0.035 in.). This is larger than the initial flaw size measured in the field failure, where a_0 was 0.0432 cm (0.017 in.). However the calculated value should be high since, as the crack grew, the width c varied from 1.12 to about 1.8, so that later stages of growth would have been faster, and the necessary initial flaw size smaller. Additionally, no account was taken in these calculations of the effect of residual stresses developed during pinch clamping which would have assisted crack propagation, especially in the early stages of growth.

Simulation tests

A squeeze clamp with 5.0-cm (1.97-in.) diameter pinch bars was used to clamp pipe sections for various lengths of time at 22 °C (72 °F) and 4.4 °C (40 °F). The interior of the pipes tested suffered extensive surface damage which occurred in two zones 180° apart, both approximately 1.0 cm (0.39 in.) wide and 5.0 cm (1.97 in.) long, running longitudinally along the inside of the pipe. They were centered on the points of maximum strain due to clamping. The zones contained folded material as well as tears and bubble-like protrusions on the surface. These bubbles were visible to the unaided eye as elliptical white spots and were observed on the pipe which failed in service as well as the laboratory clamped specimens. SEM examination of the surface features showed them to be ruptured bubbles (Fig. 7) with subsurface cracking (Fig. 8). The cracking indicated that the bubbles represented regions of microcracking just beneath the surface.

The laboratory squeeze-clamped specimens were cooled in liquid nitrogen and then broken along the line of maximum strain due to clamping. Examination of the resulting fracture surfaces disclosed the presence of features (Fig. 7) very similar in shape and morphology to the origin site of the in-service failure. These features were associated with bubble damage at the inner surface of the pipe. The size of the features was found to increase for longer times under the clamp as well as lower clamping temperatures.

It is of particular note that the clamping produced the bubble structures only in the particular sample of PE 3306 which failed in service. Similar tests on PE 2306 materials (the current commonly used material) produced a folded zone but none of the gross damage observed on this pipe.

Discussion

The diametral and wall thickness measurements indicated that the pipe was manufactured within the specifications of the applicable ASTM standards. Measurements did not indicate any significant deviation from specifications in the region of the failure. The consistent results of dynamic mechanical spectroscopy and density measurements, both of which are influenced largely by crystalline morphology, indicated that a localized manufacturing inhomogeneity (e.g., spot cooling or a coolant spill) was not a likely cause of the failure. Thus, any deficiency in the material's performance will be generally applicable to the entire pipe, not just the region of the failure.

SEM fractography of the service failure combined with evidence of damage in laboratory clamped pipes indicated that squeeze clamping introduced a flaw at the inside surface of the pipe. The large voids and long fibrils of the first zone of the fractured surface indicated a high initial rate of growth. This may have been due to greater extrusion-induced residual stresses and higher clamp stresses near the inner pipe surface. The rate of propagation decreased, leading to smaller voids and fibrils, as the crack propagated farther from the initiation site. At a critical point, the crack propagated rapidly the remaining distance through the pipe wall, leading to the nearly smooth bands in the duplex region near the outer surface of the pipe.

The stepped duplex region indicates that a meniscus type instability may have occurred at the final stages of crack growth. As the stably growing crack passes a depth of half the wall thickness, the local stress intensity K becomes an increasingly sensitive function of the local uncracked ligament width. Thus, small local variations in crack growth rate initiated by material inhomogeneity can result in continued, quite variable, local crack growth rates and locally unstable crack growth.

Fracture mechanics calculations clearly indicate that there was nearly sufficient loading available from known sources to grow flaws of the type introduced by squeeze clamping to the dimensions of the apparent critical flaw. The fact that the crack grew in a curved path along the fold on the outer surface of the pipe caused by application of a clamp suggested that some of the additional stresses were in fact due to residual stresses from clamping.

Conclusion and Recommendations

Most probable cause

The most probable cause of the failure was the introduction, by pinch clamping, of a small crack at the inner wall of the pipe. This initiated a process of slow stable crack growth until the crack reached a critical size at which time it propagated rapidly through the remaining thickness of the pipe.

Remedial action

Recommended remedial action included rerounding of pipes after clamping followed by installation of a reinforcing collar around the clamped location, and screening of materials currently in use for susceptibility to clamp-induced damage. Action taken by several utilities included total replacement of pipes of the grade involved in this failure whenever encountered during maintenance operations.

References

1. J.G. Williams, J.M. Hodgkinson, *Polym. Eng. Sci.*, Vol 21, 1981, p 822.
2. R.W. Hertzberg, *Deformation and Fracture Mechanics of Engineering Materials*, 2nd ed., John Wiley and Sons, New York, 1983, p 282.
3. D.P. Rooke and D.J. Cartwright, *Stress Intensity Factors*, Her Majesty's Stationery Office, London, 1976, p 88.
4. M.K.V. Chan and J.G. Williams, *Polymer*, Vol 24, 1983, p 234.

Feedwater Piping Erosion at a Waste-to-Energy Plant

Wendy L. Weiss, Radian Corporation, Austin, Texas
Brian McClave, Hartford Steam Boiler Inspection and Insurance Company, New York Branch Office

The carbon steel feedwater piping at a waste-to-energy plant was suffering from wall thinning and leaking after being in service for approximately six years. Metallographic examination of ring sections removed from the piping revealed a normal microstructure consisting of pearlite and ferrite. However, the internal surface on the thicker regions of the rings exhibited significant deposit buildup, where the thinned regions showed none. No significant corrosion or pitting was observed on either the internal or external surface of the piping. The lack of internal deposits on the affected areas and the evidence of flow patterns indicated that the wall thinning and subsequent failure were caused by internal erosion damage. The exact cause of the erosion could not be determined by the appearance of the piping. Probable causes of the erosion include an excessively high velocity flow through the piping, extremely turbulent flow, and/or intrusions (weld backing rings or weld bead protrusions) on the internal surface of the pipes. Increasing the pipe diameter and decreasing the intrusions on the internal surface would help to eliminate the problem.

Key Words

Piping, Mechanical properties	Carbon steels, Mechanical properties	Erosion
Wall thickness	Turbulent flow	Leakage
Diameters		

Background

Wall thinning and leaking were plaguing the feedwater piping in a waste-to energy plant (Fig. 1). Following the initial leaks, the piping was analyzed by the boiler manufacturer and by the water treatment consultant. The boiler manufacturer felt the leakage was due to feedwater chemistry problems and the water treatment consultant felt that there was no chemistry problem.

Pertinent specifications

Some piping had a 10-cm (4-in.) internal diameter with a nominal wall thickness of 0.86 cm (0.337 in.); some piping had smaller dimensions. The piping exhibited wall thinning to below 0.025 mm (0.001 in.) without an obvious source of the problem. Reportedly, it had taken approximately six years for the piping to develop the leaks.

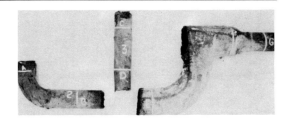

Fig. 1 The feedwater piping received for analysis. The lettered lines indicate areas from which ring sections were removed.

Performance of other parts in similar service

The same problem was occurring in all three boilers at the plant.

Testing Procedures and Results

Surface examination

Visual. The external surface of the feedwater piping contained several pad welds, but did not exhibit any significant corrosion or wear. A portion of the piping was longitudinally split for visual observation of the internal surface. The thinned areas of the piping were free of internal deposits, whereas the unaffected area had substantial deposit buildup.

Wear Patterns. Flow patterns were observed on the internal surface (Fig. 2). An eroded area on one

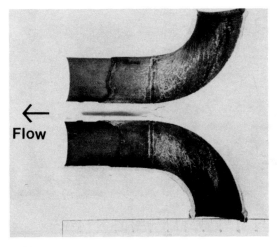

Fig. 2 A macrograph showing the internal surface of a section of piping. Notice the grooved surface in the upper section.

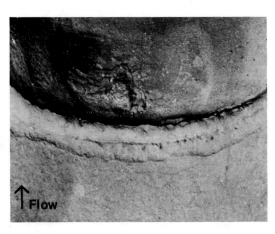

Fig. 3 The jagged weld band and erosion caused by an eddy. Approximately 0.82×

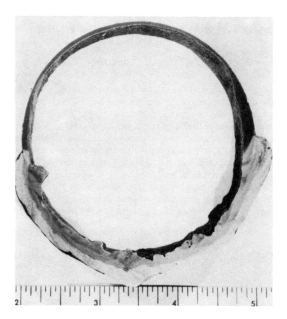

Fig. 4 The thinnest ring section removed from the feedwater piping. The lower half has been weld repaired.

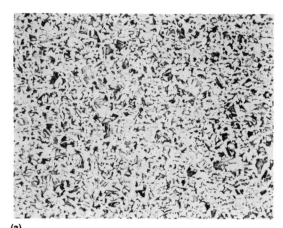

(a)

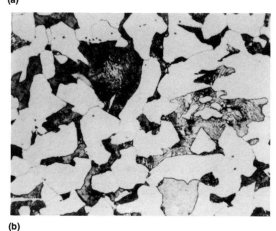

(b)

Fig. 5 Microstructure of the feedwater piping. The microstructure consists of pearlite and ferrite. Nital etch. (a) 61×. (b) 488×

side of a weld backing ring was also observed (Fig. 3).

Metallography

Microstructural Analysis. Six ring sections were removed from the feedwater piping; the areas from which they were removed are shown in Fig. 1. All six rings exhibited wall thinning. The two sections with the most severe thinning were polished for metallographic examination. The thinnest ring is shown in Fig. 4.

The microstructure observed on the ring sections was normal for carbon steel piping, consisting of a mixture of pearlite and ferrite (Fig. 5). However, the internal surface on the thicker regions of the rings exhibited significant deposit buildup while the thinner regions exhibited no deposit buildup (Fig. 6). No significant corrosion or pitting was observed on either the internal or external surfaces of the piping.

Chemical analysis/identification

Corrosion or Wear Deposits. Both bulk internal deposits and pit deposits were analyzed using energy dispersive X-ray spectrography. The deposits were found to contain a major amount of iron, and minor to trace amounts of copper, silicon, and sulfur.

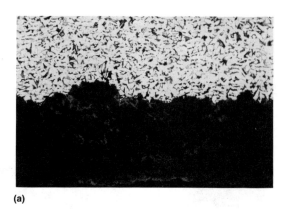

(a)

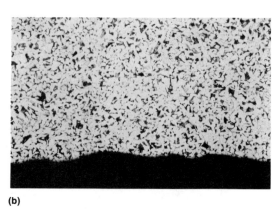

(b)

Fig. 6 Typical internal surfaces of regions on the ring sections. Nital etch. 61×. (a) Unthinned regions on the ring sections. (b) Thinned regions on the ring sections

Discussion

The lack of internal deposits on the affected areas and the evidence of flow patterns indicated that the wall thinning and subsequent failures were caused by internal erosion damage. Metal was likely removed from the inside surface to the extent that the wall could no longer support the internal water pressure. Neither general nor pitting corrosion conditions appeared to be a significant factor in the failure. However, even slightly corrosive conditions can be accelerated by erosive action, which can remove protective surface scale and provide larger amounts of corrosive species to the unprotected surface of the metal.

Conclusion and Recommendations

Most probable cause

Probable causes of the erosion include an excessively high velocity flow through the piping, extremely turbulent flow, and/or the intrusions (weld backing rings or weld bead protrusions) on the internal surface of the pipes. Suspended particles in the water or cavitation may also have aggravated the erosion, as evidenced by the jagged edge of the weld band. However, the exact cause of the erosion could not be determined by the appearance of the piping.

Remedial action

Erosion damage most likely occurs when fluid velocities exceed 2.13 m/s (7 ft/s) (Ref 1). The damage generally occurs first at locations where the direction of flow changes, such as at elbows and U-bends. Flow velocity, V, can be determined using the equation $m/3600 = \rho \cdot v \cdot A$, where m is the mass flow rate in pounds of fluid per hour, ρ is the density of water at a specific temperature in lb/ft^3, and A is the cross-sectional area of the pipe in ft^2. The flow velocity in the 10-cm (4-in.) diam pipe, with a mass flow rate of 87090 kg/h (192,000 lb/h), was estimated to be 3.23 m/s (10.6 ft/s). The use of a larger diameter pipe would reduce the flow velocities; flow through a 15-cm (6-in.) diam pipe would be approximately 1.43 m/s (4.7 ft/s).

The presence of an intrusion on the internal surface can upset the flow through the pipe and cause an eddy to form next to the intrusion, as shown in Fig. 3. The eddy erodes the pipe metal on one side (upstream) of the intrusion, but not on the other. Removing or minimizing surface discontinuities can reduce the likelihood of solid particle erosion or cavitation damage.

The plant elected to replace the failed piping with identical piping in all three boilers. The new piping has been in service for approximately one year. Since replacement, periodic ultrasonic (UT) wall thickness readings have been taken, and there have been no signs of thinning. The new piping may lack the internal intrusions that aggravated the erosion. Operating experience will determine whether or not the problem was resolved by installing the same size, material, and configured piping with possibly an improved inside surface.

Reference

1. *Metals Handbook, Vol. 10: Failure Analysis and Prevention*, 8th ed., ASM, Metals Park, OH, 1975, p 542.

Hydrogen Damage of Waterwall Tubes

Sarah Jane Hahn, Radian Corporation, Austin, Texas
Jimmy D. Wiser, The Hartford Steam Boiler Inspection & Insurance Company, Atlanta, Georgia

Waterwall tube failure samples removed from a coal- and oil-fired boiler in service for 12 years exhibited localized underdeposit corrosion and hydrogen damage. EDS and XRD revealed that bulk internal deposits collected from the tubes contained metallic copper, which can accelerate corrosion through galvanic effects and can promote hydrogen damage. Ultrasonic testing was recommended to locate tubes with severe gouging and corrosion, which are suspect locations for hydrogen damage. The source of the copper should be identified and future chemical cleaning of the boiler should address its presence in the waterwall tubes.

Key Words	Boiler tubes, Corrosion	Hydrogen embrittlement	Corrosion
	Chemical cleaning	Boilers	Galvanic corrosion

Background

Applications

The waterwall tube samples were reportedly removed from a coal- and oil-fired Riley Stoker boiler which operates at 2.0 MPa (2900 psig) and 541 °C (1005 °F) (superheater and reheater) on a swing cycle. The boiler, in service for twelve years, was chemically cleaned with acid nine years ago and one year ago.

Performance of other parts in same or similar service

The boiler had reportedly had several water-wall tube failures. Of the four boilers in the plant, this unit had had 4 to 5 times as many tube failures.

Selection of specimens

Four waterwall tube samples were received for analysis. The tube samples, labeled 1, 2, 3, and 4, were reportedly removed from a slanted section above the burners. Transverse ring sections were removed from the tube samples. The hot (fireside) side of each tube was designated as the 12:00 position.

Visual Examination of General Physical Features

The tubes were examined visually and photographed as received for analysis (Fig. 1). A transverse ring section is shown in Fig. 1(b). The internal surfaces of the tubes were rifled, and gouging on the internal surface at the 12:00 position was visible.

 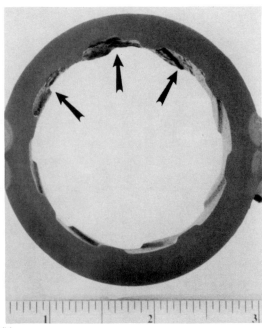

(a)   (b)

Fig. 1 Tube samples as received for analysis. (a) From top to bottom, Tubes 1, 2, 3, and 4, showing hot side (furnace side). (b) Transverse ring section removed from a tube sample for metallographic analysis. Note the internal gouging at the 12:00 position (arrows). Nital etch

Testing Procedure and Results

Metallography

Microstructural Analysis. The transverse ring sections were prepared for metallographic examination by grinding, polishing, and etching. The prepared sections were examined using a metallographic microscope to assess microstructure and internal and external surface conditions.

Copper deposits, voids, and microfissures beneath the deposits were observed along the internal surfaces of the tube samples. Metallic copper was observed on approximately half of the inner diameter (hot side). Figure 2 shows typical microfissures observed near the internal gouging. The typical microstructure of pearlite and ferrite observed in the tube samples is shown in Fig. 3. The microstructure exhibited no evidence of overheating.

Chemical analysis/identification

Corrosion or Wear Deposits. Bulk internal deposits were collected from the tubes and analyzed using energy dispersive X-ray spectroscopy (EDS) and powder X-ray diffraction (XRD). Figure 4 presents EDS results for the bulk internal deposits from Sample 1. The EDS and XRD analysis results were similar for all four tube samples. XRD identified the iron oxide magnetite (Fe_3O_4), metallic copper (Cu), the iron oxides hematite (Fe_2O_3) and wüstite (FeO), and metallic iron (Fe).

Magnetite is the normal, protective oxide found on boiler waterside surfaces. Hematite may be indicative of excessive boiler water oxygen content or minor acidic attack. Hematite is also a known product of pitting corrosion. The tube samples were obviously torch cut, and the wüstite and metallic iron most likely resulted from tube sample removal. In addition to iron and copper, EDS detected small amounts of nickel, silicon, sulfur, chromium, manganese, and zinc. Metallic copper was also visible during metallographic examination.

Mechanical properties

Hardness of the transverse ring sections ranged from 73 to 80 HRB. These hardness values are consistent with the observed microstructure.

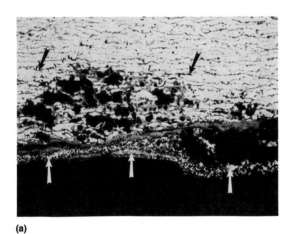

(a)

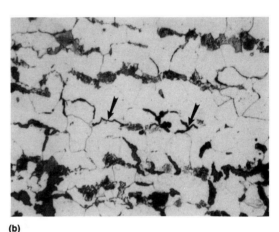

(b)

Fig. 2 Internal surface of a sample at the 12:00 position. Voids from hydrogen damage are evident beneath the internal corrosion (black arrows). Nital etch. (a) Copper deposits are visible on the internal surface (white arrows). (b) Microfissures near the internal gouging. 800×

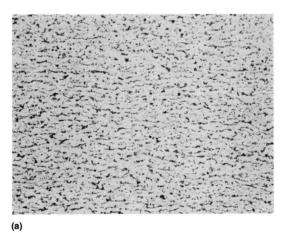

(a)

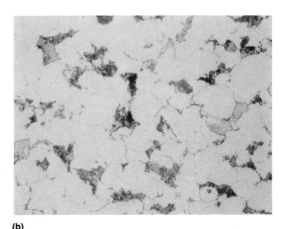

(b)

Fig. 3 Typical microstructure of pearlite and ferrite observed in the tube samples. Nital etch. (a) 100×. (b) 800×

Discussion

The metallurgical analysis of the waterwall tubes showed evidence of hydrogen damage, associated with heavy deposits and localized, underdeposit corrosion. Heavy deposits can serve as concentration sites for oxygen and acid-forming contaminants. When acidic or basic (low or high pH,

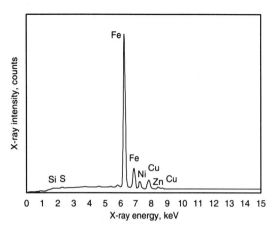

Fig. 4 EDS results for the internal deposit from Tube 1. These results are representative of all four tube samples.

respectively) conditions are present, hydrogen is generated during rapid corrosion of the inner diameter surface. Atomic hydrogen can diffuse into the metal and combine with carbon from iron carbides to form methane, or with other hydrogen atoms to form molecular hydrogen. These large gas molecules become trapped and produce very high localized stresses, leading to the formation of microfissures. When enough microfissures link up, cracks are formed. Brittle rupture can result.

The presence of copper in the internal deposits, in addition to aggravating corrosion, can promote the occurrence of hydrogen damage by helping to generate more free hydrogen at the tube/water interface. Any future chemical cleaning should address the presence of copper in the waterwall tubes. Hydrogen damage can occur in random, discrete locations in boiler tubing; it is not a general mechanism. Severe wastage on the inner diameter surface could be located by ultrasonic testing (UT) methods. (The internal rifling of the tubes must be taken into consideration.)

Conclusions and Recommendations

Most probable cause

The tube samples exhibited both localized underdeposit corrosion and hydrogen damage. The bulk internal deposits contained copper, indicating corrosion of the pre-boiler system. The presence of copper is of concern, as it is known to accelerate corrosion through galvanic effects and can promote hydrogen damage by generating more free hydrogen at the tube/water interface. The previous tube failures were most likely due to hydrogen damage caused by underdeposit corrosion and aggravated by the presence of copper.

Remedial action

Ultrasonic testing should be used to locate tubes with severe gouging/corrosion, which are suspect locations for hydrogen damage. The source of copper must be identified. Future chemical cleaning should address the presence of copper in the waterwall tubes.

Solvent-Induced Cracking Failure of Polycarbonate Ophthalmic Lenses

Edward C. Lochanski, Metalmax, Inc., Thiensville, Wisconsin

Metal-framed polycarbonate (PC) ophthalmic lenses shattered from acetone solvent-induced cracking. The lenses exhibited primary and secondary cracks with solvent swelling and crazing. A laboratory accident splashed acetone onto the lenses. The metal frames gripped approximately two-thirds of the lenses' periphery and introduced an unevenly distributed force on the lenses. To prevent future failures, it was recommended to protect PC from service environments with solvents, such as acetone; or from marking pens, adhesives or soaps which contain undesirable solvents; and to not apply excessive stress on ophthalmic lenses in the form of working or residual stresses.

Key Words

Polycarbonates,
  Mechanical properties
Acetone, Environment
Solvent-induced cracking

Safety glasses,
  Mechanical properties
Crazing
Plastic

Cracking,
  Environmental effects
Swelling
Ophthalmic lenses

Background

Applications

Metal-framed polycarbonate ophthalmic lenses were inadvertently substituted for plastic-framed polycarbonate ophthalmic lenses for use as prescription safety glasses in a chemical laboratory.

Circumstances leading up to the failure

A laboratory accident resulted in an unknown splashed solvent on metal-framed polycarbonate ophthalmic lenses. These lenses shattered, exhibiting signs of solvent swelling and crazing.

Pertinent specifications

The metal-framed ophthalmic lenses are specified to be unfilled polycarbonate which exhibits glass-like clarity. The metal frames were a custom designer series, chosen exclusively for style. These metal frames gripped approximately two-thirds of the lenses' periphery.

Performance of other parts in same or similar service

All previous solvent splashes on plastic-framed ophthalmic safety glasses did not produce any fractures. However, these lenses did exhibit signs of solvent swelling.

Selection of specimens

A new pair and a used pair of metal-framed ophthalmic lenses were examined.

Testing Procedure and Results

Surface examination

Visual. The metal-framed ophthalmic lenses exhibited primary and secondary cracks with solvent swelling and crazing.

Macrofractography. Typical new polycarbonate ophthalmic lenses exhibited well polished surfaces (Fig. 1). The failed polycarbonate ophthalmic lenses exhibited primary and secondary cracks which were associated with solvent swelling and crazing (Fig. 2). The frames are not shown in order to protect manufacturer confidentiality.

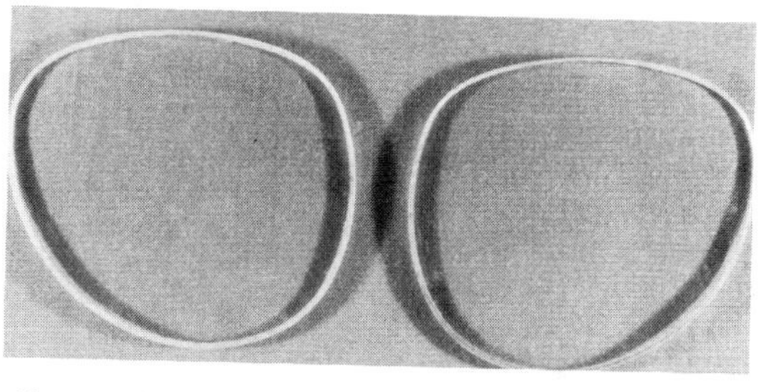

Fig. 1 Typical new polycarbonate ophthalmic lenses exhibited well polished surfaces. Approximately 0.55×

Fig. 2 The failed polycarbonate ophthalmic lenses exhibited primary and secondary cracks which were associated with solvent swelling and crazing. Approximately 0.55×

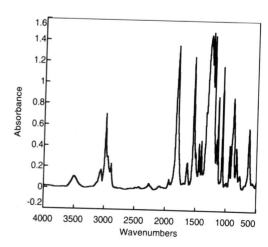

Fig. 3 The failed ophthalmic lenses were verified to be the specified polycarbonate material by Fourier-transform infrared (FTIR) analysis.

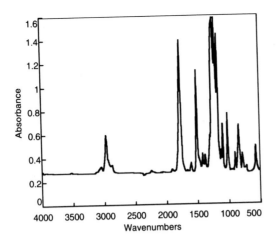

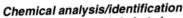

Fig. 4 A Fourier-transform infrared (FTIR) analysis of a reference polycarbonate sample

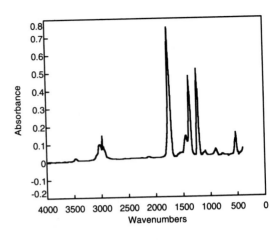

Fig. 5 The source of the splashed liquid was found to contain primarily acetone by Fourier-transform infrared (FTIR) analysis.

Chemical analysis/identification

Material. The failed ophthalmic lenses were verified to be the specified polycarbonate material by Fourier-transform infrared (FTIR) analyses (Fig. 3 and 4).

The source of the splashed liquid was found to contain primarily acetone by Fourier-transform infrared (FTIR) analyses (Fig. 5 and 6).

Discussion

Polycarbonate plastics are often used in applications which require glass-like clarity with extreme toughness. One such application is safety glasses.

Under most circumstances, when polycarbonate material is stressed in the absence of a solvent environment, no shattering will occur. The converse of this event is also true. However, when polycarbonate material is stressed in the presence of a solvent environment, solvent-induced cracking, or environmentally induced cracking, can occur, resulting in a catastrophic brittle-like failure. This phenomenon also occurs with other plastics and other solvents.

Solvent-induced cracking of the polycarbonate lenses resulted with the metal-framed ophthalmic lenses (but not the plastic-framed ophthalmic lenses) since the metal frames were exerting an uneven stress distribution on the polycarbonate

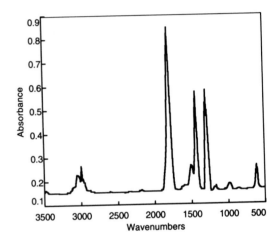

Fig. 6 A Fourier-transform infrared (FTIR) analysis of a reference acetone sample

ophthalmic lenses.

It must be emphasized that polycarbonate ophthalmic lenses are usually not designed or designated for chemical splash protection. The wearer should always consult the manufacturer's recommendations before usage.

Conclusions and Recommendations

Most probable cause

The most probable cause of failure of the metal-framed polycarbonate ophthalmic lenses is acetone solvent-induced cracking. The stress component was due to unevenly distributed force from the metal frames. The solvent component was due to an accidental splash of acetone.

Remedial action

Polycarbonate ophthalmic lenses should not be exposed to service environments with solvents, such as acetone. In addition, marking pens, adhe-

sives or soaps, which contain undesirable solvents, should never be used on the ophthalmic lenses.

Stresses on the ophthalmic lenses should be removed or reduced. Excessive stress can originate from working stress or from residual stress.

How failure could have been prevented

The failure could have been prevented by specifying and using the correct type of safety glasses, which are approved plastic-framed polycarbonate ophthalmic lenses.

Creep Failure of a Superheater Tube Promoted by Graphitization

Sarah Jane Hahn, Radian Corporation, Austin, Texas
Tom Kurtz, The Hartford Steam Boiler Inspection and Insurance Company, Concord, California

Tube 3 from a utility boiler in service for 13 years under operating conditions of 540 °C (1005 °F), 13.7 MPa (1990 psi) and 1,189,320 kg/h (2,662,000 lb/h) incurred a longitudinal rupture near its 90° bend while Tube 4 from the same boiler exhibited deformation near its bend. Metallographic examination revealed creep voids near the rupture in addition to graphite nodules. Exposure of the SA209 Grade T1A steel tubing to a calculated mean operating temperature of 530 °C (983 °F) for the 13 years resulted in graphitization and subsequent creep failure in Tube 3. The deformation in Tube 4 was likely the result of steam washing from the Tube 3 failure. Graphitization observed remote from the rupture in Tube 3 and in Tube 4 indicated that adjacent tubing also was susceptible to creep failure. In-situ metallography identified other graphitized tubes to be replaced during a scheduled outage.

Key Words			
	Carbon-molybdenum steel, mechanical properties		Boiler tubes
	Superheaters	Creep rupture	Graphitization
Alloys	Carbon-molybdenum steel—SA209 Grade T1A		

Background

Applications

The superheater tubes were from a utility boiler, a base-loaded unit that had been in service for 13 years. Normal steam conditions were reportedly 540 °C (1005 °F), 13.7 MPa (1990 psi), and 1,189,320 kg/h (2,622,000 lb/h).

Pertinent specifications

The tubes were made of SA209 Grade T1A steel tubing, 51 mm (2.0 in.) outer diameter, and 5.6 mm (0.220 in.) minimum wall thickness. The remote ring wall thicknesses ranged from 6.1 to 6.9 mm (0.241 to 0.271 in.) in Tube 3, and from 6.5 to 6.6 mm (0.255 to 0.260 in.) in Tube 4. Wall thickness near the rupture ranged from 5.8 to 6.6 mm (0.230 to 0.259 in.).

Selection of specimens

Two superheater tubes were received for analysis. The tubes were reportedly the third and fourth tubes from the front of the ninth pendant in Unit 1, and are identified as Tube 3 and Tube 4. Tube 3 contained a longitudinal rupture.

Visual Examination of General Physical Features

The tubes were examined visually and photographed as received for analysis (Fig. 1). Figure 2 shows the damaged areas observed in each tube.

Tube 3 contained a longitudinal rupture near its 90° bend, and Tube 4 exhibited deformation near its 90° bend.

Testing Procedure and Results

Metallography

Microstructural Analysis. Three transverse ring sections were cut from Tube 3; one through the rupture and one on either side of the rupture, remote from the failed area. An additional ring section was removed from Tube 4 remote from the damaged area. The ring sections were prepared for metallographic examination by grinding, polishing, and etching. The prepared sections were examined using an optical metallographic microscope to assess microstructure and internal and external surface conditions.

The typical microstructure in the failed area is shown in Fig. 3. Creep voids were evident near the rupture, in addition to graphite nodules. The typical microstructures in the remote rings from Tubes 3 and 4 are shown in Fig. 4. The remote microstructures consisted of ferrite and bainite with graphite nodules. Both tubes contained dense, uniform, internal scale approximately 76 μm (0.003 in.) thick.

Mechanical properties

Hardness was measured on the transverse ring sections. Hardness values for the remote ring sections of both tube samples ranged from 72 to 79

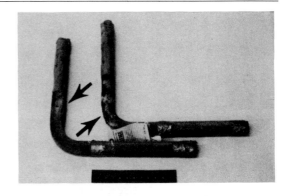

Fig. 1 The tubes as received for analysis. The arrows indicate the damaged areas. Tube 3 is to the left and Tube 4 is to the right.

HRB. Hardness values for the ruptured ring section of Tube 3 ranged from 71 to 77 HRB.

Stress analysis

Analytical. Microstructural observations suggested that the superheater tube had experienced

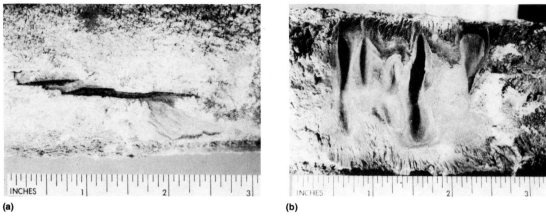

Fig. 2 Photomacrographs of the damaged areas of (a) Tube 3 and (b) Tube 4

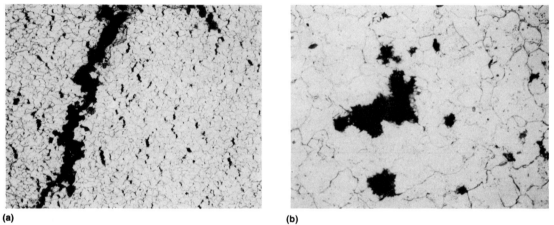

Fig. 3 Microstructure in Tube 3 at the rupture. The creep voids have linked up to form a crack. Nital etchant. (a) 177×. (b) 308×

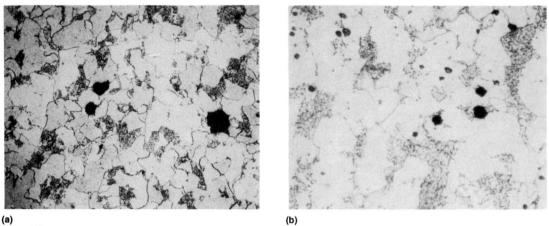

Fig. 4 Microstructure in the remote rings of (a) Tube 3 and (b) Tube 4. The microstructures consist of ferrite and bainite, and the dark spots are graphite. Nital etchant. (a) 308×. (b) 616×

temperatures from 427 to 552 °C (800 to 1025 °F). The temperature was calculated using the known wall thickness and service life on the tube, a plot of Larson-Miller parameters for the tube material, and the following assumptions:

- Only stresses due to internal pressure were present

- Operating pressure remained constant at 13.7 MPa (1990 psi)

- The original wall thickness was 6.2 mm (0.246 in.) (average wall thickness measured on ruptured ring section)

- The tube metal temperature was constant over the operating life of the tube

- The tube experienced about 8,000 hours of service each year

A mean temperature of 528 °C (983 °F) was calculated to correspond to a creep life of about 13 years using the above assumptions. This value can be considered a maximum value for constant exposure. The actual maximum temperature that the tube experienced could be higher for short-term excursions. The calculated mean operating temperature of 528 °C (983 °F) is not necessarily excessive for SA209 Grade T1A steel tubing for shorter periods of time. However, exposure to this temperature for 13 years resulted in graphitization and subsequent creep failure.

Discussion

Creep is a long-term failure mechanism depending on time, temperature, and operational stress. The creep failure of Tube 3 was aggravated by graphitization of the microstructure. Graphitization was also observed in areas remote from the rupture and in Tube 4, indicating that adjacent tubing is susceptible to creep failure.

Graphitization is the decomposition of pearlite into ferrite and carbon (graphite) that sometimes occurs in carbon or carbon-molybdenum steels subjected to moderate overheating for long periods of time. This microstructural change can embrittle steel parts, and reduce strength and creep resistance. Pearlite decomposes by either graphitization or spheroidization. Both mechanisms are temperature-dependent and occur at temperatures in the 427 to 732 °C (800 to 1350 °F) range. Graphitization usually occurs at temperatures below 550 °C (1025 °F); formation of spheroidal carbides usually occurs at higher temperatures. Tubing, piping, and other components exposed to temperatures from about 425 to 550 °C (800 to 1025 °F) for several thousand hours are most likely to be damaged by graphitization (Ref 1).

Conclusion and Recommendations

Most probable cause

Graphitization had reduced the creep resistance of the tube material, and the failure of Tube 3 was due to creep. The microstructure of the adjacent tube also exhibited graphitization. The deformation observed in Tube 4 was likely the result of steam washing from the Tube 3 failure, and was considered secondary damage.

Remedial action

If these tube samples are representative, other tubes in the superheater were probably approaching the end of their useful lives. Additional sampling or in-place metallurgical analysis was recommended to determine the extent of tubing that needed replacement. The plant subsequently requested in-place metallurgical analysis of a number of superheater pendants. The analysis identified other graphitized tubes, and allowed the plant to selectively replace damaged tubing during a scheduled outage at a significant cost savings.

Reference

1. ASM Committee on Failures of Pressure Vessels, Boilers and Pressure Piping, G.M. Slaughter, Chairman, "Failures of Boilers and Related Steam-Power-Plant Equipment," *Metals Handbook, 8th Edition, Volume 10: Failure Analysis and Prevention,* Howard E. Boyer, Ed., American Society for Metals, Metals Park, OH, 1975, p. 533

Failure of a Reformer Tube Weld by Cracking

Sergio N. Monteiro and Paulo Augusto M. Araujo, COPPE—Federal University of Rio de Janeiro, Rio de Janeiro, Brazil

An HK-40 alloy tubing weld in a reformer furnace of a petrochemical plant failed by leaking after a shorter time than that predicted by design specifications. Leaking occurred because of cracks that passed through the thickness of the weldment. Analysis of the cracked tubing indicated that the sulfur and phosphorus contents of the weld metal were higher than specified, the thickness was narrower at the weld, and the mechanical resistance of the weld metal was lower than specified. Cracking initiated at the weld root by coalescence of creep cavities. Propagation and expansion was aided by internal carburization. Quality control of welding procedures and filler metal was recommended.

Key Words		
Austenitic stainless steels Chemical processing industry Furnaces Cracking (fracturing)	Welded joints Heat-resistant steels Weld defects Chemical reactors	Crack propagation Chemical processing equipment Tubing

Alloys Austenitic stainless steel—HK-40

Background

An HK-40 alloy tubing weld in a reformer furnace of a petrochemical plant failed by leaking after a shorter time than that predicted by design specifications.

Petrochemical plants normally use reformer furnace units for a full range of processes, such as ammonia, methanol, and hydrogen production. These are gas- or oil-burning furnaces that contain heat-resistant tubing in which high-temperature reactions occur in the presence of a catalyst. The tubing construction follows a continuous arrangement, as shown in Fig. 1, typically reaching 10 to 15 m (33 to 49 ft) in height. The vertical columns in Fig. 1 are constructed by welding spun-cast pieces of tubing approximately 3 m (10 ft) in length. The connection between columns is made by welded headers. The inlet and exit of products is achieved by the use of pigtails.

In the present case, reformer tubes were made of HK-40 alloy, a 25Cr-20Ni heat-resistant austenitic stainless steel. The same stainless steel (AISI type 310-H) was used as filler metal for tungsten inert gas (TIG) welding the pieces of tubing. The tubing design life was calculated from creep data for the alloy. For instance, the American Petroleum Institute standard API-RP 530 recommends the use of Larson-Miller master curves of temperature/stress/time plots for the design of HK-40 reformer tubing. This is based on a service life of 100,000 h.

Circumstances leading to failure

After 65,000 h of operation at an average temperature of 840 °C (1545 °F) and an average internal pressure of 3.1 MPa (450 psi), a leak was detected at one of the upper weldments. The system was immediately shut down to avoid catastrophic disintegration by a torch-burning effect.

Pertinent specifications

The tubes were spun cast in ASTM A 608 grade HK-40 alloy with the specified composition limits shown in Table 1. Average dimensions of the tubes after service were:

Internal diameter	70 mm (2.8 in.)
External diameter	115 mm (4.5 in.)
Wall thickness	22.5 mm (0.89 in.)

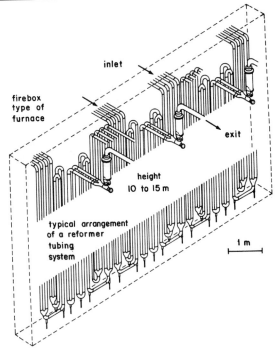

Fig. 1 Schematic of the continuous arrangement of the reformer furnace tubing

Table 1 Results of chemical analysis

Element	Composition, weight%		
	Tubing	Weld metal	HK-40 alloy requirements (ASTM 608)
Carbon	0.414	0.413	0.35-0.45
Chromium	24.2	23.4	23-28
Nickel	18.2	18.8	19-22
Molybdenum	0.39	0.22	0.5 (max)
Silicon	1.05	1.63	0.5-2.0
Manganese	1.40	1.14	1.5 (max)
Sulfur	0.03	0.09	0.04 (max)
Phosphorus	0.04	0.08	0.04 (max)

These values are within 3% of the nominal dimensions of a new tube. According to the manufacturer, the tubes were TIG welded with an AWS ER 310 H type filler metal.

Performance of other parts in same or similar service

Recorded experience from other reformer furnaces in similar industrial units indicates that tubing failure is invariably by a creep mechanism under normal operating conditions. Failure usually occurs at the tube weldments, the most vulnerable weld being the hottest one, which is subjected to greater creep damage.

Specimen selection

A set of tubing, including one failed weld and the corresponding tube pieces, was sent for examination. Specimens from the weld and adjacent tube pieces were cut and marked, as shown in Fig. 2. Smaller metallographic samples and mechanical testing specimens were prepared from the failed weldment and the tube pieces.

Visual Examination of General Physical Features	Visual examination showed that the internal and external tubing surfaces were dark and rough in appearance. This was certainly a consequence of prolonged exposure at high temperature.
Testing Procedure and Results	**Nondestructive evaluation** **Radiography.** The tubing welds were routinely inspected by X-ray techniques for crack detection during shutdowns for maintenance. Analysis of X-ray results is basically a qualitative technique that permits some evaluation of the potential of the component to continue in service throughout the next production campaign. **Liquid penetrant inspection** of all specimens received for examination revealed circumferential cracks at the weld root in the inner surface. Small cracks were also detected in the tube pieces.

Surface examination

Visual. In addition to the dark and rough appearance, close examination of the tubing inner surface revealed a large crack extending around the failed weld.

Weld Characteristics. The weld appearance is shown (Fig. 3) in a longitudinal section deep etched with aqua regia. Note the depression caused in the thickness by the welding procedure and the crack originating from the weld root toward the outer surface.

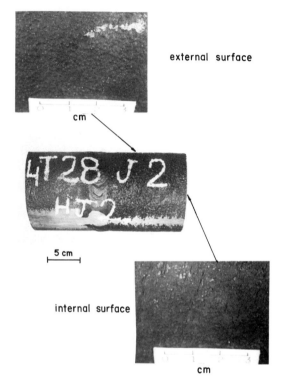

Fig. 2 Specimen taken from a failed tube weldment. Enlarged views of the outer and inner surfaces are shown.

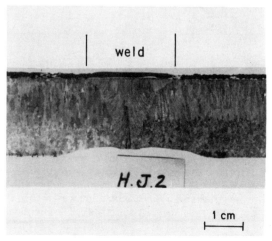

Fig. 3 Macroscopic view of a deep etched longitudinal section of the weldment.

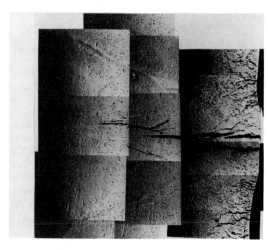

Fig. 4 Polished and lightly etched view of the longitudinal section across the failed weld.

Fig. 5 Crack formed by coalescence of creep cavities at carbide interfaces. Cavity alignment follows dendritic arms.

Fig. 6 Large crack in the weld outlined by a carburized layer.

Table 2 Results of tensile testing

Specimen	Temperature		Yield strength		Ultimate tensile strength		Uniform elongation, %	Total elongation, %
	°C	°F	MPa	ksi	MPa	ksi		
Failed weld		RT	302	43.8	369	53.5	0.8	0.8
	840	1545	134	19.4	174	25.2	6.2	15.2
Adjacent tube		RT	364	52.8	486	70.5	3.6	3.6
	840	1545	154	22.3	190	27.6	5.5	43.2
Unused tube		RT	367	53.2	510	73.9	9.5	9.5
	840	1545	275	39.9	278	40.3	3.5	16.3
ASTM specification for HK-40		RT	245-340	35.5-49.3	450-590	65.3-85.6	...	10
	840	1545	125	18.1	180	26.1	...	18

Metallography

Microstructural Analysis. The complete view of the polished and lightly etched longitudinal section across the failed weld is shown in Fig. 4. The main features are the large cracks that start at the base of the weld root and run perpendicular to the tube surface. The inner surface of the tube near the weld root also discloses a pattern of many small cracks. Another significant feature shown in Fig. 4 is the large number of cavities.

Crack Origins/Paths. The large cracks responsible for the weld failure (Fig. 4) were clearly formed by coalescence of cavities. This is actually a process of alignment, followed by interconnection of creep cavities (Fig. 5). Because the creep cavities are preferentially nucleated at carbide interfaces, the cracks tended to follow the direction of dendritic arms where carbide precipitation had occurred. Thus, the crack path tended to be perpendicular to the tube surface.

Chemical analysis/identification

Material and Weld. The results of chemical analysis performed on samples of the tube pieces are summarized in Table 1. The composition of the tube alloy was within the ASTM 608 specification for HK-40. The chemical analysis performed on the weld metal is also included in Table 1. The sulfur and phosphorus contents are significantly larger than those specified.

Surface Layers. Both the outer and inner surfaces of the tubing were covered with a carburized layer. Furthermore, the large cracks in the weld were also outlined by a carburized layer, as illustrated in Fig. 6.

Associated Environments. The reformer tubes and related welds were subjected to high-temperature exposure to carbon-rich atmospheres. Hydrocarbon vapor and combustion gases were in permanent contact with the tubing inner and outer surfaces, respectively.

Mechanical properties

Hardness. Microhardness measurements were performed along the wall thickness of the weld and adjacent pieces of tube. The weld displayed Vickers hardnesses decreasing from 330 HV at the outer surface to 250 HV at the inner surface (weld root). The tube had values from 240 HV at the outer surface to 280 HV at the inner surface. An unused HK-40 tube from the same manufacturer exhibited an almost constant hardness of 225 HV across the thickness.

Tensile Properties. Table 2 shows the values of the main tensile properties at room temperature and 840 °C (1545 °F) for the weldment and adjacent piece of tube. It is also shown that the corresponding property values specified by ASTM as well as the tested values for an unused HK-40 tube of the same manufacturer. Note that the weldment exhibited total strength that was below specification both at room and high temperature. Its low ductility at room temperature was apparently a consequence of the cracking in the weld metal.

Discussion

Figure 4 clearly shows that the HK-40 reformed tubing failed by a cracking process initiating at the weld root. The mechanism of crack nucleation was related to the coalescence of creep cavities. Under the average service conditions of 840 °C (1545 °F) and 3.1 MPa (450 psi), this alloy was subjected to diffusional creep effects that caused cavities at carbide interfaces after 65,000 h. Cracking was facilitated by several factors:

- Thickness reduction at the weld root (Fig. 3)
- Alignment of carbides in the direction of dendritic arms (Fig. 5)

- Carburized layer that surrounded the expanding cracks (Fig. 6) as a consequence of penetration of hydrocarbon vapor
- High sulfur and phosphorus content in the weld metal (Table 1), which helps to form second phases and sites for cavity nucleation
- Relatively low values of the hardness and tensile strength (Table 2), associated with a decrease in the mechanical resistance of the weld.

Consequently, large circumferential cracks developed and eventually penetrated through the weld thickness, resulting in failure by leaking.

Conclusion and Recommendations

Most probable cause

The failure of the reformer tube weld was caused by creep cracking assisted by microstructural factors and other factors related to the welding procedure and characteristics.

Remedial action

The damaged tubing column was replaced with a new one. This is a common procedure in reformer furnaces. It was also recommended that all factors shortening tube life be eliminated. This includes chemical control of the filler metal and quality assurance of the weldments, particularly through the use of good welding procedures.

How failure could have been prevented

The reformer tubing could certainly have been functional for the predicted lifetime of 100,000 h if a more rigorous quality-control program had been practiced concerning welding procedures and the selection of the filler metal.

Hot Corrosion of Stage 1 Nozzles in an Industrial Gas Turbine

Sarah Jane Hahn, Radian Corporation, Austin, Texas

The first-stage nozzles of a high-pressure turbine section of an industrial gas turbine exhibited leading-and trailing-edge deterioration. The nozzles were made of X-40, a cobalt-base alloy, and were aluminide coated. Failure analysis determined that the deterioration was the result of hot corrosion caused by a combination of contaminants, cooling-hole blockage, and coating loss.

Key Words	Cobalt-base alloys, corrosion	Nozzles, corrosion	Hot gas corrosion
	Gas turbines		
Alloys	Cobalt-base alloy—X-40		

Background

The first-stage nozzles of a high-pressure turbine (HPT) section of an industrial gas turbine exhibited deterioration of the leading and trailing edges.

Circumstances leading to failure

This particular set of nozzles had operated for 12,000 to 14,000 h. The damaged nozzles were found during an inspection for possible causes of significant power deterioration over the period of a year. Most of the power degradation appeared to be occurring in the HPT section. Damage of this type had not been seen before in this engine.

The compressor section of the engine was washed on-line with water every day. Every other day, a chemical wash solution was added to the water. The chemical wash was followed by a plain-water rinse. The water wash chemicals vaporized before they reached the stage 1 HPT nozzles. The highest-boiling-point constituent boils at roughly 315 to 370 °C (600 to 700 °F). The engine was also washed each time it was taken off-line for maintenance. The gas turbine was brought up to cranking speed with the starter and injected with a water/chemical mixture. This chemical wash was normally followed by two rinse charges.

Pertinent specifications

The temperature of the gas entering the first-stage nozzles was not measured directly, although it was estimated to be 1095 °C (2000 °F). Nozzle metal temperature was estimated to be 760 to 815 °C (1400 to 1500 °F). The gas turbine was fueled with natural gas. Both the fuel and the inlet air were filtered. The nozzles were made of X-40, a cobalt-base alloy, and aluminide coated.

Specimen selection

Three nozzle segments were submitted for failure analysis.

Visual Examination of General Physical Features

The as-received nozzle segments are shown in Fig. 1. The nozzles exhibited leading- and trailing-edge deterioration, and the coating appeared to be damaged and/or missing in several areas (Fig. 2).

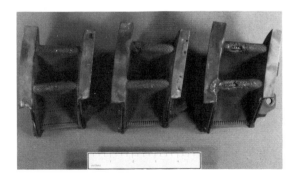 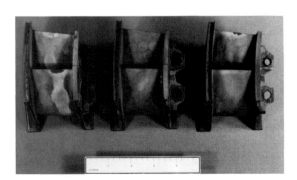

Fig. 1 As-received nozzle segments. (a) Leading edges. (b) Trailing edges

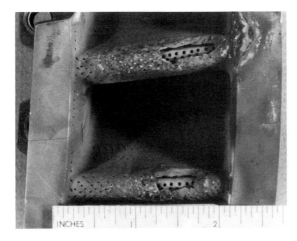

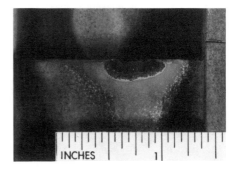

Fig. 2 Representative leading (a) and trailing (b) edges of the nozzles

Testing
Procedure
and Results

Surface examination

Scanning Electron Microscopy/Fractography. Figure 3 shows an area exhibiting missing coating. Examination of the nozzle segments by scanning electron microscopy (SEM) also revealed some blockage of the cooling holes (Fig. 4). These holes appeared to be partially filled with deposit, which was typical of all three segments.

Metallography

Microstructural Analysis. Representative cross sections were prepared for metallographic evaluation by grinding, polishing, and etching. Cooling holes are shown in cross section in Fig. 5 and 6. Metallography revealed surface corrosive attack.

Chemical analysis/identification

Material. The composition of the base metal was analyzed by quantitative chemical analysis. The chemical composition met the requirements for X-40, as shown in Table 1.

Coatings or Surface Layers. Sections were cut from deteriorated areas of the leading and trailing edges (Fig. 7). Surface deposits on some of the sections were analyzed *in situ* using energy-dispersive spectroscopy (EDS). This technique was also used on the metallographic sections to analyze surface deposits and base metal constituents.

Corrosion or Wear Deposits. Analysis by EDS detected contaminants such as silicon, phosphorus, sulfur, chlorine, and calcium in the cooling hole deposits, along with calcium, phosphorus, zinc, and chlorine in the corrosion products.

Associated Environments. A cooling tower was reportedly located near the turbine. Analysis of the cooling tower water showed that it contained calcium carbonate, sulfates, sulfites, and chlorides.

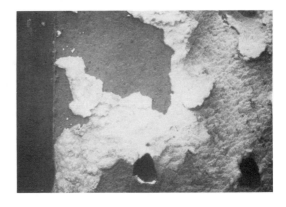

Fig. 3 SEM micrograph showing missing coating in a deteriorated area. 12.4×

Fig. 4 SEM micrographs showing partial blockage of cooling holes. (a) ~24.4×. (b) ~36.6×

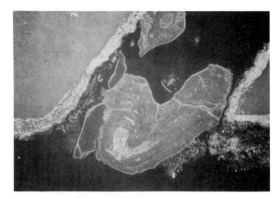

Fig. 5 Cross section through the deteriorated leading edge of a nozzle, showing cooling-hole blockage. (a) 10.37×. (b) 61×

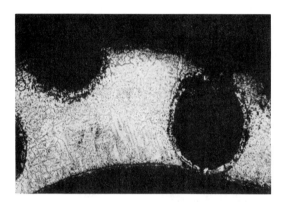
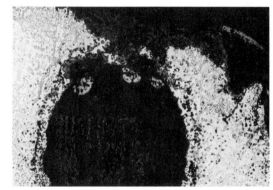

Fig. 6 Micrographs showing surface attack in a deteriorated/distressed area. Etched with Marble's reagent. (a) 30.5×. (b) 61×

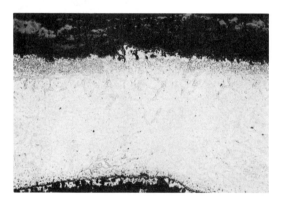
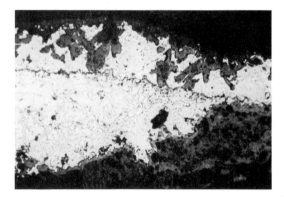

Fig. 7 Micrographs showing a deteriorated area of a nozzle. The surface corrosion is evident. Etched with Marble's reagent. (a) Leading edge. 61×. (b) Trailing edge. 244×

Discussion   Alloy surfaces exposed to high temperatures and gas environments can become coated with foreign deposits (sulfates, oxides, chlorides, etc.) from combustion air or fuel contaminants. Corrosion products can react with or dissolve into the deposits, and these surface deposits usually cause the alloys to be attacked by the gas at increased rates. The deposit-modified, gas-induced, high-temperature degradation of alloys is called hot corrosion.

Blades and vanes located in the hot gas path of a turbine are subject to hot corrosion. A layer-type corrosion (type II), characterized by an uneven scale/metal interface and the absence of subscale

sulfides, occurs from 650 to 705 °C (1200 to 1300 °F). Above 760 °C (1400 °F), a nonlayer-type corrosion (type I) occurs. Metallography revealed a layer-type corrosion and uneven scale/metal interfaces on the nozzles, corresponding to type II hot corrosion.

Cooling holes provide film cooling that allows the nozzle material to withstand gas stream temperatures. Cooling hole blockage results in excessive metal temperatures. The elevated temperatures promote cracking or spalling of the coating because of thermal stresses and lead to hot corrosion of the base metal. The durability of aluminide coatings depends on their adherence. If the coating is damaged or removed, oxides will form on the exposed base metal surface. Thermally induced stresses also cause cracking and spalling of these oxide scales.

Conclusion and Recommendations

Most probable cause

The combination of contaminants, cooling hole blockage, and coating loss resulted in hot corrosion of the nozzles. The rinse water and the nearby cooling tower were possible sources of contaminants. The nozzle cooling holes contained deposits that prevented proper film cooling and resulted in overheating. The overheating probably contributed to coating loss.

Remedial action

The source of water-borne contaminants should be identified, and their contact with the turbine reduced or prevented.

Table 1 Results of chemical analysis

Element	Composition, % Base metal	X-40 nominal requirements(a)
Carbon	0.51	0.50
Manganese	<0.01	0.75
Phosphorus	0.007	NR
Sulfur	<0.005	NR
Silicon	0.54	0.75
Nickel	12.53	10.50
Chromium	25.50	25.50
Molybdenum	<0.01	NR
Cobalt	bal	54.00
Tungsten	7.92	7.50
Titanium	0.05	NR
Aluminum	0.06	NR

(a) NR, no requirement

Corrosion Failure of Stainless Steel Components During Surface Pretreatment

R.K. Dayal and J.B. Gnanamoorthy, Metallurgy Division, and G. Srinivasan, Reactor Group, Indira Gandhi Centre for Atomic Research, Kalpakkam, India

Two AISI type 316 stainless steel components intended for use in a reducer section for sodium piping in a fast breeder test reactor were found to be severely corroded—the first soon after pickling, and the second after passivation treatments. Metallographic examination revealed that one of the components was in a highly sensitized condition and that the pickling and passivation had resulted in severe intergranular corrosion. The other component was fabricated from thick plate and, after machining, the outer surface represented the transverse section of the original plate. Pickling and passivation resulted in severe pitting because of end-grain effect. Strict control of heat treatment parameters to prevent sensitization and modification of pickling and passivating conditions for machined components were recommended.

Key Words

Austenitic stainless steels, corrosion
Fast nuclear reactors
Pickling
Finishing baths, environment

Intergranular corrosion
Nuclear reactor components
Sodium-cooled reactors
Breeder reactors

Passivation
Pitting (corrosion)
Pipe fittings

Alloys

Austenitic stainless steel—316

Background

Two stainless steel components fabricated from AISI type 316 stainless steel were found to be severely corroded—the first soon after pickling, and the second after passivation treatments.

Circumstances leading to failure

A reducer section in the second wall (double envelope) of sodium piping in a fast breeder test reactor was to be made from two halves and assembled over an inner core pipe by welding. Because of stringent weld fit-up tolerances, each half was to be made from a full conical reducer by slitting it into two unequal parts and using the larger half after machining to the required size. The conical reducers from which the two halves were made were machined from two different stocks:

- A 135 mm (5.4 in.) diam forged bar (component A)
- A round trepanned from a 110 mm (4.4 in.) thick plate (component B). In this case, the reducer was machined through the thickness of the plate

To relieve residual stresses caused by machining, the components were annealed prior to finish machining. During the annealing treatment, after a 30 min soak at 1050 °C (1920 °F), component A was furnace cooled at a rate of 120 °C/h (215 °F/h), whereas component B was air cooled.

Subsequently, the parts were pickled in a solution of 10 to 20% HNO_3, 2 to 3% HF, and water at 60 °C (140 °F) for 20 min and passivated in a solution of 10 to 20% HNO_3 and water at 60 °C (140 °F) for 30 min (Ref 1).

Soon after the surface treatment, component B developed severe pitting attack on both the inside and outside surfaces, but no pitting occurred on the cross-sectional surfaces (Fig. 1b). Component A did not suffer any pitting attack, but the surface had a dull gray finish (Fig. 1a).

(a)

(b)

Fig. 1 Components after pickling and passivation treatment. (a) Component A. (b) Component B. 0.6×

Pertinent specifications

The reducer section was fabricated from AISI type 316 stainless steel, which has a typical chemical composition of 0.054% C, 16.46% Cr, 12.43% Ni, 2.28% Mo, 0.64% Si, 1.69% Mn, 0.025% P, and 0.006% S, with the balance iron.

Testing Procedure and Results

Metallography

Optical microscopic examination of etched cross sections of the two components revealed that component A was in a sensitized condition and that intergranular corrosion attack up to about 20 μm (0.8 mil) in depth had occurred on both components, from inside and outside surfaces (Fig. 2-4). However, component B did not show any sign of sensitization (Fig. 5). It was also noted that the grain size of component B was significantly larger than that of component A. The inclusion content of component B, as measured on a cross-sectional surface per ASTM E-45 (Ref 2), was found to be low (A type = 0.11, B type = 0.31, C type = 0.05, D type = 0.63; A + B + C + D = 1.10).

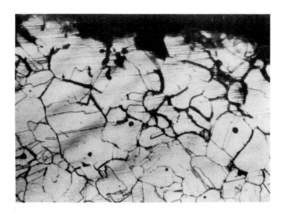

Fig. 2 Micrograph of component A on thickness cross section, near inside edge. Etched in Marble's reagent. 165×

Fig. 3 Micrograph of component A on thickness cross section, middle of thickness. Etched in Marble's reagent. 165×

Fig. 4 Micrograph of component A on thickness cross section, near outside edge. Etched in Marble's reagent. 165×

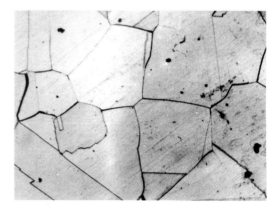

Fig. 5 Micrograph of component B on thickness cross section. Etched in Marble's reagent. 165×

Discussion

Both components were attacked by corrosion after the surface treatment, though in different forms. Component A, which was in a sensitized condition, underwent intergranular attack. The cooling rate of 120 °C/h (216 °F/h) after the postmachining annealing treatment was not adequate to prevent sensitization (Ref 3). Component B was air cooled after the annealing treatment and did not undergo sensitization. This component, which had been machined from the cross section of a thick plate, experienced severe pitting, primarily because of the exposure of smaller end grains and, perhaps, because of the larger inclusion content on the exposed surface.

Conclusion and Recommendations

Most probable cause

Component A was severely sensitized; the pickling and passivation processes resulted in severe intergranular corrosion. Component B was fabricated from a thick plate; after machining, its outer surface represented the transverse section of the original plate. Because of the end-grain effect, in which more inclusions were exposed on the surface, pickling and passivation processes resulted in severe pitting.

Remedial action Component A was reused after removing the intergranularly attacked layer and conducting re-solution annealing. Component B was reused after polishing to remove the pits and the severe machining marks. Subsequently, pickling and passivation treatments were carried out by carefully controlling composition, temperature, and time against the specified parameters (Ref 1) described in the section "Circumstances Leading to Failure."

How failure could have been prevented
First, the heat treatment parameters should be strictly controlled to avoid sensitization (air cooling, in this case, appears to provide a sufficient cooling rate to avoid sensitization in austenitic stainless steels). Second, the pickling and passivation procedures normally adopted for rolled components are not adequate for components machined through the thickness from thick plates, because the surfaces in these parts consist of end-grain material, which is generally smaller in grain size and has a larger inclusion content. Therefore, specific parameters, with respect to composition and duration of pickling and passivation, may have to be developed for such materials.

References
1. "Specification for Surface Treatments of Steel Parts—Fast Breeder Test Reactor," Specification No. 0-1732-1, Reactor Research Centre (Kalpakkam, India), 1974
2. "Recommended Practice of Determination of Inclusion Content of Steel," ASTM E-45-76, *Annual Book of ASTM Standards, Part II*, ASTM, 1977
3. R.K. Dayal and J.B. Gnanamoorthy, *Corrosion*, Vol 36, 1980, p 104

Cracking During Forging of Extruded Aluminum Alloy Bar Stock Material

A.K. Das, Aircraft Design Bureau, Hindustan Aeronautics Ltd., Bangalore, India

During the preproduction stages of forging, an initial batch of 50 mm (2 in.) diam Al-4Cu alloy (L77) extruded bar stock material was found to be cracking randomly. Failure analysis was conducted to determine the metallurgical factors underlying the phenomenon. Microexamination of sections across the defects revealed intergranular cracks tracing a path of round, segregated particles and oxide film discontinuities. The segregated particles were rich in copper. It was concluded that the cracking was the result of segregations occurring in poor-quality raw material. The source of segregation was suspected to be the use of improperly made master alloys. Use of improved melting techniques and proper master alloys was recommended.

Key Words			
Aluminum-base alloys		Segregations	Forgings
Master alloys		Cracking (fracturing)	

Alloys   Aluminum-base alloy—L77

Background

During the preproduction stages of forging, an initial batch of 50 mm (2 in.) diam Al-4Cu alloy extruded bar stock material was found to be cracking randomly. The quality of the raw material had been microstructurally checked and strength properties in the heat-treated condition had been evaluated prior to acceptance for forging.

Circumstances leading to failure

The majority of the first 50 forgings made from the same batch exhibited visible cracks and random burstings in both the preforging and finish forging stages (Fig. 1).

Pertinent specifications

The forging stock was an Indian alloy that is similar to British BSL L77 alloy.

Specimen selection

Three sections across the cracked regions of the parts (one each from the upsetting, preforging, and final forging stages) were selected for low-power optical and high-power scanning electron microscopy (SEM) examination. A specimen was cut from the supplied raw material for detailed and comparative studies.

Visual Examination of General Physical Features

Examination of the failed parts revealed several fold-type discontinuities/cracks and radial bursts (Fig. 1).

Testing Procedure and Results

Surface examination

Macrofractography. Macroexamination of the cross sections of the extruded bar and of the preforged failed sample under a low-power stereoscopic microscope showed the presence of oxide films confined to the coarse grain envelope (Fig. 2).

Metallography

Microstructural Analysis. Microexamination of sections across the defects revealed intergranular cracks tracing the path of round, segregated particles and oxide film discontinuities (Fig. 3). In addition, the structure showed pockets of gas voids in chain form, which appeared as cracklike dis-

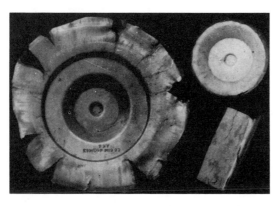

Fig. 1 Preforged and finish forged parts, showing cracks and bursts at various locations

Fig. 2 Fractured sections of the raw material (left) and the upset bar stock, showing oxide inclusions confined to coarse grain envelopes

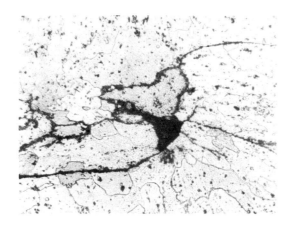

Fig. 3 Optical micrograph of a polished and etched section through the cracked zone of a finish forging, showing black, segregated particles along the intergranular cracking path

(a)

(b)

Fig. 4 Macrographs of the upset bar stock (a) and the raw material (b), showing numerous gas voids in chain form

Table 1   Results of chemical analysis

Element	Composition, %	
	Bar stock	L77 specifications
Copper	3.96-4.09	3.9-5.0
Magnesium	0.36-0.56	0.2-0.8
Silicon	0.62-0.76	0.5-0.9
Manganese	0.52-0.61	0.4-1.2
Titanium + zirconium	0.017	0.2 (max)
Iron	0.28-0.44	0.5 (max)
Nickel	0.06	0.2 (max)
Zinc	0.10-0.17	0.2 (max)
Lead	0.002	0.05 (max)
Tin	0.0023	0.05 (max)
Chromium	0.08	0.2 (max)
Aluminum	bal	bal

continuities (Fig. 4a). Microexamination of the raw material confirmed the presence of gas voids, mostly in chain form, and zones of segregation (Fig. 4b).

To identify the nature of the segregated particles and to differentiate them from other microstructural constituents that are normally present in the heat-treated Al-4Cu alloy structure (i.e., $Fe-Mn_3$, $Si-Al_{12}$, $Cu-Al_2$, Cu_2-Mg_8, and Si_6-Al_5), a differential etching test with 10% phosphoric acid was performed on the microspecimens. Although normal constituents were easily identified by changes in their respective colors, the remaining particles were unchanged, suggesting the presence of segregation.

SEM examination of the segregated particles concentrated along the crack path showed them to be dark in color against the white round particles of $Cu-Al_2$ (Fig. 5). To confirm this, an etched specimen of a satisfactorily heat-treated Al-4Cu alloy extruded material (which microexamination showed to be free of segregation) was examined by SEM; no dark-colored particles were visible (Fig. 6).

Chemical analysis/identification

Chemical analysis of the forging stock material showed it to conform to L77 specifications (Table 1).

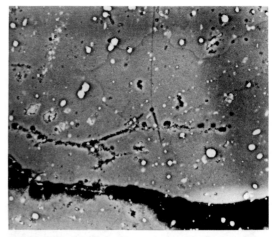

Fig. 5 SEM micrograph of the cracked area in Fig. 3, showing black, segregated particles in the crack path. White particles are normal $Cu-Al_2$ constituents.

Elemental analysis using a microprobe identified the segregations as copper-base particles (Fig. 7).

Fig. 6 SEM micrograph of a satisfactorily heat-treated Al-4Cu alloy extruded bar, showing absence of black, segregated particles

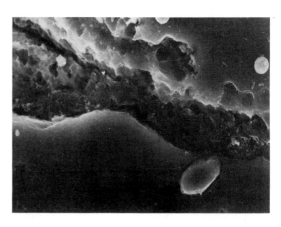

Fig. 7 X-ray map obtained with an SEM microprobe, showing copper-rich particles in the crack path

Discussion

Macroexamination and microstructural analysis clearly showed that the unusual cracking and radial bursting of the forgings was associated with the poor quality of the raw material, which contained harmful segregation of copper-rich particles, oxide film, and gas voids caused by improper melting practices. The high concentration of segregation zones was the primary cause of failure. The raw material had been examined and found acceptable prior to forging.

Stray zones of segregated copper-rich particles probably could not be differentiated from the normal constituents under low-power optical microexamination (at 100×). Therefore, in-depth SEM and microprobe examination was not thought necessary during routine acceptance testing.

Conclusion and Recommendations

Most probable cause

The basic cause of random cracking and bursting of the preforged and finish forged parts was the segregation of copper-rich particles that were present in the raw material. These segregated particles provided the matrix with sites for inherent brittleness during forging operations. The presence of oxide film, gas voids in chain form, and coarse grain envelopes also promoted cracking.

Remedial action

It was recommended that the supplier improve melting techniques, including the quality of master alloys, to avoid formation of harmful segregation, oxide films, and gas voids. It was also recommended that the forging shop perform additional quality control checks by ultrasonic testing of the extruded forging stock and by macro- and microexamination of slices cut from the front and back ends of the stock.

Failure of Compound Bow Handle Risers

George M. Goodrich, Taussig Associates, Inc., Skokie, Illinois

Compound bow handle risers that had failed in service and during assembly along with an unassembled riser were submitted for analysis. The risers were die cast from magnesium-base alloy AM60A. Inspection of the failed risers and metallurgical investigations conducted on the stock riser revealed the presence of cold shuts at the same site in all specimens. It was recommended that all risers be thoroughly inspected and that the bow company work with their die casting shop to design a mold with acceptable filling characteristics.

Key Words			
	Magnesium-base alloys	Casting defects	Die castings
	Die casting dies, design	Cracking (fracturing)	Sporting goods
Alloys	Magnesium-base alloy—AM60A		

Background

Compound bow handle risers that had failed in service and during assembly along with an unassembled riser were submitted for analysis. The risers were die cast from magnesium-base alloy AM60A (ASTM B94).

Testing Procedure and Results

Surface examination

The riser removed from stock was examined visually and using a stereomicroscope (up to 45×) to evaluate the surface of the riser at the sites where fracturing had occurred on the failed risers. Three of the broken risers had fractured at the thumb position on the riser grip. Figure 1 shows the stock riser and a broken riser. Examination of both sides of the grip revealed surface flow lines visibly present at the thumb position (Fig. 2). Flow lines can be an indication of cold shuts, which result when two solidification fronts fail to fuse during the casting process.

Metallography

A cross section was cut from the stock riser at the site of flow lines, then mounted, ground, and polished. The resulting sample was microscopically examined in both the unetched and etched conditions.

Several cold shuts were present at the surface and immediately beneath the surface. Figures 3 and 4 illustrate the solidification fronts that failed to fuse. This lack of fusion can create a plane of weakness where fracturing can initiate in a manner similar to that observed on the broken risers.

Chemical analysis/identification

Sections from two broken risers were subjected to quantitative chemical analysis. The results, presented in Table 1, showed that the compositions of both risers conformed to requirements for alloy AM60A (ASTM B94).

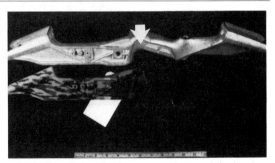

Fig. 1 Overall view of a riser removed from stock (top) and a broken riser. Arrow indicates the location of cold shuts in the stock riser.

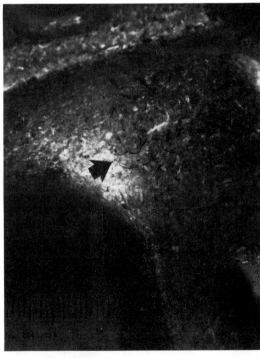

Fig. 2 Closeup view of the thumb position on the stock riser. Arrow indicates flow lines on the surface.

Table 1 Results of chemical analysis

Element	Failed riser 1	Failed riser 2	Alloy AM60A requirements (ASTM B94)
Aluminum	5.80	6.35	5.5-6.5
Silicon	<0.05	<0.05	0.50
Copper	<0.03	<0.03	0.35
Zinc	<0.05	<0.05	0.22
Iron	<0.05	<0.05	...
Nickel	<0.01	<0.01	0.03
Manganese	0.29	0.30	0.13 (min)

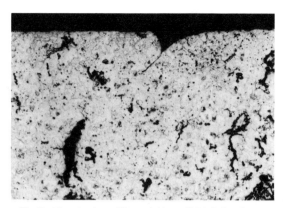

Fig. 3 Microstructure of the metal immediately below the surface flow lines shown in Fig. 2. Etched. 62×

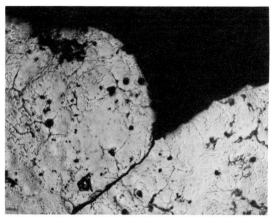

Fig. 4 Micrograph showing the plane of weakness in the structure at the cold shut shown in Fig. 3. 248×

Discussion

Examination revealed that the stock riser had cold shuts present at the site where fracturing occurred on the other risers. Cold shuts occur during the casting process as metal flows into the mold and fails to fuse at local sites. This lack of fusion can be caused by cold metal, cold molds, and/or metal flow patterns in the mold cavity. Often, one location in a mold is particularly sensitive to cold shuts, and a defect can result in all of the castings made from that mold.

Conclusion and Recommendations

It was recommended that all risers be thoroughly inspected and that the bow company work with their die casting shop to design a mold with acceptable filling characteristics.

Failure of a Steel Wire Rope From a Television Tower

Carmine D'Antonio, Department of Metallurgy and Materials Science, Polytechnic University, Brooklyn, New York

A 6 × 19 fiber core steel wire rope failed as it was being used to lower a steel television tower. Fracture of the rope occurred at a point under one of two clips used to fashion a spliced loop that was directly connected to the top of the tower. Microscopic examination of the fracture surfaces and the condition of the individual wires revealed that 59% of the wires failed by shear, 39% failed in tension, and 2% had been cut. In addition, 87% of the wires showed some degree of crushing damage, ranging from mild to severe. The failure was attributed to improper installation of the clips.

Key Words	Steels	Wire rope	Compression damage

Background	A wire rope failed as it was being used to lower a steel television tower. The rope was an integral component of a lowering mechanism on the tower. It parted when the tower was in a near horizontal position.	***Pertinent specifications*** The wire rope consisted of a 6 × 19 fiber core in the Seale arrangement, as specified.

Visual Examination of General Physical Features

Figure 1 shows the portion of the wire rope assembly submitted for analysis. The section of the system that failed is in the foreground. Parting of the rope occurred in a spliced loop that was directly connected to the top of the tower. Two wire clips were installed on the loop, one on either side of the splice. Fracture of the rope took place at the point where one of the clips was installed. This clip was recovered subsequent to the failure and is shown in Fig. 1.

Figure 2 shows a closeup view of the failure area. Fracture occurred in the leg of the loop between the clips that contained the splice. When the two clips were installed, this leg apparently was

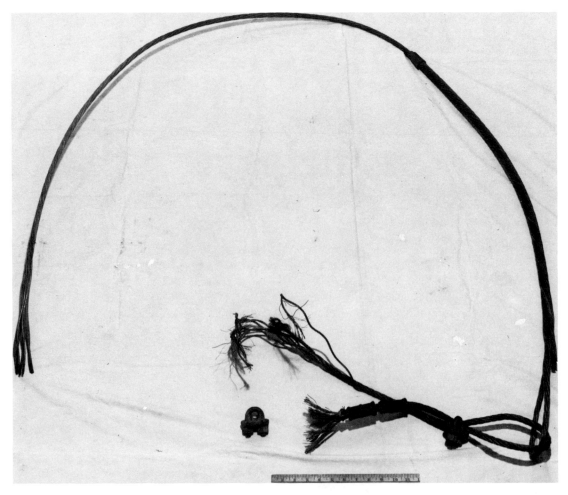

Fig. 1 Section of wire rope that contains the failure zone (foreground)

made shorter than the other, resulting in the greater part of the load (perhaps all of it) being sustained by this leg. This leg of the loop is also the one against which the U-bolt portion of the clip pressed when it was installed.

The arrow in Fig. 2 indicates the zone on the other leg of the loop where the clip was positioned. There was evidence of crushing and fracture of wires in this area.

Testing Procedure and Results

Surface examination

The fracture surfaces and the condition of the individual wires were examined microscopically to determine the mode of failure. Because the wire rope was a 6 × 19 fiber core in the Seale arrangement, it contained 114 individual wires, 60 of which were large diameter and 54 of which were small diameter. Examination revealed that 67 of the wires (59%) failed by shear, 45 (39%) failed in tension, and 2 (2%) had been cut. In addition, 99 of the individual wires (87%) showed some degree of crushing damage, ranging from mild to severe. Figures 3(a) to (c) show many of the individual wires, their mode of fracture, and the crushing damage. Arrows marked "T" indicate tensile failures, arrows marked "S" indicate shear failures, and arrows marked "C" indicate crushing damage.

Discussion

Crushing damage was evident in 87% of the individual wire fracture zones. This damage was caused by overtightening of the clip, which seriously reduced the load-carrying capacity of the rope. Failure of the rope occurred in the leg of the loop at the point where the U-bolt side of the clip pressed against it. Consequently, this side was the most severely damaged by overtightening of the clip, because the circular cross section of the bolt tended to dig into the individual wires. The pressure of the U-bolt also tended to cause kinking of the individual wires.

Overtightening of the clip was also evidenced

Fig. 2 Closeup view of failure zone

(a)

Fig. 3 Micrographs showing individual wires and their failure modes. T, tension failure; S, shear failure; C, crushing damage. ~6.1×

(continued)

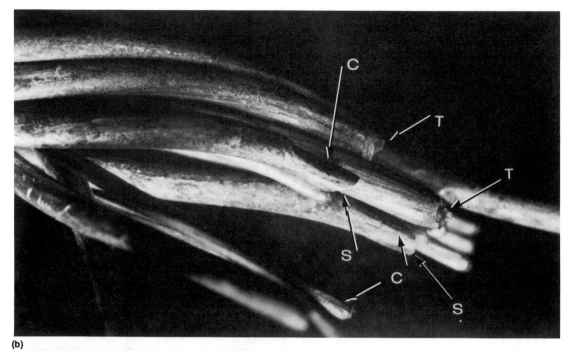

(b)

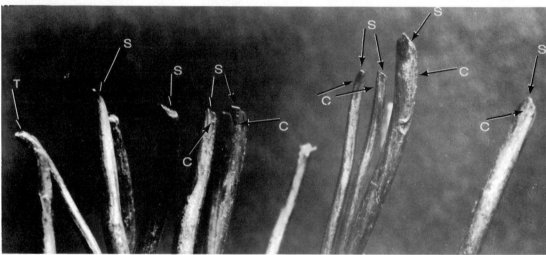

(c)

Fig. 3 Micrographs showing individual wires and their failure modes. T, tension failure; S, shear failure; C, crushing damage. ~6.1×

by the crushing and fracturing of wires on the other leg of the loop (see arrow, Fig. 2). This leg was less severely damaged, because the rope lay in the flat saddle portion of the clip. Because the failure took place in the spliced leg of the loop between the clips, this leg was probably shorter than the other and therefore sustained the greater part, or all, of the load during service.

In uninhibited failure of wire rope in tension, failure of individual wires would exhibit a charac-

teristic cup-cone fracture. Only 39% of the wires in the rope showed this mode of failure. On the other hand, 59% of the wires showed shear failures. Shear failure will occur when individual wires are subjected to transverse compression (crushing) and are kinked. Because metals are substantially weaker in shear than in tension, the effective strength of the rope was greatly reduced when the wires were induced to fail by shearing.

Conclusion and Recommendations

Failure of the wire rope was the direct result of improper installation of the clips. Severe overtightening of one of the clips caused extraordinary compressive forces and kinking of individual wires, thus inducing them to fail in shear. Shear failure required substantially lower service loads. In addition, overtightening of the clip caused

physical crushing damage in a majority of the individual wires, thereby further, and again substantially, reducing the load-carrying capacity of the rope. It is also quite likely that the legs between the clips were uneven in length, causing the shorter leg (the failed leg) to sustain most, or all, of the service loads.

The crushing damage was, in all probability, exacerbated by improper installation of the clip under which failure occurred. Proper installation requires that the saddle portion of the clip bear against the load-bearing segment of the rope—in this instance, the shorter leg. This arrangement reduces the likelihood of damage, because the saddle represents a much less severe condition for crushing than the U-bolt. Because the clip under which failure occurred separated from the system, there is no unequivocal proof that it was improperly used. However, an indirect indication that this was the case can be inferred from the other clip, which was installed improperly.

Failure of a Transmission Brake Disc

Edward C. Lochanski, Metalmax, Thiensville, Wisconsin

Failure of AISI 1015 steel brake discs used in power transmissions in emergency winches was investigated using various testing methods. The failed discs were stampings that had replaced cast discs. Residual stresses in the fillets of new cast and new stamped brake discs were measured by X-ray diffraction. The results indicated that the stamped brake discs had failed by fatigue caused by a tensile residual stress pattern in the fillet. The residual stress pattern was attributed to the change in manufacturing process from casting to stamping. Use of a manufacturing process that yields a compressive residual stress in the fillet, appropriate heat treatment of stamped discs, or redesign of the disc and/or transmission assembly was recommended.

Key Words			
	Residual stress	Carbon steels	Stampings
	Brake discs	Fatigue failure	Stress concentration
	Mechanical transmissions	Winches	
Alloys	Carbon steel—1015		

Background

AISI 1015 steel brake discs used in power transmissions in emergency winches began failing after the design of the discs was changed from a casting to a stamping.

Applications

Friction elements are pressed against the brake disc to provide a cable position lock or to take up slack in an outgoing cable in the emergency winch. A microprocessor-based controller determines the times and rates for the brake application and release operations. The brake assembly has a very low duty cycle, estimated at less than 1%.

Circumstances leading to failure

After a cost-driven change in the manufacturing process of the brake discs, numerous catastrophic field failures began to occur. In each case, either the cable did not brake properly, usually spilling cargo, or the cable locked up into an unwanted position, making the cargo completely inaccessible.

The brake discs were formerly castings that received subsequent heat treatment and a light 6.4 mm (0.25 in.) radius fillet roll. The new-style brake discs were stampings that did not receive subsequent heat treatment or a fillet roll.

Pertinent specifications

Both the cast and the stamped brake discs had similar dimensions. The discs were approximately 300 mm (12 in.) in diameter. The friction surfaces were approximately 50 mm (2 in.) wide, with a wall thickness of approximately 13 mm (0.5 in.). All fillets were specified to a 5 mm (0.20 in.) radius, except the failed fillet, which was specified to a 6.4 mm (0.25 in.) radius.

The cast brake disc was specified to be an ASTM A395 60-40-18 ductile iron casting with a subsequent anneal heat treatment. A mostly ferritic microstructure with no massive carbides was specified. In addition, graphite formation was required to contain a minimum of 90% types I and II graphite. A hardness range of 143 to 187 HB, anywhere on the component, was specified. The 6.4 mm (0.25 in.) radius was specified to be lightly rolled.

The stamped brake disc was specified to be an AISI/SAE 1015 steel stamping with no subsequent heat treatment. A mostly ferritic microstructure yielding a hardness range of 121 to 173 HB, anywhere on the component, was specified. The 6.4 mm (0.25 in.) radius was not specified to be lightly rolled.

Visual Examination of General Physical Features

New cast brake discs and new stamped brake discs appear similar to one another with the unaided eye, except for different inscribed manufacturing codes. Old cast brake discs exhibited no signs of cracking in the 6.4 mm (0.25 in.) radius fillet. The fillet regions were smooth, with a gradual radius contour. Failed stamped brake discs exhibited partial cracking or complete cracking in the 6.4 mm (0.25 in.) radius fillet. The fillet regions were smooth, with a gradual radius contour. Completely cracked brake discs displayed a mostly smooth and flat fracture, except for a small shear lip opposite the 6.4 mm (0.25 in.) radius fillet. The fracture area is shown schematically in Fig. 1.

Testing Procedure and Results

Surface examination
Scanning Electron Microscopy/Fractography. A scanning electron microscope (SEM) traverse of the entire fracture section revealed three distinct fracture zones: the origin zone, the intermediate zone, and the termination zone. These zones are depicted schematically in Fig. 2.

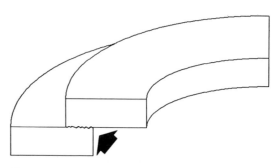

Fig. 1 Schematic of fracture area

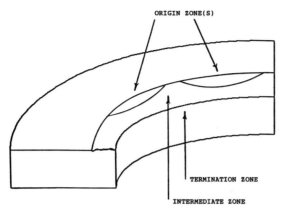

Fig. 2 An SEM traverse of the entire fracture section revealed three distinct fracture zones, shown here schematically.

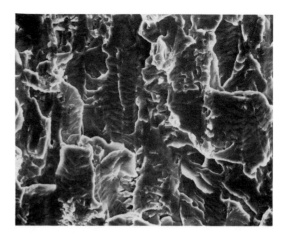

Fig. 3 Areas adjacent to each fatigue origin exhibit fatigue striation artifacts, the result of interaction between slip lines and crystallographic cleavage planes in the plastically deformed region. 630×

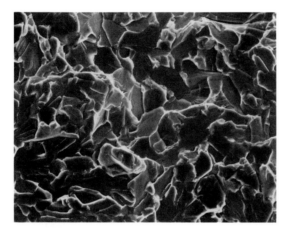

Fig. 4 Intermediate zone, exhibiting primarily cleavage facets. 630×

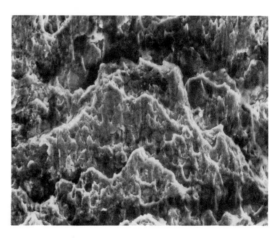

Fig. 5 Termination zone, exhibiting typical shear lip facets. 630×

The origin zone exhibited multiple fatigue origins at the 6.4 mm (0.25 in.) radius surface. The microfacets of the fatigue areas were completely obliterated because of rubbing in service and were not documented. The areas adjacent to each fatigue origin displayed fatigue striation artifacts, the result of interaction between slip lines and crystallographic cleavage planes in the plastically deformed region (Fig. 3). The intermediate zone exhibited primarily cleavage facets (Fig. 4). The termination zone exhibited typical shear lip dimple facets (Fig. 5).

Metallography
Microstructural Analysis. Optical microscope traverses of transverse metallographic cross sections of a used cast brake disc confirmed that the microstructure in the 6.4 mm (0.25 in.) radius fillet was similar to the microstructure in the disc wall, which is typical of a cast material. A typical unetched cross section of the 6.4 mm (0.25 in.) radius fillet displayed small, round manganese sulfide inclusions (average diameter ~5 µm), no free massive carbides, and more than 90% types I and

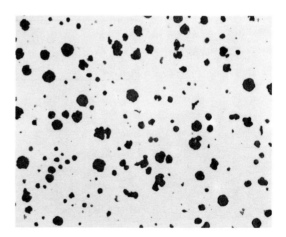

Fig. 6 Typical unetched metallographic cross section of a used cast brake disc 6.4 mm (0.25 in.) radius fillet, displaying small, round manganese sulfide inclusions with no free massive carbides and more than 90% types I and II graphite formation. 63×

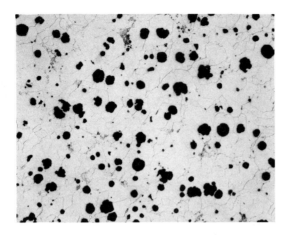

Fig. 7 Corresponding etched view of Fig. 6, illustrating uniform-size grains of mostly ferrite with some pearlite. 63×

Fig. 8 Typical unetched metallographic cross section of a failed stamped brake disc 6.4 mm (0.25 in.) radius fillet, displaying small, round manganese sulfide inclusions. 63×

Fig. 9 Corresponding etched view of Fig. 8, illustrating highly cold-worked grains of mostly ferrite with some pearlite. 63×

Fig. 10 Etched view of the disc wall of a failed stamping, illustrating uniform-size grains of mostly ferrite with some pearlite. 63×

Table 1 Results of chemical analysis

	Composition, %			
Element	Cast brake disc	ASTM A395 60-40-18 requirements(a)	Stamped brake disc	AISI-SAE 1015 requirements(a)
Carbon	3.21	3.00 (min)	0.14	0.13-0.18
Manganese	0.56	NR	0.52	0.30-0.60
Phosphorus	0.054	0.080 (max)	0.025	0.040 (max)
Sulfur	0.049	NR	0.030	0.050 (max)
Silicon	2.38	2.50 (max)	0.04	NR
Nickel	0.09	NR	0.02	NR
Chromium	0.14	NR	0.03	NR
Molybdenum	0.06	NR	0.01	NR
Copper	0.02	NR	0.01	NR
Aluminum	0.046	NR	0.021	NR

(a) NR, no requirement

II graphite formation (Fig. 6). The corresponding etched cross section showed uniform-size grains (average ASTM grain size 5) of mostly ferrite with some pearlite (Fig. 7). Thus, the used cast brake disc met the specified ASTM A395 60-40-18 microstructural requirements.

Optical microscope traverses of transverse metallographic cross sections of a failed stamped brake disc confirmed that the microstructure in the 6.4 mm (0.25 in.) radius fillet was not similar to the microstructure in the disc wall, which is typical of a stamped material. A typical unetched cross

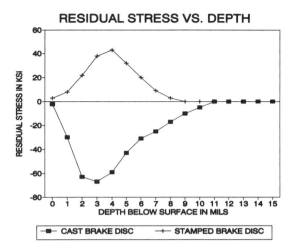

Fig. 11 Residual stresses in the 6.4 mm (0.25 in.) radius fillets of new cast brake discs and new stamped brake discs, as measured by X-ray diffraction

section of the 6.4 mm (0.25 in.) radius fillet exhibited small, round manganese sulfide inclusions (average diameter ~2 μm) (Fig. 8). The corresponding etched cross section showed highly cold-worked grains of mostly ferrite with some pearlite (Fig. 9). An etched view of the disc wall revealed uniform-size grains (average ASTM grain size 8) of mostly ferrite and some pearlite (Fig. 10). Thus, the failed stamped brake disc met the specified AISI-SAE 1015 microstructural requirements.

It should be noted that cold work can yield fatigue striation artifacts on a fracture surface. Although pearlite lamellae and serpentine glide steps are well-known examples of fatigue striation artifacts, cleavage fractures of highly cold-worked material, which yield fatigue striation artifacts, are not well known. The highly cold-worked areas of the stamped brake disc fracture are good examples of this phenomenon.

Chemical analysis/identification

Optical emission spectroscopy and combustion (carbon and sulfur) results verified that the used cast brake disc met the specified ASTM A395 60-40-18 chemical composition requirements and that the failed stamped brake disc met the specified AISI-SAE 1015 chemical composition requirements (Table 1).

Mechanical properties

Hardness. A hardness traverse of the used cast brake disc yielded a hardness reading of 161 to 175 HB anywhere on the sample. Thus, the used cast brake disc met the specified ASTM A395 60-40-18 hardness requirements.

A hardness traverse of a failed stamped brake disc wall yielded a reading of 135 to 152 HB. A hardness traverse of the same brake disc in the 6.4 mm (0.25 in.) radius fillet yielded values of 148 to 171 HB. Thus, the failed stamped brake disc met the specified AISI-SAE 1015 hardness requirements.

Stress analysis

Experimental. Residual stresses in the 6.4 mm (0.25 in.) radius fillets of a new cast brake disc and a new stamped brake disc were measured by X-ray diffraction (Fig. 11). The peak residual stress of the fillet in the cast brake disc was approximately 463 MPa (67.1 ksi) in radial compression, at a distance of approximately 3 mils from the surface. The peak residual stress of the fillet in the stamped brake disc was approximately 298 MPa (43.2 ksi) in radial tension, at a distance of approximately 4 mils from the surface.

Discussion

Residual stress analysis revealed that a new cast brake disc contained a desirable compressive residual stress pattern in the 6.4 mm (0.25 in.) radius fillet, whereas a new stamped brake disc contained an undesirable tensile residual stress pattern in the fillet. The different manufacturing processes used to produce the brake discs resulted in different polarities and magnitudes of residual stress.

The new cast brake disc exhibited a desirable compressive residual stress pattern because of fillet rolling operations. The tensile service stress must overcome the compressive residual stress before any brake disc tensile loading occurs.

On the other hand, the new stamped brake disc exhibited an undesirable tensile residual stress pattern in the fillet because of upsetting operations. The tensile service stress is added to the tensile residual stress. In effect, the stamped brake disc is always under tensile load.

Most properly operating disc brakes are usually not subjected to fatigue loading in the 6.4 mm (0.25 in.) radius fillet. However, long idle periods would cause the caliper centering assembly of the brake disc to rust solid. Subsequent application of the brake would engage only the friction pad on the caliper piston, resulting in brake disc fatigue failures. An interim solution included a rolled 6.4 mm (0.25 in.) radius fillet, introducing compressive residual stress at the fillet surface to prevent fatigue initiation. The problem was finally circumvented by redesigning the original all-metal caliper centering assembly with high-performance composite bushings.

A major product liability concern was that a failed brake disc may not be replaced along with the revised caliper centering assembly. Despite the fact that all owners of the emergency winch were notified of the caliper centering assembly recall, a review of field inspection reports verified that more than 50% of all replaced brake discs were installed by unauthorized service personnel. None of the unauthorized service personnel replaced the caliper centering assemblies.

Conclusion and Recommendations

Most probable cause

The most likely cause of the transmission brake disc failure was the occurrence of an undesirable tensile residual stress pattern in a tensile-loaded 6.4 mm (0.25 in.) radius fillet. Fatigue initiated at these areas of high tensile residual stress, culminating in catastrophic overload.

The tensile residual stress pattern was attributed to the change in manufacturing process. The cast brake disc exhibited a desirable compressive residual stress pattern in a tensile-loaded 6.4 mm (0.25 in.) radius fillet. However, the stamped brake disc exhibited an undesirable tensile residual stress pattern in a tensile-loaded 6.4 mm (0.25 in.) radius fillet.

Remedial action

One long-term solution to the problem is to select a manufacturing process that produces a compressive residual stress in the 6.4 mm (0.25 in.) radius fillet of the brake disc, such as rolling or shot peening. Another option is to select a manufacturing process that produces almost no residual stress in the 6.4 mm (0.25 in.) radius fillet of the brake disc, such as annealing or stress relieving. A final recommendation involves revision of the overall design of the brake disc and/or the transmission assembly so that the tensile stresses in the 6.4 mm (0.25 in.) radius fillet of the brake disc are minimized in service.

Explosion of an Oxygen Line

David O. Leeser, Consulting Engineer, Scottsdale, Arizona

An oxygen line that was part of a mobile, truck-mounted oxygen-acetylene welding unit exploded in service. Analysis revealed that the failure occurred at the flexible hose-to-valve connection. It was further determined that a steel adapter had been installed at the point of failure to make the connection. Use of the adapter, which joined with a brass nipple, created an unacceptable dissimilar metal joint. The steel also provided a source for the generation of sparks. Loctite, a hydrocarbon sealant that is highly flammable and explosive in contact with pure oxygen, had been used to seal the threaded joint. It was recommended that only brass fittings be used to assemble removable joints and that use of washers, sealants, and hydrocarbon lubricants be strictly avoided.

Key Words   Welding machines   Oxyacetylene welding   Explosions
Dissimilar metals

Background

An oxygen line that was part of an oxygen-acetylene welding setup exploded in service.

Applications

The equipment involved was a mobile unit with an oxygen-acetylene welding setup on the bed of a pickup truck used to perform various on-site jobs and repairs.

Circumstances leading to failure

On a hot summer day (<48 °C, or 118 °F), when both steel bottles of gas were exposed to the sun, the welder opened the valve on a fully pressurized (>17.2 MPa, or 2500 psig) bottle of oxygen gas. Spontaneously, the connection between the valve and the flexible hose exploded with such violence that the flexible hose completely disappeared. The noise produced by the explosion caused the welder to suffer tinnitus, a condition of constant ringing in the ears that results from eardrum damage.

Visual Examination of General Physical Features

Abuse had been suffered by the brass nipple that connected the gas bottle to the flexible line. The marks were not recent, because they were not shiny and bright (Fig. 1). Burn marks on the steel adapter were evidence of the elevated temperatures created during the explosion.

Loctite adhesive was used as a sealant to prevent loosening of the joint between the two dissimilar metals as they expanded and contracted at different rates in the hot sun and during cool nights. Close examination of the threaded interface (Fig. 2) clearly indicated use of this adhesive, which is unacceptable for oxygen systems.

Evidence of spikes of incipient melting at the seat of the metal-to-metal steel fitting is shown in Fig. 3. Because steel melts at temperatures above 1540 °C (2800 °F), three scenarios were possible. First, the adhesive may have loosened and finally melted enough to ooze into contact with the oxygen gas. Second, the worker who replaced the empty bottle with a fully recharged bottle may have used some lubricant to facilitate the reassembly of the

connection fittings. Third, the fittings may have been dropped unintentionally into some oil or wiped with an oily rag for the purpose of cleaning them before reassembly. These scenarios are reinforced by the fact that the explosion occurred the instant the valve on the fully recharged bottle was opened for the first time.

Fig. 1 Bottle-to-flexible hose fitting used on mobile unit. Note damage to rim of steel fitting caused by explosion. 0.88×

Fig. 2 Closeup view of brass-to-steel joint, showing heavy wrench marks on nipple and puddled adhesive at threaded joint. 4.6×

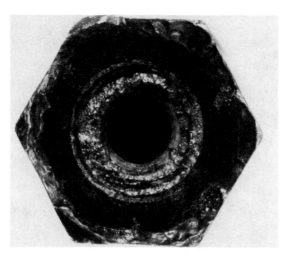

Fig. 3 Closeup view of base at inner section of steel fitting, focusing on evidence of incipient melting upon detonation. 3.5×

Fig. 4 Magnified angular view at opened end of steel fitting shown in Fig. 2. Note torn metal, burn marks, and incipient melting at base. Hexagonal rim did not flare out. 2.5×

The remains of the steel fitting, after the steel head was pushed through it, showed that the load-bearing threads were completely gone. Striations illustrated how material was violently dug out by the explosive force behind the head (Fig. 4). Blackened burn marks were evidence of the high temperatures created by the resulting friction that was generated. The loss of shrapnel-type bits of steel from the lip of the fitting occurred as the head was ejected.

Testing Procedure and Results

This investigation was conducted in a noninvasive manner; that is, all evidence remained intact.

Chemical analysis/identification

Associated Environments. Oxygen is about 20% of ambient air. Although it encourages and supports fire and combustion, it does not itself burn. However, it will appear to explode when instantaneously depressurized from 17.6 MPa (2550 psig) to empty at 0.1 MPa (1 atm). This condition results when the oxygen in the fully pressurized steel bottle used for welding is suddenly released. The thrust produced can propel objects for great distances. Moreover, when oxygen combines with hydrocarbon materials, such as oils and certain plastics, the combustion, or burning, rate is instantaneous (about 0.001 s). The reaction also is explosive. In addition to the two explosive forces that are generated, tremendous amounts of heat are produced, as evidenced by spectacular temperature increases.

Conclusion and Recommendations

Remedial action

It is important to avoid each of the possible conditions that caused the explosion. Dissimilar materials should not be used at removable joint positions. Brass fittings, rather than spark-prone steel fittings, should be used to assembly removable joints. Copper tubing can be used for the hard line.

Washers or sealants should not be used for seating or on threaded joints. If leaks occur that cannot be sealed using metal-to-metal brass joints, then the fittings should be replaced with new, tight-sealing brass fittings.

Hydrocarbon lubricants, such as oils or greases, and oily rags should not be used to assist in the mating or cleaning of fittings. The regulator next to the gas bottle should be maintained using high-pressure fittings without modifications.

How failure could have been prevented

Problem areas can be detected by examining damaged parts that bear evidence of thread sealants and abusive forces used in attempts to maintain seals between dissimilar metals.

At the time of the accident, the regulator on the mobile unit was located at the end of the truck bed. Since then, it has been moved back and now resides next to the bottle for safety reasons. The regulator cannot prevent hammerlike high pressuring in both the flexible and hard gas lines. Although this arrangement is not as convenient for the welder, it will help to prevent similar accidents.

Fatigue Failure of a Carbon Steel Piston Shaft on an Extrusion Press Billet-Loading Tray

Philip Green, University of Hull, United Kingdom

A recurring piston shaft failure problem on the billet-loading tray of an extrusion press was investigated. Two shafts fractured within a period of 10 days. The shaft was machined from normalized EN3 (AISI C1022) steel stock without further treatment. Visual, microstructural, chemical, and mechanical (hardness and tensile properties) analyses of failed shaft specimens were conducted. The examinations showed that the shafts had failed by fatigue. It was recommended that a low-alloy steel (e.g., 3% Ni-Cr) in the hardened and tempered condition and subjected to shot-peening surface-hardening treatment be used. The provision of a stop to reduce bending stresses was also recommended.

'Key Words

Carbon steels
Extruders

Pistons
Fatigue failure

Shafts (power)

Alloys   Carbon steels—C1022, EN3

Background

A piston shaft on the billet-loading tray of an extrusion press experienced recurring failure. The shaft, which was subjected to repeated tensile, compressive, and bending forces, had been machined from normalized stock without further treatment.

Circumstances leading to failure

The billet-loading tray on an industrial extrusion press is operated by two pneumatic rams. The pistons are mounted beneath the billet-loading tray in a vertical position (Fig. 1). Every 6 to 12 months, the shafts that connect the air cylinders to the billet-loading tray fracture. Replacement of a shaft takes a minimum of 4 h, although usually the work of an entire shift is lost. Two shafts fractured within a period of 10 days, and an investigation was subsequently requested to determine the cause of the failure.

Specimen selection

One shaft, which had fractured in service, was available for examination. This shaft had been in operation for approximately four shifts per week for 2 months, after which the work rate was increased to three shifts per day until the occurrence of the failure 10 months later.

Pertinent specifications

The billet-loading tray is of the split type, with one piston to each half. Air pressure is applied si-

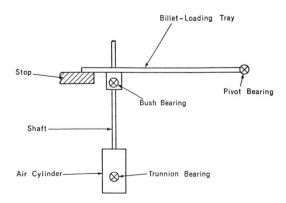

Fig. 1 Schematic side view of billet-loading tray, shaft, and air cylinder

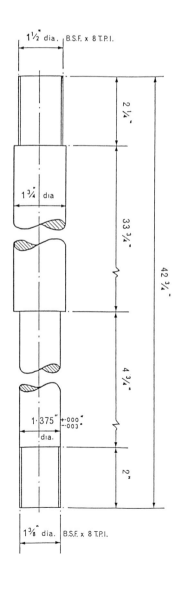

Fig. 2 Dimensions of shaft

multaneously to coordinate the lift. Each air cylinder is mounted on trunnion bearings to allow free rotation during operation.

The dimensions of the shaft are shown in Fig. 2. The shaft has a nominal diameter of 44 mm (1 3/4 in.) for most of its total length of 1090 mm (42 3/4 in.). Both ends are threaded. The 35 mm (1 3/8 in.) diam British Standard Fine (BSF) thread is at the lower end of the shaft and screws into the piston of the air cylinder. The larger 38 mm (1 1/2 in.) diam BSF thread is at the upper end and is screwed into a tongue that is free to rotate on a bush bearing mounted on the billet-loading tray (Fig. 3). A holding pin passes through the tongue and shaft at the thread to prevent unscrewing during operation. The end of the upper 38 mm (1 1/2 in.) diam thread, where it meets the bar surface, is undercut to the depth of the thread. The width of the undercut is 3.2 mm (1/8 in.) and the nose radius is 1.6 mm (1/16 in.). Thus, at this point the diameter of the section is reduced to 34.0 mm (1.34 in.). The most recent shaft failures occurred by fracture across the section at this point. On some shafts, although not on the one examined, the lower thread is also undercut, in which case the diameter is reduced to 30.86 mm (1.215 in.).

The air cylinders are 200 mm (8 in.) in diameter. Air pressure of 550 kPa (80 psi) is applied for the duration of both the lift period and, in reverse, for the downstroke. The application of air pressure to assist the downstroke is necessary to ensure that the tray is out of the path of the advancing press. The downward movement of the billet-loading tray is halted by a stop.

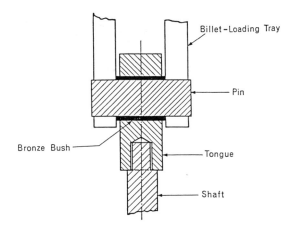

Fig. 3 Schematic view of the shaft, tongue, and bush

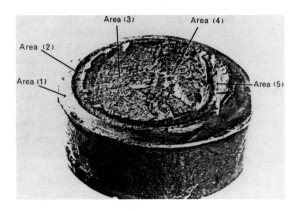

Fig. 4 Fracture surface. 1.04×

Visual Examination of General Physical Features

The fracture surface of the shaft is shown in Fig. 4. The mating surface of this fracture was not available for examination. In Fig. 4, five areas can be distinguished:

1. A beveled and polished outer annulus with a lip of metal around the circumference
2. An inner incomplete annulus with a rougher rusted surface on which machining marks are visible
3. A semicircle approximately 33 mm (1.3 in.) in diameter with a flat surface of rough texture on which striations are visible
4. A semicircle of the same approximate diameter as area 3 with an uneven fibrous surface
5. An area with a relatively smooth appearance

Areas 3 and 4 were separated by cracks running across the section, parallel to the axis of the bar.

The fracture occurred across the shaft at the undercut at the end of the 38 mm (1 1/2 in.) diam thread. The thread was situated at the upper end of the bar, where it screwed into the tongue.

Testing Procedure and Results

Metallography

Microstructural Analysis. A longitudinal section was made parallel with the axis of the bar and perpendicular to the line of cracks separating areas 3 and 4. The cracks were small, with a depth of less than 0.81 mm (0.032 in.). The microstructure consisted of α-iron, with pearlite occupying approximately 25% of the surface area. Manganese sulfide particles were observed, which were elongated in the direction of the longitudinal axis of the bar. No decarburization was observed.

Table 1 Chemical composition of the shaft

	Composition, %							
	C	Si	Mn	S	P	Ni	Cr	Fe
Steel shaft	0.16	0.22	0.60	0.11	0.05	0.175	0.055	bal
EN 3, nominal specification	0.15-0.25	0.05-0.35	0.60-1.0	0.06 max	0.06 max	...	...	bal

Table 2 Tensile testing results

Specimen identification(a)	Reduction of area, %	Elongation, %	Yield stress		0.2% proof stress		Ultimate tensile strength	
			MPa	tonf/in.²	MPa	tonf/in.²	MPa	tonf/in.²
L1	52	20	426.1	30.9	470.2	34.1	541.9	39.3
L2	58	20	456.4	33.1	481.2	34.9	510.2	37.0
L3	57	22	434.3	31.5	462.0	33.5	488.2	35.4
T1	29	13	391.6	28.4	415.1	30.1	485.4	35.2
T2	27	13	372.3	27.0	398.5	28.9	475.8	34.5
T3	28	14	380.6	27.6	405.4	29.4	484.0	35.1

(a) L, longitudinal; T, transverse

Chemical analysis/identification

The chemical composition of the shaft is given in Table 1.

Mechanical properties

The hardness of the bar was measured on a transverse section. The hardness was 169 HV near the surface and 175 HV in the central region.

Tensile Properties. Six tensile specimens were machined from the shaft. Three specimens were machined with their axis parallel to the longitudinal axis of the shaft, and the others were machined perpendicular to this direction. The results of the tensile tests, which were carried out at room temperature and at a crosshead speed of 0.5 mm/min (0.02 in./min) are presented in Table 2. As expected, there is a difference between the mechanical properties measured in the transverse and in the longitudinal directions. However, the results are reproducible and typical for steel of this composition.

Discussion

When the billet-loading tray is lifted by the pneumatic ram, the shaft is subjected to a compressive stress. The downstroke is also powered by air pressure. However, in this case, when the downward motion of the billet-loading tray is halted by the stop, a sudden tensile stress is applied to the shaft. The air pressure of 550 kPa (80 psi) acts on a piston diameter of 200 mm (8 in.); consequently, the tensile force on the shaft will be approximately 18 kN (2 tonf). The design of the billet-loading unit is such that the tray moves through an arc of approximately 90° during the lift. During the same movement, the air cylinders, which are bottom heavy, are tilted through an angle of about 20°. It is clear that on the downstroke, when the billet-loading tray is halted by the stop and the shaft is subjected to a tensile stress, the momentum of each air cylinder will apply a bending force to its respective shaft.

The appearance of the fracture surface examined in this investigation is indicative of a fatigue failure. Beach marks are visible on area 3, depicting the advance of the fatigue crack. The final abrupt fracture of the shaft occurred over area 5 when the reduced cross-sectional area was no longer able to sustain the applied tensile stress.

From the area of this part of the fracture surface, 4.5 mm² (0.07 in.²), and assuming a maximum force of 18 kN (2 tonf), an approximate final fracture stress of 400 MPa (29 tonf/in.²) may be calculated. This is comparable to the values measured in the tensile tests. The rougher surface of area 4 presumably reflects the influence of the increasing tensile and bending stresses accelerating the fatigue failure.

Areas 1 and 2 are located around the circumference of the fracture. It is clear that area 2 is the original surface of a portion of the side of the undercut. Area 1 has a polished appearance and has been plastically deformed so that a lip of metal protrudes around the entire outside diameter. This suggests that, after fracture on the downstroke, the air pressure has been reapplied at least once, and the end of the shaft has collided with the inner rim of the tongue.

From the chemical analysis and tensile strength, the steel of the shaft corresponds most closely to EN 3 (AISI C1022), the nominal specification of which is given in Table 1. The microstructure is typical of this material in the normalized condition.

Conclusion and Recommendations

Most probable cause

The examination revealed that the growth of cracks by fatigue accounted for approximately 95% of the surface area of the fracture. Furthermore, the shaft had been machined from normalized stock and had not been subsequently heat treated or surface hardened. In addition to the tensile and compressive stresses applied to the shaft, bending forces were present.

Surface finish and sharp changes in surface geometry can be more important than the nominal cross-sectional area with regard to fatigue suscep-

tibility. This is highlighted by the fact that fracture of these shafts did not always occur at the smallest section. It may not be merely the presence of the undercut that initiated failure but its surface finish. However, since all recent shaft failures have occurred at the upper undercut, it is probable that the bending stresses are greater at this point.

Remedial action

The provision of a stop to absorb the momentum of the swing of the air cylinder during the downstroke would help to reduce the bending forces. If possible, the undercuts at the end of the threads should be omitted; alternatively, their surface finish should be good.

In selecting a material for applications involving repeated stressing, it is well established that there is an increase in fatigue limit with tensile strength to about 1100 MPa (80 tonf/in.^{2}). In addition, fatigue resistance is improved by a microstructure of tempered martensite and by a surface-hardening treatment. Hence, a hardened and tempered steel is recommended, with a minimum tensile strength of 410 kPa (60 psi), such as 3% Ni-Cr steel, and the surface should be hardened by shot peening. If possible, the shaft should be roll threaded.

Fracture of the Bottom Platen of an 800 Ton Hydraulic Press

I.B. Eryürek and M. Capa, Istanbul Technical University, Istanbul, Turkey

The side supporting flange of the bottom platen of an 800 ton hydraulic press fractured after 9×10^5 cycles under a maximum load of 530 tons. The platen material specified in the design was cast steel 52. Metallographic examination of the fracture surface indicated that the platen had failed in fatigue as a result of a high stress concentration in a sharp 0.6 mm (0.02 in.) radius fillet. Stress analysis and fracture mechanics predictions revealed that there was also danger of fatigue failure for platens with the design radius of 10 mm (0.4 in.) if the press operates at 800 tons. It was recommended that the remaining life of similar presses be assessed periodically controlling the cracks, their dimensions, and their propagation rates. An increase in the radius of the fillet was also recommended.

Key Words	Carbon steels	Hydraulic presses	Castings
	Fatigue failure		
Alloys	Cast steel—52		

Background

The side supporting flange of the bottom platen of an 800 ton hydraulic press fractured after 9×10^5 cycles under a maximum load of 530 tons.

Applications

The press was installed in a ceramic factory. The geometry and the dimensions of the press platen are shown in Fig. 1, along with the location of the crack initiation zone.

Circumstances leading to failure

The press had operated for 809 h under a load of 530 tons when the failure was observed. During this period, the number of impacts (repeated load) were as follows:

70 h	15 impacts/min	(63,000 total impacts)
739 h	19 impacts/min	(842,460 total impacts)

The total number of impacts was 905,460, or ~9 × 10^5 impacts.

Pertinent specifications

The platen material specified in the design was cast steel 52. The fillet radius measured 0.6 mm (0.02 in.), which was much smaller than the specified value of 10 mm (0.4 in.) (Fig. 1).

Performance of other parts in same or similar service

No damage has yet been observed on other presses of the same make and capacity operating in the same workshop.

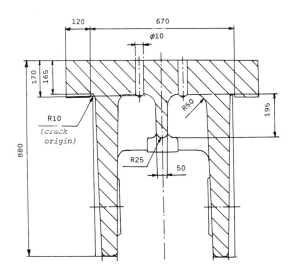

Fig. 1 Dimensions (in millimeters) of the press platen. Crack origin is also shown.

Testing Procedure and Results

Surface examination

Visual examination of the fracture surface of the platen indicated that the crack had initiated at the sharp fillet (Fig. 1) and had followed a curved path into the main body of the platen (Fig. 2), which is characteristic of fatigue fracture at sharply filleted changes in section. No indications of fatigue beach marks were found on the fracture surface.

Metallography

Microstructural Analysis. The microstructure of the platen material is shown in Fig. 3. It is apparent that the structure is a coarse-grained cast structure with a ferrite grain size of ASTM 2 to 3. The presence of some Widmanstätten structure is also visible.

Mechanical properties

Mechanical properties of the platen material are given in Table 1. The endurance limit (σ_{eo}) of cast steels in reversed bending is generally as-

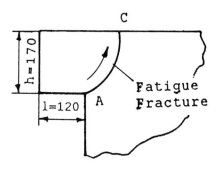

Fig. 2 Curved path of the fatigue crack

Table 1 Mechanical properties of the platen

Property	
Yield strength, MPa (ksi)	284 (41.2)
Tensile strength, MPa (ksi)	478 (69.3)
Ductility (σ_5), %	12
Reduction of area, %	15
Hardness, HB	143

sumed to be equal to 40% of the tensile strength of steel. Then,

$$\sigma_{eo} = 0.4 \, \sigma_{TS} = 0.4 \times 478 = 191 \, MPa$$

Stress analysis

Analytical. The platen-supporting flange was considered a built-in cantilever beam (Fig. 4). The table weight, residual stresses, and friction forces between the flange and its supporting surface are neglected in the calculation. The reaction force on the bottom of the flange was considered as a single force acting midway from the end of the built-in cantilever. Maximum bending stress, σ, at point A (Fig. 4) is:

$$\sigma = (\tfrac{3}{2})(P \cdot {}^l\!/_{Bh^2}) \qquad \text{(Eq 1)}$$

where $l = 0.12$ m (0.4 ft), $h = 0.17$ m (0.6 ft), and $B = 0.82$ m (2.7 ft) are the length, width, and thickness of the beam, respectively (Fig. 4).

For repeated loading between $P_{min} = 0$ and $P_{max} = 530$ tons, maximum, minimum, mean, and alternating normal stresses at point A (Fig. 4), respectively, are:
$\sigma_{max} = 40$ MPa (5.8 ksi), $\sigma_{min} = 0$ MPa, $\sigma_m = 20$ MPa (2.9 ksi), $\sigma_a = 20$ MPa (2.9 ksi). For the geometry shown in Fig. 4 and in bending, the notch factor, K_t, is (Ref 1):

$$K_t = 1 + 0.5 \, (h/2r)^{1/2} \qquad \text{(Eq 2)}$$

For $h = 0.17$ m (0.4 ft) and $r = 6 \times 10^{-4}$ m (measured fillet radius), $K_t = 7$. The fatigue safety factor of the press platen n can be written in the form:

$$n = \sigma_e \cdot C_{su} \cdot C_{sz}/\sigma_a \cdot C_{im} \cdot K_f \qquad \text{(Eq 3)}$$

where σ_e is the endurance limit of the material corresponding to the mean stress of 20 MPa (2.9 ksi), σ_a is the alternating stress, C_{su} is the correction factor for the surface roughness, C_{sz} for the part thickness, C_{im} for the loading rate, and K_f for the notch effect. The numerical values adopted in Eq 3 are:

$\quad \sigma_e = 183$ MPa (26.5 ksi) (from Goodman's rule)
$\quad \sigma_a = 20$ MPa (2.9 ksi) (calculated)
$\quad C_{su} = 0.8$ (machined surface, Ref 2)
$\quad C_{sz} = 0.7$ ($h = 170$ mm and bending, Ref 2)
$\quad C_{im} = 1.5$ (medium shock, Ref 2)
$\quad K_f = 0.5 \, K_t = 0.5 \times 7 = 3.5$ ($r = 0.6$ mm (0.2 in), Ref 3)

Then, $n = 0.98$ is obtained. Thus, the fatigue life of the platen is limited to 530 tons, because the safety factor is less than 1.

Fracture Mechanics Predictions. The data

Fig. 3 Microstructure of the table material. 83×

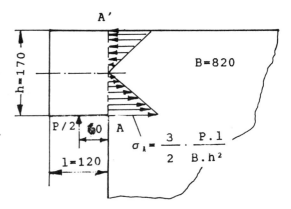

Fig. 4 Stress distribution at the fillet

published by various investigators on fatigue crack propagation threshold (ΔK_{th}) values of steels show that, for stress ratios smaller than 0.1, ΔK_{th} is a constant equal to 6 MPa√m (5.5 ksi√in.). If $\Delta K_I > \Delta K_{th}$, an existing crack will propagate under the stress-intensity factor fluctuation, ΔK_I.

For the geometry and the loading in question, the stress-intensity factor, K_I, has been obtained from the stress-concentration factor, as follows:

$$K_I = 0.46 \, {}^{Pl}\!/_{Bh}{}^{3/2} \qquad \text{(Eq 4)}$$

Substituting the values $l = 0.12$ m (0.4 ft), $B = 0.82$ m (2.7 ft), $h = 0.17$ m (0.6 ft), and $P = 530$ tons (5.2 MN) into Eq 4, $K_I = 5$ MPa√m (4.5 ksi√in.) is obtained. Because the stress ratio is equal to zero, $\Delta K_I = K_I$. Then the safety factor, neglecting the effect of impact loading, is:

$$n = \Delta K_{th}/K_I = 6/5 = 1.2 \qquad \text{(Eq 5)}$$

Thus, fatigue cracking will propagate at a platen fillet of 0.6 mm (0.02 in.) radius, because the safety factor is not large enough.

Discussion   Examination of the fracture surface of the platen indicated that the crack had initiated at the sharp fillet and had followed a curved path into the main body of the platen, which is characteristic of fatigue fracture at sharply filleted changes in section. No fatigue beach marks were visually observed on the fracture surface, because the press had operated continuously under the same repeated loading and the same environmental conditions. However, calculations using two different methods have shown that the press platen is not safe against fatigue failure at loads greater than 500 tons. Additional calculations showed that there is also a danger of fatigue failure at the design radius of 10 mm (0.4 in.) if the press operates at 800 tons. For this reason, the fillet radius should be increased to 30 mm (1.2 in.). This provides the safety factors of 2.4 and 1.6 for 530 and 800 tons, respectively.

Conclusion and Recommendations	***Most probable cause*** The platen failed in fatigue that resulted from a high stress concentration. ***Remedial action*** Results obtained from these studies show that to minimize losses, periodic assessment of the remaining life of the working presses should be made by controlling the cracks, their dimensions, and propagation rates. Also, the radius of the fillet should be increased to 30 mm (1.2 in.) to reduce the notch effect.

References

1. F.A. McClintock and A.S. Argon, *Mechanical Behaviour of Materials*, Addison-Wesley, 1966, p 411-412
2. H. Tauscher, *Dauerfestigkeit von Stahl und Gusseisen*, Archimedes Verlag, 1971
3. M. Klesnil and P. Lukas, *Fatigue of Metallic Materials*, Elsevier, 1980, p 196

Hydrogen Embrittlement Cracking in a Batch of Steel Forgings

A.K. Das, Aircraft Design Bureau, Hindustan Aeronautics Ltd., Bangalore, India

The repeated occurrence of random cracks in the fillet radius portion of low-alloy steel (38KhA) end frame forgings following heat treatment was investigated. Microstructural analyses were carried out on both the failed part and disks of the rolled bar from which the part was made. Subsurface cracks were found to be zigzag and discontinuous as well as intergranular in nature. A mixed mode of fracture involving ductile and brittle flat facets was observed. Micropores and rod-shaped manganese sulfide inclusions were also noted. The material had a hydrogen content of 22 ppm, and cracking was attributed to hydrogen embrittlement. Measurement of hydrogen content in the raw material prior to fabrication was recommended. Careful control of acid pickling procedures for descaling of the hot-rolled bars was also deemed necessary.

Key Words	Chromium steels Cracking (fracturing)	Forgings	Hydrogen embrittlement
Alloys	Chromium steels—38KhA		

Background

The repeated occurrence of random cracks in the fillet radius portion of low-alloy steel and frame forgings following heat treatment was investigated.

Applications

A batch of 120 end frame parts forged and machined from a 200 mm (8 in.) diam medium-carbon chromium steel rolled bar was heat treated in accordance with specification requirements. These parts are used for critical applications in helicopters, and thus utmost care is necessary to ensure quality from the melting stage onward.

Circumstances leading to failure

Magnetic-particle inspection of the initial batch of 50 heat-treated parts after cadmium plating revealed several tiny radial cracks (Fig. 1).

Pertinent specifications

Russian alloy 38KhA was used for the fabrication of the parts.

Sample selection

One cracked sample was selected for low-power optical and high-power scanning electron microscopic (SEM) examination. Two 20 mm (0.8 in.) thick disks sliced from the rolled bar stock used for forging were also examined.

Visual Examination of General Physical Features

Examination of the parts that had been inspected by magnetic-particle testing showed several tiny radial cracks at random locations, but primarily confined to the base edge and fillet radius zones (Fig. 1).

Testing Procedure and Results

Surface examination

Scanning Electron Microscopy/Fractography. One of the tiny cracks was carefully opened. Because the depth of the crack was negligible, it was difficult to retain the fracture zone intact. The frac-

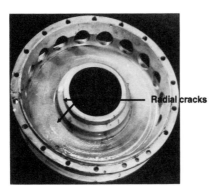

Fig. 1 Heat-treated end frame component after cadmium plating. Several tiny cracklike indications, primarily in the base fillet radial zones, were detected by magnetic-particle testing.

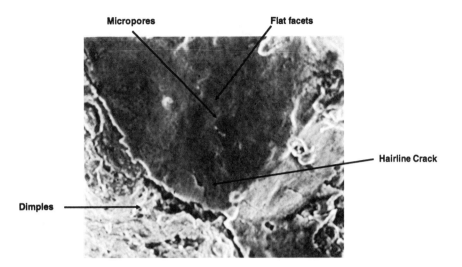

Fig. 3 High-magnification SEM micrograph of the fracture zone of a subsurface crack in the bore, showing brittle flat facets, ductile dimples, micropores, and hairline cracks—all indicative of hydrogen embrittlement

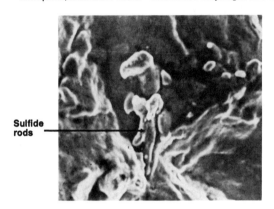

Fig. 4 High-magnification SEM micrograph of the same zone shown in Fig. 3. Rod-shaped manganese sulfide inclusions are visible, normally a preferred site for hydrogen accumulation.

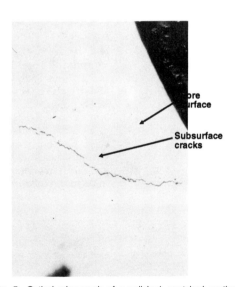

Fig. 5 Optical micrograph of a polished, unetched section across a crack indication shown in Fig. 2, showing the clear presence of subsurface cracks

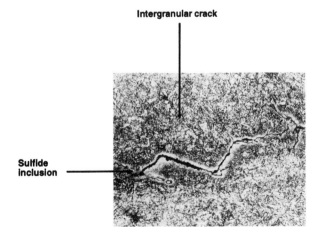

Fig. 6 High-magnification optical micrograph showing intergranular cracks with characteristic manganese sulfide and oxide inclusions in the crack path of a section across the radial cracks in the heat-treated part

ture zone was examined by SEM, which showed the predominantly dimple structure of a ductile failure. No clue as to the nature and type of cracking could be established.

It was suspected that the hairline cracks could be due to hydrogen embrittlement. Two disks were sliced from the rolled bar, one in the annealed (as-supplied) condition and the other in the heat-treated condition. A 50 mm (2 in.) diam hole similar to the one in the finished end frame part was bored in the center of each disk. Magnetic-particle inspection of both disks revealed tiny discontinuous subsurface cracklike indications in the bore surface (Fig. 2). One of the subsurface cracks was carefully opened so that the very small zone of cracking could be retained undamaged for SEM examination. The fracture clearly exhibited dim-

Table 1 Chemical specifications for 38KhA steel

Element	Composition, % Min	Max
Carbon	0.34	0.42
Manganese	0.50	0.80
Silicon	0.17	0.37
Sulfur	...	0.03
Phosphorus	...	0.03
Nickel	...	0.40
Chromium	0.80	1.1

pled structures associated with brittle (cleavage) flat facets, ductile hairline cracks, micropores (Fig. 3), and rod-shaped manganese sulfide inclusions (Fig. 4). All of these fracture characteristics are typical of hydrogen embrittlement.

Metallography

Microstructural Analysis. Microexamination of sections taken across the defect indications in both disks revealed subsurface discontinuous hairline (zigzag) cracks (Fig. 5), which were intergranular in nature with no evidence of branching. The microstructure of the defective part at higher magnification also showed discontinuous intergranular cracks, with typical manganese sulfide inclusions along the path of cracking (Fig. 6).

Chemical analysis

The chemical composition of the end frame material was found to conform to specifications (Table 1). The cracked end frame part was also subjected to instrumental gas analysis; the hydrogen gas content was found to be on the order of 22 ppm.

Mechanical properties

Impact Toughness. Charpy impact properties determined in locations both near and remote from the defective zone in the longitudinal and transverse directions were generally low—44 and 34 J, or 32 and 25 ft · lbf, respectively—compared with the specification requirement of 49 J (36 ft · lbf).

Discussion

SEM fractography clearly established that the random occurrence of tiny cracks was caused by hydrogen embrittlement. This finding was also supported by the presence of a high hydrogen content in the material prior to heat treatment.

Initially, from the nature and location of the cracks (mostly in the fillet radius zones), faulty heat treatment was thought responsible. However, when subsequent heat treatment of the second batch resulted in identical cracks in the same location (despite satisfactory chemistry, microstructure, and hardness), the random cracking was strongly suspected to be associated with a hydrogen-induced phenomenon. Because the tiny cracks in the part could not be successfully opened, a subsurface crack in the heat-treated disk was forced open. The fracture was examined by SEM, which revealed characteristic hydrogen embrittlement features (Fig. 3 and 4).

A 50 mm (2 in.) diam hole had been bored into the disk to develop residual stresses sufficient to cause subsurface cracking. Under high stresses, hydrogen gas pressure increases rapidly, causing flakes and fissures. A gas content on the order of 22 ppm present in the raw material was high enough to cause hydrogen embrittlement in high-strength steel without any external stresses. The low impact properties measured on the failed part also indicated the brittle nature of the material caused by hydrogen embrittlement.

Conclusion and Recommendations

Most probable cause

The basic cause of the development of tiny cracks was hydrogen embrittlement.

Remedial action

In the absence of records related to the manufacturing history of the supplied raw material, hydrogen absorption was suspected to have occurred during improper pickling of the hot-rolled bars. Careful control of the acid pickling operation within specified process parameters was recommended. Measurement of hydrogen content in the raw material prior to fabrication was also suggested.

INDEX

B

C

Fracture mechanics

Fracture mechanics analysis

Fractures

Fracture toughness

Fracturing

Frames

Fretting

Furnace conveyors

Furnaces

Fused salts

Fused salts, environment

G

Gallionella
iron bacteria found in hydrotests, **Vol. 1•**179, **Vol. 1•**180

Galvanic corrosion
cap screws (alloy steel) in a refrigeration compressor, fracture of, **Vol. 1•**324-327
helicopter tail rotor blade fatigue fracture due to field-induced corrosion, **Vol. 2•**30-32
nickel-base superalloy heat-exchanger tube failure in a black liquor heater, **Vol. 2•**95-98
pipe flange assemblies of austenitic stainless steel, **Vol. 2•**197-200
waterwall tube hydrogen damage, **Vol. 2•**490-492

Galvanized steels
ACSR electrical transmission cable aluminum connector failure, **Vol. 1•**428-430
anchor bolt fractures in chemical plant construction project, **Vol. 1•**328-331

Galvanized steels, mechanical properties
large grain storage bin brittle fracture, **Vol. 2•**470-477

Gamma radiography
tanks (stainless steel type 304L), corrosion failure of, **Vol. 1•**194-195

Gas metal arc welding
exhaust diffuser assembly failure analysis, **Vol. 2•**61
truck cross members, fatigue failure of, **Vol. 1•**95-101

Gas pipelines
electrostatic discharge attack on a thrust bearing face in a power turbine, **Vol. 1•**225-227

Gas tungsten arc welding (GTAW)
aluminum alloy 5083-O piping, mercury liquid embrittlement failure, **Vol. 2•**207, **Vol. 2•**208
bellows failure (AISI Type 347 stainless steel), **Vol. 2•**259
oil tank assembly procedure, helicopters, **Vol. 1•**40

Gas turbine engines
aero engine compressor discs, fatigue fracture of, **Vol. 1•**241-250
diaphragm fatigue cracking, **Vol. 2•**295-298
spot weld (stainless steel type 321) mode III fatigue crack growth following heat-affected zone curvature, **Vol. 1•**39-46

Gas turbines
first-stage nozzles in an industrial gas turbine, hot corrosion of, **Vol. 2•**502-505
impeller shaft fracture, **Vol. 1•**297-298
turbine blade failure analysis, **Vol. 2•**289-294

Gas wells
P-110 couplings for mating 180mm casing in oil field production, hydrogen embrittlement of, **Vol. 1•**396-400

Gate valves
cast iron, in oleum and sulfuric acid service, brittle fracture, **Vol. 1•**202-209

Gears
connecting bolts in a truck crane, fatigue failures of, **Vol. 2•**419-424
pressurized heavy-water reactor refueling machine, gear failure of, **Vol. 1•**231-233
segment failures in cast steel, **Vol. 2•**45-52
transfer gear shaft surface damage, **Vol. 1•**299-300

Gear teeth
carburized steel gear from a helicopter transmission, fatigue failure, **Vol. 1•**228-230
pressurized heavy-water reactor refueling machine, gear failure of, **Vol. 1•**231-233

Girders
tension flange of a steel box-girder bridge, brittle fracture, **Vol. 1•**369-377

Grain boundary segregation
solenoid valve seats (Zn-Al alloy), intergranular corrosion failure, **Vol. 1•**421-423

Grain size
femoral prosthesis of cast stainless steel, fracture of, **Vol. 2•**448-450

Graphitization
superheater tube promoted by graphitization, creep failure of, **Vol. 2•**495-497

Graphitization, heating effects
superheater tube of low-alloy steel, graphitization-related failure, **Vol. 2•**201-203

Gray iron
valve body castings failure, **Vol. 2•**364-366
valve (cast iron) in oleum and sulfuric acid service, brittle fracture of, **Vol. 1•**202-209
Yankee dryer roll on a modified paper machine, corrosion fatigue and subsequent rupture, **Vol. 1•**132-135

Gray iron, specific types
247 type A
valve in oleum and sulfuric acid service, brittle fracture, **Vol. 1•**202-209

Grinding damage
aircraft main landing gear sliding struts, cracking of, **Vol. 2•**7-10

Gusset plates
welded truss, hydrogen-induced cracking, **Vol. 1•**382-384

H

Hammers
ball-peen hammer spalling, **Vol. 2•**467-469
sledgehammer chipping failure, **Vol. 1•**417-420
sledgehammer head composition, **Vol. 1•**418
spike maul failure, **Vol. 2•**367-369

F.E. Harris equation, gas carburizing, Vol. 1•109

Hardness
missile launcher detent spring, failure of, **Vol. 2•**455-458

Heat-affected zone (HAZ)
ACSR electrical transmission cable aluminum connector failure, **Vol. 1•**429-430
axles from a prototype urban transit vehicle, fatigue failure, **Vol. 1•**93
carrier shafts, **Vol. 1•**287, **Vol. 1•**288, **Vol. 1•**289, **Vol. 1•**290
C-Mn steel in CO_2 absorber in a chemical plant, stress-corrosion cracking, **Vol. 1•**191-193
crane bolster (cast steel) failure, **Vol. 1•**440-442
electrical resistance welded production tubing (Grade J-55) failure, **Vol. 1•**393-395
feedwater line break, single-phase erosion corrosion, **Vol. 1•**183-186
Ferralium (E-Brite-clad) tube sheet in nitric acid service, intergranular corrosion of, **Vol. 1•**124-125
marine riser clamp, brittle fracture of, **Vol. 1•**385-388
mode III fatigue crack growth following curvature of stainless steel type 321 spot weld, **Vol. 1•**39-46
nickel-base superalloy heat-exchanger tube failure in a black liquor heater, **Vol. 2•**95-98
pipe welds intergranular stress-corrosion cracking in a Kamyr continuous digester equalizer line, **Vol. 1•**168, **Vol. 1•**169, **Vol. 1•**170
stainless steel pipe reducer section in bleached pulp stock service, intergranular corrosion, **Vol. 1•**164-167
stainless steel tank used for storage of heavy water/helium, failure of, **Vol. 2•**253-255
tension flange of a steel box-girder bridge, brittle fracture, **Vol. 1•**372
vanes (stainless steel) from a closed riveted impeller, intergranular cracking and failure, **Vol. 1•**278-283
welded helium tank failure, **Vol. 2•**249-252
welded truck cross members, fatigue failure of, **Vol. 1•**95-101
welded truss gusset plates, hydrogen-induced cracking, **Vol. 1•**382-384

Heat cracking
stainless steel pipe (76mm, 3in.) and liner from a hydrogen plant quench pot vessel, **Vol. 1•**107-109

I

K

L

J

Lubricants

toilet-tank floats (plastic) failure, **Vol. 1**•434-436

Lubrication

trunnion bolt used to couple railway cars, fracture of, **Vol. 2**•394-395

Lubrication systems

heat exchanger tubes (brass), failure of, **Vol. 1**•115-117

oil tank stainless steel type 321 spot weld mode III fatigue crack growth, **Vol. 1**•39-46

springs (stainless steel) used in an oil ring lip seal, failure of, **Vol. 1**•401-405

M

Machine tools

draw-in bolts of 4340 steel, hydrogen embrittlement delayed failure, **Vol. 2**•401-404

Macroexamination

copper tubing leaks from cooling coils of a large air-conditioning unit, **Vol. 2**•204

Macrofractography

ACSR electrical transmission cable aluminum connector failure, **Vol. 1**•429

actuator (cast A356 aluminum) premature torquing failures, **Vol. 1**•47

admiralty brass condenser tubes, failure analysis, **Vol. 1**•111

aircraft landing gear fracture, **Vol. 1**•5-6

aircraft main landing gear sliding struts, cracking of, **Vol. 2**•8-9

aircraft nose landing gear strut, cracking in, **Vol. 2**•11-12

aircraft (transport) crankshaft fatigue fracture during flight, **Vol. 2**•36-38

aircraft wing main spar cracking at a bolt hole, **Vol. 1**•10

airplane wing component (17-4PH) hydrogen assisted fracture, **Vol. 1**•31

anchor bolt fractures in chemical plant construction project, **Vol. 1**•329

axles from a prototype urban transit vehicle, fatigue failure, **Vol. 1**•93

bronze rupture disc liquid metal embrittlement, **Vol. 2**•130

C130 aircraft main landing gear wheel flange, fatigue failure, **Vol. 1**•27, **Vol. 1**•29

cap screws (alloy steel) in a refrigeration compressor, fracture of, **Vol. 1**•325

closed-die forgings parting line, cracking, **Vol. 2**•463

copper condenser dashpot failure, **Vol. 2**•100, **Vol. 2**•101

coupling in a line-shaft vertical turbine pump, fracture of, **Vol. 2**•354, **Vol. 2**•355

die insert (tool steel D2) service failure, **Vol. 1**•413, **Vol. 1**•415

diesel engine bearing cap bolts, anomalous fractures of, **Vol. 2**•374

downcomer expansion joint, stress-corrosion cracking, **Vol. 2**•223

extruded aluminum alloy bar stock material, cracking during forging, **Vol. 2**•509

fan blade fatigue fracture, **Vol. 2**•301, **Vol. 2**•302

fighter aircraft high-strength steel frame stress-corrosion cracking, **Vol. 1**•52

gas turbine diaphragm fatigue cracking, **Vol. 2**•296

heat-resistant sinter belt failure, **Vol. 1**•345

helical compression springs (Cr-Si steel) residual stresses, stress-corrosion cracking, **Vol. 1**•407

helicopter main rotor blade failure, **Vol. 1**•21

helicopter tail rotor blade fatigue fracture due to field-induced corrosion, **Vol. 2**•30-31

helicopter tail rotor blade processing-induced fatigue failure, **Vol. 2**•33, **Vol. 2**•34

high-pressure water-line plug in a fire sprinkler system, failure of, **Vol. 1**•159-160

hoist chain link tensile failure, **Vol. 1**•452-453, **Vol. 1**•454

hydroturbine shaft failure analysis, **Vol. 2**•338

I-beam failure, **Vol. 1**•379

impeller blades (stainless steel) in a circulating water pump, flow-induced vibration fatigue, **Vol. 1**•254

large grain storage bin brittle fracture, **Vol. 2**•471-473

load cells (stainless steel) in a milk storage tank, corrosion fatigue failure, **Vol. 2**•413

main boiler feed pump impeller cracking, **Vol. 2**•324

P-110 couplings for mating 180mm casing in oilfield production, hydrogen embrittlement of, **Vol. 1**•397

pipe nipple (brass) failure, **Vol. 1**•157

pipe reducer section in bleached pulp stock service, **Vol. 1**•165

polycarbonate ophthalmic lenses, solvent-induced cracking failure, **Vol. 2**•493

pressure vessel (carbon steel) hydrotest failure, **Vol. 1**•144

pressurized heavy-water reactor refueling machine, gear failure of, **Vol. 1**•231

relay valve guide failure in transport aircrafts, **Vol. 2**•28, **Vol. 2**•29

screws (stainless steel type 316L) employed for surgical implanting, fatigue fracture, **Vol. 1**•315

silver solid-state bonds of uranium, stress corrosion fracture, **Vol. 1**•148, **Vol. 1**•149

spike maul failure, **Vol. 2**•367-368

spiral-welded water line (1830mm diam), catastrophic failure, **Vol. 1**•150

squeeze-clamped polyethylene natural gas pipeline failure, **Vol. 2**•483

steam superheater 800H tube failure, **Vol. 2**•218, **Vol. 2**•219

steam turbine rotor disk cracking, **Vol. 2**•277-278

tempered glass panels, delayed fracture due to nickel sulfide inclusions, **Vol. 1**•432

titanium nitride-coated high speed steel hob, brittle failure of, **Vol. 2**•459, 460

toilet-tank floats (plastic) failure, **Vol. 1**•435

trailer wheel fatigue failure at the bolt holes, **Vol. 1**•86

train wheel fracture due to thermally induced fatigue and residual stress, **Vol. 2**•71-72

tricycle agricultural field chemical applicator steering spindle, fatigue failure of, **Vol. 1**•89

trunnion bolt used to couple railway cars, fracture of, **Vol. 2**•395

tube (stainless steel) thermal fatigue, **Vol. 1**•363

turbine blade failure analysis, **Vol. 2**•290

turbine impeller of aluminum, fatigue failure of, **Vol. 2**•329, **Vol. 2**•330

turntable rail quench cracking, **Vol. 2**•80, **Vol. 2**•82

valve body castings failure, **Vol. 2**•364-365

welded truss gusset plates, hydrogen-induced cracking, **Vol. 1**•383

wires (stainless steel type 316L) in an electrostatic precipitator at a paper plant, fatigue fracture of, **Vol. 1**•219

electrostatic discharge attack on a thrust bearing face in a power turbine, **Vol. 1**•225-226

Macrofractography/scanning electron fractography

aircraft components, accident caused by explosive sabotage, **Vol. 2**•4

Macro (optical) fractography

trailer kingpin failure caused by overheating during forging, **Vol. 2**•53, **Vol. 2**•54, **Vol. 2**•55, **Vol. 2**•57

Macrophotography

arcing fault burndown in low voltage residential service entrance with aluminum conductors, **Vol. 2**•409-410

heat exchanger tubes (brass), failure of, **Vol. 1**•115

vanes (stainless steel) from a closed riveted impeller, intergranular cracking and failure, **Vol. 1**•280

Magnesium-base alloys

compound bow handle riser failure, **Vol. 2**•512-513

Magnesium-base alloys, s.t.

AM 60A

compound bow handle riser failure, **Vol. 2**•512-513

Magnetic particle inspection

aircraft (transport) crankshaft fatigue fracture during flight, **Vol. 2**•36, **Vol. 2**•38

bolts from an Army tank recoil mechanism, failure of, **Vol. 2**•384, **Vol. 2**•385, **Vol. 2**•387

bull gear contact fatigue failure, **Vol. 2**•39, **Vol. 2**•40, **Vol. 2**•43, **Vol. 2**•44

carburized steel gear from a helicopter transmission, fatigue failure, **Vol. 1**•229

circulating water pump shaft fatigue failure, **Vol. 2**•353

closed-die forgings parting line, cracking, **Vol. 2**•463, **Vol. 2**•465

compressor turbine impeller cracking, metallurgical failure analysis of, **Vol. 2**•333

crane long-travel worm drive shaft failure, **Vol. 2**•343

diesel engine bearing cap bolts, anomalous fractures of, **Vol. 2**•374

N

O

S

Scanning electron microscopy/energy-dispersive spectroscopy (SEM/EDS)

Scanning electron microscopy/fractography

T

W

X

X-ray analysis
helicopter main rotor bolt failure analysis, **Vol. 2•**392-393
hydroturbine shaft failure analysis, **Vol. 2•**338, **Vol. 2•**339

X-ray diffraction (XRD) analysis
aluminum/refrigerant reaction resulting in the failure of a centrifugal compressor, **Vol. 2•**320, **Vol. 2•**321-323
anchor bolt fractures in chemical plant construction project, **Vol. 1•**330
bellows failure (AISI Type 347 stainless steel), **Vol. 2•**261
boiler tube failures induced by phosphate water treatment, alkaline-type, **Vol. 2•**143, **Vol. 2•**144
brass tube in a generator air cooler unit, stress-corrosion cracking, **Vol. 2•**107, **Vol. 2•**109
chemical process piping cross-tee assembly, corrosion failure, **Vol. 2•**158
diesel engine bearing cap bolts, anomalous fractures of, **Vol. 2•**379
heat exchanger corrosion in a chlorinated solvent incinerator, **Vol. 1•**122
heat-resistant alloy (metal dusting), gaseous corrosion of, **Vol. 1•**352
helical compression springs (Cr-Si steel) residual stresses, stress-corrosion cracking, **Vol. 1•**408, **Vol. 1•**409
nickel anodes in a heavy water upgrading plant, failure of, **Vol. 2•**417
reactor tube failure, **Vol. 2•**189, **Vol. 2•**191

refinery boiler tube hydrogen-induced failure, **Vol. 2•**146
stainless steel grate bars in taconite indurators, hot corrosion, **Vol. 1•**359, **Vol. 1•**360-361
superheater tube U-bend caustic gouging and caustic-induced stress-corrosion cracking, **Vol. 2•**149
transmission brake disc failure, **Vol. 2•**518, **Vol. 2•**521
waterwall tube hydrogen damage, **Vol. 2•**490, 491
welded helium tank failure, **Vol. 2•**252

X-ray energy spectroscopy
helicopter tail rotor blade fatigue fracture due to field-induced corrosion, **Vol. 2•**31
helicopter tail rotor blade processing-induced fatigue failure, **Vol. 2•**34

X-ray fluorescence
steam superheater 800H tube failure, **Vol. 2•**220

X-ray fluorescence and combustomeric infrared (IR) absorption
bond between a cobalt alloy prosthetic casting and a sintered porous coating, **Vol. 1•**449, **Vol. 1•**451

X-ray photoelectron microscopy
turbine blade failure analysis, **Vol. 2•**292

X-ray photoelectron spectroscopy (XPS)
bronze rupture disc liquid metal embrittlement, **Vol. 2•**131, **Vol. 2•**134
steam turbine rotor disk cracking, **Vol. 2•**277, **Vol. 2•**278

X-ray spectroscopy
boiler feed pump second-stage impeller failure, **Vol. 1•**270-271
helicopter main rotor blade failure, **Vol. 1•**21-22
stainless steel grate bars in taconite indurators, hot corrosion, **Vol. 1•**359
to examine regenerator screens in Stirling engines, **Vol. 1•**66, **Vol. 1•**67-68, **Vol. 1•**69

Y Z

Yield strength
missile launcher detent spring, failure of, **Vol. 2•**455-458

Zinc base alloys
solenoid valve seats (Zn-Al alloy), intergranular corrosion failure, **Vol. 1•**421-423

Zinc plating
wheel studs, carburized and zinc plated SAE Grade 8, hydrogen-assisted stress cracking of, **Vol. 2•**396-400